Charting New Pathways to C_4 Rice

Charting New Pathways to C_4 Rice

Edited by

J. E. Sheehy
P. L. Mitchell
B. Hardy

International Rice Research Institute, The Philippines

World Scientific

NEW JERSEY • LONDON • SINGAPORE • BEIJING • SHANGHAI • HONG KONG • TAIPEI • CHENNAI

Published by

World Scientific Publishing Co. Pte. Ltd.

5 Toh Tuck Link, Singapore 596224

USA office: 27 Warren Street, Suite 401-402, Hackensack, NJ 07601

UK office: 57 Shelton Street, Covent Garden, London WC2H 9HE

The International Rice Research Institute (IRRI) was established in 1960 by the Ford and Rockefeller Foundations with the help and approval of the Government of the Philippines. Today, IRRI is one of the 15 nonprofit international research centers supported by the Consultative Group on International Agricultural Research (CGIAR – www.cgiar.org).

IRRI receives support from several CGIAR members, including the World Bank, European Union, Asian Development Bank, International Fund for Agricultural Development, Rockefeller Foundation, Food and Agriculture Organization of the United Nations, and agencies of the following countries: Australia, Brazil, Canada, Denmark, France, Germany, India, Iran, Japan, Malaysia, Norway, People's Republic of China, Republic of Korea, Republic of the Philippines, Sweden, Switzerland, Thailand, United Kingdom, United States, and Vietnam.

The responsibility for this publication rests with the International Rice Research Institute.

CHARTING NEW PATHWAYS TO C_4 RICE

Mailing address: DAPO Box 7777, Metro Manila, Philippines
Phone: +63 (2) 580-5600
Fax: +63 (2) 580-5699
Email: irri@cgiar.org
Web: www.irri.org.
Rice Knowledge Bank: www.knowledgebank.irri.org
Courier address: Suite 1009, Security Bank Center
6776 Ayala Avenue, Makati City, Philippines
Tel. +63 (2) 891-1236, 891-1174, 891-1258, 891-1303

Suggested Citation:
Edited by J.E. Sheehy, P.L. Mitchell, and B. Hardy, editors. 2008. Charting new pathways to C_4 rice. Singapore: World Scientific Publishing, and Los Baños (Philippines): International Rice Research Institute. 412 p.

ISBN-13 978-981-270-951-6
ISBN-10 981-270-951-7

Cover design: Juan Lazaro IV
Page makeup and composition: Ariel Paelmo
Figures and illustrations: Ariel Paelmo

Printed in Singapore by World Scientific Printers

Foreword

Agriculture is the indispensable base of human society and the nature and productivity of agriculture are determined by water and climate and largely directed by the products of agricultural research. Today, the world's population is 6.6 billion, and 5.1 billion live in the developing world where most of the world's existing poverty is concentrated. Currently, a billion people live on less than one dollar a day and spend half their income on food, 854 million people are hungry, and each day about 25,000 people die from hunger-related causes. The United Nations Millennium Declaration, agreed upon in September 2000, commits the world's nations to "eradicate extreme poverty and hunger." Solving the current problem would be sufficiently challenging, but what makes it even more daunting is that several aggravating features are magnifying. Over the next 50 years, the world population will increase by about 50% and climate change will probably result in more extreme variations in weather and cause adverse shifts in the world's existing climatic patterns. Water scarcity will grow and the demand for biofuels will result in competition between grain for fuel and grain for food, resulting in price increases. Furthermore, 75% of the world's people will live in cities, whose populations will need to be largely supported by a continuous chain of intensive food production and delivery. All of these adverse factors are occurring at a time when the developed nations are both reducing their investments in agricultural research and turning their remaining research investments away from productivity gains. If all of this weren't bad enough, the elite rice cultivars that dominate the food supplies of the millions of poor people in Asia have approached a yield barrier and production growth is slowing.

Each hectare of land used for rice production in Asia currently provides food for 27 people, but by 2050 that land will have to support at least 43 people. Feeding the 5.6 billion Asians in the 21st century will require a second Green Revolution to boost yields by 50% using less water and fertilizer. Theoretical models have been used to examine this problem and they suggest that this can be done only by increasing the efficiency with which photosynthesis uses solar energy. Fortunately, evolution has provided an example of a much more efficient photosynthetic system (C_4) than that possessed by rice or wheat (C_3). Maize, for example, is one of these C_4 plants. Boosting the photosynthetic efficiency of rice by changing it from C_3 to C_4 photosynthesis will be like supercharging a car's engine by fitting a new fuel injection system. Until the era of modern plant breeding, including genetic engineering, this was thought to be an intractable problem; now, there are many reasons for being optimistic about finding a solution. The chapters in this book are written by world-renowned experts and each of them offers special insights into the various forms of C_4 photosynthesis and how they might be introduced into rice.

The imperative for this project is necessity rather than curiosity. It will take an international consortium of research institutions to make C_4 rice a reality over the next 10 to 15 years. To that end, IRRI has formed a C_4 Rice Consortium to stimulate and conduct the research needed to invent C_4 rice. I am delighted to be able to use IRRI's resources to provide initial funds and we hope that donors will provide the support necessary to complete this important task.

ROBERT S. ZEIGLER
Director General
International Rice Research Institute

Preface

Rice displays a biphasic growth pattern: half the first phase of vegetative growth precedes the second phase of reproductive growth. The rates at which the phases proceed are strongly influenced by temperature, which largely accounts for different crop durations in temperate and tropical environments. The second phase begins when the rate of vegetative growth reaches a maximum and it reaches a maximum when the rate of growth of the vegetative phase falls to zero. During the first phase, full light interception is reached and the reservoir of nutrients for use in the second phase reaches a maximum. During the second phase of growth, the reservoirs in the vegetative portions of the crop are depleted and the second phase comes to a halt when the crop is mature: the time when most grains are filled and the fewest have been shed. Both empirical and theoretical investigations suggest that the maximum fraction of a crop's total biomass that can be grain is about 50%.

Solar energy captured in photosynthesis over the duration of a crop gives it the capacity to grow. The upper limit to crop biomass is determined by the laws of thermodynamics and mass conservation. At the limit, the total biomass is simply a function of the total quantity of solar energy captured and the efficiency with which that energy is made available for synthetic processes. Total solar energy absorption is largely a function of canopy architecture and crop duration. The efficiency of energy use is largely determined by photorespiration, dark respiration, and losses of biomass that occur owing to senescence. Canopy architecture is usually thought of in terms of leaf erectness and, given that plant breeders have selected for erectness over the past 30 years, little more can be gained in that direction. The opportunities for reducing dark respiration are very limited and senescence is essential in terms of recycling essential nutrients from the vegetative portions of the crop to the reproductive ones. There are many evolutionary examples of plants that have eliminated photorespiration by concentrating CO_2 around the photosynthetic enzyme Rubisco using a four-carbon acid (C_4) cycle. Plants such as rice that do not have a concentrating CO_2 mechanism fix CO_2 into three carbon acids (C_3 plants); their photosynthetic rates in hot environments are about half that of C_4 plants. C_4 plants have double the water-use efficiency of C_3 plants, and use about 40% less nitrogen to achieve 50% higher yields. Evolution has made it clear that photorespiration can be eliminated; therefore, it is the obvious candidate for work aimed at significant increases in yields.

The repeated evolution of C_4 photosynthesis indicates that it should be feasible to create C_4 rice plants by engineering C_4 genes into C_3 rice and replicating strong selection pressure for C_4 traits that we think exist in nature. The development of the C_4 system can be seen as an addition to the C_3 system and it is now clear that the C_3 and C_4 syndromes are not as rigidly separated as was first thought. The enzymes that are prominent in the C_4 pathway also exist in C_3 leaves although with very low activity. More surprisingly, there is a well-developed C_4 pathway in certain locations in C_3 plants: in the green tissue around vascular bundles, and probably in rice spikelets. In the opposite direction, maize, a thoroughly C_4 plant, has patches of C_3 tissue wherever a mesophyll cell is not adjacent to a bundle sheath cell, particularly in leaf sheaths and husk leaves. Some of the wild relatives of rice have C_4-like anatomical features and others may have CO_2 compensation points usually associated with C_3-C_4 intermediates. When maize C_4 genes are inserted in rice, they work; the rice genome has been sequenced and sequencing of the maize genome is nearing completion. A large number of genetic resources are available for use in screening programs aimed at detecting genes associated with C_4-ness: 6,000 wild relatives and 500,000 rice mutants. It has been suggested that *Arabidopsis* (C_3) can be used as a test system for transferral of genes from its closest C_4 relative, *Cleome gynandra*. The advantages of this are that all the knowledge of *Arabidopsis* can be used and *Cleome* has a short life cycle.

There are, of course, differences of opinion (contrasting hypotheses) between scientists as to which form of C_4 photosynthesis (single-cell and dual-cell systems) can be achieved most rapidly in rice and the ultimate effectiveness of the different forms in delivering significant increases in yield. This book explores those differences, but begins with a broad perspective of the economic problems surrounding rice and the potential impact on the poor of failing to contain upward pressure on food prices. It continues setting the scene by describing how the rice crop works and the consequences of supercharging photosynthesis.

In the second section of the book, Jane Langdale and her coauthors describe progress in various genetic approaches to understanding chloroplast development and then speculate on solutions to solving the problem of how to convert C_3 systems to C_4 ones.

The chapter of Richard Leegood examines metabolite transport and some of the structural and physiological changes that might be required when adding C_4 systems to C_3 ones.

Susanne von Caemmerer and her coauthors use models of diffusion to explore the effects of leaf anatomy and leakiness of cells on the efficiencies of the two-cell and single-cell forms of C_4 photosynthesis. Finally, they turn their attention to the anatomical and physiological requirements for C_4 rice.

Julian Hibberd advocates dual-track and fast-track approaches to the challenge of producing C_4 rice by inserting genes from *Cleome gynandra* into *Arabidopsis thaliana*.

John Evans and coauthors address the question of how the correct amounts of NADPH (biochemical reducing power) and ATP are provided in C_4 plants by two sorts of chloroplast in two types of cell. They also point out briefly the advantages of the single-cell C_4 system and identify its particular weakness (carbon dioxide leakage from the chloroplast).

The chapter by Rowan and Tammy Sage is a tour de force. It begins by identifying the essential features of a C_4 system and then examines its diversity in *Flaveria* before turning to an examination of the evolutionary factors critical to the emergence of C_4 systems. The chapter ends with a skeleton proposal on how to combine biotechnology and screening to produce C_4 plants from C_3 rice.

The genus *Flaveria* contains not only C_3 and C_4 species, but also several intermediate C_3-C_4 species. Udo Gowik and Peter Westhoff discuss the use of *Flaveria* as a model system for studying the evolution of genes involved in C_4 photosynthesis and the subtle differences between C_3 and C_4 orthologous genes.

D.M. Jiao speculates on ways a C_4 rice could be constructed by various genetic engineering approaches.

The third section of the book is devoted to an examination of single-cell C_4 systems, how they work, and what they might deliver if engineered into rice.

Jim Burnell opens this section by reviewing the early history of attempts to increase the rate of photosynthesis by manipulating the expression of foreign genes and moves on to suggest critical issues that might be examined.

Gerry Edwards and his coauthors briefly review critical features of C_4 plants, paying particular attention to chloroplast position and differentiation in Kranz types and single-cell aquatic types. They provide a more detailed description of single-cell terrestrial C_4 mechanisms and finally suggest some single-cell models for C_4 rice.

John Raven and coauthors describe lessons relevant to C_4 to be learned from diatoms. They provide evidence of high-capacity, low-leakage carbon-concentrating mechanisms in single cells and conclude that single-cell C_4 is a viable aim in engineering C_4 rice.

Continuing with the single-cell C_4 theme, George Bowes and his colleagues describe their work with *Hydrilla verticillata*, an aquatic monocot that operates a facultative, single-cell C_4 system. Their studies suggest that, to design a single-cell C_4 rice, transporter and permeability issues as well as the nuances of enzyme regulation need to be better understood.

Christoph Peterhänsel and his coauthors suggest a novel approach to improving photosynthesis by engineering a bypass of photorespiration in the chloroplast.

The fourth section of the book covers the background of C_4 rice and how it can be delivered.

C_4 physiology is a syndrome of interrelated developmental, anatomical, cellular, and biochemical traits that almost unavoidably must rely on regulatory networks. Tim Nelson and coauthors suggest that laser microdissection of cell types and microarray profiling can provide the comprehensive data for a systems biology approach to understanding differences between rice and C_4 leaf development.

Erik Murchie and Peter Horton draw on experiences of measuring rice photosynthesis in the field and suggest that acclimation to irradiance can result from a signal provided by mature leaves, but the nature of the signal is unknown. They also explore issues surrounding the use of nitrogen for photosynthesis in Rubisco and the conflicting demands for nitrogen in the form of protein in the grain.

D.S. Brar and J.M. Ramos discuss wild species of *Oryza* as an important reservoir of useful genes. Some of these genes have been introduced into indica and japonica rice for resistance to major diseases and insects and for tolerance of various abiotic stresses. It has been suggested that the wild types may contain aspects of C_4-ness and should be screened for anatomical, biochemical, and physiological features associated with the C_4 syndrome.

Parminder Virk and Shaobing Peng explore the consequences of inventing C_4 rice from a plant breeder's perspective. An early step would be to assess the effect of the C_4 syndrome on various agronomic traits such as resistance to pests, emergence of new pests, physical properties of the rice grains, and cooking and eating quality. Second, it would be important to evaluate the amount of expression of the syndrome in different genetic backgrounds and to identify the most promising transgenic event.

Philippe Hervé takes a genetic engineering approach and suggests that improved photosynthesis in rice can probably be achieved by engineering alleles involved in biochemical pathways and plant development. Another suggested strategy may consist of growing transgenic rice plants with C_4 features in different environments and screening for newly acquired C_4 features.

The fifth section leads the way into the formation of a C_4 Rice Consortium.

Richard Bruskiewich and Samart Wanchana deal with the role of bioinformatics in the construction of C_4 rice. They make general observations about sequenced genomes and describe a framework for gene discovery, before brainstorming on possible ways of using genomics information and bioinformatics to introduce C_4 photosynthesis into rice.

This project promises to be a universally important voyage of discovery about the most important of all plant mechanisms: photosynthesis. It will take a consortium of international institutions to make this a reality over the next 10 to 15 years. It is most encouraging that all the authors in this book have agreed to become founding members of a C_4 Rice Consortium. The next task is to build a long-term funding flow that is essential to sustaining research over the one and a half decades we estimate it will take to develop a fully functioning C_4 rice.

The book closes with a critical discussion and evaluation of the new pathways to C_4 rice. In it, all the authors highlighted important points and possibilities for success.

Contents

FOREWORD v
PREFACE vii

SECTION 1: SETTING THE SCENE **1**
How the rice crop works and why it needs a new engine **3**
J.E. Sheehy, A.B. Ferrer, P.L. Mitchell, A. Elmido-Mabilangan, P. Pablico, and M.J.A. Dionora
The case for C_4 rice **27**
P.L. Mitchell and J.E. Sheehy
Agricultural research, poverty alleviation, and key trends in Asia's rice economy **37**
D. Dawe
Catching up with the literature for C_4 rice: what we know now and didn't then **55**
P.L. Mitchell

SECTION 2: C_4 RICE FROM THEORY TO PRACTICE **79**
C_4 photosynthesis: minor or major adjustments to a C_3 theme? **81**
R.C. Leegood
C_4 photosynthesis and CO_2 diffusion **95**
S. von Caemmerer, J.R. Evans, A.B. Cousins, M.R. Badger, and R.T. Furbank
Nuclear regulation of chloroplast development in C_4 and C_3 plants **117**
J.A. Langdale, M. Waters, E.C. Moylan, and A. Bravo-Garcia
Balancing light capture with distributed metabolic demand during C_4 photosynthesis **127**
J.R. Evans, T.C. Vogelmann, and S. von Caemmerer
Redesigning C_4 rice from limited C_4 photosynthesis **145**
D.M. Jiao
Overexpression of C_4 pathway genes in the C_3 dicots potato, tobacco, and *Arabidopsis*: experiences and future challenges **163**
C. Peterhänsel, H.-J. Hirsch, and F. Kreuzaler
Molecular evolution of C_4 photosynthesis in the dicot genus *Flaveria*: implications for the design of a C_4 plant **175**
U. Gowik and P. Westhoff
Learning from nature to develop strategies for the directed evolution of C_4 rice **195**
R. Sage and T.L. Sage

The regulation of genes in C_3 plants that have been co-opted into C_4 photosynthesis, and implications for making a C_4 rice 217
J.M. Hibberd

SECTION 3: SINGLE-CELL C_4 SYSTEMS 233
C_4 rice: early endeavors and models tested 235
J. Burnell
Breaking the Kranz paradigm in terrestrial C_4 plants: does it hold promise for C_4 rice? 249
G.E. Edwards, E. Voznesenskaya, M. Smith, N. Koteyeva, Y.-I. Park, J.H. Park, O. Kiirats, T.W. Okita, and S.D.X. Chuong
Hydrilla: retrofitting a C_3 leaf with a single-cell C_4 NADP-ME system 275
G. Bowes, S.K. Rao, J.B. Reiskind, G.M. Estavillo, and V.S. Rao
The ecology and evolution of single-cell C_4-like photosynthesis in diatoms: relevance to C_4 rice 297
J.A. Raven, K. Roberts, E. Granum, and R.C. Leegood

SECTION 4: THE BACKGROUND AND HOW C_4 RICE CAN BE DELIVERED 315
The promise of systems biology for deciphering the control of C_4 leaf development: transcriptome profiling of leaf cell types 317
T. Nelson, S.L. Tausta, N. Gandotra, T. Liu, T. Ceserani, M. Chen, Y. Jiao, L. Ma, X.-W. Deng, N. Sun, M. Holfold, N. Li, and H. Zhao
Toward C_4 rice: learning from the acclimation of photosynthesis in the C_3 leaf 333
E.H. Murchie and P. Horton
Wild species of *Oryza*: a rich reservoir of genetic variability for rice improvement 351
D.S. Brar and J.M. Ramos
C_4 rice: a plant breeder's perspective 361
P.S. Virk and S. Peng
From allele engineering to phenotype 371
P. Hervé

SECTION 5: SETTING UP THE CONSORTIUM 379
C_4 rice: brainstorming from bioinformaticians 381
R. Bruskiewich and S. Wanchana
Surveying the possible pathways to C_4 rice 399
P.L. Mitchell and J.E. Sheehy

INDEX 413

Section 1:

Setting the scene

How the rice crop works and why it needs a new engine

J.E. Sheehy, A.B. Ferrer, P.L. Mitchell, A. Elmido-Mabilangan, P. Pablico, and M.J.A. Dionora

Rice is the most important crop in the world for human food. Over the past 40 years, its production has kept pace with the increase in population. However, it is clear that the gains of the first Green Revolution are largely exhausted. Rice with C_4 photosynthesis could make a major contribution to a second Green Revolution. To assess how that change could affect rice, it is necessary to understand how the rice crop works.

In this paper, we examine the properties of individual rice plants both as single individuals and as members of dense crop communities. To estimate the potential of C_4 rice, we compare the yields and radiation-use efficiencies of maize, rice, and a C_4 weed. In that context, the properties of rice canopies with respect to the interception of solar radiation and its effect on leaf temperature are examined. The influence of sink size with respect to source strength is also discussed. It is possible that wild rice types have some of the anatomical features peculiar to C_4 plants and that the wild types may contain C_3-C_4 intermediates. Consequently, we report results obtained from an examination of C_4 characteristics in the 22 species of wild rice.

Keywords: Rice, C_4 photosynthesis, radiation-use efficiency, leaf temperature, wild rice types (*Oryza* species)

Of the three major cereals that feed most of the world's population, rice is arguably the most important. Almost all of the 600 million tons produced each year are consumed directly by humans, unlike wheat and maize, of which much is used for animal feed or for industry. About half the world's population has rice as the staple cereal. For humans, rice production is about providing food in a manner that is sustainable economically, socially, and environmentally. For scientists, rice production should be about converting the maximum fraction of solar energy into the maximum amount of chemical energy in grain in the shortest possible time; that conversion should be achieved using the smallest amount of land, water, and fertilizer.

Table 1. Required percentage increases in rice yield likely to be required by 2050, relative to 2004, as a consequence of projected increases in population and projected climate change. The effect of population is an increase in yield in proportion to population increase. The carbon dioxide and temperature component takes into account an increase in concentration of carbon dioxide of up to 150 ppm, a change in temperature of up to 2 °C, and the effects on rice yields. The increase in yields to allow for disasters is calculated from the data for rice production over the past 40 years from the FAOSTAT database (2005).

Country	Percentage yield increase required			Total
	Population	CO_2 + temp.	Disaster	
Philippines	62.0	4.2	30.3	96.5
China	7.7	2.4	9.4	19.5
India	51.2	5.1	35.3	91.6
Bangladesh	77.7	4.3	18.4	100.4

Source: Sheehy et al (2006).

Ninety percent of rice is grown and consumed in Asia, where more than the combined populations of the United States and Europe live on less than US$2 a day (Cline 2004). Those Asians spend as much as 50% of their wages on rice (Dawe 2000). Over the next 50 years, it is predicted that the population of Asia will rise from 3.9 billion to 5.3 billion (UNFPA 2005). Climates are changing and many aspects of climate change such as higher temperatures and weather extremes are likely to have negative impacts on crops. Future increases in rice production will have to occur with less water, less fertilizer, and less land (Hossain and Pingali 1998, Tilman et al 2001, Evans 1998). Given the shortage of land for rice production, it is rice yields that will have to increase as a consequence of increased population and climate change (Table 1). The Green Revolution was built on breeding semidwarf cultivars that could be managed intensively with large inputs of fertilizer. The ability of that model to provide further yield increases is doubtful given that yields in many Asian countries have reached a plateau (Cassman 1999, Dawe, this volume). Indeed, yields in breeders' trials at IRRI have not increased for 30 years (Sheehy 2001a) and it has been suggested that a yield barrier has been reached (Kropff et al 1994).

In the absence of universally accepted scientific theories describing yield and its limits, disagreements about what precisely determines both biomass and grain yield are commonplace. Climate, weather, crop duration, cultivar, and management are the principal determinants of yield. Initially, the seedlings of a crop spend a brief period as individuals not limited by competition for resources from their neighbors. As a more dense community is established, the individual seedling becomes a family of tillers each bearing its own leaves and roots. For most of their existence, tillers are members of a dense community in which they compete with their family members and neighbors for resources. The properties of that community limit the expression of an individual's potential growth and yield. For meaningful crop growth, each productive individual tiller must capture and use sufficient solar energy to produce an acceptable quantity of grain. Nonetheless, the plant density must be high enough to prevent undue loss of solar energy to the soil. Sharing the incident solar energy among the individuals limits each one of them, but allows the community to intercept the available solar energy with greatest efficiency. The attributes of the individuals themselves contribute to the rapidity with which the competitive community is developed, as well as to the productivity of the community. Crop management is about balancing the attributes of the individual with the properties and requirements of the community to produce a yield acceptable to farmers. It is solar energy captured in photosynthesis that gives individual plants the capacity to synthesize, organize, and maintain a range of structural units housing a myriad of metabolic processes (Sheehy 2001b). The flow and use of energy captured by an individual is directed by control mechanisms, some of which must ultimately be peculiar to the genome of an individual species. Those mechanisms dictate plant morphology, anatomy, physiology, and the pattern of growth in a given crop microclimate. The mechanisms are the product of evolution and natural selection and must have guaranteed survival in a world of competition for resources.

Much is made of the potential of the ongoing identification of plant regulatory genes following the sequencing of the rice genome. Currently, the desire to manipulate plant morphology, anatomy, and function in the interests of crop improvement and environmental protection is intense. Higher, more nutritious yields, shorter growing seasons, and greater synchrony in development and maturity are desirable. Traits guaranteeing "survival of the fittest" may not be most suitable for high productivity in intensively managed crop communities of fairly homogeneous, weak individuals. However, ignoring the possibility of transferring traits associated with high productivity across sexually incompatible crop species, such as maize and rice, would seem perverse (Brown et al 2005). In hot climates, eliminating photorespiration while simultaneously reducing nitrogen use and increasing water-use efficiency means converting from C_3 to C_4 photosynthesis (Sheehy 2001b). It would be astonishing if yield improvements in modern cultivars, of fixed duration, were unaccompanied by improvements in canopy photosynthesis (Robson 1982, Long 1999a,b). Consequently, in this paper, we present a brief analysis of how the rice crop works and we attempt to answer the question, What would be the impact of installing C_4 photosynthesis on the future of rice production?

Growth phases of rice

Rice is a weak perennial with two strong phases of logistic growth: vegetative growth followed by reproductive growth (Sheehy et al 2004a). Grain yield was shown to be strongly dependent on weather during the second phase, whereas the maximum weight of the vegetative portion of growth was shown to be largely independent of weather. Using ^{15}N as a tracer, Sheehy et al (2004b) showed that, halfway through grain filling, nitrogen was diverted to the developing "ratoon" tillers, stimulating a possible "third"

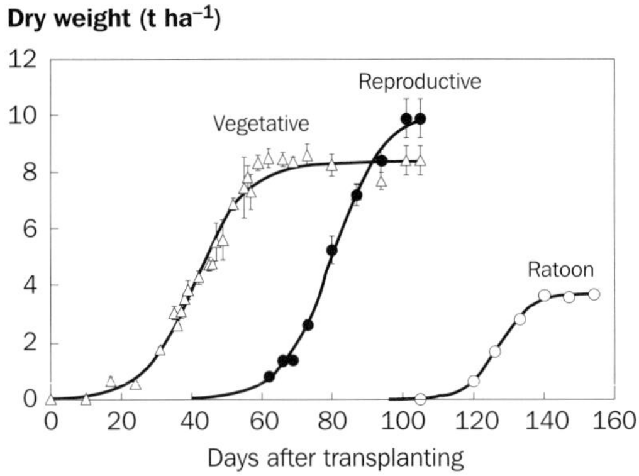

Fig. 1. The triphasic crop growth pattern for irrigated rice in the dry season at IRRI, Los Baños, Philippines. Each phase is represented by a sigmoid curve of the form y = a/(1 + exp[–(x – b)/c]), where y is dry weight, x is time (days after transplanting, DAT), a is the asymptote (t ha^{-1}), b (DAT) is the time at which y is half the asymptote, and c controls the steepness of the middle portion of the curve. Coefficient values (a, b, c) are vegetative, 8.37, 41.28, and 7.45, $r^2 = 0.98$, for $x \geq 0$; reproductive, 10.20, 80.48, and 7.30, $r^2 = 0.99$, for $x \geq 40$; reproductive ratoon generation, 3.71, 127.07, and 4.56, $r^2 = 0.99$, for $x \geq 95$.

phase of growth. A triphasic crop growth pattern for irrigated rice in the dry season is shown in Figure 1; the third phase was achieved without additional fertilizer.

The influence of temperature, on the efficiency of solar energy capture and use, is difficult to describe simply because not all yield-shaping processes respond equally to temperature. Temperature can have different effects on the acquisition of resources, the loss of resources, and the efficiency with which acquired resources are transformed into products. It can also severely damage key mechanisms such as floret fertility. As a crude but useful generalization, there are three phases to the response of yield to mean daily air temperature: (1) 16–22 °C—yields rise from zero to an optimum determined by nutrients and solar irradiance (Horie et al 1995), (2) 22–32 °C—yields decline by about 0.6 t ha^{-1} °C^{-1} (Sheehy et al 2006), and (3) 32–42 °C—floret fertility falls to zero and there is a logistic decline in yield (Satake and Yoshida 1978, Sheehy et al 2006). At temperatures greater than 32 °C, additional carbon dioxide in the atmosphere accelerates spikelet sterility (Matsui et al 1997). In the temperature range of 22–32 °C, additional carbon dioxide in the atmosphere increases crop yields by about 0.5 t ha^{-1} per 75 ppm CO_2 (Baker and Allen 1993).

The grain yield equation

That mutual shading of plants in dense crop canopies leads to low yields has long been an attractive misconception (Sheehy et al 2004c). Ultimately, solar radiation is the energy source for fueling growth and it has to be intercepted by the leaves of the canopy. There is a linear relationship between accumulated intercepted photosynthetically active solar radiation (PAR, 400–700 nm wavelength) and accumulated shoot dry weight (Monteith 1977). The slope of that linear relationship is known as the radiation-use efficiency (ε, g DW MJ^{-1}, where DW is dry weight). Radiation-use efficiency is not strictly a constant and is a conservative quantity, in part, because of the relative crudeness of its estimation (Mitchell et al 1998). Nonetheless, ε is a useful rule-of-thumb for comparing yields achieved from crops with different durations and gives some insight into whether or not there are any differences in the intrinsic physiological efficiencies of such crops. Using that concept, a simple model of grain yield can be written:

$$Y_g = H \varepsilon \int_{t_i}^{t_f} I_{int}(t)\, dt \qquad (1)$$

where Y_g is grain yield, H is harvest index (unless otherwise stated, calculated as the fraction of aboveground dry weight that is grain weight), t_i is the day of transplanting and t_f is the day of harvest, and I_{int} is the total amount of PAR intercepted by the crop. Using the data of Sheehy et al (1998), it can be shown that the harvest index of well-managed rice crops is nearly constant ($H = -0.004\ Y_{sb} + 0.59$, $P < 0.01$) over a wide range of shoot biomass (Y_{sb}). The yields of crops are simply proportional to their radiation-use efficiencies when they have similar crop durations, harvest indices, and root weight ratios (Mitchell et al 1998). Given that the ε value of maize is about 50% greater than that of rice, Sheehy et al (2001b) suggested that maize should outyield rice by 50%.

Plasticity: properties of individuals and community members

The definition of harvest index contains no description of the individual units of production, which are the tillers. The relationship between tillering and yield in rice has been studied for over a hundred years (Inagaki 1898). Jacobson (1916) reported that increased tillering was accompanied by decreasing numbers of grains per panicle. Fifty-six years later, Yoshida and Parao (1972) observed the same inverse relationship for modern cultivars. Tillers arise from buds that develop in the axils of leaves (Robson et al 1988). For rice, the increase in numbers of tillers (with at least one visible leaf) per leaf number interval (phyllochron) on the main stem can be described by a Fibonacci series up to a given leaf number. Shading, light quality, stem elongation, and the development of the panicle as a competing sink for assimilates all lead to a reduction and an eventual cessation in the predicted rate of tiller production. For a tiller to survive, the carbon inflow must equal the carbon lost in maintenance respiration. To fully develop, the individual tiller has to be able to meet the additional carbon requirements associated with the synthesis of its various organs such as leaves and panicles (Ziska et al 1997, Baker et al 1992).

The number of tillers produced by a single seedling is strongly influenced by the density of transplanting. However, for a given cultivar, a common tiller density is often achieved independently of sowing or transplanting density (Harper 1977, Zhong et al 1999). In a field experiment, using high-yielding cultivar IR72 grown at a density of one plant m^{-2}, there were 149 tillers at maturity, giving a yield of 3 t ha^{-1} (Fig. 2A). In the same experiment, when IR72 was transplanted at 25 plants m^{-2}, there were 35 tillers per plant (875 m^{-2}) but only 27 of them were productive, giving a yield of 11.6 t ha^{-1} (Sheehy et al 2000). In contrast, new plant type (NPT) transplanted at 50 plants m^{-2} produced about 8 tillers per plant (400 m^{-2}) and the same yield of 11.6 t ha^{-1} (Fig. 2B). Clearly, the size of a tiller family depends on transplanting density and, at very low density, the number of tillers per unit ground area can influence yield. Nonetheless, in crops not limited by nutrients and transplanted at the densities used in practice, differences in tiller number per unit ground area influence yield only through differences in light interception. The later a tiller appears in the sequence of tillers that comprise a rice hill, the smaller its contribution to crop yield (Fig. 2C).

What is the link between yield, leaf area, and tiller density? At full light interception, the average leaf area of an individual tiller or plant, l_i, is given by

$$l_i = L_{max}/N \qquad (2)$$

where N is the number of identical tillers per unit ground area and L_{max} is the maximum leaf area index for full light interception and yield (Sinclair and Sheehy 1999).

Using equation 2, we can calculate l_i for IR72 and the NPT when $L_{max} = 11.2$. Assuming there are approximately 675 productive tillers per square meter (T m^{-2}) in IR72, l_i would be 166 cm^2. For the NPT with 350 T m^{-2}, l_i would be 320 cm^2. By contrast, for maize with the same LAI and at 10 plants m^{-2}, l_i would be 11,200 cm^2. Furthermore, if there are approximately 4 live leaves per tiller in rice and 10 in maize, an individual leaf of maize must be about 14–27 times larger than a rice leaf. In addition, for the same grain yield, the size of the cob in maize has to be about 35–68 times larger than a rice panicle. The detailed anatomical and biomechanical implications of changing plant size are not immediately obvious, although Niklas (1992) discussed many of the principles. In maize, Kranz anatomy and vein density may have valuable biomechanical properties in addition to those associated with the concentrating mechanism for CO_2.

The challenge of integrating known mechanical principles into growth models was addressed by Silk (1984), who described the advantages of hollow panicles for flexural rigidity. Flexural rigidity (F) depends on Young's modulus of elasticity and that is a function of the composition of the plant tissue. Flexural rigidity is also a function of the moment of inertia and that depends on the geometrical arrangement of the material. Broader issues of mechanical strength have been addressed by Niklas (1994). However, as yet, such principles have not been consciously used to design desirable plant types. Maurice et al (1997) addressed problems relating to the biomechanics of an individual grass leaf, describing form and mass distribution. If yields are to be

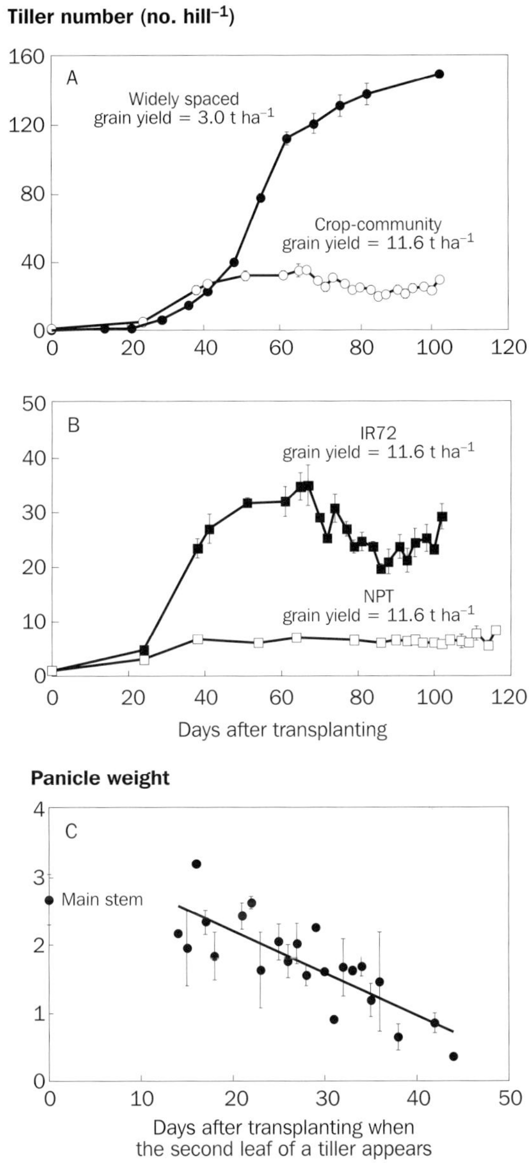

Fig. 2. (A) Tiller number of widely spaced and crop-community plants of irrigated rice IR72 in 1997 dry season; (B) tiller number of crop-community plants of irrigated rice IR72 and NPT in 1997 dry season; IR72 has the same data as in (A) plotted at higher resolution. Grain yields at 14% moisture content are shown for reference in both figures; error bars are standard error where n = 4. Tiller number is given as number per hill, where the hill is a planting position, normally with a single plant in experimental crops; (C) the relationship between the weight of a panicle at maturity (y) and the time after transplanting that the second leaf of its supporting tiller appeared (x); $y = -0.062 + 3.43$, $r^2 = 0.65$. In addition, the weight of the panicle of the main stem is shown.

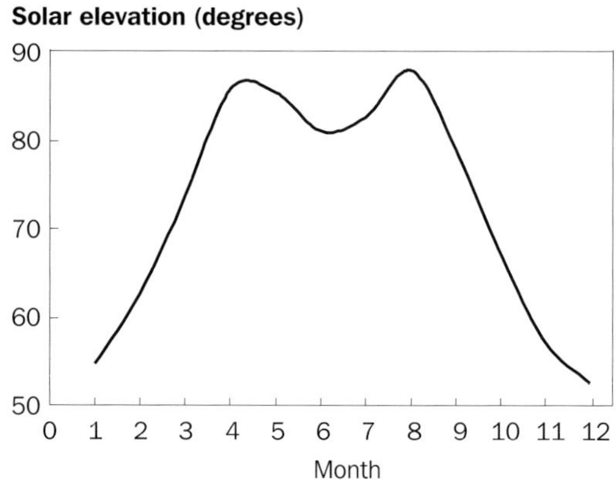

Fig. 3. The annual pattern of maximum solar elevation (mid-month) at IRRI, Los Baños, Philippines (14°11′N). Source: NOAA Solar Position Calculator at www.srrb.noaa.gov/highlights/sunrise/azel.html.

increased by 50% in a single growing season, the mechanical strength of stems and roots must be increased.

Solar radiation and canopy architecture as drivers of canopy photosynthesis

The annual pattern of solar elevation at mid-day, measured in the middle of each month, is bimodal in the tropics (Fig. 3). The monsoon climate at such a location ensures that solar irradiance can be highly variable from day to day and across seasons and years. Solar irradiances (400–2,400 nm wavelength), for the same date, in different years can vary from about 2 to 30 MJ m^{-2} day^{-1} (Fig. 4A). Temperatures at such locations are less variable, but mean values for a given day can vary by about 6 °C (Fig. 4B). The structure of a rice canopy is not uniform in the sense that each hill resembles an inverted cone with the tillers widely spaced at the top and tightly bunched near the bottom. The irradiance experienced by individual leaves depends on solar elevation, leaf depth in the canopy, erectness, and orientation to the sun. Consequently, the leaves of individual tillers experience continuous fluctuations in the energy and matter fluxes peculiar to their location in the canopy and solar elevation. Furthermore, the leaves of a rice canopy are rarely still and canopy architecture is not as uniform as theoretical models often suggest. Nonetheless, simple models of PAR distribution in canopies and its consequences for canopy photosynthesis are valuable. Two models are of interest. The first is the Bouguer–Lambert law (Monsi and Saeki 1953):

$$I/I_o = \exp(-k_{par}L) \qquad (3)$$

where I_o is the irradiance (PAR) above the canopy, I is the irradiance (PAR) at some level in the canopy beneath a leaf area per unit ground area of L, and k_{par} is the extinc-

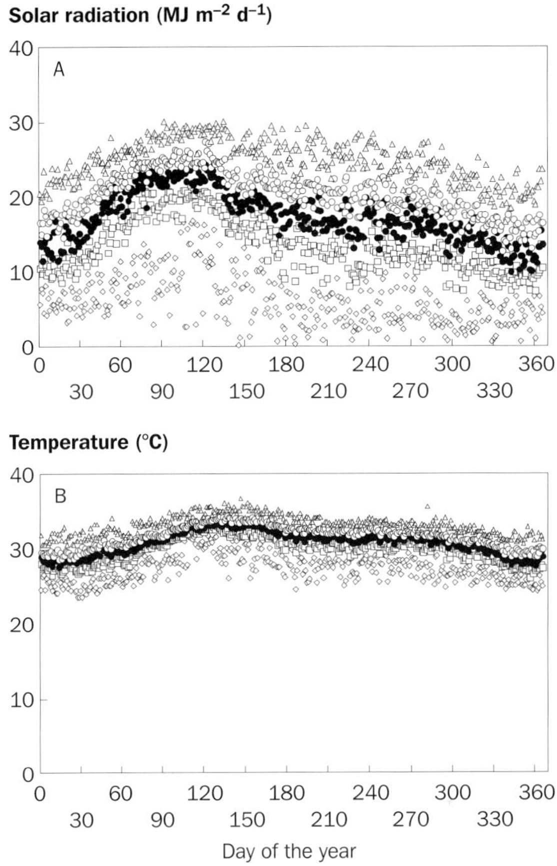

Fig. 4. Annual patterns of (A) solar radiation and (B) maximum temperature at IRRI, Los Baños, Philippines; data extracted from the IRRI weather database (1979-2005); daily median, •; third quartile, ○; first quartile, □; absolute maximum, △; absolute minimum, ◇.

tion coefficient for PAR. Measurements of the extinction coefficient in a rice canopy show that k_{par} varies with solar elevation as shown in Figure 5; the variation is more marked in clear conditions (Fig. 5A) than in overcast conditions (Fig. 5B).

Another useful model for calculating light (PAR) distribution and canopy photosynthesis was described by Monteith (1965) in terms of the fraction of light transmitted through unit leaf area index without interception (s). The leaf area of the canopy is divided into sunlit, once-shaded, and twice-shaded leaves, that is, those receiving direct sunlight, those receiving light transmitted through one leaf, and those receiving light after transmission through two leaves. The extinction coefficient, k, and s are related by the equation

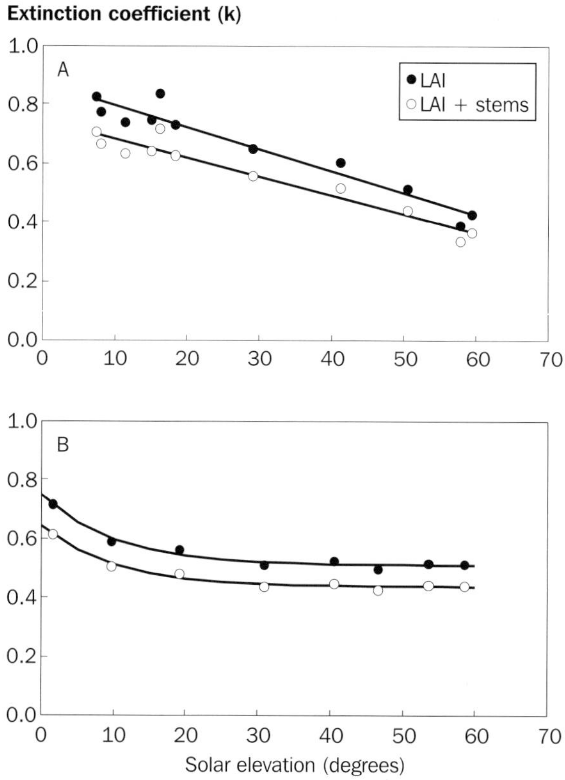

Fig. 5. Values for the extinction coefficient at different solar elevations calculated for LAI and LAI + stem area index (shoot) under (A) direct (LAI: $y = -0.0075x + 0.87$, $r^2 = 0.93$; LAI + stems: $y = -0.0064x + 0.75$, $r^2 = 0.93$) and (B) diffuse light conditions (LAI: $y = 0.51 + 0.24 \exp(-x/9.94)$, $r^2 = 0.98$; LAI + stems: $y = 0.44 + 0.21 \exp(-x/9.94)$, $r^2 = 0.98$).

$$s = [\exp(-k_{par}) - m]/[1 - m] \quad (4)$$

where m is the fraction of light transmitted through a leaf (Sheehy and Johnson 1988).

The s values for IR72 were calculated (equation 4) using the extinction coefficients for different solar elevations (Fig. 5) and a value for m of 0.1. The irradiance (PAR) above the canopy was measured on a horizontal surface at different solar elevations for clear and overcast conditions. The irradiances (PAR) of sunlit, once-shaded, and twice-shaded leaves, in the canopy, were calculated as a function of solar elevation and are shown in Figure 6. These irradiances are the values of PAR that would be measured on a surface at the same orientation as the leaf surface. In

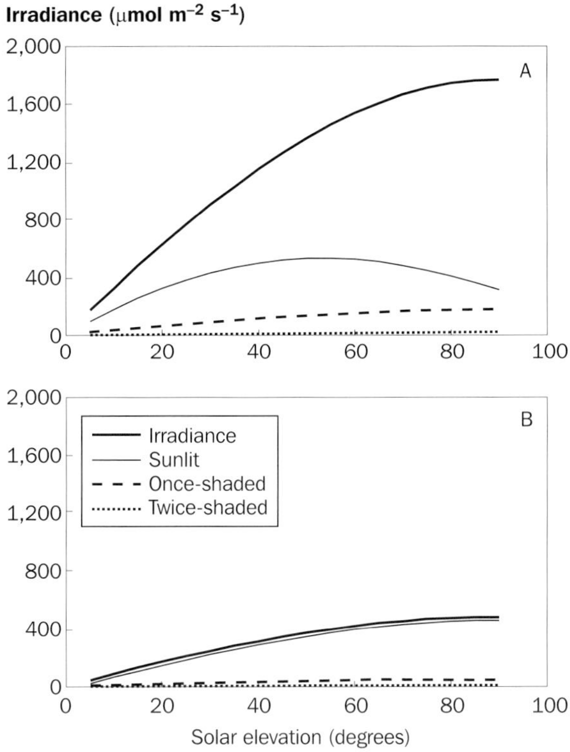

Fig. 6. The irradiances of sunlit, once-shaded, and twice-shaded leaves of an IR72 rice canopy under (A) direct and (B) diffuse light conditions for irradiance conditions in mid-April at Los Baños, Philippines (14°11′N, 121°15′E, altitude 21 m). Note that incident PAR is for a horizontal surface, but the PAR for leaves is that calculated for a surface at the same orientation as the leaf.

clear conditions (Fig. 6A), irradiance (PAR) above the canopy reached a maximum of 1,766 μmol m^{-2} s^{-1} at a solar elevation of 90°. The maximum irradiance (PAR) of sunlit leaves was 530 μmol m^{-2} s^{-1} at a solar elevation of about 55° and it decreased to 312 μmol m^{-2} s^{-1} at 90°. For overcast conditions (Fig. 6B), the irradiance (PAR) above the canopy and of nonshaded (sunlit) leaves was almost exactly the same at all solar elevations; the maximum value shown at 90° was 482 μmol m^{-2} s^{-1}. The irradiances of the nonshaded (sunlit) leaves in clear and overcast conditions were surprisingly similar given the differences in the amount of incident PAR above the canopy for those conditions. The main difference between overcast and clear conditions was found in the PAR received on the once-shaded leaves. The maximum PAR experienced by once-shaded and twice-shaded leaves in a rice canopy was estimated to be 177 and 18 μmol m^{-2} s^{-1} for clear conditions and 48 and 5 μmol m^{-2} s^{-1} for overcast conditions, respectively. Over that range, differences in the maximum rate of individual leaf photosynthesis, and to a lesser extent quantum yield at low PAR, will determine the actual rate of leaf photosynthesis. At an irradiance (PAR) of 500 μmol m^{-2} s^{-1}, the rate of leaf photosynthesis of rice is about 25 μmol CO_2 m^{-2} s^{-1} and the rate for maize is up to double that value (Mitchell and Sheehy 2000).

Canopy architecture is important for canopy photosynthesis, but we have to take into account the effects of solar elevation, as well as whether the irradiance is direct or diffuse, before the magnitude of its importance can be calculated. Furthermore, canopy architecture and solar elevation result in leaves that are not light saturated even at the highest irradiances observed on clear days.

Relationship between leaf photosynthesis, canopy photosynthesis, and yield

Can leaves of C_3 plants photosynthesize at the same rates as those of C_4 plants? Evans and von Caemmerer (2000) showed that the maximum rate of leaf photosynthesis per unit leaf area for both C_3 and C_4 plants was a linear function of leaf N content; the slope of the relationship for C_4s was greater than for C_3s. At high leaf N contents, the maximum rate of individual leaf photosynthesis per unit leaf area in C_3 plants can be as high as that in C_4 plants with lower N contents. So, leaves of individual C_3 plants can have rates of photosynthesis comparable to those of C_4 leaves. Sheehy et al (1980) showed that, in a population of individually spaced alfalfa plants, maximum photosynthetic rates per unit leaf area varied from about 13 to 51 μmol CO_2 m^{-2} s^{-1}. In addition, there was a good relationship between whole-plant photosynthesis and plant N content, but there was no relationship between individual leaf photosynthesis per unit leaf area and whole-plant photosynthesis. Whole-plant photosynthesis depends on both the rate per unit leaf area and the total leaf area of the whole plant and these can be somewhat independent of each other. Pearce et al (1969) showed that leaf photosynthesis in alfalfa depended on specific leaf weight (dry weight of leaf for unit area), so small thick leaves could have much higher rates of photosynthesis than larger thinner leaves; specific leaf weight was shown to be largely governed by growing conditions. Therefore, it was not surprising that there was no relationship between individual leaf photosynthesis and whole-plant photosynthesis in the experiments of Sheehy et al (1980). Furthermore, the maximum rate of photosynthesis of successive youngest fully expanded leaves in grass canopies declined because they developed inside the canopy in increasingly shaded conditions (Sheehy 1977, Woledge 1973). The photosynthetic rate of leaves developing on plants grown as spaced individuals differs from the photosynthetic rate of leaves developing on plants growing in dense communities (Sheehy 2001b). As an interesting aside, in the experiments of Sheehy et al (1980), the rate of biological nitrogen fixation in alfalfa depended on whole-plant photosynthesis, which in turn depended on whole-plant N content.

In crops, leaves dilute their nitrogen, and consequently their photosynthetic machinery, as they reduce their specific leaf weight and expand their leaf area owing to competition for light (Sheehy 2001b, Lemaire et al 2007). Indeed, Greenwood et al (1990) showed that, for optimally fertilized C_3 and C_4 crops, the relationship between %N and plant dry matter per unit ground area had the same form; the relationship

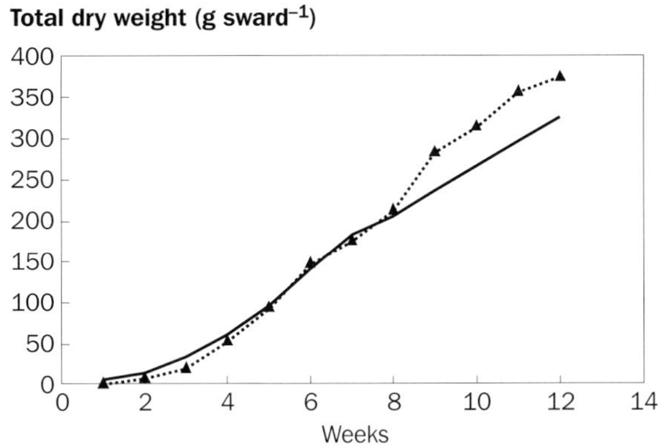

Fig. 7. The relationship between cumulative biomass production (▲) and cumulative measured canopy gross photosynthesis (——, hexose equivalent dry weight) in swards of perennial ryegrass (after Robson 1973).

declined with increasing biomass according to a power law. Under optimal supply of nitrogen, C_4 crops contained a lower concentration of N than C_3 crops at the same biomass, probably as a consequence of their higher rates of photosynthesis per unit of N.

Canopy and leaf photosynthesis were measured in grass canopies and the maximum rate of canopy photosynthesis per unit leaf area was correlated with the maximum rate of leaf photosynthesis per unit leaf area (Sheehy 1977). In those experiments, the relationship between canopy photosynthesis and irradiance was described using a simple hyperbolic relationship. In a more rigorous theoretical analysis of the relationship between canopy photosynthesis, leaf photosynthesis, and irradiance, Sheehy and Johnson (1988) showed that the maximum quantum yield of the grass crop depended on the fractional light interception, leaf transmissivity, and the maximum quantum yield of an individual leaf. They also showed that the maximum rate of canopy photosynthesis depended on LAI and the maximum rate of leaf photosynthesis. At a given temperature and concentration of atmospheric CO_2, canopy photosynthesis is completely governed by irradiance, canopy architecture, and leaf photosynthesis. Robson (1973) showed a very close relationship between cumulative biomass production and cumulative measured canopy gross photosynthesis in swards of perennial ryegrass (Fig. 7). The relationship is not surprising because the carbon content of plants is approximately 40%. In conclusion, there are clear relationships between leaf photosynthesis, canopy photosynthesis, and biomass.

Leaf and canopy temperature

Long (1999a,b) predicted daily rates of canopy photosynthesis for C_3 and C_4 canopies and suggested that C_3s have temperature optima close to 23 °C whereas rates for C_4s were still increasing at 35 °C. Leaf temperature is influenced by meteorological conditions as well as stomatal and boundary layer resistances. For a nontranspiring leaf, the energy balance equation can be written as

$$\rho c_p (T_d - T_a)/r_b = \alpha R_s - R_{dl} \quad (5)$$

where ρ is the density of the air, c_p is the specific heat capacity, T_d is the temperature of the nontranspiring leaf, T_a is the temperature of the air, r_b is the boundary layer resistance of the leaf, α is the fractional absorption of solar energy of a leaf, R_s is the irradiance of the leaf ($R_s = I_o(1 - s)$), and R_{dl} is the net emission of long-wave radiation by the leaf. By rearranging the equation, the difference between the temperature of a nontranspiring leaf and air temperature can be written as

$$T_d - T_a = r_b \alpha R_s/\rho c_p - r_b \alpha R_{dl}/\rho c_p \quad (6)$$

To evaluate the parameters of equation 6, leaves of the NPT were smeared with petroleum jelly to prevent transpiration. The temperatures of smeared and nonsmeared leaves were measured using an infrared thermometer. The difference between the temperature of the nontranspiring leaves and air temperature was plotted against the PAR incident on the leaves; the relationship was significant, albeit not impressively ($P<0.01$; Fig. 8A). Assuming that the leaf is a black body ($\alpha = 1$) and using equation 6, the boundary layer resistance was calculated to be 53 s m^{-1} and R_{dl} to be 70 W m^{-2}; both are reasonable values for crops (Woodward and Sheehy 1983, Monteith 1973). The energy balance for a transpiring leaf can be written as

$$\rho c_p (T_a - T_l)/r_b + (\alpha R_s - R_{wl}) = \lambda E \quad (7)$$

where T_l is the temperature of the transpiring leaf, R_{wl} is the net emission of long-wave radiation by the transpiring leaf, λ is the latent heat of vaporization, and E is the rate of transpiration. In theory, there is no simple relationship between leaf temperature and absorbed radiation, but in practice there was a very good correlation ($P<0.001$; Fig. 8B). In addition, leaf temperature was less than air temperature ($T_l = 0.96T_a$; $P < 0.001$).

Ku et al (2000) suggested that, when the maize PEPC gene was inserted in rice, leaf conductance increased. It is interesting to ask, Would changing rice from being a C_3 to a C_4 have any effect on leaf temperature via altered leaf conductance? Clearly, this is not an easy question to answer in the absence of C_4 rice growing in the field. At the time of writing this paper, the best we could do to provide a clue was to compare photosynthesis, transpiration, and leaf temperatures in rice (IR72), a C_4 weed (*Echinochloa glabrescens*), and maize growing as well-watered individual plants in a screenhouse. Measurements made with a Licor 6400 showed that the rate of leaf photosynthesis of the C_4 leaves was approximately 29% greater than that of rice (Table 2). Conversely, the rate of leaf transpiration of the C_4s was approximately 68% of the rate for rice and the value of leaf conductance for C_4s was 45% of the value for rice. The temperature measurements showed that the C_4 leaves were warmer (33 °C) than

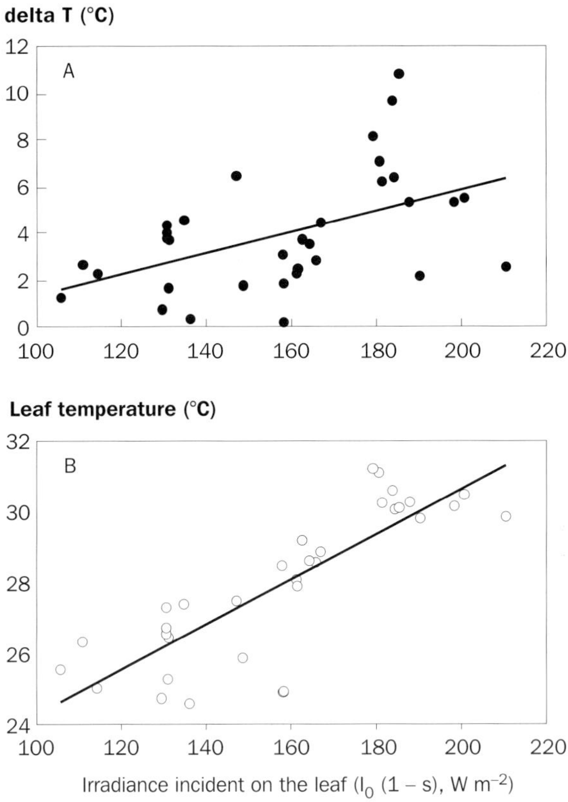

Fig. 8. (A) The relationship between the difference between leaf and air temperature and irradiance incident on the leaf for nontranspiring leaves of rice ($y = 0.045x - 3.17$, $r^2 = 0.24$). (B) The relationship between leaf temperature and irradiance incident on the transpiring leaf in rice ($y = 0.064x + 17.92$, $r^2 = 0.68$).

the C_3 leaves (31 °C), but both were less than air temperature (34.5 °C). The results suggest that leaf temperatures in rice could increase by about 2 °C as a consequence of changing the photosynthetic pathway from C_3 to C_4. When transpiration was prevented by smearing the leaves with petroleum jelly, there were differences in the leaf temperatures (T_d) of the different species (Table 2). This suggested that the properties of the leaves (other than stomatal conductance) governing heat exchange were different. Those properties are probably associated with the absorption of radiation (Davies and Buttimor 1969) or boundary layer resistance or both.

The result of Ku et al (2000) suggested that C_4 rice leaves might be cooler than C_3 rice leaves. However, the comparison between rice and the C_4 species used here

Table 2. Measurements of gas exchange and leaf temperature made on the first fully expanded leaves of plants growing in well-watered pots in a screenhouse at IRRI. The PAR in the Licor 6400 chamber was 1,000 μmol m^{-2} s^{-1} and in the screenhouse was 1,140 μmol m^{-2} s^{-1}; the air temperature in the screenhouse was 34.5 °C. Leaves were coated with petroleum jelly to prevent transpiration. The standard errors are in parentheses, n = 6.

Item	C_4 plants		C_3 plants
	Zea mays	*E. glabrescens*	*O. sativa* (IR72)
Rate of photosynthesis (μmol m^{-2} s^{-1})	31.3 (0.05)	28.8 (0.61)	23.6 (2.20)
Stomatal conductance (μmol m^{-2} s^{-1})	0.2 (0.01)	0.2 (0.01)	0.4 (0.00)
Rate of transpiration (μmol m^{-2} s^{-1})	4.1 (0.22)	3.9 (0.09)	5.9 (0.56)
Leaf temperature in Licor chamber (°C)	33.5 (0.20)	32.8 (0.30)	31.2 (0.20)
Leaf temperature in screenhouse (°C)	32.8 (0.10)	32.9 (0.10)	31.1 (0.10)
Temperature for leaves coated with petroleum jelly (°C)	39.1 (0.20)	37.3 (0.20)	35.4 (0.20)

suggests the opposite. The effect of C_4-ness on leaf and canopy temperature in rice remains unclear.

Radiation-use efficiency: two strategies for C_4 rice

Early theoretical work at IRRI suggested that grain yields of 15 t ha^{-1} were possible. This suggestion rested on erroneous values for the efficiency of radiation conversion used by Yoshida (1981). The instantaneous value of the radiation conversion factor (or radiation-use efficiency), ε, expressed as g DW MJ^{-1} (dry weight above ground, intercepted PAR in energy terms), can be written (Sheehy 2001a) as

$$\varepsilon = \frac{0.64\, P_g(t) - m_T\, W_s(t) - D_s(t)}{I_{int}} \tag{8}$$

where P_g is canopy gross photosynthesis (shoot net photosynthesis plus shoot respiration for the daylight hours), W_s is shoot weight and D_s is the rate of detachment of shoot weight, m_T is the maintenance respiration coefficient at temperature T, I_{int} is the daily total of intercepted PAR (MJ m^{-2} d^{-1}), assuming that each variable has been measured for a day, and t is time. $D_s(t)$ is negligible during vegetative growth and so it can be seen that ε is strongly influenced by photosynthesis.

An experiment to compare yields and radiation-use efficiencies of rice, maize, and the C_4 weed *Echinochloa glabrescens* was conducted at IRRI in the dry season of 2006.

The crops were sown and transplanted so that they started to grow and intercept PAR at about the same time. All the crops received 340 kg N ha^{-1}, 50 kg P ha^{-1}, and 340 kg K ha^{-1} and all of P and K fertilizers were applied as basal fertilizer incorporated a day before planting. For IR72 and maize, the N was split and applied weekly as follows: 60% at 14–50 DAT, 30% at 50–70 DAT, and 10% at 70–90 DAT. For *E. glabrescens*, the N was split and applied weekly as follows: 60% at 0–21 DAT and 40% at 21–50 DAT. The rice and weed were grown as irrigated crops (flood irrigation) and maize was irrigated every other day (i.e., grown in soil without surface water but kept well watered). Interception of PAR was measured twice weekly using a Delta T Sunscan probe and total aboveground biomass was measured weekly in the standard way (Sheehy et al 2004b, Cassman et al 1993). Both rice and maize were followed to maturity (rice, 98 days; maize, 101 days), whereas measurements ceased in *E. glabrescens* when the seeds started to shatter (42 days after transplanting). It can be seen in Figure 9A that the weed closed its canopy earlier than maize, and rice was the slowest to close. Monteith (1977) defined the slope of the relationship between shoot biomass and cumulative intercepted PAR as ε (Fig. 9B). In the above experiment, the values of ε were 4.4 g DW MJ^{-1} for maize, 4.0 g DW MJ^{-1} for *E. glabrescens*, and 2.9 g DW MJ^{-1} for rice. At maturity, the total aboveground biomass of rice was 17.9 $\pm$ 0.38 t ha^{-1} and that of maize, on the same day, was 28.8 $\pm$ 2.2 t ha^{-1}. The ratio of the values of ε were maize: rice 1.52 and *E. glabrescens*:rice 1.38. At 14% moisture content, the grain yield for maize was 13.9 $\pm$ 0.13 t ha^{-1} and for rice was 8.3 $\pm$ 0.13 t ha^{-1}.

The ratios of the radiation-use efficiencies and the ratios of the grain yields for maize and rice strongly suggest that C_4 rice would be substantially more productive than C_3 rice. From these results, we suggest two strategies that could be adopted for the crop duration of C_4 rice: a maize-like duration (100 days) and a weed-like duration (50 days). With the maize-like duration, C_4 rice biomass would be 50% greater than C_3 rice biomass so the plants would have to be correspondingly larger. With the weed-like duration, the biomass of C_4 rice would be comparable with that of a 100-day-duration C_3 rice, but it would be achieved in about 60 days.

Is the sink in rice big enough for C_4 productivity?

Table 3 shows the number of juvenile spikelets 10–15 days before panicle emergence, the number of spikelets at maturity, and the number of filled spikelets (grains) measured for IR72 in two dry seasons (Sheehy et al 2001). It can be seen that the capacity of rice crops for spikelet production (more than 100,000 m^{-2}) is more than double the final number of grains at maturity (less than 50,000 m^{-2}). Increased photosynthesis for 33 days prior to heading as a consequence of carbon dioxide enrichment was the probable cause of the 30% increase in grain yields observed by Yoshida (1973). In his experiments, the treatment increased mature spikelet number and the improved yield was the result of an increased filling percentage and individual grain weight.

The sink is much larger than required for C_3 rice and the evidence suggests that another 50% of the juvenile spikelets could be converted into grains in C_4 rice.

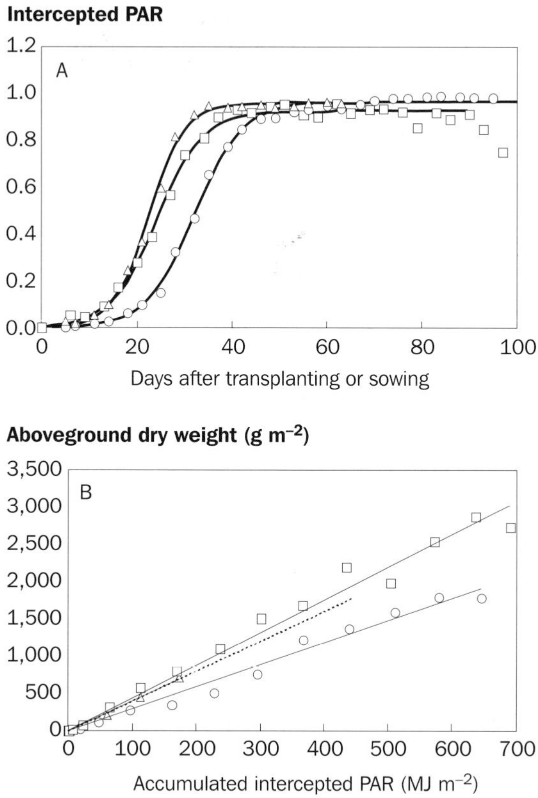

Fig. 9. (A) The time courses of fractional intercepted PAR during crop growth following transplanting (rice (IR72), ○; *Echinochloa glabrescens*, △) or sowing (maize, □). The fitted curves are of the form y = a/(1 + exp(–(x – b)/c)) and are used simply to highlight the differences; note that interception in maize declined after day 90. (B) The relationship between accumulated intercepted PAR (MJ m^{-2}) and aboveground dry weight for rice (IR72), *Echinochloa glabrescens*, and maize; symbols as in (A).

Searching for C_4-ness in wild rice

Rice belongs to the tribe *Oryzeae*, which consists of 12 genera (Vaughan 1994). The genus *Oryza* contains 24 species, two are cultivated and the others are "wild" rice; there are about 6,000 wild rice accessions in the IRRI germplasm collection. The wild types have not been studied in detail, but past work has suggested that some of the wild types have intermediate C_3-C_4 characteristics. Some accessions of *O. rufipogon* were reported to have CO_2 compensation points of about 30 µmol mol^{-1} and PEP carboxylase activity of about 3 µmol min^{-1} mg^{-1} chlorophyll (Yeo et al 1994). Furthermore, those authors observed that the photorespiration rates of *O. rufipogon*

Table 3. The mean number of juvenile spikelets, spikelets at maturity, filled spikelets, and 1,000-grain weight in crops of IR72 in the dry seasons of 1997 and 1999.

Item	Mean number
Number of juvenile spikelets (m^{-2})	113,848
Spikelets at maturity (m^{-2})	51,372
Filled spikelets at maturity (m^{-2})	38,793
1,000-grain wt. (g, 14% moisture content)	24.0

Source: Sheehy et al (2001).

were 25% lower than the rates of *O. sativa*. In addition, there is evidence for some C_4 characteristics in rice spikelets (Imaizumi et al 1997). The conclusion of those authors was that the lemmas carried out mainly C_3 photosynthesis, but also fixed some carbon dioxide by PEPcase, a mixture of routes not typically C_3 or C_4 but perhaps adapted to re-fix abundant carbon dioxide from respiration; Kranz anatomy was lacking. As a preliminary investigation, it was decided to screen a representative sample of 130 accessions drawn from the 6,000 wild relatives of rice (WRS) for aspects of anatomy and physiology associated with C_4ness. A representative from each species in the IRRI collection was included in the subsample. A small collection of C_4 plants was used to characterize some of the attributes of their leaves: *Digitaria ciliaris*, *Echinochloa colona*, *E. crus-galli*, *E. glabrescens*, another species of *Echinochloa* identified to genus only, *Panicum maximum*, and *Rottboellia cochinchinensis*.

C_4 plants discriminate less than C_3 plants against the heavier isotope ^{13}C, and the ratio of ^{13}C to ^{12}C ($\delta^{13}C$) is used to identify plants with a C_4 pathway (Cerling 1999). The $\delta^{13}C$ values for the WRS ranged from –32% to –25.1% and showed there were no C_4 types in the subsample (Fig. 10). The number of veins in a youngest fully expanded leaf (N_v) was significantly correlated with leaf width (L_w). In C_4 leaves, the relationship was $N_v = 11\ L_w$ ($P < 0.001$, leaf width in mm) and in WRS leaves it was $N_v = 5\ L_w$ ($P < 0.001$); C_4 leaves contain twice as many veins per unit leaf width than C_3 leaves. The interveinal spacing (at the middle of the blade) for the WRS ranged from 113 to 322 μm (the value for IR72 was 170 μm). The range for the C_4 weeds was 93 to 136 μm. Currently, mesophyll cell size is being estimated for the subsample and C_4 weeds. The total number of bundle sheath cells (BSC) and the number containing chloroplasts in small veins were counted. The percentage of the BSC containing chloroplasts and the percentage of BSC plan area occupied by chloroplasts were estimated; Table 4 shows some of the values for selected species. Interestingly, 100% of the BSC contained chloroplasts in *O. longistaminata* and 48% of its plan area was occupied by chloroplasts; in the C_4 species, more than 80% of the BSC plan area was occupied.

Wild rice types probably have some of the anatomical features peculiar to C_4 plants and the wild types may contain C_3-C_4 intermediates. In 2007, an enclosure technique for mass screening rice seedlings will be developed and the whole wild rice collection will be screened for photosynthetic efficiency.

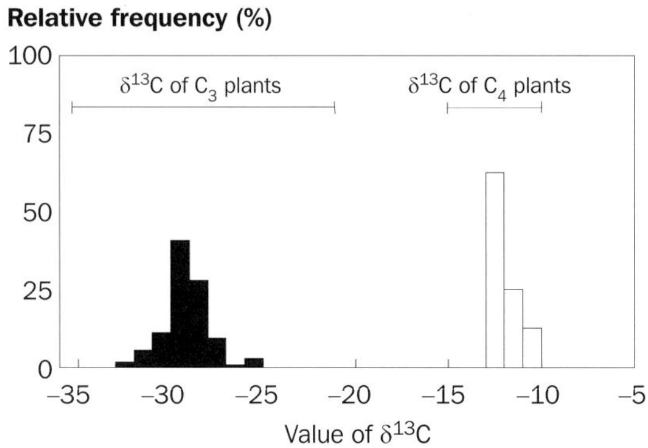

Fig. 10. Distribution of $\delta^{13}C$ values of the representative subsample of wild rice (■) and a small collection of C_4 plants (□). The range of $\delta^{13}C$ values for C_3 and C_4 plants shown is from Cerling (1999).

Table 4. Percentage of bundle sheath cells (BSC) containing chloroplasts in small vascular bundles and percentage of BSC plan area apparently occupied by chloroplasts, determined using confocal microscopy.

Species	Source	BSC with chloroplast (%)	Chloroplast area in the BSC (%)
C_4	Our collection	100	>80
C_3	Cultivated rice		
	Oryza sativa (IR64)	100	21
	Some wild rice		
	O. alta	50	52
	O. australiensis	88	41
	O. barthii	64	41
	O. longistaminata	100	48
C_3-C_4	*Panicum milioides*	100	50

Conclusions

The reason for converting the photosynthetic system in rice from C_3 to C_4 is necessity rather than curiosity. It is not good enough to be optimistic that "business as usual" will solve the problem of increasing future rice yields. New and possibly radical approaches need to be explored urgently. Using fuel more efficiently in a car with a nearly emissions-free engine is undoubtedly part of the future of motoring. Large sums of money have been invested, and the hybrid engines and fuel cells of today represent the legacy of yesterday's research. Using sunlight, land, water, and other resources more efficiently to produce food is an even greater imperative in the face of increasing populations, climate change, and economic uncertainties of the future. Can there really be any doubt that research aimed at providing the very best engine for the rice plant should be the highest priority of the International Rice Research Institute? Evolution has invented such an engine (C_4 photosynthesis) and we need to install it in the world's most important food crop.

It will take an international consortium of research institutions to make this a reality over the next 10 to 15 years. To that end, IRRI formed a C_4 Rice Consortium involving scientists from both advanced institutions and the developing countries. The Consortium will chart and conduct the research needed to invent C_4 rice and will seek financial support from donors.

References

Baker JT, Allen LH Jr, Boote KJ. 1992. Effects of CO_2 and temperature on growth and yield of rice. J. Exp. Bot. 43:959-964.

Baker JT, Allen LH Jr. 1993. Effects of CO_2 and temperature on rice: a summary of five growing seasons. J. Agric. Meteorol. 48:575-582.

Brown NJ, Parsley K, Hibberd JM. 2005. The future of C_4 research: maize, *Flaveria* or *Cleome*? Trends Plant Sci. 10:215-221.

Cassman KG, Kropff MJ, Gaunt J, Peng S. 1993. Nitrogen use efficiency of rice reconsidered: What are the key constraints? Plant Soil 155/156:359-362.

Cassman KG. 1999. Ecological intensification of cereal production systems: yield potential, soil quality, and precision agriculture. Proc. Natl. Acad. Sci. USA 96:5952-5959.

Cerling TE. 1999. Paleorecords of C_4 plants and ecosystems. In: Sage RF, Monson RK, editors. C_4 plant biology. San Diego, Calif. (USA): Academic Press. p 455-469.

Cline WR. 2004. Trade policy and global poverty. Washington, D.C.: Center for Global Development and Institute for International Economics.

Davies JA, Buttimor PH. 1969. Reflection coefficients, heating coefficients and net radiation at Simcoe, Southern Ontario. Agric. Meteorol. 6:373-386.

Dawe D. 2000. The contribution of rice research to poverty alleviation. In: Sheehy JE, Mitchell PL, Hardy B, editors. Redesigning rice photosynthesis to improve yield. Amsterdam (The Netherlands): Elsevier. p 3-12.

Evans JR, Von Caemmerer S. 2000. Would C_4 rice produce more biomass than C_3 rice? In: Sheehy JE, Mitchell PL, Hardy B, editors. Redesigning rice photosynthesis to improve yield. Amsterdam (The Netherlands): Elsevier. p 53-71.

Evans LT. 1998. Feeding the ten billion: plants and population growth. Cambridge (UK): Cambridge University Press.

Greenwood DJ, Lemaire G, Gosse G, Cruz P, Draycott A, Neeteson JJ. 1990. Decline in percentage N of C_3 and C_4 crops with increasing plant mass. Ann. Bot. 66:425-436.

Harper JL, 1977. The population biology of plants. New York, N.Y. (USA): Academic Press.

Horie T, Nakagawa H, Centeno HGS, Kropff MJ. 1995. The rice crop simulation model SIMRIW and its testing. In: Matthews RB, Kropff MJ, Bachelet D, van Laar HH, editors. Modeling the impact of climate change on rice production in Asia. Wallingford (UK): CAB International, and Manila (Philippines): International Rice Research Institute. p 51-66.

Hossain M, Pingali PL. 1998. Rice research, technological progress, and impact on productivity and poverty: an overview. In: Pingali PL, Hossain M, editors. Impact of rice research. Manila (Philippines): International Rice Research Institute. p 1-26.

Imaizumi N, Samejima M, Ishihara K. 1997. Characteristics of photosynthetic carbon metabolism of spikelets in rice. Photosyn. Res. 52:75-82.

Inagaki I. 1898. On the number of rice shoots. Bull. Coll. Agric. Tokyo Imp. Univ. 3:416-420.

Jacobson HO. 1916. Correlative characters of the rice plant. Philipp. Agric. Rev. 9:74-119.

Kropff MJ, Cassman KG, Peng S, Matthews RB, Setter TL. 1994. Quantitative understanding of yield potential. In: Cassman KG, editor. Breaking the yield barrier. Proceedings of a workshop on rice yield potential in favorable environments. Manila (Philippines): International Rice Research Institute. p 21-38.

Ku MSB, Cho D, Ranade U, Hsu T-P, Li X, Jiao D-M, Ehleringer J, Miyao M, Matsuoka M. 2000. Photosynthetic performance of transgenic rice plants overexpressing maize C_4 photosynthesis enzymes. In: Sheehy JE, Mitchell PL, Hardy B, editors. Redesigning rice photosynthesis to improve yield. Amsterdam (The Netherlands): Elsevier. p 193-204.

Lemaire G, van Oosterom E, Sheehy JE, Jeuffroy MH, Massignam A, Rossato L. 2007. Is crop N demand more closely related to dry matter accumulation or leaf area expansion during vegetative growth? Field Crops Res. 100:91-106.

Long SP. 1999a. Understanding the impacts of rising CO_2: the contribution of environmental physiology. In: Press MC, Scholes JD, Barker MG, editors. Physiological plant ecology. Oxford (UK): Blackwell Science Ltd. p 263-282.

Long SP. 1999b. Environmental responses. In: Sage RF, Monson RK, editors. C_4 plant biology. London (UK): Academic Press. p 215-249.

Matsui T, Namuco OS, Ziska LH, Horie T. 1997. Effects of high temperature and CO_2 concentration on spikelet sterility in indica rice. Field Crops Res. 51:213-219.

Maurice I, Gastal F, Durand JL. 1997. Generation of form and associated mass deposition during leaf development in grasses: a kinematic approach for non-steady growth. Ann. Bot. 80:673-683.

Mitchell PL, Sheehy JE. 2000. Performance of a potential C_4 rice: overview from quantum yield to grain yield. In: Sheehy JE, Mitchell PL, Hardy B, editors. Redesigning rice photosynthesis to improve yield. Amsterdam (The Netherlands): Elsevier. p 145-163.

Mitchell PL, Sheehy JE, Woodward FI. 1998. Potential yields and the efficiency of radiation use in rice. Manila (Philippines): IRRI Discussion Paper Series No. 32. 62 p.

Monsi M, Saeki T. 1953. Über den Lichtfaktor in den Pflanzengesellschaften und seine Bedeutung für die Stoffproduktion. Jpn. J. Bot. 14:22-52.

Monteith JL. 1973. Principles of environmental physics. London (UK): Edward Arnold Limited. 241 p.

Monteith JL. 1977. Climate and the efficiency of crop production in Britain. Philos. Trans. Royal Soc. London B281:277-294.

Monteith JL. 1965. Light distribution and photosynthesis in field crops. Ann. Bot. 29:17-37.

Niklas KJ. 1992. Plant biomechanics: an engineering approach to plant form and function. Chicago, Ill. (USA): The University of Chicago Press. 607 p.

Niklas KJ. 1994. Plant allometry: the scaling of form and process. Chicago, Ill. (USA): The University of Chicago Press. 395 p.

Pearce RB, Carlson GE, Barnes DK, Hart RH, Hanson CH. 1969. Specific leaf weight and photosynthesis in alfalfa. Crop Sci. 9:423-426.

Robson MJ. 1973. The growth and development of simulated swards of perennial rye grass. I. Leaf growth and dry weight changes as related to the ceiling yield of a seedling sward. Ann. Bot. 37:487-500.

Robson MJ. 1982. The growth and carbon economy of selection lines of *Lolium perrene* cv. 323 with differing rates of dark respiration. I. Grown as simulated swards during a regrowth period. Ann. Bot. 49:321-329.

Robson MJ, Ryle GJA, Woledge J. 1988. The grass plant: its form and function. In: Jones MB, Lazenby A, editors. The grass crop: the physiological basis of production. London (UK): Chapman and Hall Ltd. p 25-83.

Satake T, Yoshida S. 1978. High temperature-induced sterility in indica rices at flowering. Jpn. J. Crop Sci. 47(1):6-17.

Sheehy JE. 1977. Microclimate, canopy structure and photosynthesis in canopies of three contrasting temperate forage grasses. III. Canopy photosynthesis, individual leaf photosynthesis and the distribution of current assimilate. Ann. Bot. 41:593-604.

Sheehy JE, Fishbeck KA, Phillips DA. 1980. Relationships between apparent nitrogen fixation and carbon exchange rate in alfalfa. Crop Sci. 20:491-495.

Sheehy JE, Johnson IR. 1988. Physiological models of grass growth. In: Jones MB, Lazenby A, editors. The grass crop: the physiological basis of production. London (UK): Chapman and Hall. p 243-275.

Sheehy JE, Dionora MJA, Mitchell PL, Peng S, Cassman KG, Lemaire G, Williams RL. 1998. Critical nitrogen concentrations: implications for high yielding rice (*Oryza sativa* L.) cultivars in the tropics. Field Crops Res. 59:31-41.

Sheehy JE, Mitchell PL, Dionora MJA, Tsukaguchi T, Peng S, Khush GS. 2000. Unlocking the yield barrier in rice through a nitrogen-led improvement in the radiation conversion factor. Plant Prod. Sci. 3:372-374.

Sheehy JE, Dionora MJA, Mitchell PL. 2001. Spikelet numbers, sink size and potential yield in rice. Field Crops Res. 71:77-85.

Sheehy JE. 2001a. Will yield barriers limit future rice production? In: Nösberger J, Geiger HH, Struik PC, editors. Crop science: progress and prospects. Hamburg (Germany): CAB International. p 281-305.

Sheehy JE. 2001b. Future food requirements: are improvements in photosynthesis required? Proceedings of the 12th International Congress on Photosynthesis 2001. Melbourne (Australia): CSIRO Publishing.

Sheehy JE, Mitchell PL, Ferrer AB. 2004a. Bi-phasic growth patterns in rice. Ann. Bot. 94:811-817.

Sheehy JE, Mnzava M, Cassman KG, Mitchell PL, Pablico P, Robles RP, Ferrer AB. 2004b. Uptake of nitrogen by rice studied with a point-placement technique. Plant Soil 259:259-265

Sheehy JE, Peng S, Dobermann A, Mitchell PL, Ferrer A, Jianchang Yang, Yingbin Zou, Xuhua Zhong, Jianliang Huang. 2004c. Fantastic yields in the system of rice intensification: fact or fallacy? Field Crops Res. 88:1-8.

Sheehy JE, Mitchell PL, Allen LH, Ferrer AB. 2006. Mathematical consequences of using various empirical expressions of crop yield as a function of temperature. Field Crops Res. 98:216-221.

Silk WK. 1984. Quantitative descriptions of development. Annu. Rev. Plant Physiol. 35:479-518.

Sinclair TR, Sheehy JE. 1999. Erect leaves and photosynthesis in rice. Science 283:1456-1457.

Tilman D, Fargione J, Wolff B, D'Antonio C, Dobson A, Howarth R, Schindler D, Schlesinger WH, Simberloff D, Swackhamer D. 2001. Forecasting agriculturally driven global environmental change. Science 292:281-284.

UNFPA (United Nations Population Fund). 2005. State of world population 2005. New York (USA).www.unfpa.org.
Vaughan DA. 1994. The wild relatives of rice: a genetic resources handbook. Manila (Philippines): International Rice Research Institute. 137 p.
Woledge J. 1973. The photosynthesis of ryegrass leaves grown in a simulated sward. Ann. Appl. Biol. 73:229-237.
Woodward FI, Sheehy JE. 1983. Principles and measurements in environmental biology. London (UK): Butterworth & Co. Ltd. 263 p.
Yeo ME, Yeo AR, Flowers TJ. 1994. Photosynthesis and photorespiration in the genus *Oryza*. J. Exp. Bot. 45:553-560.
Yoshida S. 1973. Effects of CO_2 enrichment at different stages of panicle development on yield components and yield of rice (*Oryza sativa* L.). Soil Sci. Plant Nutr. 19:311-316.
Yoshida S. 1981. Fundamentals of rice crop science. Manila (Philippines): International Rice Research Institute. 269 p.
Yoshida S, Parao FT. 1972. Performance of improved rice varieties in the tropics with special reference to tillering capacity. Exp. Agric. 8:203-212.
Zhong X, Peng S, Sheehy JE, Liu H, Visperas RM. 1999. Parameterization, validation and comparison of three tillering models for irrigated rice in the tropics. Plant Prod. Sci. 2(4):258-266.
Ziska LH, Namuco O, Moya T, Quilang J. 1997. Growth and yield response of field grown tropical rice to increasing carbon dioxide and air temperature. Agron. J. 89:45-53.

Notes

Authors' addresses: J.E. Sheehy, A.B. Ferrer, A. Elmido-Mabilangan, P. Pablico, and M.J.A. Dionora, Crop and Environmental Sciences Division, International Rice Research Institute, DAPO Box 7777, Metro Manila, Philippines, e-mail: j.sheehy@cgiar.org; P.L. Mitchell, Department of Animal and Plant Sciences, University of Sheffield, Sheffield S10 2TN, UK.
Acknowledgment: We gratefully acknowledge the scientific and financial support given for this work by Dr. Robert S. Zeigler, without whose enthusiasm the C_4 conference would not have taken place.

The case for C_4 rice

P.L. Mitchell and J.E. Sheehy

The case for C_4 rice is summarized as a chain of argument, starting from the need to produce more rice, through higher yield potential, to the consequent improvements in growth and photosynthesis. We conclude that it is essential to make Rubisco, the key enzyme of photosynthesis, work harder, by concentrating carbon dioxide around Rubisco, thus raising the light-saturated rate of photosynthesis and greatly reducing photorespiration, as occurs in C_4 photosynthesis. The additional agricultural benefits of shifting from C_3 to C_4 rice are that C_4 photosynthesis requires (1) less Rubisco and hence less nitrogen, and (2) less water, since a steeper concentration gradient for carbon dioxide diffusion can be maintained through partly closed stomata.

The chain of argument also shows that research should be planned from the perspective of what we wish to achieve, that is, from the top downward, and quantitatively as far as possible. Developments in screening phenotypes, plant breeding, molecular biology, and genetic engineering are proceeding rapidly and need to be directed toward the applied goal. The pathway to success cannot be seen completely but it is very likely that techniques will arise to enable the construction of C_4 rice. The certainties of population growth, climate change, and future shortages of water for agriculture mean that it is essential to start research now.

Keywords: C_4 photosynthesis, C_4 rice, Rubisco, top-down planning, yield potential

We started thinking seriously about C_4 rice in 1998, and we were far from being the first (see Burnell, this volume; Akita 1994). Three pieces of information were influential (Mitchell et al 1998).

1. Crop biomass and yield are directly proportional to the solar radiation intercepted by the crop (Monteith 1977) and the constant of proportionality, a measure of how effectively radiation is converted to energy in biomass, is the radiation conversion factor, so-called radiation-use efficiency (RUE).
2. There is a consistent large difference in RUE between C_4 maize and C_3 rice.

3. Ultimately, RUE depends on the surplus of photosynthesis over respiration; little can be done to decrease respiration so higher RUE, biomass, and yield require increased photosynthesis.

Evidence is also accumulating to show that the C_4 mechanism is added to the basic system of C_3 photosynthesis, and has been added by evolution on many occasions (Mitchell and Sheehy 2006); nonetheless, C_4 rice has not evolved naturally so making it is an ambitious project. For nearly a decade, the idea has continued to seem attractive to us, almost inevitable for a substantial gain in rice productivity, and we have refined our arguments for C_4 rice. There are two parts to the case. The first is a chain of argument leading from the need to reduce hunger and poverty (Millennium Development Goals—IRRI 2006), through an increase in rice productivity, the need for a substantial increase in yield potential, and a corresponding increase in photosynthesis, to the features of C_4 photosynthesis that appear so advantageous. The second part is the rapid progress made in screening methods for desirable phenotypes, plant breeding with biotechnology, molecular biology, and genetic engineering. The chain of argument would remain as armchair speculation if it were not for the astonishing rates of progress in these fields.

The chain of argument

Most of the justification for the chain of argument, set out in Table 1, can be found in Dawe (2000), Sheehy (2000), and Mitchell and Sheehy (2000); it is also summarized in Mitchell and Sheehy (2006). A completely independent analysis by Long et al (2006) for grain crops reaches the same conclusion: that improved photosynthesis is necessary to increase yields, and they explore several methods for this, overlapping with ideas in Table 1. We will comment briefly on a few links in the chain.

For Requirement 5, there are several reasons for selecting a 50% increase in yield potential. First, yield potential in rice (tropical, growth duration around 110 days) has stagnated for 30 years at about 10 t ha^{-1} (Cassman et al 2003); it is time to catch up. Second, rice yields of 15 t ha^{-1} do occur in temperate regions with growth durations of 5–6 months (IRRI 2002) so rice culms (stems) can be strong enough for panicles that are 50% heavier (assuming the same number of culms per unit area). Third, the RUE of maize, a typical productive C_4 plant, is 50% greater than that of rice (Mitchell and Sheehy 2000), recently confirmed by experiments at IRRI (Sheehy et al, this volume).

An alternative method of raising yield potential (Requirement 5) would be to take a smaller yield from each of two crops of shorter duration, for example, 7.5 t ha^{-1} in 50 days, both fitted into the tropical dry season when solar radiation is the highest in the year. This is prompted by the observation (Sheehy et al, this volume) that the rapid growth of the C_4 weed *Echinochloa glabrescens*, RUE 38% higher than rice, was channeled into a much shorter growth duration: 42 days from transplanting of seedlings to seed dispersal. A very short duration rice of this high productivity (150 kg ha^{-1} day^{-1}) would also permit more diverse cropping patterns, which could include

Table 1. The chain of argument leading to C_4 rice. Abbreviations are given at the end of the table.

Requirement	Reasons
1. Alleviation of hunger and poverty in global regions dependent on rice	Simple humanity; contribution to Millennium Development Goals and to the missions of IRRI and the CGIAR.
2. Rice crops	Rice, rather than other crops, for cultural, nutritional, and agro-climatic preferences.
3. Increased productivity of rice crops	Provides several routes to alleviation of poverty through lower price of rice, higher profits for farmers, increased demand for agricultural labor.
4. Increase in yield per hectare	From increased productivity, and because suitable arable area in rice regions is decreasing slowly but steadily.
5. Increase yield potential of new cultivars by 50%	To increase yields while maintaining yield gap between yield potential and farm yields. Increase from 10 t ha^{-1} to 15 t ha^{-1} for which rice culms are strong enough (110-day growth duration, yield from 90 to 135 kg ha^{-1} day^{-1}). Or a yield of 7.5 t ha^{-1} over a duration of 50 days (150 kg ha^{-1} day^{-1}), to fit two crops into the dry season with high solar radiation.
6. Increase crop biomass by 50%	Cannot increase harvest index beyond 0.5, which is the best currently obtainable.
7. Increase in number of spikelets filled by 50%	Many more spikelets are initiated on the panicle than appear as filled grains so there is unused capacity for yield.
8. Increase RUE by 50%	Crop growth is driven by intercepted PAR, and cannot increase crop duration, incident PAR, or fraction of PAR intercepted.
9. Increase canopy photosynthesis by 50%	Daily RUE is balance of photosynthesis and respiration, and there are no obvious ways of reducing respiration.
10. Increase in leaf photosynthesis	To contribute to canopy photosynthesis.
11. Higher rate of leaf photosynthesis when saturated by PAR	For leaves in the canopy receiving high PAR, i.e., high in the canopy, at appropriate angles, when incident PAR is high.
12. Higher quantum yield	For leaves in the canopy receiving low PAR, i.e., in canopies with PAR well distributed through the canopy, or when incident PAR is low.
13. Better performance from Rubisco	No alternative to Rubisco for continuous net fixation of carbon dioxide into carbohydrate, but suffers from oxygenation activity, hence occurrence of photorespiration.

Continued on next page

Table 1 continued.

Requirement	Reasons
Options	
(a) More Rubisco in thicker leaves	More photosynthesis from more Rubisco (but requires more nitrogen).
(b) Better Rubisco: higher specificity for carbon dioxide	Reduced photorespiration (but also brings lower rate of catalytic activity so may not increase rate of leaf photosynthesis by much).
(c) Protect Rubisco from oxygen: confine to chloroplasts without PSII	Reduced photorespiration (but not necessarily a large increase in rate of leaf photosynthesis).
(d) Make Rubisco work harder by increasing concentration of carbon dioxide around Rubisco	Absence of photorespiration and more effective use of Rubisco in higher concentration of carbon dioxide.
14. From 13, option (d): C_4 photosynthesis as productive as maize	Maize RUE 50% higher than rice RUE; both maximum rate of leaf photosynthesis and quantum yield higher than in rice. Better use of nitrogen and water.
15. Features of C_4 photosynthesis in maize	
I. Concentration of carbon dioxide around Rubisco raised 3–8 times higher than in C_3 plants	Raises maximum rate of leaf photosynthesis.
II. Low leakage from the compartment with Rubisco, by low conductance for carbon dioxide	Minimizes futile cycling and minimizes reduction in quantum yield from leakage.
Options	
(a) C_4 photosynthesis with Kranz anatomy	By far the commonest C_4 system, high productivity in many cases (maize, sugar cane, *Miscanthus*).
(b) Single-cell C_4 photosynthesis with low leakage	Avoids complexity of anatomical modification; requires changes within existing photosynthetic cells only; allows design of systems not occurring naturally.
16. From 15, option (a): C_4 photosynthesis with Kranz anatomy, as in maize	
I. Kranz anatomy	Spatial separation of initial fixation in mesophyll cells from final fixation in bundle sheath cells but connection by abundant plasmodesmata for rapid diffusion of metabolites. Reduction in leakage of carbon dioxide from bundle sheath cells to maintain high concentration around Rubisco. Close vein spacing so each mesophyll cell is adjacent to a bundle sheath cell.

Continued on next page

Table 1 continued.

Requirement	Reasons
II. Changed pattern of expression of photosynthetic enzymes	In mesophyll cells, need PEPcase of C_4 type in large amounts, and CA and PPDK. In bundle sheath cells, need NADP-ME (or other decarboxylating enzyme) and Rubisco, absence of CA, absence of PSII. Appropriate changes to transporters across the chloroplast envelope.
17. From 15, option (b): single-cell C_4 photosynthesis with low leakage	
I. Chloroplast envelope with reduced conductance to carbon dioxide	Need to reduce leakage from the compartment in which carbon dioxide concentration is raised to minimize futile cycling and maximize quantum yield.
II. Changed pattern of expression of photosynthetic enzymes	In cytosol, need PEPcase of C_4 type in large amounts, and CA and PPDK (if carboxylase is NADP-ME). In chloroplast, need decarboxylase (NADP-ME or PEPCK), absence of CA. Appropriate changes to transporters across the chloroplast envelope.

Abbreviations in Table 1: CA = carbonic anhydrase, CGIAR = Consultative Group on International Agricultural Research, NADP-ME = NADP-dependent malic enzyme, PAR = photosynthetically active radiation, PEPcase = phospho*enol*pyruvate carboxylase, PEPCK = phospho*enol*pyruvate carboxykinase, PPDK = pyruvate, orthophosphate dikinase, PSII = photosystem II, Rubisco = ribulose 1,5-bisphosphate carboxylase–oxygenase, RUE = radiation-use efficiency.

extra vegetable crops or crops for biofuel, addressing the alleviation of hunger and poverty (Requirement 1) in other ways.

Requirement 7 tackles the problem of matching source and sink in the formation of grain yield. Sheehy et al (2001) showed that many more spikelets (florets) are initiated on the panicle than appear as filled grains, that is, there is spare capacity in the sink, which can use photosynthate from a larger source.

The key role of Rubisco (Requirement 13) arises from its being the only carboxylase to provide a continuous net gain in fixed carbon, given biochemistry based on carbohydrate, particularly trioses, pentoses, and hexoses. Rubisco can be made to work harder (Option d of Requirement 13) by providing it with conditions more like the primitive atmosphere in which it evolved, an atmosphere higher in carbon dioxide and much lower in oxygen. Experiments with elevated concentrations of carbon dioxide show that Rubisco in rice can indeed respond with increased photosynthesis (Baker and Allen 1993, Ziska et al 1997, Kobayashi et al 2005). The objective of constructing C_4 rice is to make the plant itself provide a higher concentration of carbon dioxide around Rubisco; in C_4 plants, this concentration is 3–8 times higher than in C_3 plants (Requirement 15, I; Kanai and Edwards 1999).

Requirement 15, II refers to the sensitivity of quantum yield in C_4 plants to leakage from the compartment in which carbon dioxide concentration is raised (bundle sheath cells in Kranz anatomy, chloroplasts in single-cell C_4). When carbon dioxide is lost from this compartment, the energy required to regenerate phospho*enol*pyruvate (PEP) from pyruvate has been used in vain (since carbon dioxide has been fixed and then re-appears after decarboxylation—futile cycling). The advantage in quantum yield that C_4 plants have over C_3 plants is because the absence of photorespiration outweighs the cost of regenerating PEP (at temperatures above 21 °C) but the advantage is easily squandered if leakage is high (Mitchell and Sheehy 2000).

Requirements 16 and 17 sketch the pathways to C_4 rice, with or without Kranz anatomy, respectively.

We have focused on the route to C_4 photosynthesis in Table 1 but there are branching points from the chain. For example, from Requirement 8, in principle it could be worth lengthening the growth duration of the wet-season crop in the humid tropics when only one crop is to be taken, to make better use of a growing season of 7 months. Possibly there are changes in the timing of canopy growth and in canopy architecture that could produce improved canopy photosynthesis that would lead to higher yields (Requirement 9). Option b in Requirement 13 has been explored by Zhu et al (2004), as summarized in Mitchell (this volume). Accepting the inevitable oxygenase activity of Rubisco but arranging that the photorespiratory mechanism releases carbon dioxide in the chloroplast, for recapture by Rubisco, has been tried by Peterhänsel et al (this volume). Long et al (2006) make several suggestions, some mentioned above (canopy architecture) or in Table 1 (better Rubisco), but in addition a better rate of recovery from photoprotection so that leaves experiencing a decrease in light below saturation resume photosynthesis more rapidly, or faster regeneration of ribulose 1,5-bisphosphate (RuBP) to increase rates of light-saturated photosynthesis. Raines (2006) also discusses some of these possible improvements to photosynthesis, and others such as improving the heat stability of Rubisco activase above 30 °C, or using a cyanobacterial enzyme for the accumulation of inorganic carbon, introduced into tobacco and *Arabidopsis* with some encouraging results although the precise function of the enzyme is unknown.

Any or several of these methods may be effective in increasing rice photosynthesis, perhaps in the range of 10–25%. But only C_4 photosynthesis offers the prospect of a substantial increase of 50% plus more effective use of nitrogen and water (summarized in Mitchell and Sheehy 2006). Although significant amounts of enzymes in the C_4 pathway are required, they are more than offset by the decreased amount of Rubisco in C_4 plants so that overall the requirement for nitrogen is smaller (Greenwood et al 1990). The combination of carbonic anhydrase and PEP carboxylase in the mesophyll cells of C_4 plants takes up carbon dioxide very effectively, producing a low concentration in the air spaces of the leaf. This steepens the concentration gradient so that a rapid flux of carbon dioxide can be maintained even with a smaller stomatal conductance (reduced aperture), thus reducing transpiration. This will be a direct benefit to upland rice and to rainfed rice when the soil is dry. For irrigated rice, for which most water is used for soil preparation and flood irrigation, the increased yield per hectare will ensure higher water productivity for the crop.

Progress in techniques

Improvements in screening for phenotypes and modern plant breeding using biotechnology are covered by Hervé (this volume) and by Virk and Peng (this volume). What we have noted with interest, as nonpractitioners, is the progress in molecular biology and genetic engineering that appears relevant to the task of making C_4 rice. It is now routine to insert genes for C_4 photosynthetic enzymes into a C_3 plant and have them expressed successfully (Häusler et al 2002). The rice genome and most of the maize genome have been sequenced and are being annotated (Bruskiewich and Wanchana, this volume). There is increasing use of specific promoters so that genes are switched on only in particular tissues, or in response to environmental cues, or at particular stages of plant development. Many enhancer trap lines have been identified in rice (Wu et al 2003, Johnson et al 2005, Liang et al 2006) and these can be used for analysis of the genome since it is possible to control where in a plant the transgene is expressed. Plastids are being transformed, that is, genes can be inserted into the chloroplast and be expressed there (Maliga 2002, Lee et al 2006). We conclude that genetic engineering will provide a method for introducing C_4 photosynthesis into C_3 plants once the key genes are identified, not immediately but in the medium-term future.

As an example of the rate of progress, consider the speculation in August 2005 of one of us (P.L.M.), unhindered by expertise in this field, on possible methods of keeping together several genes introduced to rice by genetic engineering. "Perhaps the necessary genes could be inserted into the chromosome together so that segregation is minimized. In the future, will it be possible to introduce genes to particular positions in a selected chromosome? Are there genes that prevent crossover, perhaps of specific chromosomes? Could the genes be introduced on two small extra chromosomes, which pair and segregate but do not cross over, as sex chromosomes do?" In less than 18 months, we learned (Hervé, this volume) that several of these suggestions are already in use (site-specific recombination technology, chromosome-based engineering, artificial chromosomes).

The top-down approach

The top-down approach is appropriate for any complex task, certainly for one as difficult as constructing C_4 rice. It is essential to specify the final objective and work backward to plan the stages required to reach the objective. However, research so far on producing C_4 rice seems to have been pushed by technology: it is possible to introduce genes for a few enzymes of the C_4 pathway, thought to increase the rate of photosynthesis, so it has been tried (e.g., Ku et al 1999). This tendency seems to be common where genetic engineering is applied to crops. Sinclair et al (2004) pointed out that the weakness of this approach is that improvements made at the molecular level are dissipated when scaled up through biochemical and physiological levels to the response of crops in the field. In their calculations for soya bean, a 50% increase in messenger RNA for Rubisco declined to a 33% increase in maximum rate of leaf photosynthesis, then to an 18% increase in crop biomass, and finally to a 6% increase in grain yield. The impetus for this kind of research generally comes from a newly developed technique in genetic engineering, so from the bottom, with the hope that the improvement will be worthwhile when scaled up.

In contrast, to construct C_4 rice will require a top-down approach, specifying the ultimate objective and then the steps required to achieve it, as in Table 1. The pathway to C_4 rice may not be short and direct: there are likely to be some surprises along the way, and additional refinements found to be necessary once the C_4 mechanism is introduced into rice. Some of these have been alluded to, such as the mechanical strength of culms and crop duration. The top-down approach also encourages a quantitative view since a numerical requirement at one step has implications for the next step down. For example, we need to state a desired and quantified improvement in yield potential, and then specify the changes in photosynthesis required. This then leads to seeking the solutions in terms of biochemistry, or biochemistry and anatomy, either in natural variation or in the genetic engineering required, or very likely both. Of course, to elaborate and quantify all the steps there is much work to be done, which will include mathematical modeling, experiments, and fieldwork, and probably comparative work on other species such as C_4 weeds of rice.

Envoi

It is easy to suggest that the construction of C_4 rice will be especially difficult given current knowledge, or that the cost might be unusually high for agricultural research. But the need is correspondingly high: global population continues to increase, climate change will alter cropping patterns and probably reduce yields, and water available for agriculture will become scarce or more expensive (IRRI 2006). The chain of argument (Table 1) leads us to the inescapable conclusion that we must make C_4 rice in order to achieve a 50% increase in yield potential while using nitrogen and water more efficiently. Constructing C_4 rice is a high-reward, high-risk venture, likely to take at least 15 years to complete. It will require the ingenuity and skills of researchers from around the world, hence the formation of a Consortium for C_4 Rice. We must start now, confident that developments in plant breeding, including genetic engineering, will provide the techniques required.

References

Akita S. 1994. Eco-physiological aspects of raising the yield plateau of irrigated rice in the tropics. In: Cassman KG, editor. Breaking the yield barrier. Proceedings of a workshop on Rice Yield Potential in Favorable Environments, 29 November-4 December 1993, at IRRI. Manila (Philippines): International Rice Research Institute. p 85-89.

Baker JT, Allen LH. 1993. Effects of CO_2 and temperature on rice: a summary of five growing seasons. J. Agric. Meteorol. 48:575-582.

Cassman KG, Dobermann A, Walters DT, Yang H. 2003. Meeting cereal demand while protecting natural resources and improving environmental quality. Annu. Rev. Environ. Res. 28:315-358.

Dawe D. 2000. The contribution of rice research to poverty alleviation. In: Sheehy JE, Mitchell PL, Hardy B, editors. Redesigning rice photosynthesis to improve yield. Proceedings of a workshop on The Quest to Reduce Hunger: Redesigning Rice Photosynthesis, 30 November-3 December 1999, at IRRI. Amsterdam (Netherlands): Elsevier, and Makati City (Philippines): International Rice Research Institute. p 3-12.

Greenwood DJ, Lemaire G, Gosse G, Cruz P, Draycott A, Neeteson JJ. 1990. Decline in percentage N of C_3 and C_4 crops with increasing plant mass. Ann. Bot. 66:425-436.

Häusler RE, Hirsch HJ, Kreuzaler F, Peterhänsel C. 2002. Overexpression of C_4-cycle enzymes in transgenic C_3 plants: a biotechnological approach to improve C_3 photosynthesis. J. Exp. Bot. 53:591-607.

IRRI (International Rice Research Institute). 2002. Rice almanac. Third edition. In: Maclean JL, Dawe DC, Hardy B, Hettel GP, editors. Manila (Philippines): International Rice Research Institute. 253 p.

IRRI (International Rice Research Institute). 2006. Bringing hope, improving lives: strategic plan 2007-2015. Manila (Philippines): IRRI. 61 p.

Johnson AAT, Hibberd JM, Gay C, Essah PA, Haseloff J, Tester M, Guiderdoni E. 2005. Spatial control of transgene expression in rice (*Oryza sativa* L.) using the GAL4 enhancer trapping system. Plant J. 41:779-789.

Kanai R, Edwards G.E. 1999. The biochemistry of C_4 photosynthesis. In: Sage RF, Monson RK, editors. C_4 plant biology. London (UK): Academic Press. p 49-87.

Kobayashi K, Kim H, Lieffering M, Miura S, Okada M. 2005. Paddy rice response to free-air CO_2 enrichment. J. Agric. Meteorol. 60:475-479.

Ku MSB, Agarie S, Nomura M, Fukayama H, Tsuchida H, Ono K, Hirose S, Toki S, Miyao M, Matsuoka M. 1999. High-level expression of maize phospho*enol*pyruvate carboxylase in transgenic rice plants. Nature Biotechnol. 17:76-80.

Lee SM, Kang K, Chung H, Yoo SH, Xu XM, Lee S-B, Cheong J-J, Daniell H, Kim M. 2006. Plastid transformation in the monocotyledonous cereal crop, rice (*Oryza sativa*) and transmission of transgenes to their progeny. Mol. Cells 21:401-410.

Liang D, Wu C, Li C, Xu C, Zhang J, Kilian A, Li X, Zhang Q, Xiong L. 2006. Establishment of a patterned GAL4-VP16 transactivation system for discovering gene function in rice. Plant J. 46:1059-1072.

Long SP, Zhu X-G, Naidu SL, Ort DR. 2006. Can improvement in photosynthesis increase crop yields? Plant Cell Environ. 29:315-330.

Maliga P. 2002. Engineering the plastid genome of higher plants. Curr. Opin. Plant Biol. 5:164-172.

Mitchell PL, Sheehy JE, Woodward FI. 1998. Potential yields and the efficiency of radiation use in rice. Discussion Paper No. 32. Manila (Philippines): International Rice Research Institute.

Mitchell PL, Sheehy JE. 2000. Performance of a potential C_4 rice: overview from quantum yield to grain yield. In: Sheehy JE, Mitchell PL, Hardy B, editors. Redesigning rice photosynthesis to improve yield. Proceedings of a workshop on The Quest to Reduce Hunger: Redesigning Rice Photosynthesis, 30 November-3 December 1999, at IRRI. Amsterdam (Netherlands): Elsevier, and Makati City (Philippines): International Rice Research Institute. p 145-163.

Mitchell PL, Sheehy JE. 2006. Supercharging rice photosynthesis to increase yield. New Phytol. 171:688-693.

Monteith JL. 1977. Climate and the efficiency of crop production in Britain. Philos. Trans. Royal Soc. London B 281:277-294.

Raines CA. 2006. Transgenic approaches to manipulate the environmental responses of the C_3 carbon fixation cycle. Plant Cell Environ. 29:331-339.
Sheehy JE. 2000. Limits to yield for C_3 and C_4 rice: an agronomist's view. In: Sheehy JE, Mitchell PL, Hardy B, editors. Redesigning rice photosynthesis to improve yield. Proceedings of a workshop on The Quest to Reduce Hunger: Redesigning Rice Photosynthesis, 30 November-3 December 1999, at IRRI. Amsterdam (Netherlands): Elsevier, and Makati City (Philippines): International Rice Research Institute. p 39-52.
Sheehy JE, Dionora MJA, Mitchell PL. 2001. Spikelet numbers, sink size and potential yield in rice. Field Crops Res. 71:77-85.
Sinclair TR, Purcell LC, Sneller CH. 2004. Crop transformation and the challenge to increase yield potential. Trends Plant Sci. 9:70-75.
Wu C, Li X, Yuan W, Chen G, Kilian A, Li J, Xu C, Li X, Zhou D-X, Wang S, Zhang Q. 2003. Development of enhancer trap lines for functional analysis of the rice genome. Plant J. 35:418-427.
Zhu X-G, Portis AR, Long SP. 2004. Would transformation of C_3 crop plants with foreign Rubisco increase productivity? A computational analysis extrapolating from kinetic properties to canopy photosynthesis. Plant Cell Environ. 27:155-165.
Ziska LH, Namuco O, Moya T, Quilang J. 1997. Growth and yield response of field grown tropical rice to increasing carbon dioxide and air temperature. Agron. J. 89:45-53.

Notes

Authors' addresses: P.L. Mitchell, Department of Animal and Plant Sciences, University of Sheffield S10 2TN, U.K.; J.E. Sheehy, Crop and Environmental Sciences Division, International Rice Research Institute, Los Baños, Philippines.
Acknowledgments: Use of facilities at the Department of Animal and Plant Sciences and support from Professor F.I. Woodward are gratefully acknowledged by P.L.M.

Agricultural research, poverty alleviation, and key trends in Asia's rice economy

D. Dawe

Rice is the most widely planted crop in Asia and it is the dominant source of calories for Asians living in poverty. Since a majority of the world's poor live in Asia, improving the livelihoods of rice farmers and consumers is therefore critical to global poverty alleviation. Poor farmers need high profits from growing rice, and poor consumers need lower prices so that they can increase the quality and quantity of food consumption and still have money left over for investments in education. Agricultural research is perhaps the most important way to achieve both high profits for farmers and low prices for consumers.

Although some trends in the rice economy give cause for optimism regarding the future path of rice prices, other trends raise concerns that prices will increase in the next 20 years. Uncertainties abound regarding future oil prices, demand for biofuels, water scarcity, climate change, and the pace of slowdowns in population growth and dietary diversification. In the face of such uncertainty, it seems prudent to invest in research for C_4 rice. Without such investments in productivity-enhancing technologies, it is the poor who will suffer the most from adverse shocks that put upward pressure on food prices.

Keywords: rice, agricultural research, poverty alleviation

Rice is by far the most widely planted crop in the developing world, with average harvested area of 147 million hectares from 2003 to 2005 (wheat and maize were next, each with slightly less than 100 million hectares). It is also the most important source of calories for consumers in the developing world, providing 655 calories per capita per day, or 25% of total intake in 2003. Wheat is next, at 17% (all data in this paragraph are from FAO 2006a).

Although rice production and consumption are growing rapidly in Africa (from a small base), approximately 90% of the world's rice is produced and consumed in a belt ranging from Pakistan in the west to Japan in the east. This part of Asia, which will be referred to in this paper as rice-producing Asia (RPA), is also home to a majority of the world's poor. In 2001, more than 700 million Asians still lived on less than US$1 a day, more than the entire population of sub-Saharan Africa at that time (FAO 2006b). To some extent, this is because Asia already has such a large population.

But even in relative terms, malnutrition appears to affect a substantially larger share of the population in South Asia than in Africa, the two poorest regions in the world (Svedberg 1999). As of 2000-02, there were still 548 million undernourished people in the developing and transition economies of Asia and the Pacific (FAO 2006b).

Rice is by far the most important commodity for the Asian poor. In many of the poorest countries, it accounts for more than 60% of caloric intake and more than half of protein consumption (Bangladesh, Cambodia, Lao PDR, Myanmar, and Vietnam), and it is usually more than 50% of the crop area harvested in those countries (all data are from FAO 2006a). For Asia and the Pacific as a whole, it accounted for 23% of total crop area harvested in 2004 (more than any other crop) and more than 30% of total caloric intake among Asian developing countries. In fact, while the countries of South, Southeast, and East Asia are very diverse in terms of religion, political structure, and stage of economic development, they are all united by the importance of rice in food, agriculture, and culture.

Given the dominant role of rice in the lives of Asia's poor, and the fact that most of the world's poor reside in Asia, rice research has the potential to make an important contribution to global poverty alleviation.

Pathways from agricultural research to poverty alleviation: theory[1]

Direct, short-term benefits

How does rice research help alleviate poverty? In some ways, this is an easy question to answer. Higher standards of living for rice farmers can be sustained only if farmers are able to produce more rice per unit of input. This higher productivity leads to higher profits from farming and a reduction in poverty. Thus, one way that rice research helps alleviate poverty is by increasing the productivity of farmers. This is the "direct" contribution to poverty alleviation. It is an important one, and it applies primarily to farmers that own land. This short-term effect can also lead to longer-term effects, as farmers with increased profits invest in education for their children, helping to keep subsequent generations out of poverty.

Indirect, medium-term benefits

If this direct contribution were the only one, research priorities would be relatively simple to set. The only goal would be to help the poorest farmers directly, and this would mean working with farmers on marginal lands without access to roads and irrigation. But rice research also makes an important indirect contribution to poverty alleviation, a contribution that is often overlooked. This indirect contribution makes itself felt in both the medium term and the long term through lower prices that are the natural outcome of the law of supply and demand: increases in production reduce prices relative to what they would have been without the increase in production.

---

[1] Parts of this section draw heavily on Dawe (2000).

Low prices are important because several classes of poor people do not have surplus rice to sell, and must buy their rice on markets. Such people often do not get enough to eat. In fact, the World Health Organization (WHO) in a recent issue of its World Health Report (2002) identified "being underweight" as the most burdensome health risk in the world, being responsible for the annual loss of 138 million disability-adjusted life years (DALYs). Lower food prices can contribute to reducing the incidence of underweight by increasing the affordability of food for the poor. Lower prices for basic foodstuffs also increase the effective income that the poor have at their disposal for other important expenditures such as education, health, and consumption of nutritious foods such as fish, meat, and dairy products.

Who are the poor people who are net consumers of rice? One increasingly important group is the urban poor. As of 2005, urban dwellers accounted for 38% of the total population in the developing countries of rice-producing Asia (RPA). Although this means that most of the population still lives in rural areas, urbanization is increasing. From 2000 to 2005, Asia's rural population grew at just 0.26% per annum, while the urban population grew more than ten times as fast at 2.73% per annum. In absolute terms, the urban population increased by 167 million people from 2000 to 2005, whereas the rural population increased by just 29 million. Population projections indicate that the rural population will be declining in absolute terms before 2015, and the urban population will equal the rural population in RPA by 2025 (all data from UN 2006). Average levels of income are surely higher in urban areas than in rural areas, but not all urban dwellers are well-to-do. There is little doubt that, while there is more poverty in rural areas, the number of urban poor is increasing, both in absolute terms and as a share of the total (Haddad et al 1999).

A second important group of poor rice consumers is the rural landless or near landless who derive most of their income from agricultural labor. Landless agricultural workers are most common in South Asia, Indonesia, and the Philippines. They are less common in Thailand (where population density is lower), China, and Vietnam (due to comprehensive land reforms). In the Philippines, they constitute 13% of the agricultural labor force, and are one of the poorest groups in the countryside, with income 30% lower than that of rice farmers (Dawe et al 2006). Although some of these laborers work on rice farms and are occasionally paid in rice, surveys show that they do not earn enough rice to sell a surplus on the market. Instead, they need to purchase rice on markets and are likely to benefit from lower prices.

A third important group of poor rice consumers is rural dwellers who own land, but use it to grow nonrice crops. They would benefit from cheaper rice prices. In Indonesia, many farmers plant maize, cassava, and soybeans. In the Philippines, maize and coconut are important crops grown by poor smallholders, with maize farmers being particularly poor.

Thus, many poor Asians are net purchasers of rice. Rice constitutes an important part of their daily expenditures, so the contribution of lower prices is not trivial. For example, for the poorest 10% of the urban population in Bangladesh, rice accounts for half of total expenditures. Even for the poorest 60% of the urban population, nearly 40% of expenditures go to rice (Bangladesh Bureau of Statistics 1998). In Indonesia, rice accounts for 20% of total expenditures for the poorest quarter of the urban population (about 20 million people). For the poorest 5% of the urban population, the share of rice is about 25% (Biro Pusat Statistik 1998).

What are the effects of high rice prices for those who spend a substantial proportion of their income on rice? Senauer and Sur (2001) estimated that, if there were a 20% increase in food prices between then and 2025 (due, for example, to a decline in funding for international rice research), the number of undernourished people in Asia would increase by 158 million. An interesting study using data from rural Central Java in Indonesia (Block et al 2004) found that, when rice prices increased in the late 1990s, mothers in poor families responded by reducing their caloric intake in order to better feed their children, leading to an increase in maternal wasting. Furthermore, purchases of more nutritious foods were reduced in order to afford the more expensive rice. This led to a measurable decline in blood haemoglobin levels in young children (and in their mothers), increasing the probability of developmental damage. A negative correlation between rice prices and nutritional status has also been observed in Bangladesh (Torlesse et al 2003).

Indirect, long-term benefits

While the short- and medium-term effects described above are important, poverty alleviation is ultimately a long-term, broad-based process, and higher farm productivity makes several important long-term indirect contributions to poverty alleviation. These effects operate through lower rice prices.

One pathway is that lower rice prices lead to increased calorie consumption and improved nutrition for the poor. Lower retail prices allow some people to consume larger quantities of rice (especially the very poor who are calorie-deficient), while also allowing people to spend less money on rice, which in turn frees up income to spend on other, more nutritious foods that can reduce "hidden hunger" by increasing the intake of important vitamins and minerals (Welch and Graham 1999). There is some evidence that increased calorie consumption increases the efficiency and productivity of laborers, especially those who work in jobs requiring physical strength (Strauss and Thomas 1998). Improved nutritional status can increase cognitive abilities (e.g., through increased consumption of iron-rich foods, see Horton and Ross 2003), which has the potential to affect worker productivity in jobs throughout the economy. Finally, the increased worker productivity due to greater caloric intake and improved nutritional status leads to economic growth, which is a necessary (although not sufficient) condition for increasing the living conditions and incomes of the poor.

The effects of lower retail prices for rice are reinforced by farm diversification in response to lower farm prices for rice. As rice prices decline, more farmers will be encouraged to produce other crops and livestock that are more nutritious. Not all farmers will shift crops because the production of other crops and agricultural products is often riskier than rice, but the increased production from some farmers will in turn serve to lower prices of these items, thus making them more affordable for the poor. The second important long-term indirect contribution of lower rice prices to poverty alleviation is that this helps to accelerate the structural transformation of the economy.

Long-term poverty alleviation requires the creation of jobs in the relatively higher-productivity industrial and service sectors of the economy. This is obviously true if one is trying to reach the urban poor. It is also true for the long-term reduction in rural poverty, because no country in history has managed to grow rich while keeping a large share of its population in agriculture. This is a striking empirical regularity of the development process (Timmer 1988). In other words, long-term poverty alleviation requires a structural transformation of the economy away from agriculture and toward industry and services.

Low rice prices that are the result of higher productivity induced by agricultural research contribute to this structural transformation of the economy. Low rice prices allow nominal wages to be lower without sacrificing any welfare on the part of the workers, because their effective purchasing power increases with lower rice prices. In conjunction with other factors (e.g., a stable macroeconomic environment, an efficient and fair legal system), these lower wages stimulate the job creation and growth that are necessary for sustainable poverty alleviation. If rice prices are high, workers will legitimately demand higher wages. But these higher wages will discourage investment, both domestic and foreign, and the growth process will be retarded.

But it is not possible to accelerate the growth process simply by pursuing policies that arbitrarily depress rice (food) prices through subsidies. This strategy has been tried, and it has not worked. If low retail prices are accompanied by low farm prices in the absence of rising productivity, farmers have little incentive to produce, leading to reduced supplies of food. In conjunction with high levels of consumption encouraged by low prices, this results in black markets and high prices for those without privileged access to cheap food. Alternatively, if low retail prices are accompanied by high farm prices through the use of subsidies in the absence of rising productivity, the typical result in developing countries is large government fiscal deficits that cannot be sustained or that reduce funds available for other critical investments such as education, health, and infrastructure. Thus, it is critical to have low prices that are the result of higher productivity, not low prices achieved through other means.

Negative effects of low prices

Of course, there can also be negative effects from the lower prices that are the inevitable result of increases in agricultural productivity in a market economy. It is true that lower rice prices, *holding all else constant*, adversely affect poor rice farmers who produce a surplus of rice. However, it is misleading to couch the issue in terms of lower prices with all else held constant, because the lower prices are the consequence of higher yields and increased multiple cropping that serve to increase supplies faster than demand. These higher yields and increases in harvested area lead to sharply increased production that compensates farmers for the lower prices. Lower prices do hurt farmers with a surplus to sell who have not adopted innovations or diversified their cropping systems into higher-value crops. Thus, it will be important to ensure that even poor farmers have the ability to adopt innovations such as C_4 rice. This should not be a problem if C_4 rice is developed in the public sector, which will likely make it available to all farmers without licensing fees. Even if C_4 rice were developed in the private sector, such companies do not want to charge prices so high that no farmers can adopt the innovation. But private companies may still charge prices that are high enough that small farmers are indeed excluded. With this possibility in mind, it will be important to develop arrangements such as those that were developed for "golden rice," for which the private sector has donated intellectual property so that poor farmers will be able to save their own seed (IRRI 2005).

As another example of a possible negative effect, lower rice prices can reduce the wages of poor farm laborers by reducing the demand for farm labor, although this effect will not always be present (FAO 2006b). Working against this adverse outcome are rising farm yields due to increased productivity, which will tend to increase the demand for farm labor and raise wages for landless laborers. The net effect of these two changes is ambiguous.

To summarize, the main point of this section is that agricultural research has complex effects that affect all poor people in the entire economy, not just rice farmers (who are a minority of the poor). Further, the vast majority of these effects are beneficial, as will be seen in the next section.

Agricultural research and poverty alleviation: evidence

The main papers that directly measure the effects of agricultural research on poverty reduction are by Shenggen Fan and colleagues at the International Food Policy Research Institute (IFPRI). For the case of India, Fan et al (2000a) show that government expenditures on agricultural research and development (R&D) have a larger impact on rural poverty in India than any other expenditures aside from roads. If urban poverty is included, then agricultural R&D have a larger impact on total poverty than any other government expenditure (Fan 2002). Fan et al (2000b) show that agricultural R&D also have a large impact on rural poverty reduction in China, second only to the effect of education.

These studies show not only that agricultural research contributes to poverty alleviation, but that the marginal effects on poverty of a dollar spent on research are larger than the effects of a dollar spent on other public goods such as irrigation, roads, electricity, or education (except in the case of China for the latter). This is not to say that these other investments are not important—they are. However, money spent on agricultural research appears to do more for poverty reduction than money spent on other investments.

Other studies show results that are generally consistent with these findings, although the analysis is not as direct. For example, Thirtle et al (2003), using cross-country data on poverty and yields, find that a 10% increase in yields is associated with a 4.8% decline in poverty in Asia. Datt and Ravallion (1998) show that a 10% increase in farm yields in India reduces rural poverty by 9%, with the effect on incomes being strongest for the poorest of the poor (i.e., those far below the poverty line). From these studies, it is just a short step to conclude that agricultural research (without which higher yields are difficult to achieve) is important for poverty reduction.

A large group of other studies examines the effects of economic growth on poverty alleviation, and in particular the effects of agricultural growth (as distinguished from nonagricultural growth). Nearly all studies agree that economic growth contributes to poverty alleviation, and many present evidence showing that agricultural growth in particular contributes more to poverty alleviation than nonagricultural growth, mainly because a given dose of agricultural growth has a larger effect on poverty than the same dose of nonagricultural growth (Timmer 1997, Ravallion and Chen 2004, Christiaensen et al 2006). In some respects, this is not surprising because of the greater importance of agriculture for the incomes of the poor and the greater importance of food in the expenditures of poor consumers. In addition, agricultural growth also seems to generate more nonagricultural growth through various backward and forward linkages than the other way around.

Not all studies agree with these conclusions on agricultural growth versus growth in other sectors. For example, Ravallion and Datt (1996, 2002) find that growth in the service sector does more to alleviate poverty than growth in agriculture. Bravo-Ortega and Lederman (2005) find that nonagricultural growth raises incomes for the poorest quintile more than does agricultural growth. Nevertheless, these studies do find that agricultural growth reduces poverty.

Key trends in the Asian rice economy

Rice prices

As late as 1981, world market rice prices were in excess of $1,000 per ton (in inflation-adjusted 2005 US dollars), similar to the average that prevailed from 1950 to 1980. Since then, however, there have been two episodes of major price declines: 1981 to 1986, when prices declined by 66%; and 1998 to 2001, when prices declined by 48% (from the new lower base). Both of these declines were due (in part) to large increases in production (Dawe 2002). Since 1980, rice prices have declined more rapidly than maize and wheat prices. Although most Asian governments do not allow changes in world prices to be directly transmitted to domestic markets, domestic rice prices (after adjusting for inflation) are also lower in many Asian countries than those 40 years ago.

The new lower prices are due in large part to the lower per ton production costs made possible by the Green Revolution. These lower prices have allowed the poor improved access to calorie supplies, and the proportion of undernourished Asians has declined sharply.

By 2001, world prices reached a record low of $190 per ton for Thai 100B, a high-quality indica grain commonly traded in world markets. Four years later, however (2005), world rice prices had increased to $288 per ton, an increase of 51% from the record low in 2001. (Despite this increase, prices in 2005 were still below the average of $391 per ton from 1986 to 1998.) An important question, then, is what will be the evolution of future rice prices? It is not possible to predict commodity prices with much precision, but some key trends that will affect future prices will now be discussed.

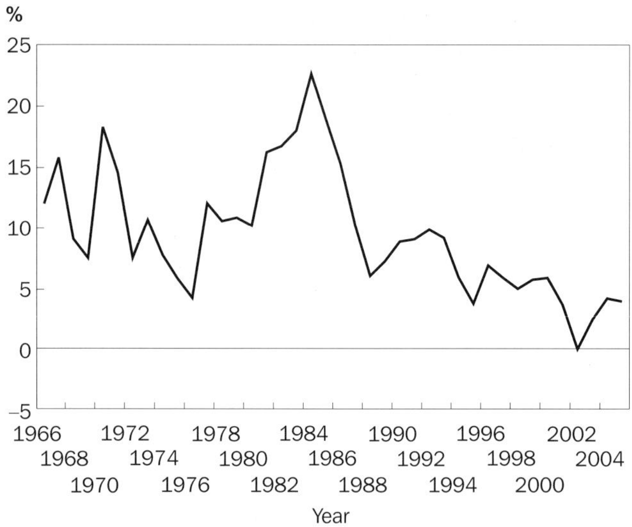

Fig. 1. Cumulative percent change in rice yield during the previous 5 years in rice-producing Asia. Source of raw data: FAO (2006a). Rice-producing Asia refers to Bangladesh, Bhutan, Brunei, Cambodia, China, Democratic People's Republic of Korea, India, Indonesia, Japan, Laos, Malaysia, Myanmar, Nepal, Pakistan, Philippines, Republic of Korea, Sri Lanka, Thailand, Timor-Leste, and Vietnam.

Trends in area, yield, and the adoption of modern varieties

Much of the 139% increase in rice production in Asia between 1965 and 1999 came from higher yields. Perhaps surprisingly, however, the area harvested to rice increased by 23% during this period despite the loss of some land to urbanization,[2] as irrigation and shorter-duration varieties allowed many farmers to grow multiple crops of rice per year. Since 1999, however, rice area harvested has been declining from its peak. This process has been most rapid in East Asia, where area harvested in 2004 was 10% below its peak five years earlier. However, even for the rest of Asia combined (South and Southeast Asia), rice area harvested in 2004 was 3% below the 1999 level (data for the calculations in this paragraph come from FAO 2006a). Thus, it seems unlikely that expanded area will be a major source of future growth in rice production (unless prices were to increase substantially and encourage farmers to shift into rice).

Rice yields continue to increase, but this rate of increase has slowed tremendously in recent years. While cumulative increases of 10% every five years were typical up until the early 1990s, yields in 2004 were just 3% higher than in 1999 (see Fig. 1). Some of the dramatic slowdown in yield growth (and the decline in area harvested) may be due to the low level of prices, as farmers find that growing other crops, or

[2]These losses are much smaller than is often supposed, however. See FAO (2006b) for some brief discussion.

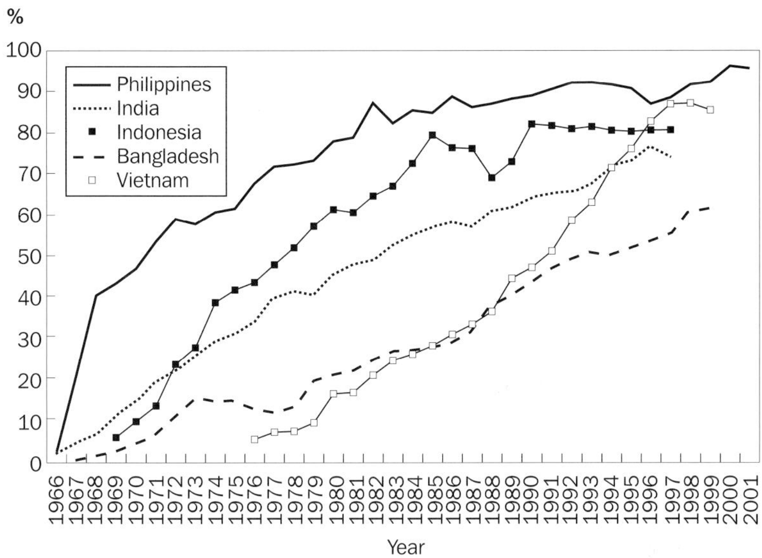

Fig. 2. Percent adoption of modern rice varieties, selected Asian countries, 1966 to 2001. Source of data: IRRI (2006a).

leaving agriculture altogether, is more profitable, and they devote less attention to management of their rice crops. On the other hand, growth in the adoption of modern varieties has slowed down as adoption rates reach plateaus of 75% to 90% in many countries (see Fig. 2; further adoption is probably constrained because not all rice-growing areas have adequate water supplies). Furthermore, rice yields have remained stagnant in recent years in Indonesia (an increase from 4.43 tons per hectare in 1997 to just 4.52 tons per hectare in 2004), despite large increases in domestic rice prices during that period due to government restrictions on imports. This suggests that higher prices may not bring forth higher yields in the current environment. Instead, it may be that yields are constrained by a lack of post-Green Revolution yield-enhancing technologies that are available and are being communicated effectively to farmers. If true, this situation is much more worrisome than if slow yield growth were due only to low prices for farmers.

Hybrid rice is one possible solution to this lack of new technologies, and adoption is increasing in several countries. Putting aside the case of China, where it has been widely adopted for decades, hybrid rice has made the most progress in recent years in northern Vietnam, the Philippines, and India (see Fig. 3). At this stage, it is not clear how much hybrid rice will be able to boost rice yields in the medium to longer term. Adoption has been quite rapid in the Philippines, but it may have been fueled to some extent by government subsidies. It is not clear whether farmers will return to inbred modern varieties once these subsidies are withdrawn.

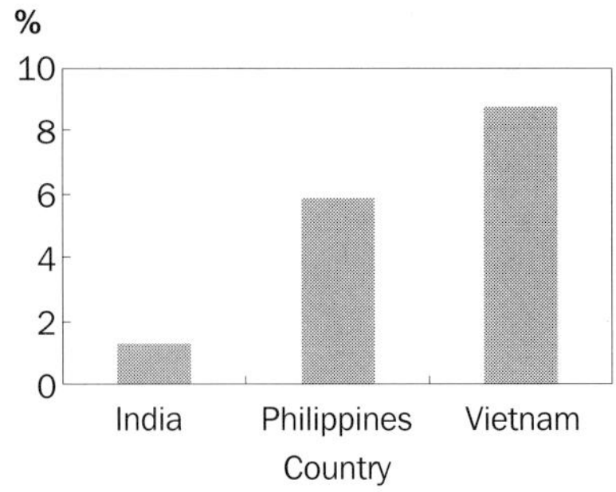

Fig. 3. Percent adoption of hybrid rice, selected Asian countries, 2004. Source of data: IRRI (2006b). Data for Philippines refer to 2005.

Future rice demand growth

While slowing yield growth may portend future price increases, population growth continues to slow in the region, which will help to reduce pressure on future rice prices. After growing by 2.4% per year in the early 1970s in rice-producing Asia, the population expanded by just 1.2% per year from 2000 to 2005. The most recent medium variant forecast from the United Nations projects that total population in these countries will grow just 29% between 2005 and 2050 (UN 2006), after which the population will soon begin to decline in absolute terms in the region.

Reinforcing the trend of slower population growth is the decline in per capita rice consumption that occurs as countries become wealthier. This decline is due to the universal human desire to diversify diets noted by Bennett's Law, which states that the proportion of calories in the diet coming from starchy staple foods (including cereals and roots and tubers) declines as incomes increase. In addition to diversifying away from cereals toward livestock, dairy products, fruits, and vegetables, Asian diets are also diversifying within the class of cereals. Since most Asians eat rice as their primary staple, this typically means shifting from rice toward wheat (rice remains dominant, but less so).[3] The net result can be seen in Japan, for example, where rice consumption has declined from 111 kg of milled rice per capita per year in 1961 to just 57 kg in 2003. Similar declines have taken place in the Republic of Korea (141 kg per capita per year in 1978 to 78 in 2003) and Malaysia (127 kg per capita per year in 1974 to 71 in 2003). As more Asian countries continue on the path of economic growth, per capita rice consumption will continue to decline.

[3] However, there are other cases where consumers who eat diets based on wheat (northern India, northern China) or maize (southern Philippines) are shifting from those staples toward rice.

Fertilizer prices, petroleum prices, and biofuels

World market prices in 2005 for urea, the most important source of nitrogen for rice in Asia, were nearly triple those in 1999 due to higher prices for petroleum and natural gas. Nevertheless, prices are still slightly below their level of ten years earlier after adjusting for inflation. And these prices are substantially below the levels reached during the late 1970s and early 1980s in the face of repeated shocks on world petroleum markets.

It is extremely difficult to forecast how petroleum and urea prices will evolve during the next 20 years. While there are legitimate concerns that higher prices are here to stay, the world economy has weathered previous disruptions to petroleum markets. For example, prices rose sharply from 1972 to 1981 due to a succession of problems in the Middle East, but then declined sharply to very low values. But this time may be different, because of increased demand from China and other rapidly developing economies that is likely to be more permanent.

Permanently higher urea prices are not likely to cause major disruptions on the world rice market, however, because urea represents only about 10% of the gross value of production. Thus, even if prices were to double again from their elevated current levels, an increase in rice prices of 10% would be enough to keep farm profitability constant and keep land in production.

A more worrisome possibility is that high petroleum prices will lead to substantially increased demand for biofuels from sugar and maize. If more ethanol plants are built and production continues to expand, demand for sugar and maize will increase, leading to higher prices of these crops if sufficient additional production is not forthcoming through either expansion of area or higher yields.[4] Higher prices will substantially increase the profitability of growing these crops, and Asian farmers may be induced to substitute land out of rice, for example, shifting from a rice-rice rotation to rice-maize. The resulting decrease in rice supply would put upward pressure on rice prices.

The possibility for substantial disruptions in rice markets would seem to be nontrivial. A report from OECD (2006) estimates that ethanol from sugarcane in Brazil and from maize in the United States is a competitive source of transport fuel at oil prices of $44 per barrel ($29 in the case of Brazil), which was below the price of oil in most of 2005 and the first half of 2006. Further, at current levels of agricultural and ethanol production technology, the report estimates that, if oil prices are sustained at $60 per barrel, prices for sugar in 2014 will increase by about 80% relative to a scenario of constant (at the levels prevailing in 2004) biofuel production, with maize prices increasing by 20% (no calculations were done for rice). This calculation ignores the possibility of bringing fallow land into production and using international trade, so it is only vaguely indicative of how world markets will evolve, and clearly more research is needed in this area. Nevertheless, these increases are substantial, especially since the increased biofuel production that would result will raise the share of biofuels in

---

[4] Demand will also grow for bio-diesel made from oilseed crops.

total EU-15 transport fuel consumption to just 6%, and less than 3% in the U.S. and Canada. Thus, there would appear to be substantial upside potential if biofuels were to become even more important as a source of transport fuel. Indeed, if biofuels were to account for 10% of total transport fuel, this would require nearly one-third of all land currently harvested for cereals, oilseeds, and sugar in the U.S. and Canada, and more than 70% in the EU-15 (OECD 2006).[5] Recalling Senauer and Sur's (2001) estimate that a 20% increase in food prices would increase the number of undernourished in Asia by 158 million, the impact of biofuel demand is worrisome indeed.

Higher petroleum prices may also lead to higher production costs in terms of fuel for tractors and combine-harvesters. However, most rice production in Asia is not highly mechanized, with labor being the main component of production costs. Thus, higher fuel prices will not substantially affect rice production costs through this channel. But production costs of maize and wheat in developed economies could be substantially affected, and, to the extent that this drives up prices for wheat and maize on world markets, rice markets can be affected through crop substitution on the supply side as noted above.

Water scarcity

Water scarcity is also a potential constraint to future rice production. At a macro level, Rosegrant et al (2002) project that consumptive water use for domestic and industrial purposes in Asia will increase by 98% and 88%, respectively, between 1995 and 2025. During the same period, consumptive use for irrigation will increase just 1%. But this approximately constant amount of irrigation water will have to support 50% higher levels of cereal production. In addition, because it is cost-prohibitive to move water long distances (the south-to-north water conveyance projects in China will be an exception), the role of markets and prices in allocating water to its highest-value use will be limited. In other words, while it may be possible to shift water to higher-valued uses within a basin, it will not be possible, for example, for the Philippines to sell surplus water to northern India.

Although many areas will continue to have adequate supplies of water for decades to come, there will be shortages localized in space and/or time. For example, in the Zhang He irrigation system in Hubei, China, competition from domestic and industrial uses has substantially reduced the volume of water available for irrigation during the past 40 years (Hong et al 2001; see Fig. 4). Although there were increases in water productivity during this time, rice production started to decline during the 1990s (although other factors such as labor scarcity may have also contributed to this outcome).

Periodic widespread water shortages may also become more common in the future if global warming leads to more variable weather patterns. One example is the El Niño Southern Oscillation (ENSO) phenomenon that has been shown to affect rice

---

[5] This calculation assumes current agricultural yields and current levels of transport fuel consumption. Both are likely to increase, so it is not clear if the reported calculation is biased up or down.

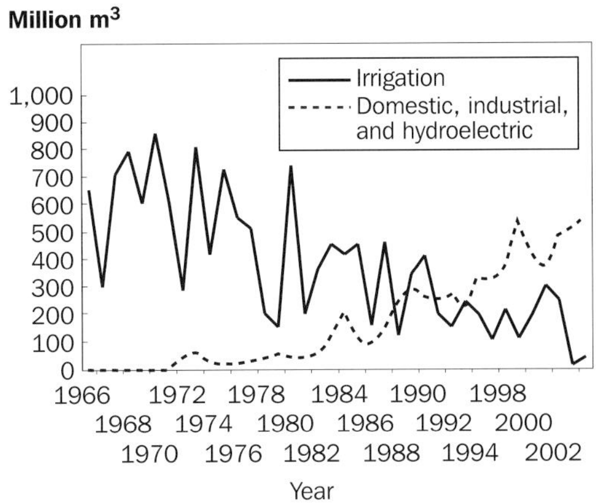

Fig. 4. Water allocations by sector, Zhanghe Reservoir, Hubei, China, 1966 to 2003. Source of data: Hong et al (2001) updated with unpublished data from Zhanghe Irrigation District.

production in India, Indonesia, the Philippines, and Sri Lanka (Selvaraju 2003, Naylor et al 2001, Falcon et al 2004, Zubair 2002). For example, rice production in the dry season declined by 27% in the Philippines during the El Niño of 1997-98. Since dry-season production accounts for nearly half of annual production, this drought required substantial additional quantities of imports.

Given increased future water scarcity, the implications for rice prices are potentially severe if water management is not improved. Under the "Crisis" scenario of Rosegrant et al (2002), rice prices in 2025 are higher by 80% than in a "Business as usual" scenario.

Diversification of farm household income

In addition to scarcity of natural resources, farmers will also experience a scarcity of time for farm management in the future. Farms throughout Asia are diversifying into nonfarm activities such as trading, construction, services, and transportation, among many others. Pingali (2006) states that, on average, 32% of farm household income in Asia comes from nonfarm activities. Furthermore, this share is increasing over time. Hossain et al (2000) found that the share of rural household income coming from nonagriculture in four selected Philippine villages increased from 36% to 60% between 1985 and 1997. Nearly all—96%—of the households surveyed in 1997 had income from agriculture, so these rural households are not abandoning farming entirely.

As farmers find less time to devote to farm management, it will become critical to develop technologies that do not require large investments of knowledge or time on the part of farmers; otherwise, farmers may not adopt the technologies. C_4 rice could be particularly useful in this regard, because the knowledge is embodied in the seed. Farmers may need to make some adjustments in cultivation techniques with C_4 rice, but these are likely to be minor.

The difficulty of part-time farmers applying knowledge-intensive technologies in Asia is compounded by the fact that farm sizes are very small, typically 1 to 2 hectares compared with farm sizes of more than 100 hectares in countries such as the United States and Australia. Farmers with small amounts of land under their control have less incentive to learn sophisticated crop management methods, because the benefits of higher yields on a small plot of land are not likely to compensate for the fixed cost of knowledge acquisition. The benefits of increased yield on large plots of land have a much greater chance of covering the fixed costs.

Land consolidation could remedy this problem, and farm sizes will eventually increase as development proceeds in Asia. But this process of consolidation could turn out to be quite slow. For example, the average farm size in Japan increased from 1.00 ha in 1960 to 1.57 ha in 2002. In Korea, average farm size rose from 0.88 ha in 1970 to 1.46 ha in 2002 (Fan and Chan-Kang 2003). These increases are not very large, less than a hectare in 30 to 40 years.

Conclusions

The main objective of the paper has been to show that agricultural research aimed at raising productivity has an important role to play in poverty alleviation. If C_4 rice is technically feasible, it would make a very important contribution in that regard. Higher yields would increase rice supplies, helping to lower rice prices for the poor and allowing farmers to diversify into other important crops without sacrifices in aggregate rice production at the country level. Higher yields would also contribute to higher farm profits, although the net effect would be ambiguous because of lower prices in the long run.

C_4 rice that increased grain yield per unit of water transpired would make an important contribution to managing future water scarcity in Asia, which would contribute to both poverty alleviation and environmental goals. C_4 rice would also increase nitrogen-use efficiency, reducing production costs per ton for farmers and helping to reduce nitrogen loads in the environment.

Furthermore, although some trends in the rice economy give cause for optimism regarding the future path of rice prices, other trends raise concerns that prices will increase in the next 20 years. Clearly, a great deal of uncertainty surrounds any projections that are made. The possible effects of climate change have also not been considered, adding a further layer of uncertainty.

In such circumstances, it seems prudent to provide solid funding for additional research on C_4 rice. If the optimistic scenario holds (e.g., oil prices return to low levels, demand for biofuels does not increase substantially, population growth declines rapidly, diets diversify away from rice quickly, water management improves), the added productivity from C_4 rice will still contribute to making poverty alleviation even more rapid than it otherwise would have been. In other words, the faster that cereal prices decline, the faster that undernourishment will become a distant memory. But, if a more pessimistic scenario unfolds, many poor people will suffer severely from the resultant high prices. C_4 rice would provide an important buffer to make sure this eventuality does not come to pass.

References

Bangladesh Bureau of Statistics. 1998. Household expenditure survey 1995-96. 364 p.

Biro Pusat Statistik. 1998. Statistical yearbook of Indonesia. Jakarta, Indonesia. 594 p.

Block S, Kiess L, Webb P, Kosen S, Moench-Pfanner R, Bloem MW, Timmer CP. 2004. Macro shocks and micro outcomes: child nutrition during Indonesia's crisis. Econ. Human Biol. 2(1):21-44.

Bravo-Ortega C, Lederman D. 2005. Agriculture and national welfare around the world: causality and international heterogeneity since 1960. World Bank Policy Research Working Paper 3499. Washington, D.C. (USA): World Bank.

Christiaensen L, Demery L, Kuhl J. 2006. The role of agriculture in poverty reduction: an empirical perspective. Unpublished paper.

Datt G, Ravallion M. 1998. Farm productivity and rural poverty in India. J. Dev. Studies 34(4):62-85.

Dawe D. 2000. The contribution of rice research to poverty alleviation. In: Sheehy JE, Mitchell PL, Hardy B, editors. Redesigning rice photosynthesis to increase yield. Proceedings of the Workshop on The Quest to Reduce Hunger: Redesigning Rice Photosynthesis, 30 Nov.-3 Dec. 1999, Los Baños, Philippines. Makati City (Philippines): International Rice Research Institute and Amsterdam (The Netherlands): Elsevier Science B.V. p 3-12.

Dawe D. 2002. The changing structure of the world rice market, 1950-2000. Food Policy 27(4):355-370.

Dawe D, Moya P, Casiwan C, editors. 2006. Why does the Philippines import rice? Meeting the challenge of trade liberalization. Los Baños (Philippines): International Rice Research Institute and Philippine Rice Research Institute. 166 p.

Falcon WP, Naylor RL, Smith WL, Burke MB, McCullough EB. 2004. Using climate models to improve Indonesian food security. Bull. Indonesian Econ. Studies 40(3):355-377.

Fan S. 2002. Agricultural research and urban poverty in India. Environment and Production Technology Division Discussion Paper No. 94. Washington, D.C. (USA): International Food Policy Research Institute.

Fan S, Hazell P, Thorat S. 2000a. Government spending, growth and poverty in rural India. Am. J. Agric. Econ. 82(4):1038-1051.

Fan S, Zhang L, Zhang X. 2000b. Growth, inequality, and poverty in rural China: the role of public investments. IFPRI Research Report No. 125. Washington, D.C. (USA): International Food Policy Research Institute.

Fan S, Chan-Kang C. 2003. Is small beautiful? Farm size, productivity and poverty in Asian agriculture. Proceedings of the 25th International Conference of Agricultural Economists, Durban, South Africa, 16-22 August.

FAO (Food and Agriculture Organization). 2006a. FAOSTAT (FAO statistical databases). Available at www.fao.org/waicent/portal/statistics_en.asp.

FAO (Food and Agriculture Organization). 2006b. The state of food and agriculture in Asia and the Pacific. Bangkok (Thailand). FAO. Also available at www.fao.org/es/esa/en/pubs_sofa.htm.

Haddad L, Ruel MT, Garrett JL. 1999. Are urban poverty and undernutrition growing? Some newly assembled evidence. World Dev. 27(11):1891-1904.

Hong L, Li YH, Deng L, Chen CD, Dawe D, Barker R. 2001. Analysis of changes in water allocations and crop production in the Zhanghe irrigation system and district, 1966-98. In: Barker R, Li YH, Tuong TP, editors. Water Saving Irrigation for Rice: Proceedings of an International Workshop held in Wuhan, China, 23-25 March 2001. Colombo (Sri Lanka): International Water Management Institute.

Horton S, Ross J. 2003. The economics of iron deficiency. Food Policy 28:51-75.

Hossain M, Gascon F, Marciano E. 2000. Income distribution and poverty in rural Philippines: insights from a repeat village study. Econ. Political Weekly 35(52-53):4650-4656.

IRRI. 2005. Golden Rice Fact Sheet available at www.goldenrice.org/PDFs/fs_GR_IRRI_2005.pdf.

IRRI. 2006a. World rice statistics on-line database. Available at www.irri.org/science/ricestat/index.asp.

IRRI. 2006b. Data presented during planning workshop on "Climate change and rice," 20-24 March 2006, Los Baños, Philippines.

Naylor R, Falcon WP, Rochberg D, Wada N. 2001. Using El Niño-Southern Oscillation climate data to predict rice production in Indonesia. Climate Change 50:255-265.

OECD (Organization for Economic Cooperation and Development). 2006. Agricultural market impacts of future growth in the production of biofuels. AGR/CA/APM(2005)24/FINAL. Available at www.oecd.org/dataoecd/58/62/36074135.pdf.

Pingali P. 2006. Agricultural growth and economic development: a view through the globalization lens. Presidential address to the 26th International Conference of Agricultural Economists, 12-18 August 2006, Gold Coast, Australia.

Ravallion M, Chen S. 2004. China's (uneven) progress against poverty. World Bank Policy Research Working Paper 3408. Washington, D.C. (USA): World Bank.

Ravallion M, Datt G. 1996. How important to India's poor is the sectoral composition of economic growth? World Bank Econ. Rev. 10(1):1-25.

Ravallion M, Datt G. 2002. Why has economic growth been more pro-poor in some states of India than others? J. Dev. Econ. 65:381-400.

Rosegrant MW, Cai X, Cline SA. 2002. World water and food to 2025. Washington, D.C. (USA): International Food Policy Research Institute.

Selvaraju R. 2003. Impact of El Niño-Southern Oscillation on Indian foodgrain production. Int. J. Climatol. 23:187-206.

Senauer B, Sur M. 2001. Ending global hunger in the 21st century: projections of the number of food insecure people. Rev. Agric. Econ. 23(1):68-81.

Strauss J, Thomas D. 1998. Health, nutrition and economic development. J. Econ. Lit. XXXVI(2):766-817.

Svedberg P. 1999. 841 million undernourished? World Dev. 27(12):2081-2098.

Thirtle C, Lin L, Piesse J. 2003. The impact of research-led agricultural productivity growth on poverty reduction in Africa, Asia and Latin America. World Dev. 31(12):1959-1975.

Timmer CP. 1988. The agricultural transformation. In: Chenery H, Srinivasan TN, editors. Handbook of development economics. Vol. 1. Amsterdam (Netherlands): North-Holland (Elsevier Science Publishers). p 275-331.

Timmer CP. 1997. How well do the poor connect to the growth process? CAER II Discussion Paper No. 17. Cambridge, Mass. (USA): Harvard Institute of International Development. 29 p.

Torlesse H, Kiess L, Bloem MW. 2003. Association of household rice expenditure with child nutritional status indicates a role for macroeconomic food policy in combating malnutrition. J. Nutr. 133(5):1320-1325.

UN (United Nations). 2006. World population prospects: the 2004 revision population database. Available at http://esa.un.org/unpp/.
Welch RM, Graham RD. 1999. A new paradigm for world agriculture: meeting human needs with productive, sustainable, nutritious food. Field Crops Res. 60:1-10.
WHO (World Health Organization). 2002. World health report. Geneva (Switzerland): WHO.
Zubair L. 2002. El Niño-Southern Oscillation influences on rice production in Sri Lanka. Int. J. Climatol. 22:249-260.

Notes

Author's address: Senior food systems economist, Food and Agriculture Organization Regional Office for Asia and the Pacific, Bangkok, Thailand.

Catching up with the literature for C_4 rice: what we know now and didn't then

P.L. Mitchell

Developments in understanding C_4 photosynthesis since 1999 are reviewed to provide a starting point for the conference on Supercharging the Rice Engine in 2006. It is now clear that all C_3 plants have the enzymes for the C_4 pathway and some plants employ them to capture carbon dioxide in particular tissues, although not as full C_4 photosynthesis with reduced or absent photorespiration. Conversely, parts of C_4 plants, mesophyll cells not close to bundle sheath cells, carry out C_3 photosynthesis. Several types of single-cell C_4 photosynthesis have now been studied; none seem to be highly productive but instead are adaptive for these plants in environments conducive to high rates of photorespiration. Introduction of one or a few genes for C_4 enzymes into rice has become routine but full C_4 photosynthesis or greatly increased growth has not resulted. Kranz anatomy appears to be essential for productive C_4 photosynthesis because there is no other way of confining carbon dioxide at high concentrations around ribulose 1,5-bisphosphate carboxylase–oxygenase (Rubisco), which is the key to success. At least 45 independent origins of C_4 photosynthesis across 19 angiosperm families are known. Study of C_3-C_4 intermediates has led to the definition of seven phases of evolution from C_3 to C_4. Given low atmospheric concentrations of carbon dioxide and high rates of photorespiration, C_4 photosynthesis seems to evolve readily, although not in the subfamily of grasses containing rice. Nevertheless, these recent advances in knowledge are encouraging for those intending to produce C_4 rice.

Keywords: C_3-C_4 intermediate, C_4 enzymes, evolution, Kranz anatomy, single-cell C_4

In 1999, a conference was held at the International Rice Research Institute on redesigning rice photosynthesis, with the proceedings published in October 2000. Apart from a few cases where earlier work was overlooked or incompletely understood, Sheehy et al (2000) is a complete statement of research relevent to C_4 rice up to the end of 1999. That conference also had the benefit of a book summarizing current knowledge, *C_4 Plant Biology*, which had appeared earlier in the year (Sage and Monson 1999).

This paper covers developments in our understanding of C_4 photosynthesis since then, to provide a starting point for the conference in July 2006 called Supercharging the Rice Engine. It is based on a literature review, prepared when there was renewed serious interest in C_4 rice, finished in August 2005, and now selectively updated. Inevitably, parts of this review are rendered incomplete or out of date by developments reported at the conference, to be found in papers in this volume, but it records a state of knowledge that was seen as sufficiently encouraging to merit the funding of the 2006 conference.

For readers new to the field, it may be helpful to mention some key reviews. They are Sage (2004) on the evolution of C_4 photosynthesis (with a concise introduction to the biochemistry and anatomy), Edwards et al (2004) on single-cell C_4 photosynthesis, Leegood (2002) and Raines (2006) on the prospects for introducing the C_4 pathway into C_3 plants, and Matsuoka et al (2001) and Miyao (2003) on the molecular biology and genetic engineering of C_4 photosynthesis.

Current understanding of C_4 photosynthesis

Photosynthesis with C_4 features in petioles and stems

Hibberd and Quick (2002) discovered that the green tissue around vascular bundles in celery and tobacco petioles could fix carbon supplied in solution through the xylem as carbon dioxide or as malate. They used carbon sources labeled with ^{14}C and found that the fixed carbon was exported to growth points. The enzymes neeed for decarboxylation of malate (all three used in the three main subtypes of C_4 photosynthesis) occurred in the green tissue with activities 9–13 times higher than in leaf tissue (expressed relative to unit amount of chlorophyll). The enzyme pyruvate, orthophosphate dikinase (PPDK), which produces phospho*enol*pyruvate (PEP) from pyruvate, was also present, at 18 times the activity in leaf tissue. Green cells around vascular tissue are widespread: they found them in species from 30 families. These photosynthetic cells are in organs with few stomata, several cell layers from the surface, and with few intercellular spaces so that carbon dioxide direct from the atmosphere must contribute negligibly to photosynthesis. Hibberd and Quick (2002) concluded that this was a system akin to C_4 photosynthesis where carbon dioxide was fixed twice: initially in roots by phospho*enol*pyruvate carboxylase (PEPcase), where carbon dioxide came from respiration or by diffusion from the soil, and finally by ribulose 1,5-bisphosphate carboxylase–oxygenase (Rubisco) in green cells adjacent to vascular tissue once the malate had traveled up in the xylem sap and been decarboxylated in the cells. Thus, the components of C_4 photosynthesis and the regulation of the necessary genes occurred already in C_3 plants, which helps to explain why the C_4 system has evolved so many times (at least 45 times in 19 families—Sage 2004).

Photosynthesis with C_4 features in developing fruit

Imaizumi et al (1997) studied photosynthesis in rice spikelets because around 25% of the carbohydrate in grains comes from panicle photosynthesis. They carried out ^{14}C-pulse, ^{12}C-chase experiments on the flag leaf (for reference as typically C_3) and on the lemma (one of the bracts surrounding the spikelet), in cultivar Nipponbare, grown outside. Substantial fixation was found in malate and aspartate, followed by transfer to sugar phosphates, then to sucrose and amino acids. They established that there was direct fixation by PEPcase, and found that about half the C_4 acids were decarboxylated and refixed by Rubisco; the other half probably entered metabolism through the usual anaplerotic route (topping up the Krebs cycle after intermediates are drawn off for synthesis of amino acids). The conclusion was that the lemmas carried out mainly C_3 photosynthesis but also fixed some carbon dioxide by PEPcase, a mixture of routes not typically C_3 or C_4 but perhaps adapted to refix abundant carbon dioxide from respiration. Kranz anatomy was lacking, however, and more work was needed on the localization of enzymes. A critical experiment would be to examine rates of photosynthesis by lemmas in 21% and 2% oxygen to find out whether photorespiration is reduced at all, as would be required in any C_4 photosynthesis that was of interest for making C_4 rice.

King et al (1998) found that the seeds and endocarp (inner wall of the siliqua, the pod) of developing fruits of rape (canola) had high activities of PEPcase and Rubisco that refixed carbon dioxide from respiration. Inside the pod, the carbon dioxide concentration rose to 2.5% since there was rapid respiration during oil formation, and the outer wall contained precursors of lignin and suberin that might decrease gas diffusion. The activity of PEPcase was especially high in the seeds, indicating a large flux through the anaplerotic pathway, and the rate of fixation by Rubisco would certainly be increased by the much higher concentration of carbon dioxide inside the pod, with reduced photorespiration. Again, this does not seem to be C_4 photosynthesis as such but local adjustment to abundant carbon dioxide inside a structure, and a need for carbon skeletons for the production of storage compounds in seeds.

Profiles of photosynthetic enzymes in rice

Prompted by findings of C_3 photosynthesis in parts of C_4 plants and vice versa, including Imaizumi's work, Tsuchida et al (2005) examined soluble proteins from six parts of rice: lamina, leaf sheath, stem, root, rachis-branch, and lemmas plus paleas (the bracts around the spikelet). Eleven enzymes involved in photosynthesis and carbohydrate and nitrogen metabolism were identifed by antisera. The overall pattern for rachis-branch was similar to that of the lamina, including abundant Rubisco and enzymes of the Calvin cycle. The stem and lemmas plus paleas had features in common, for example, the absence of carbonic anhydrase (CA) of the chloroplast type, that were "not inconsistent with operation of the C_4-like pathway."

Wang et al (2006) studied photosynthesis in two recent super-hybrids of rice and a current widely-grown F_1 hybrid. They measured the activities of PEPcase, NADP-dependent malate dehydrogenase (NADP-MDH), NADP-dependent malic enzyme (NADP-ME), and PPDK in addition to Rubisco and concluded that the higher activities of the C_4 enzymes in the flag leaf and lemma of one super-hybrid indicated "that the C_4 pathway works more efficiently" than in the current hybrid. However, no measurements of carbon dioxide uptake with 2% oxygen were made to assess the effect of photorespiration, nor was carbon dioxide compensation point recorded, so there is no evidence for C_4 photosynthesis as properly understood. The high activity of these enzymes is probably indicative of a large flux through the anaplerotic pathway or energetically inefficient fixation of carbon dioxide more than once in the photosynthetic cells.

Photosynthesis in the sedge genus *Eleocharis*

Recent publications are extending knowledge of the range of C_4 photosynthesis in species of *Eleocharis*, leafless sedges that can grow with the culms submerged or in air. It is clear from Barroca et al (2005) and Ueno (2004) that some species are strongly C_4, in both terrestrial and submerged culms, and others are C_4-like, C_3-like, or typical C_3. One technique used was inhibition of PEPcase to see how much photosynthesis was affected, in theory greatly for the strongly C_4 species, less so if C_4 is weakly expressed (Ueno and Ishimaru 2002). *Eleocharis vivipara* has C_4 terrestrial culms with Kranz anatomy and submerged culms that are C_3 without Kranz anatomy; it has been studied in detail by Agarie et al (2002a) and Ueno (2001). Of particular interest is the switch between C_3 and C_4 modes according to the culm habitat.

Vein spacing

It has long been known that veins are closer in the leaves of C_4 grasses than in C_3 grasses and that this relates to the differentiation of mesophyll and bundle sheath cells. Ogle (2003) assembled a large set of data to test the idea that quantum yield would be related to interveinal distance (IVD): it was and she discusses the minimum IVD (56 μm) expected for the highest quantum yield for a C_4 grass and the maximum IVD (196 μm) at which C_4 grasses have an advantage over C_3 ones. The reason for the expected relationship between IVD and quantum yield is the proportion of bundle sheath tissue in the leaf. Since the bundle sheath is the tissue with photosynthetic carbon reduction (PCR, i.e., Rubisco and the Calvin cycle) in C_4 photosynthesis, the more there is, the higher the quantum yield. The model she constructs helps to explain why shade-tolerant C_4 grasses are so rare. For positive carbon gain, such grasses must have a very small IVD, close to the limit set by the maximum quantum yield possible, and this is possessed by very few species.

The relationships are entirely empirical and nowhere does Ogle mention the number of mesophyll cells or their sizes. In the light of the discovery that maize photosynthetic tissue is purely C_4 in the leaves when there are only four cells between adjacent vascular bundles, and elsewhere it is a mix of C_4 and C_3 (see below), the variation in quantum yield could be interpreted as representing the proportion of C_4 tissue in the leaf. For reference (Ogle 2003), in C_3 grasses the mean IVD was 267 μm and the median 259 μm ($n = 89$); for C_4 grasses, the mean was 118 μm and the median 111 μm ($n = 147$). In round figures, the IVD in rice is 230 μm but in maize and also *Echinochloa* it is 100 μm.

Anatomy of the bundle sheath

Understanding the fine details of bundle sheath anatomy may be critical for making C_4 rice. The vascular bundles in C_3 grasses have two layers of cells sheathing them:

an inner layer called the mestome sheath and the outer bundle sheath proper (Dengler and Nelson 1999). The cells of the mestome sheath do not contain chloroplasts and are sclerenchymatous but the cells of the bundle sheath are parenchymatous and contain chloroplasts. There is not always a continuous mestome sheath around the smallest vascular bundles. Presumably, the conspicuous bundle sheath cells in rice are bundle sheath proper since they contain a few chloroplasts.

In C_4 grasses of the NAD-dependent malic enzyme (NAD-ME) and phospho-*enol*pyruvate carboxykinase (PEPCK) subtypes, both sheaths are present around the vascular bundles but the outer one is modified for C_4 photosynthesis (PCR), the bundle sheath in the functional sense. However, in the NADP-ME subtype (including maize), there is only one sheath evident, modified for C_4 photosynthesis (PCR), and it is the mestome sheath, and "the outer parenchymatous bundle sheath layer is missing" (Dengler and Nelson 1999, p 139).

This means that the bundle sheath cells in maize are not homologous to the bundle sheath cells in rice that we hope to convert to C_4 photosynthesis. In addition, the mestome sheath develops from the procambial strand so is part of the vascular tissue but the bundle sheath arises from ground tissue (Sage 2004, p 345). This difference in origin may also mean that mestome sheath cells will be elongated longitudinally but that bundle sheath cells will be more isodiametric. The signal in C_4 plants that modifies a cell type to become functional bundle sheath (PCR) may well be different in the NADP-ME subtype from that in the other subtypes. In the NADP-ME subtype, the signal is directed at the mestome sheath, part of the vascular tissue, and it turns it from a potentially sclerenchymatous cell to a parenchymatous one as well as inducing all the other changes associated with acquiring PCR activity. Is the lack of true bundle sheath cells in maize a significant detail, one affecting understanding of how C_4 anatomy develops, and affecting how C_4 rice might be constructed?

Differentiation of bundle sheath and mesophyll cells

One aspect of C_4 development that was known by 1999 but that was not picked up by Mitchell and Sheehy (2000) was the work of Jane Langdale on maize (discussed in Taylor 2000). She pointed out (Langdale and Nelson 1991) that the C_3 condition is the default development state in maize: it is evident in tissues grown in the dark before they respond to light by becoming green, and cells with C_3 photosynthesis remain in that state if they are too far away from a vein. Consequently, while leaf laminas are completely C_4, since veins are separated by only four cells, in leaf sheaths and husk leaves (modified leaf sheaths) where veins are wider apart, there are mesophyll cells carrying out C_3 photosynthesis in between. The repeated pattern of cell types in a transverse section can be summarized with abbreviations: V = vein, BS = bundle sheath cell of C_4 type, M = mesophyll cell of C_4 type, C_3M = mesophyll cell of C_3 type, DC = distinctive cell (see below). The standard C_4 pattern is V–BS–M–M–BS– repeated across the lamina. The repeated pattern in leaf sheaths is V–BS–M–$[C_3M]_n$–M–BS–, and presumably rates of photosynthesis and occurrence of photorespiration are intermediate between C_3 and C_4 values in proportion to the amount of C_3 and C_4 tissues. These patterns arise because the signal to switch to C_4 development (in the light in maize, not in *Amaranthus*—Edwards et al 2001) appears to move out from the vein, working on the cells next to a vein (becoming C_4 bundle sheath cells) and on the adjacent cells (becoming C_4 mesophyll cells) but not on cells further out.

Additional evidence for the next-to-vein restriction of C_4 development comes from the genus *Arundinella* (C_4 panicoid grasses), whose members have 7–8 cells between veins (Langdale and Nelson 1991, Wakayama et al 2003). Longitudinal strands of specialized cells develop between the veins, known as distinctive cells, which have PCR activity just as bundle sheath cells do. The cells adjacent to distinctive cells develop as C_4 mesophyll cells. The repeated pattern across the leaf is then V–BS–M–M–DC–M–M–BS–; again, with all mesophyll cells (C_4) adjacent to a cell that is carrying out PCR whether bundle sheath or distinctive cell. As the cells in the leaf develop, the pattern of enzyme accumulation is identical in BS and DC, and identical in the two types of mesophyll cell, next to BS or next to DC (Wakayama et al 2003). This is evidence that functionally there is no difference between BS and DC, and the mesophyll cells are identical irrespective of whether they are adjacent to BS or DC. The distinctive cells occur in the location where minor veins would be expected in other C_4 grasses, but they differentiate from ground tissue, not from vascular tissue or its precursor the procambial strand. This suggests that it is position, not cell lineage, that determines differentiation as bundle sheath equivalents (Dengler and Nelson 1999, p 157, Wakayama et al 2003). Yet there is a lingering doubt over whether DCs "might be procambium descendants that failed to correctly differentiate into vascular tissues with [bundle sheath cells]," because of the alignment of DC files and minor veins in certain places in the leaf (Wakayama et al 2003, p 1339).

Some species of the sedge *Eleocharis* have two layers of mesophyll cells called inner and outer with reference to the vein they surround. The outer mesophyll cells have weaker development of C_4 mesophyll characteristics: they contain less PEPcase and more Rubisco than the inner mesophyll cells (Ueno 2004). Again, this indicates the importance of vicinity to the bundle sheath cell or to the vein for full differentiation of C_4 characteristics. Whatever the factor that influences differentiation may be, it seems to be able to discriminate between cells that are next to a bundle sheath and cells further away. (In *Eleocharis*, the bundle sheath cells are actually cells within the vascular bundle so that the mestome sheath of smaller, thick-walled, colorless cells separates the cells with bundle sheath function [Ueno (2004) calls them Kranz cells] from the inner mesophyll cells.)

Evolution of C_4 photosynthesis

Sage (2004) divides the evolution of C_4 photosynthesis into seven phases. These are summarized in Table 1. Plants at Phase 1 or 2 would be recognized as standard C_3 plants. Those at Phase 3 might be described as C_3-like; the C_3-C_4 intermediates that restrict the location of glycine decarboxylase to bundle sheath cells are at Phase 4; Phase 5 is C_4-like; and by Phases 6 and 7 plants exhibit the full set of C_4 characteristics.

The C_4 pathway cannot have arisen as a single change but rather as a series of steps, some of them clearly having to precede certain other steps. Moreover, each intermediate stage had to be selected for, so must be at least neutral to survival of the

Table 1. Phases of evolution of C_4 photosynthesis according to Sage (2004).

Phase	Explanation
1. General preconditioning	Gene duplication, especially favored in organisms that reproduce sexually and have short life cycles.
2. Anatomical preconditioning	Tendency toward close spacing of veins in the leaf, for example, for structural or hydraulic reasons.
3. Increase in organelles in bundle sheath cells	With closer veins, bundle sheath cells are larger fraction of leaf tissue so become more active to maintain photosynthetic capacity, acquiring more chloroplasts and mitochondria.
4. Changes in photorespiration	Scavenging the products of oxygenation by Rubisco, that is, the photorespiration pathway, normally occurs in each photosynthetic cell. At one point, glycine decarboxylase combines two glycine molecules to produce serine and carbon dioxide. In C_3-C_4 intermediate plants, this reaction is confined to bundle sheath cells so glycine diffuses there and carbon dioxide concentration becomes slightly higher.
5. Increase in PEPcase in mesophyll	Leakage of carbon dioxide back to mesophyll leads to enhanced activity of PEPcase, with four-carbon acids being decarboxylated in bundle sheath cells, where decarboxylases already occur to process organic acids from the vascular tissue. In time, as PEPcase activity increases, more carbon dioxide from intercellular spaces is fixed by PEPcase in the mesophyll cells than by Rubisco. Enhanced activity of PPDK, to regenerate PEP, seems to occur late in the evolution of C_4 photosynthesis.
6. Integration of C_3 and C_4 pathways	Changes in expression patterns of genes for photosynthetic enzymes, especially restriction to mesophyll or bundle sheath.
7. Optimization and adjustmentof the whole plant	Changes in kinetic properties of enzymes, appropriate to different metabolic conditions; including changes in Rubisco, now in high carbon dioxide, low oxygen surroundings so acquires faster rate of catalysis, even if at the expense of specificity for carbon dioxide. Changes in stomatal behavior.

plant, if not beneficial. We are far from understanding this aspect of C_4 evolution. For single-cell C_4 photosynthesis, our ignorance is complete: there are too few examples known and no intermediates so how it evolved is obscure.

Single-cell C_4 photosynthesis

Four species are now known to operate C_4 photosynthesis within a single cell: the submerged water plants *Hydrilla verticillata* and *Egeria densa* and the terrestrial chenopods (family Chenopodiaceae) *Borszczowia aralocaspica* and *Bienertia cycloptera*. Original publications are listed in Table 2. There is an excellent brief account in Sage

Table 2. Original accounts of single-cell C_4 photosynthesis.

Species and reference	Comments
Hydrilla verticillata	
Spencer et al (1994)	Field and laboratory measurements; ecological significance.
Reiskind et al (1997)	Induction of C_4 photosynthesis.
Magnin et al (1997)	Appearance and location of C_4 enzymes.
Rao et al (2002)	Three types of phospho*enol*pyruvate carboxylase in *Hydrilla*.
Rao et al (2005)	Attempts to isolate different forms of carbonic anhydrase.
Egeria densa	
Casati et al (2000)	First report of induction of C_4 photosynthesis.
Lara et al (2002)	Full account of C_4 photosynthesis.
Borszczowia aralocaspica	
Freitag and Stichler (2000)	First report of C_4 photosynthesis.
Voznesenskaya et al (2001)	Full account of C_4 photosynthesis.
Voznesenskaya et al (2004)	Development of C_4 photosynthesis in cotyledons.
Bienertia cycloptera	
Freitag and Stichler (2002)	First report of C_4 photosynthesis.
Voznesenskaya et al (2002)	Full account of C_4 photosynthesis.

(2004) with new drawings to make clear the spatial separation of initial and final fixation of carbon dioxide within the cells of the terrestrial examples. Single-cell C_4 photosynthesis has been reviewed by Edwards et al (2004), Bowes et al (2002), and Sage (2002a); see also Bowes et al (this volume) and Edwards et al (this volume).

Hydrilla and *Egeria* switch to C_4 mode when it is advantageous, in conditions of low carbon dioxide and bicarbonate concentrations in the water, high oxygen concentration, and high temperature, which all tend to produce high rates of photorespiration. The quantum yield of *Hydrilla* in C_4 mode is half that of C_3 mode (Spencer et al 1994), indicating that the C_4 mode is expensive in terms of energy. The quantum yield figures are given in the discussion, with little detail and in unfamiliar units, so it is not possible to compare them with other published figures, but the difference is large and unambiguous in direction (but see Bowes et al, this volume).

In *Borszczowia*, the photosynthetic cells are elongated and arranged radially around the vascular tissue. Most chloroplasts, containing Rubisco, are in a dense mass at the inner end of the cell. Initial fixation by PEPcase occurs at the outer end, where there are intercellular air spaces, and the C_4 acids diffuse to the opposite end for decarboxylation and refixation of carbon dioxide. The barrier to leakage is apparently the cell vacuole and lack of intercellular air spaces at the inner end of the cells. It resembles Kranz anatomy in all but the absence of a transverse cell wall midway along the cell.

The photosynthetic cells in *Bienertia* have a peculiar arrangement of cytoplasm, and also seem to use the vacuole as the barrier to carbon dioxide diffusion away from the site of concentration around Rubisco. There is a central portion of cytoplasm, containing chloroplasts and Rubisco, connected by strands across the vacuole to peripheral cytoplasm. Initial fixation by PEPcase takes place at the periphery, metabolites diffuse along the strands, and final fixation occurs in the center.

Using a model given in full by von Caemmerer and Furbank (2003), who emphasized the importance of a high resistance to carbon dioxide between mesophyll and bundle sheath in C_4 plants, von Caemmerer (2003) examined what would happen if the C_4 pathway could be added to C_3 photosynthesis in a single cell. She concluded that such a system would have high leakiness in most conditions, that is, carbon dioxide would not be effectively concentrated around Rubisco, and some energy necessary for the initial fixation (for the regeneration of PEP) would be wasted, leading to low quantum yield. The leakiness was worse when capacity to fix carbon dioxide was high and photosynthetic rates were highest. The system would have some advantages when intercellular carbon dioxide concentration was very low in the leaf, for example, when stomata were almost closed in water-limited conditions.

Given that the resistance to carbon dioxide diffusion across the cell wall and plasmalemma must be low, for adequate diffusion of carbon dioxide into the cell for initial fixation, any barrier to carbon dioxide diffusion would have to be in the chloroplast envelope, where in C_3 photosynthesis there is every advantage in having a low resistance at this point. None of the single-cell C_4 systems rely on a high resistance to carbon dioxide across the chloroplast envelope. This is the point where it would be required in *Hydrilla* but *Hydrilla* seems to be a leaky system with low quantum yield (see above) because energy efficiency is less important than making some photosynthetic gain in conditions conducive to very high photorespiration (but see Bowes et al, this volume). In *Borszczowia*, the resistance to diffusion along 40 µm of vacuole was calculated to be as high as the resistance found in C_4 plants between mesophyll and bundle sheath (von Caemmerer 2003).

From the details given above, it is possible to agree with Sage (2002a) that "C_4 photosynthesis in terrestrial plants does not require Kranz anatomy." But the terrestrial plants with single-cell C_4 photosynthesis are chenopods from arid, saline habitats where growth rates are slow. Their rates of photosynthesis (when nearly saturated for photosynthetically active radiation, PAR, expressed per unit weight of chlorophyll) are low (von Caemmerer 2003): about 15% (*Bienertia*) or 31% (*Borszczowia*) of the average for C_3 plants (but see Edwards et al, this volume). I conclude that we should now say "*productive* C_4 photosynthesis requires Kranz anatomy." Edwards et al (2004) speculate that there may be more single-cell C_4 systems to be discovered, but I find it hard to believe that it has been overlooked in any highly productive plant or in any crop. They also point out that the *Bienertia* system, apparently free of anatomical constraints and possibly switching between C_3 and C_4 modes, could have advantages in a crop. However, we are probably more ignorant of the controls of cellular development necessary to make a compartment within the cell in which to concentrate carbon dioxide around Rubisco than we are of the controls of cell differentiation that result in mesophyll and bundle sheath cells with high resistance to diffusion of carbon dioxide between them.

A variety of carbon-concentrating mechanisms in algae and in cyanobacteria may provide new ideas for converting a C_3 plant to C_4 photosynthesis—see references cited by von Caemmerer (2003) and see Raven et al (this volume). It seems likely that these mechanisms will be adaptations to particular environments (so not energetically efficient), rather than highly productive as is required for C_4 rice.

The questions about *Hydrilla* and *Borszczowia* posed in Mitchell and Sheehy (2000) have been answered or superseded. It is evident that quantum yield is low in *Hydrilla* from Spencer et al (1994), a paper not known to anyone at the 1999 conference. A model to simulate the single-cell system has been constructed (von Caemmerer and Furbank 2003) and used to see whether a C_3 plant transformed to C_4 in this way would be productive (von Caemmerer 2003). The answer is no, not for mainstream rice, at best perhaps for upland rice, where growth was continually limited by a shortage of water.

Molecular biology and genetic engineering of C_4 photosynthesis

Enzymes of C_4 photosynthesis

Matsuoka et al (2001) published a review of molecular biology and genetic engineering of C_4 enzymes, covering work up to 2000. The aims were to understand the mechanisms of C_4 photosynthesis as well as, ultimately, to improve C_3 crops. Their Table 1 is a useful summary of the locations and functions of C_4 enzymes. It had recently become clear that the C_4 enzymes were not specific to C_4 plants but that C_3 plants also possessed them, in two forms. One was for housekeeping and the other was similar to the C_4 form but produced in very small amounts. From this finding arose the proposal that the C_4 genes evolved from a set already present in C_3 plants, acquiring modifications in the pattern of expression and in kinetic properties of the enzymes. The bulk of the paper is a comprehensive review of the genes introduced, the promoters used, and the effects on plant physiology. The enzymes included are PEPcase, PPDK, NADP-MDH, aspartate aminotransferase, NADP-ME, and PEPCK (their Table 3), each of which has been transferred to rice except NADP-MDH. The main conclusions were that the C_4 system was finely tuned with precise expression of genes combined with anatomical modifications, and that techniques to produce transgenic plants with C_4 enzymes were well established: genes could be expressed in large amounts in specified locations. This was the starting point for introducing C_4 photosynthesis to C_3 plants but further work was needed to obtain the required precision of gene expression in the correct place in the plant, including auxiliary enzymes such as CA and pyrophosphatase, and transporters for metabolites across membranes. It was unclear whether this would work without Kranz anatomy. There was the example of *Hydrilla* but doubts were emerging that its C_4 system was a survival mechanism rather than a productive one.

The review by Miyao (2003) covers much the same ground, with references up to 2002. The emphasis is on the evolution of C_4 genes and the results of overproducing

Table 3. Original papers on rice transgenic for C_4 enzymes.

Enzyme	References
PEPcase from maize	Ku et al (1999, 2000, 2001)
	Agarie et al (2002b)
	Jiao et al (2002)
	Fukayama et al (2003)
	Suzuki et al (2006)
PEPCK	Suzuki et al (2000)
PPDK	Fukayama et al (2001)
NADP-ME	Takeuchi et al (2000)
	Tsuchida et al (2001)

C_4 enzymes in transgenic rice, complementing Häusler et al (2002), who cover other species. The gene for PPDK is taken as an example and its structure and evolution are shown in diagrams. The *Pdk1* gene has two promoters: one for the cytosol form of the enzyme, the other for the chloroplast form with a transit peptide for transfer into the chloroplast. There is a list of C_4 genes transferred to rice with their promoters, enzyme activity, and references. The general conclusion was that overproduction of a single C_4 enzyme in C_3 plants can alter metabolism but there were no definite positive effects on photosynthesis. Transgenic rice with more than one C_4 enzyme is now being produced but whether it will have improved photosynthesis is unclear.

Monson (2003) reviewed the properties of the genes for C_4 enzymes in the context of explaining the pattern of evolution of C_4 photosynthesis (i.e., many monocots and rather few dicots, no large trees). Although a low concentration of atmospheric carbon dioxide is the main driver, genetic factors are important in preventing or favoring the successful appearance of C_4 plants. Given the importance of first accumulating a reservoir of duplicated C_3 genes, before separate genes become specialized for different functions, the features that favor C_4 evolution are reproduction mainly by sexual means, large populations, and short generation times.

Carbonic anhydrase is the first enzyme of the C_4 mechanism, for which it ensures the supply of bicarbonate for PEPcase (reviewed by Badger 2003). That CA is essential for C_4 photosynthesis has been demonstrated in work with transgenic *Flaveria bidentis*, a C_4 dicot, which also showed that there was ample CA activity to support the highest rates of photosynthesis (von Caemmerer et al 2004). Further work on the role of CA in C_4 photosynthesis is required to increase the number of species studied and to explore any relationship between growth conditions, CA activity, and photosynthetic capacity.

Phospho*enol*pyruvate carboxylase has naturally received much attention: its properties are reviewed by Izui et al (2004), and Engelmann et al (2003) discuss the different forms found across the C_3, C_3-C_4, and C_4 species of *Flaveria*. Christin et al (n.d.) have studied the phylogeny of the genes for PEPcase in grasses and found that the four lineages each included rice and a panicoid C_4 grass. This is a preliminary result because their sample of DNA was unavoidably biased toward genes expressed in green leaves, and did not include any PEPcase genes from C_3 panicoid grasses.

Leegood (2002) and Edwards et al (2001) drew attention to a pleiotropic effect that occurred when the gene for NADP-ME was introduced into rice chloroplasts and strongly expressed. The activity of the enzyme was inversely related to the amount of chlorophyll and the activity of photosystem II; in addition, the chloroplasts lacked grana (Takeuchi et al 2000). These are precisely the changes seen in the chloroplasts of C_4 bundle sheath cells (NADP-ME subtype). There were other deleterious effects of high activity of NADP-ME (Tsuchida et al 2001); nevertheless, this positive pleiotropy is encouraging.

Leaf development

The pattern of maize leaf development was reviewed by Langdale and Nelson (1991) and Hall and Langdale (1996): bundle sheath cells develop next to a vein (Rubisco in chloroplasts without grana), mesophyll cells next to bundle sheath cells (no Rubisco, chloroplasts with grana, abundant PEPcase), and any mesophyll cells further away from the bundle sheath remain in the default C_3 state (Rubisco in chloroplasts with grana, no more than housekeeping amounts of PEPcase). The switch from the default state requires light and then it is position relative to a vein, not cell lineage, that determines the development of C_4 characteristics in bundle sheath and mesophyll cells.

In three related papers, the genetic control of this process is unraveled. First, a mutant gene in maize, *Golden2* (*G2*), was identified (Rossini et al 2001); second, four mutant alleles of the gene were used to discover more about the gene function (Cribb et al 2001); and finally these genes were put in a wider context once similar ones were found in *Arabidopsis* (Fitter et al 2002). The *G2* gene is a general regulator of chloroplast development in C_3 tissue, that is, the default state and in any mesophyll cells not adjacent to a bundle sheath cell. Once the C_4 program is switched on, *G2* is a specific regulator of chloroplast development in bundle sheath cells, giving them their C_4 characteristics. There is a second gene in maize, *ZmGlk1* (*Zea mays G2-like1*), which regulates chloroplast development in the mesophyll cells that become C_4.

There are orthologous genes in rice: *OsGlk2* pairs with *G2* and *OsGlk1* with *ZmGlk1*. These genes in rice are involved in chloroplast development in overlapping sets of plant parts. Consequently, there is at least some redundancy, that is, one gene can substitute for the other at least partly.

Equivalent genes have now been found in *Arabidopsis*. All these genes are *GLK* genes and are part of the GARP superfamily of genes, and they have been identified from their structure as transcription factors. The occurrence of pairs of genes suggests that duplication took place before the divergence of rice and maize. Subsequently, the maize genes have been able to acquire specialized and complementary functions, controlling the development of C_4 characteristics in chloroplasts of bundle sheath or mesophyll cells. Recently, it has been shown that *GLK* genes are ancient, occurring in moss, so preceding the split between mosses and flowering plants more than 400 million years ago (Yasumura et al 2005).

An enhancer trapping system in rice

Control of where transgenes are expressed has relied on finding a promoter that allows expression only at a particular developmental stage or in a particular type of cell or tissue or organ. There are rather few promoters, and they tend to be specific for individual species, that is, promoters show little conservation across species. An alternative approach is to use enhancers that exist already in the genome of the experimental plant. (An enhancer controls expression of its associated gene in a specific tissue [Lawrence 2000]. It is a control sequence [of DNA] in the gene where a gene-regulating protein can bind and produce a large increase in transcription.) Johnson et al (2005) used an enhancer trapping system to obtain a set of enhancer trap lines in rice that transactivate genes in many specific cell types of root, leaf, flower, and seed. (Transactivate means that the gene activated need not be close to its promoter or enhancer.) They produced hundreds of lines and identified which types of cell or tissue showed expression of an indicator gene (green fluorescent protein). The range of patterns of expression was from broad, for example, mesophyll, vascular bundles, to very narrow, for example, coleorhiza, guard cells in lamina, lodicules in spikelet. Work continues to find lines with expression patterns specific to developmental stages, or after exposure to high salinity or other changes in the environment. Many uses are envisaged for these lines in investigations of specific processes in particular locations in the plant because it is now possible, by selecting the appropriate line, to control where in a plant a transgene will be expressed. Other enhancer trap lines in rice have been produced by Wu et al (2003), and a transactivation system for rice has been described by Liang et al (2006).

Use of *Cleome*

Brown et al (2005) argue that progress on understanding the details of C_4 development would be faster using *Cleome gynandra* as the model C_4 plant, since it is the C_4 plant most closely related to *Arabidopsis thaliana*, and like *Arabidopsis* is a small plant with a short life cycle and a small genome. They concede (footnote to Table 1) that work on dicots may be of less direct use for monocots but the advantages of *Cleome* over maize (large plant, long life cycle, large genome albeit well known) are evident.

Rice transgenic for C_4 photosynthetic enzymes

There has been some controversy over whether transgenic rice with C_4-type PEPcase from maize exhibits improved photosynthesis or not, and what is the explanation for any changes in photosynthesis. The original papers are listed in Table 3, along with those on other C_4 enzymes. The problem is discussed by Miyao (2003), by von Caemmerer and Furbank (2003), and briefly by Leegood (2002). The conclusion now seems to be that the early results suggesting improved photosynthesis (Ku et al 1999, 2000) have not been supported by subsequent work; indeed, photosynthesis is probably somewhat reduced. If it was ever seriously thought that simply producing abundant PEPcase in leaves of C_3 plants would boost photosynthesis (either by fixing carbon dioxide in parallel to Rubisco or in somehow pumping carbon dioxide toward

Rubisco or both), then it is clear now that this strategy was not successful. To obtain the benefits of C_4 photosynthesis, it is necessary to introduce the full system, as carefully analyzed by Leegood (2002).

An independent line of research was taken by Suzuki et al (2000), who attempted to construct a single-cell system for concentrating carbon dioxide around Rubisco in rice, using what Leegood (2002) described as the simplest possible pathway, although one that does not exist in any C_4 plant. If large amounts of PEPCK were active inside the chloroplast, this could decarboxylate any oxaloacetic acid (OAA) that entered the chloroplast as a result of PEPcase fixing carbon dioxide in the cytosol. Decarboxylation would increase the concentration of carbon dioxide around Rubisco, and produce PEP for diffusion back to the cytosol and further use by PEPcase. The beauty of this set of enzymes is that PPDK is not required to regenerate PEP so the cost is only one ATP for each carbon dioxide molecule pumped into the chloroplast (Leegood 2002).

Suzuki et al (2000) succeeded in constructing rice transgenic with PEPCK from a panicoid C_4 grass (PEPCK subtype) using a maize promoter and the rice transit peptide from the small subunit of Rubisco to ensure that the protein reached the chloroplast. They showed in their transgenic lines that PEPCK in the chloroplasts was decarboxylating OAA, made by endogenous PEPcase in the cytosol, the carbon dioxide was refixed by Rubisco, and the PEP was used by the cytosolic PEPcase. However, this C_4-like cycle made a "very limited" contribution to photosynthesis (not statistically significant differences between control and transgenic plants in leaf photosynthesis or carbon dioxide compensation point for the leaf), which was attributed to the low activity of PEPCK in the chloroplasts and the low activity of endogenous PEPcase in the cytosol. They proposed that the system might work better if greater activity of PEPCK could be achieved, and with a C_4-type PEPcase introduced to the cytosol, as well as with attention to metabolite transporters across the chloroplast envelope. Suzuki et al (2006) added the maize gene for PEPcase but found that photosynthesis was 10% less than in the control plants, and the chloroplasts showed abnormal development. See Burnell (this volume) for further information.

This ingenious and simple system seems to have been comparatively easy to implement, at least as a first attempt. As for other single-cell systems, I suspect that the weakness is the lack of high resistance to carbon dioxide diffusion across the chloroplast envelope. High leakage of carbon dioxide out of the chloroplast is probable, leading to futile cycling of carbon dioxide where carbon fixed by PEPcase is not captured by Rubisco, which makes the system expensive even if it requires only half the amount of ATP per carbon dioxide as the NADP-ME system.

Better Rubisco for improved photosynthesis

Photosynthesis of C_3 plants is reduced by the occurrence of photorespiration, which is a consequence of the properties of Rubisco. But there is some variation in Rubisco across the range of photosynthetic organisms in which it is found. A Rubisco with better specificity for carbon dioxide ought to reduce the amount of photorespiration. Unfortunately, Rubiscos with higher specificity tend to have lower rates of catalysis.

Zhu et al (2004) examined the inverse relationship between specificity and rate of catalysis, and simulated canopy photosynthesis for Rubisco of various specificities for carbon dioxide with corresponding rates of catalysis from the relationship. The combination of average values for specificity and rate of catalysis for C_3 plants was found to be optimal for atmospheric carbon dioxide concentration of 200 ppm, which is about the average for the past 400,000 years. The current specificity is too high for carbon dioxide of 370 ppm; if it were reduced to the optimal value for this concentration (with rate of catalysis increasing), canopy photosynthesis could be increased, although by only 3%. (Their simulation was for a day of photosynthesis at 44°N, for 19 July [day 200 in the year], and leaf temperature 25 °C.) A few Rubiscos do not lie close to the curve relating specificity and rate of catalysis, mostly from nongreen algae. Using Rubiscos with the best available properties increased canopy photosynthesis by 17% (Rubisco from *Amaranthus*, a C_4 plant) or 27% (*Griffithsia*, a red alga).

A general conclusion was that Rubisco in vascular plants had evolved to the best working compromise, with a large affinity for carbon dioxide and adequate rate of catalysis. If it were easy to obtain higher specificity without rate of catalysis decreasing, surely it would have turned up. (Mutant forms of Rubisco found so far have always had worse combinations of specificity and rate of catalysis.) Instead, the C_4 system had evolved many times, despite the complexities of Kranz anatomy and an extra biochemical pathway. Interestingly, Rubisco appears to have evolved optimally for 200 ppm carbon dioxide, close to the range of values prevailing for most of the time that life has been on land, and below the preindustrial (280 ppm) or current values.

Zhu et al (2004) pointed out that there was no interest in increasing the amount of Rubisco. It already accounted for up to 50% of soluble protein in C_3 crop leaves and the work of Pyke and Leech (1987) suggested that there was little physical space in photosynthetic cells or chloroplasts to add more Rubisco. They concluded that better Rubisco was the answer. However, recent analysis suggests that the trade-off between specificity for carbon dioxide and rate of catalysis is inherent in the chemistry of Rubisco and its reactants, and that Rubiscos are finely adapted for the temperature and concentrations of carbon dioxide and oxygen in the chloroplast of their particular species (Tcherkez et al 2006, Gutteridge and Pierce 2006).

Progress in transferring a selected Rubisco into a different plant was reviewed briefly by Zhu et al (2004) and by Maliga (2002), citing Kanevski et al (1999), Whitney and Andrews (2001a), and Whitney et al (2001) but not always agreeing in interpretation. The large subunit of Rubisco in tobacco was replaced with a bacterial version; the tobacco plants grew successfully but only in higher concentrations of carbon dioxide, as expected from the kinetic properties of the bacterial Rubisco (Whitney and Andrews 2001b). Study of these papers may shed more light on the differing interpretations; also relevant is a review of Rubisco properties (Spreitzer and Salvucci 2002), cited by Zhu et al (2004) for the inverse relationship between specificity and rate of catalysis, and see also Sage (2002b).

Reflections on constructing C_4 rice

General discussion of findings

The C_4 mechanism is an addition to the basic C_3 system of photosynthesis, just as a supercharger can be added to an internal combustion engine. Apart from a few intermediate species, plants are categorized as C_3 or C_4 but it is now clear that the C_3 and C_4 syndromes are not as rigidly separated as was first thought. The enzymes that are especially important in the C_4 mechanism also exist in C_3 leaves although with very low activity (Matsuoka et al 2001). Unexpectedly, there is a well-developed C_4 pathway in certain locations in C_3 plants: in the green tissue around vascular bundles (Hibberd and Quick 2002), and possibly in developing fruits generally (Imaizumi et al 1997, King et al 1998). In contrast, maize, a thoroughly C_4 plant, has patches of C_3 tissue wherever a mesophyll cell is not adjacent to a bundle sheath cell, particularly in leaf sheaths (Langdale and Nelson 1991). This information helps in understanding why the C_4 system has evolved repeatedly. It appears that there is lots of preconditioning and the C_4 system readily emerges whenever there is a premium on productive photosynthesis in conditions with high photorespiration.

Single-cell systems of C_4 photosynthesis are now much better known than in 1999 when we discussed only *Hydrilla* at the conference. They appear to have evolved as survival mechanisms in particular environments. The work of von Caemmerer (2003) makes it clear that a single-cell system is unlikely to be appropriate for productive C_4 rice, because it is impossible to have a high resistance to diffusion of carbon dioxide within a cell in order to concentrate carbon dioxide around Rubisco effectively.

Similarly, rice transgenic with one, two, or three enzymes of the C_4 pathway is not the answer. There may be some additional fixing of carbon dioxide but it does not produce a consistent and substantial increase in the rate of photosynthesis.

An alternative route to increased photosynthesis is to use a better Rubisco. Predictions from a model gave mixed results (Zhu et al 2004). The best available C_3 Rubisco produced only a 3% increase in canopy photosynthesis. The more speculative introduction of Rubisco from a red alga gave a 27% increase, which would be well worth having in rice. Zhu et al (2004) have established a good procedure with the use of a model to assess the effects of changes in Rubisco properties when scaled up to canopy photosynthesis and integrated over the day. (The effects of higher specificity for carbon dioxide or higher rate of catalysis vary according to whether leaf photosynthesis is PAR-limited or close to saturation for PAR, and consequently depend on canopy structure and the distribution of values of incident PAR during the day.) There is scope for further modeling: for instance, they used a leaf temperature of 25 °C but 30 °C would be more typical of tropical rice, with corresponding higher photorespiration so greater benefits expected from an improved Rubisco (as implied by results in their Figure 4d). Another approach would be to select a 10%, 20%, or 50% increase in canopy photosynthesis and find out what properties of Rubisco were required, and then whether any such Rubisco were known. If C_4 rice were constructed, it would probably include insertion of a C_4-type Rubisco so developments in the technology

Table 4. The five main subfamilies of grasses, here in order from C_3 to C_4, not in taxonomic order. Information from Mabberley (1997) and Sage et al (1999).

Subfamily	Characteristics
Pooideae	Entirely C_3; mainly north temperate, including wheat, barley, oats, rye.
Bambusoideae	Entirely C_3; bamboo, rice and relatives; mainly tropical and subtropical, often in shady or wet habitats.
Arundinoideae	Mostly C_3: 92% of genera are C_3; cosmopolitan.
Panicoideae	Mostly C_4: 80% of genera are C_4; tropical to warm temperate; including maize, sorghum, sugar cane.
Chloridoideae	Entirely C_4; mainly tropical and subtropical, and dry climates; a single species has apparently reverted to C_3.

of Rubisco manipulation are worth watching. Various other routes to improved photosynthesis are reviewed by Long et al (2006) and Raines (2006).

If the case made by Brown et al (2005) is accepted, then we could expect faster progress in dicots in comparing C_3 and C_4 genomes. Perhaps *Arabidopsis* will be the first C_3 species converted to C_4 photosynthesis. Much of the knowledge is likely to be transferable to monocots since molecular biology seems to be well conserved among angiosperms (e.g., Fitter et al 2002).

Why has C_4 rice not evolved?

Given the multiple origins of C_4 photosynthesis—45 independent lineages across 19 families of angiosperms, 11 lineages in grasses alone (Sage 2004)—why does C_4 photosynthesis not occur in rice or any of its close relatives? The distribution of C_4 photosynthesis across the main subfamilies of grasses is shown in Table 4. The lack of C_4 in the pooid grasses is understandable since they occur mainly in cool temperate regions where C_4 photosynthesis would be of no advantage. But the largely tropical bambusoids "ought" to be C_4.

Sage (2004) analyzed the geographical origins of C_4 dicots; no corresponding analysis was possible for monocots because they are too numerous and arose too early for the origins to be discernible now. The origins of C_4 dicots are thought to be in (1) southwestern North America, (2) east-central South America, (3) eastern and southern Africa, and (4) south-central Asia and the Middle East. For what it is worth, these centers of origin do not include the rice regions of South, East, and Southeast Asia (India eastward to Japan and southeast to New Guinea, for *Oryza sativa*) or West Africa (for *O. glaberrima*).

There are two main drivers for the evolution of C_4 photosynthesis (Sage 2004). One is the need to compensate for high rates of photorespiration and the other is simple carbon deficiency, when the atmospheric concentration of carbon dioxide is low, especially when rather close to the long-term compensation point where respiration balances photosynthesis over the lifetime of the plant. Various combinations of low atmospheric concentration of carbon dioxide, high temperature, low humidity, drought, and high salinity produce conditions of carbon deficiency and high photorespiration, thus determining where and when evolution of C_4 photosynthesis is likely to occur. This does not preclude evolution in wet habitats if high temperature promoted high photorespiration, as has occurred with sedges of tropical wetlands; and there are plenty of wetland C_4 grasses, for example, species of *Echinochloa*.

I speculate that one reason why C_4 photosynthesis has not arisen in the bambusoid grasses is their occurrence predominantly in shady habitats. Shade-tolerant C_4 grasses are rare and Ogle (2003) has proposed an empirical explanation. In addition, for bamboos, the semelparous habit, that is, they flower only once and die, after several to many years of vegetative growth, is not conducive to rapid evolution. Rapid evolution, especially the accumulation of duplicated genes (Monson 2003), is part of the general preconditioning envisaged as Phase 1 of C_4 evolution (Sage 2004).

Rice crops, of course, are not grown under shade but in the open for the sake of high yields. Moreover, where irrigation permits, crops are grown in the dry season under high solar radiation. These are the conditions that promote high rates of photorespiration, where C_4 photosynthesis is perceived as an advantage, and certainly one that is exploited by C_4 weeds of rice. Since in nature rice and its relatives grew in shade, or completed their life cycles during the wet season with lower solar radiation and lower temperatures, perhaps these species did not suffer from high photorespiration during the periods of low atmospheric concentrations of carbon dioxide when C_4 photosynthesis evolved in the grasses, starting 30 million years ago (Sage 2004). The question is: Has evolution of rice anatomy and biochemistry since that time proceeded along lines that now preclude the changes required for C_4 photosynthesis?

Phases of C_4 evolution

The phases of C_4 evolution may guide the construction of C_4 rice but need not constrain it. It should be possible to nurture any poorly adapted intermediates so that components of the system need not be added in order, and need not achieve a working system at each stage.

The characteristics of Phases 1, 2, and 3 do suggest features to look for in rice. Can duplicate genes be identified for all the enzymes required? Are there rice varieties or closely related species that have closer vein spacing in the leaves, or bundle sheath cells that are especially large or metabolically active?

The changes in photorespiration in Phase 4 are a plausible mechanism for the evolution of C_4 photosynthesis. The C_3-C_4 intermediates with glycine decarboxylase confined to bundle sheath cells have reached this phase, but they are not likely to be especially productive. It does not appear necessary in constructing C_4 rice to pass through this phase, and it does not appear to be a useful end-point at which to aim.

I think that designing C_4 rice can concentrate on Phases 5, 6, and 7. We now have a clear idea of the requirements (Leegood 2002, von Caemmerer and Furbank 2003, Miyao 2003) and, once a design is specified, techniques for achieving it can be sought.

Reasons for hope

Constructing C_4 rice is an ambitious project but there are several reasons to start work optimistically. The frequency and apparent ease with which C_4 photosynthesis has evolved is encouraging. There are C_4 features already present in C_3 plants and much is understood about the different forms of the relevant enzymes. Genes for many of these enzymes can be inserted into rice and made to work, sometimes in precise locations in the plant or in the cell. The enhancer trap lines now available for rice are a promising resource. One case of pleiotropy in a helpful direction has been noted. Progress in understanding leaf development is slower but there is still hope that genes will be found, perhaps only a few, with far-reaching effects on controlling vein spacing and differentiation of C_4 bundle sheath and mesophyll cells.

References

Agarie S, Kai M, Takatsuji H, Ueno O. 2002a. Environmental and hormonal regulation of gene expression of C_4 photosynthetic enzymes in the amphibious sedge *Eleocharis vivipara*. Plant Sci. 163:571-580.

Agarie S, Miura A, Sumikura R, Tsukamoto S, Nose A, Arima S, Matsuoka M, Miyao-Tokutomi M. 2002b. Overexpression of C_4 PEPC caused O_2-insensitive photosynthesis in transgenic rice plants. Plant Sci. 162:257-265.

Badger M. 2003. The roles of carbonic anhydrases in photosynthetic CO_2 concentrating mechanisms. Photosynthesis Res. 77:83-94.

Barroca J, Murphy LR, Franceschi VR, Lee R, Roalson E, Edwards GE, Ku MS. 2005. Diversification and plasticity of C_4 photosynthetic pathway in *Eleocharis* (Cyperaceae). In: van der Est A, Bruce D, editors. Photosynthesis: fundamental aspects to global perspectives. Proceedings of the 13th International Congress on Photosynthesis, Montreal, Canada, August-September 2004). Lawrence, Kansas (USA): Alliance Communications Group, for International Society of Photosynthesis Research. p 646-649.

Bowes G, Rao SK, Estavillo GM, Reiskind JB. 2002. C_4 mechanisms in aquatic angiosperms: comparisons with terrestrial C_4 systems. Funct. Plant Biol. 29:379-392.

Brown NJ, Parsley K, Hibberd JM. 2005. The future of C_4 research—maize, *Flaveria* or *Cleome*? Trends Plant Sci. 10:215-221.

Casati P, Lara MV, Andreo CS. 2000. Induction of a C_4-like mechanism of CO_2 fixation in *Egeria densa*, a submersed aquatic species. Plant Physiol. 123:1611-1621.

Christin PA, Salamin N, Savolainen V, Besnard G. n.d. A phylogenetic study of the phospho*enol*pyruvate carboxylase multigene family in Poaceae: understanding the molecular changes linked to C_4 photosynthesis evolution. Kew Bulletin. (In press.)

Cribb L, Hall LN, Langdale JA. 2001. Four mutant alleles elucidate the role of the G2 protein in the development of C_4 and C_3 photosynthesizing maize tissues. Genetics 159:787-797.

Dengler NG, Nelson T. 1999. Leaf structure and development in C_4 plants. In: Sage RF, Monson RK, editors. C_4 plant biology. London (UK): Academic Press. p 133-172.

Edwards GE, Furbank RT, Hatch MD, Osmond CB. 2001. What does it take to be C_4? Lessons from the evolution of C_4 photosynthesis. Plant Physiol. 125:46-49.

Edwards GE, Franceschi VR, Voznesenskaya EV. 2004. Single-cell C_4 photosynthesis versus the dual-cell (Kranz) paradigm. Annu. Rev. Plant Biol. 55:173-196.

Engelmann S, Blasing OE, Gowick U, Svensson P, Westhoff P. 2003. Molecular evolution of C_4 phospho*enol*pyruvate carboxylase in the genus *Flaveria*: a gradual increase from C_3 to C_4 characteristics. Planta 217:717-725.

Fitter DW, Martin DJ, Copley MJ, Scotland RW, Langdale JA. 2002. *GLK* gene pairs regulate chloroplast development in diverse plant species. Plant J. 31:713-727.

Freitag H, Stichler W. 2000. A remarkable new leaf type with unusual photosynthetic tissue in a central Asiatic genus of Chenopodiaceae. Plant Biol. 2:154-160.

Freitag H, Stichler W. 2002. *Bienertia cycloptera* Bunge ex Boiss., Chenopodiaceae, another C_4 plant without Kranz tissues. Plant Biol. 4:121-132.

Fukayama H, Tsuchida H, Agarie S, Nomura M, Onodera H, Ono K, Lee BH, Hirose S, Toki S, Ku MSB. 2001. Significant accumulation of C_4-specific pyruvate, orthophosphate dikinase in a C_3 plant, rice. Plant Physiol. 127:1136-1146.

Fukayama H, Hatch MD, Tamai T, Tsuchida H, Sudoh S, Furbank RT, Miyao M. 2003. Activity regulation and physiological impacts of maize C_4-specific phospho*enol*pyruvate carboxylase overproduced in transgenic rice plants. Photosynthesis Res. 77:227-239.

Gutteridge S, Pierce J. 2006. A unified theory for the basis of limitations of the primary reaction of photosynthetic CO_2 fixation: was Dr Pangloss right? Proc. Nat. Acad. Sci. USA 103:7203-7204.

Hall LN, Langdale JA. 1996. Molecular genetics of cellular differentiation in leaves. New Phytol. 132:533-553.

Häusler RE, Hirsch HJ, Kreuzaler F, Peterhänsel C. 2002. Overexpression of C_4-cycle enzymes in transgenic C_3 plants: a biotechnological approach to improve C_3 photosynthesis. J. Exp. Bot. 53:591-607.

Hibberd JM, Quick WP. 2002. Characteristics of C_4 photosynthesis in stems and petioles of C_3 flowering plants. Nature 415:451-454.

Imaizumi N, Samejima M, Ishihara K. 1997. Characteristics of photosynthetic carbon metabolism of spikelets in rice. Photosynthesis Res. 52:75-82.

Izui K, Matsumura H, Furumoto T, Kai Y. 2004. Phospho*enol*pyruvate carboxylase: a new era of structural biology. Annu. Rev. Plant Biol. 55:69-84.

Jiao D, Huang X, Li X, Chi W, Kuang T, Zhang Q, Ku MSB, Cho D. 2002. Photosynthetic characteristics and tolerance to photo-oxidation of transgenic rice expressing C_4 photosynthetic enzymes. Photosynthesis Res. 72:85-93.

Johnson AAT, Hibberd JM, Gay C, Essah PA, Haseloff J, Tester M, Guiderdoni E. 2005. Spatial control of transgene expression in rice (*Oryza sativa* L.) using the GAL4 enhancer trapping system. Plant J. 41:779-789.

Kanevski I, Maliga P, Rhoades DF, Gutteridge S. 1999. Plastome engineering of ribulose-1,5-bisphosphate carboxylase/oxygenase in tobacco to form a sunflower large subunit and a tobacco small subunit hybrid. Plant Physiol. 119:133-141.

King SP, Badger MR, Furbank RT. 1998. CO_2 refixation characteristics of developing canola seeds and silique wall. Aust. J. Plant Physiol. 25:377-386.

Ku MSB, Agarie S, Nomura M, Fukayama H, Tsuchida H, Ono K, Hirose S, Toki S, Miyao M, Matsuoka M. 1999. High-level expression of maize phospho*enol*pyruvate carboxylase in transgenic rice plants. Nature Biotechnol. 17:76-80.

Ku MSB, Cho D, Ranade U, Hsu T-P, Li X, Jiao D-M, Ehleringer J, Miyao M, Matsuoka M. 2000. Photosynthetic performance of transgenic rice plants overexpressing the maize C_4 photosynthesis enzymes. In: Sheehy JE, Mitchell PL, Hardy B, editors. Redesigning rice photosynthesis to improve yield. Proceedings of a workshop on The Quest to Reduce Hunger: Redesigning Rice Photosynthesis, 30 November-3 December 1999, at IRRI. Amsterdam (Netherlands): Elsevier, and Makati City (Philippines): International Rice Research Institute. p 193-204.

Ku MSB, Cho D, Li X, Jiang D, Pinto M, Mitsue M, Matsuoka M. 2001. Introduction of genes encoding C_4 photosynthesis enzymes into rice plants: physiological consequences. In: Chadwick D, Goode JA, editors. Rice biotechnology: improving yield, stress tolerance and grain quality. Chichester, Sussex (Wiley; Novartis Foundation Symposium no. 236, March 2000). p 100-111.

Langdale JA, Nelson T. 1991. Spatial regulation of photosynthetic development in C_4 plants. Trends Genet. 7:191-196.

Lara MV, Casati P, Andreo CS. 2002. CO_2-concentrating mechanisms in *Egeria densa*, a submersed aquatic plant. Physiol. Plant. 115:487-495.

Lawrence E, editor. 2000. Henderson's dictionary of biological terms. Harlow, Essex (UK): Pearson.

Leegood RC. 2002. C_4 photosynthesis: principles of CO_2 concentration and prospects for its introduction into C_3 plants. J. Exp. Bot. 53:581-590.

Liang D, Wu C, Li C, Xu C, Zhang J, Kilian A, Li X, Zhang Q, Xiong L. 2006. Establishment of a patterned GAL4-VP16 transactivation system for discovering gene function in rice. Plant J. 46:1059-1072.

Long SP, Zhu X-G, Naidu SL, Ort DR. 2006. Can improvement in photosynthesis increase crop yields? Plant Cell Environ. 29:315-330.

Mabberley DJ. 1997. The plant-book: a portable dictionary of the higher plants. Second edition. Cambridge (UK): Cambridge University Press.

Magnin NC, Cooley BA, Reiskind JB, Bowes G. 1997. Regulation and localization of key enzymes during the induction of Kranz-less, C_4-type photosynthesis in *Hydrilla verticillata*. Plant Physiol. 115:1681-1689.

Maliga P. 2002. Engineering the plastid genome of higher plants. Curr. Opin. Plant Biol. 5:164-172.

Matsuoka M, Furbank RT, Fukayama H, Miyao M. 2001. Molecular engineering of C_4 photo synthesis. Annu. Rev. Plant Physiol. Plant Mol. Biol. 52:297-314.

Mitchell PL, Sheehy JE. 2000. Discussion: opportunities for redesigning rice photosynthesis. In: Sheehy JE, Mitchell PL, Hardy B, editors. Redesigning rice photosynthesis to improve yield. Proceedings of a workshop on The Quest to Reduce Hunger: Redesigning Rice Photosynthesis, 30 November-3 December 1999, at IRRI. Amsterdam (Netherlands): Elsevier, and Makati City (Philippines): International Rice Research Institute. p 269-288.

Miyao M. 2003. Molecular evolution and genetic engineering of C_4 photosynthetic enzymes. J. Exp. Bot. 54:179-189.

Monson RK. 2003. Gene duplication, neofunctionalization, and the evolution of C_4 photosynthesis. Int. J. Plant Sci. 164(3 Supplement):S43-S54.

Ogle K. 2003. Implications of interveinal distance for quantum yield in C_4 grasses: a modeling and meta-analysis. Oecologia 136:532-542.

Pyke KA, Leech RM. 1987. The control of chloroplast number in wheat mesophyll cells. Planta 170:416-420.

Raines CA. 2006. Transgenic approaches to manipulate the environmental responses of the C_3 carbon fixation cycle. Plant Cell Environ. 29:331-339.
Rao VS, Magnin NC, Reiskind JB, Bowes G. 2002. Photosynthetic and other phospho*enol*pyruvate carboxylase isoforms in the single-cell, facultative C_4 system of *Hydrilla verticillata*. Plant Physiol. 130:876-886.
Rao VS, Rao SK, Reiskind JB, Bowes G. 2005. Carbonic anhydrase isoforms in the C_4 CCM of *Hydrilla*. In: van der Est A, Bruce D, editors. Photosynthesis: fundamental aspects to global perspectives. Proceedings of the 13th International Congress on Photosynthesis, Montreal, Canada, August-September 2004. Lawrence, Kansas (USA): Alliance Communications Group, for International Society of Photosynthesis Research. p 954-955.
Reiskind JB, Madsen TV, van Ginkel LC, Bowes G. 1997. Evidence that inducible C_4-type photosynthesis is a chloroplastic CO_2-concentrating mechanism in *Hydrilla*, a submersed monocot. Plant Cell Environ. 20:211-220.
Rossini L, Cribb L, Martin DJ, Langdale JA. 2001. The maize *Golden2* gene defines a novel class of transcriptional regulators in plants. Plant Cell 13:1231-1244.
Sage RF, Monson RK, editors. 1999a. C_4 plant biology. London (UK): Academic Press.
Sage RF, Li M, Monson RK. 1999b. The taxonomic distribution of C_4 photosynthesis. In: Sage RF, Monson RK, editors. C_4 plant biology. London (UK): Academic Press. p 551-584.
Sage RF. 2002a. C_4 photosynthesis in terrestrial plants does not require Kranz anatomy. Trends Plant Sci. 7:283-285.
Sage RF. 2002b. Variation in the k_{cat} of Rubisco in C_3 and C_4 plants and some implications for photosynthetic performance at high and low temperature. J. Exp. Bot. 53:609-620.
Sage RF. 2004. The evolution of C_4 photosynthesis. New Phytol. 161:341-370.
Sheehy JE, Mitchell PL, Hardy B, editors. 2000. Redesigning rice photosynthesis to improve yield. Proceedings of a workshop on The Quest to Reduce Hunger: Redesigning Rice Photosynthesis, 30 November-3 December 1999, at IRRI. Amsterdam (Netherlands): Elsevier, and Makati City (Philippines): International Rice Research Institute.
Spencer WE, Teeri J, Wetzel RG. 1994. Acclimation of photosynthetic phenotype to environmental heterogeneity. Ecology 75:301-314.
Spreitzer RJ, Salvucci ME. 2002. RUBISCO: structure, regulatory interactions, and possibilities for a better enzyme. Annu. Rev. Plant Biol. 53:449-475.
Suzuki S, Murai N, Burnell JN, Arai M. 2000. Changes in photosynthetic carbon flow in transgenic rice plants that express C_4-type phospho*enol*pyruvate carboxykinase from *Urochloa panicoides*. Plant Physiol. 124:163-172.
Suzuki S, Murai N, Kasaoka K, Hiyoshi T, Imaseki H, Burnell JN, Arai M. 2006. Carbon metabolism in transgenic rice plants that express phospho*enol*pyruvate carboxylase and/or phospho*enol*pyruvate carboxykinase. Plant Sci. 170:1010-1019.
Takeuchi Y, Akagi H, Kamasawa N, Osumi M, Honda H. 2000. Aberrant chloroplasts in transgenic rice plants expressing a high level of maize NADP-dependent malic enzyme. Planta 211:265-274.
Taylor WC. 2000. C_4 rice: what are the lessons from developmental and molecular studies? In: Sheehy JE, Mitchell PL, Hardy B, editors. Redesigning rice photosynthesis to improve yield. Proceedings of a workshop on The Quest to Reduce Hunger: Redesigning Rice Photosynthesis, 30 November-3 December 1999, at IRRI. Amsterdam (Netherlands): Elsevier, and Makati City (Philippines): International Rice Research Institute. p 87-96.

Tcherkez GGB, Farquhar GD, Andrews TJ. 2006. Despite slow catalysis and confused substrate specificity, all ribulose bisphosphate carboxylases may be nearly perfectly optimized. Proc. Nat. Acad. Sci. USA 103:7246-7251.

Tsuchida H, Tamai T, Fukayama H, Agarie S, Nomura M,Onodera H, Ono K, Nishizawa Y, Lee BH, Hirose S, Toki S, Ku MSB, Matsuoka M, Miyao M. 2001. High level expression of C_4-specific NADP-malic enzyme in leaves and impairment of photoautotrophic growth in a C_3 plant, rice. Plant Cell Physiol. 42:138-145.

Tsuchida H, Fukayama H, Miyao-Tokutomi M. 2005. Proteomic analysis of six different organs of rice: comparison of enzymes involved in photosynthetic and primary metabolism. In: van der Est A, Bruce D, editors. Photosynthesis: fundamental aspects to global perspectives. Proceedings of the 13th International Congress on Photosynthesis, Montreal, Canada, August-September 2004. Lawrence, Kansas (USA): Alliance Communications Group, for International Society of Photosynthesis Research. p 949-951.

Ueno O. 2001. Environmental regulation of C_3 and C_4 differentiation in the amphibious sedge *Eleocharis vivipara*. Plant Physiol. 127:1524-1532.

Ueno O, Ishimaru K. 2002. Effects of an inhibitor of phospho*enol*pyruvate carboxylase on photosynthesis of the terrestrial forms of amphibious *Eleocharis* species. Photosynthesis Res. 71:265-272.

Ueno O. 2004. Environmental regulation of photosynthetic metabolism in the amphibious sedge *Eleocharis baldwinii* and comparisons with related species. Plant Cell Environ. 27:627-639.

von Caemmerer S. 2003. C_4 photosynthesis in a single C_3 cell is theoretically inefficient but may ameliorate internal CO_2 diffusion limitations of C_3 leaves. Plant Cell Environ. 26:1191-1197.

von Caemmerer S, Furbank RT. 2003. The C_4 pathway: an efficient CO_2 pump. Photosynthesis Res. 77:191-207.

von Caemmerer S, Quinn V, Hancock NC, Price GD, Furbank RT, Ludwig M. 2004. Carbonic anhydrase and C_4 photosynthesis: a transgenic analysis. Plant Cell Environ. 27:697-703.

Voznesenskaya EV, Franceschi VR, Kiirats O, Freitag H, Edwards GE. 2001. Kranz anatomy is not essential for terrestrial C_4 plant photosynthesis. Nature 414:543-546.

Voznesenskaya EV, Franceschi VR, Kiirats O, Artyusheva EG, Freitag H, Edwards GE. 2002. Proof of C_4 photosynthesis without Kranz anatomy in *Bienertia cycloptera* (Chenopodiaceae). Plant J. 31:649-662.

Voznesenskaya EV, Franceschi VR, Edwards GE. 2004. Light-dependent development of single-cell C_4 photosynthesis in cotyledons of *Borszczowia aralocaspica* (Chenopodiaceae) during transformation from a storage to a photosynthetic organ. Ann. Bot. 93:177-187.

Wakayama M, Ueno O, Ohnishi J. 2003. Photosynthetic enzyme accumulation during leaf development of *Arundinella hirta*, a C_4 grass having Kranz cells not associated with veins. Plant Cell Physiol. 44:1330-1340.

Wang Q, Zhang Q, Fan D, Lu C. 2006. Photosynthetic light and CO_2 utilization and C_4 traits of two novel super-rice hybrids. J. Plant Physiol. 163:529-537.

Whitney SM, Baldett P, Hudson GS, Andrews TJ. 2001. Form I Rubiscos from non-green algae are expressed abundantly but not assembled in tobacco chloroplasts. Plant J. 26:535-547.

Whitney SM, Andrews TJ. 2001a. The gene for ribulose-1,5-bisphosphate carboxylase/oxygenase (Rubisco) small subunit relocated to the plastid genome of tobacco directs the synthesis of small subunits that assemble in Rubisco. Plant Cell 13:193-205.

Whitney SM, Andrews TJ. 2001b. Plastome-encoded bacterial ribulose-1,5-bisphosphate carboxylase/oxygenase (RubisCO) supports photosynthesis and growth of tobacco. Proc. Nat. Acad. Sci. USA 98:14738-14743.

Wu C, Li X, Yuan W, Chen G, Kilian A, Li J, Xu C, Li X, Zhou D-X, Wang S, Zhang Q. 2003. Development of enhancer trap lines for functional analysis of the rice genome. Plant J. 35:418-427.

Yasumura Y, Moylan EC, Langdale J.A. 2005. A conserved transcription factor mediates nuclear control of organelle biogenesis in anciently diverged land plants. Plant Cell 17:1894-1907.

Zhu X-G, Portis AR, Long SP. 2004. Would transformation of C_3 crop plants with foreign Rubisco increase productivity? A computational analysis extrapolating from kinetic properties to canopy photosynthesis. Plant Cell Environ. 27:155-165.

Notes

Author's address: Department of Animal and Plant Sciences, University of Sheffield, Sheffield S10 2TN, U.K.

Acknowledgments: I am grateful to Peter Horton, Colin Osborne, and conference participants for information on relevant papers, and to Professor F.I. Woodward and the Department of Animal and Plant Sciences for general support and encouragement in this work.

Section 2:

C_4 rice from theory to practice

C_4 photosynthesis: minor or major adjustments to a C_3 theme?

R.C. Leegood

This article explores some of the structural and physiological changes that may be required for the introduction of C_4 photosynthesis into the leaf of a C_3 plant, such as rice. Transport, both between cells and within cells, is essential to the operation of a C_4 photosynthetic system. In this article, I discuss some of the relationships between mesophyll and bundle sheath cells, and how transport of photosynthetic intermediates via the plasmodesmata is driven by large concentration gradients of metabolites. In turn, the properties of the enzymes of carbohydrate synthesis are modified to accommodate the high concentrations of metabolite precursors. The properties of intracellular metabolite transporters are also modified in C_4 photosynthesis. It is also likely that metabolites moving between the mesophyll and bundle sheath act as the messengers that ensure that individual enzymes are regulated so that photosynthetic fluxes are coordinated between the two cell types.

Keywords: enzyme regulation, metabolite transport, plasmodesmata

The most outstanding feature of high-capacity C_4 photosynthetic systems is the presence of two cell types, mesophyll cells and bundle sheath cells, which cooperate in carbon fixation. This anatomical feature of C_4 plants was recognized over a century ago by Haberlandt (1884), who drew attention to the distinctive Kranz (German for "wreath") arrangement of the chlorenchymatous bundle sheath cells in C_4 leaves. The metabolic cooperation between mesophyll and bundle sheath cells (Fig. 1) is perhaps only paralleled by that between the vegetative cells and heterocysts of cyanobacteria. If we are to introduce C_4 photosynthesis into the leaves of C_3 plants, one of the major challenges is to understand what mechanisms are used to coordinate the activities of the two cell types. In addition, we need to understand which of the enzymes of C_4 photosynthesis are already expressed in the bundle sheath and how metabolism has been altered compared to the bundle sheath of C_3 plants. An almost equally valid question would be to ask what the bundle sheath does in C_3 plants. What processes in the C_3 bundle sheath might be disrupted by introducing C_4 photosynthesis? However,

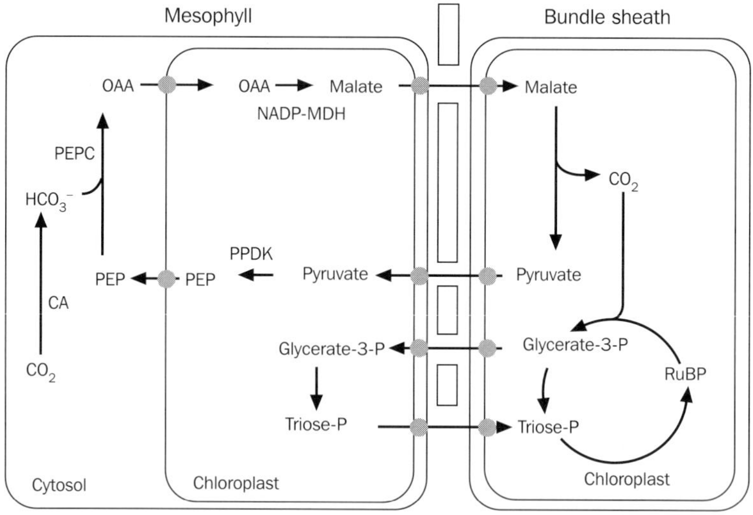

Fig. 1. The simplest form of the C_4 pathway, found in NADP-malic enzyme species such as sugar cane, showing the necessity for both intercellular transport between the mesophyll and bundle sheath, and for metabolite transport across the chloroplast envelope. It also shows the glycerate 3-P/triose-P shuttle. Transporters are depicted by shaded circles. CA = carbonic anhydrase, NADP-MDH = NADP malate dehydrogenase, NADP-ME = NADP-malic enzyme, PPDK = pyruvate, phosphate dikinase, PEPC = PEP carboxylase.

this article will concentrate on some of the changes in metabolic regulation that have occurred in C_4 plants compared with C_3 plants.

Intercellular metabolite transport in C_4 plants

Photosynthetic CO_2 fixation in C_4 plants depends on metabolite transport, which occurs by symplastic diffusion between the bundle sheath and mesophyll cells. This transport must occur at rates equivalent to rates of photosynthesis. Transport depends on a number of factors, including plasmodesmatal frequency, number, location, and properties of the various organelles, and the properties of metabolite transporters. The necessity for diffusive metabolite transport between the mesophyll and bundle sheath cells (Fig. 1) requires close contact between the two cell types and therefore sets limits on the number of mesophyll cells that can be functionally associated with bundle sheath cells. Thus, the interveinal distance (i.e., the number of mesophyll cells between adjacent bundle sheaths) is usually much smaller than in the leaves of C_3 plants and this also influences leaf thickness in C_4 plants. There are also appreciable differences between the various C_4 decarboxylation types (see Hattersley 1992). One particularly interesting anatomical mutant is the *lis1* genotype in *Panicum maximum*, in which the interveinal distance is increased from 2 to between 6 and 7 cells (Fladung 1994). This results in a lower rate of photosynthesis and an increase in the CO_2

compensation point, partly because it must affect intercellular transport, but perhaps also because it influences the development of C_4 metabolism in the mesophyll (Nelson and Langdale 1992). However, the mutant is no longer extant.

An important requirement for intercellular diffusion of metabolites is the presence of plasmodesmata, which permit the rapid exchange of solutes between cells. The plasmodesmata exclude large molecules, such as cytosolic proteins, with a size exclusion limit in bundle sheath strands of about 900 Da (Weiner et al 1988). The plasmodesmatal frequency is related to the CO_2 assimilation rate (Botha 1992). Extensive pit-fields in the wall between mesophyll and bundle sheath cells provide symplastic connections between the two cell types. Plasmodesmata are frequent only in primary pit-fields at the areas of contact between mesophyll and bundle sheath cells. In two C_4 grasses, *Themeda triandra* and *Panicum maximum*, 56% and 77%, respectively, of all the plasmodesmata in vascular bundles were present at this interface, compared with only 32% in *Bromus unioloides*, a C_3 grass (Botha and Evert 1988, Botha 1992). Mesophyll/bundle sheath plasmodesmatal frequencies in C_4 plants are about three to five times those in C_3 plants (Botha 1992).

Intracellular metabolite transport in C_4 plants

Intracellular metabolite transport between organelles is as important to the operation of C_4 photosynthesis as is intercellular metabolite exchange and it must proceed at comparable fluxes (again much higher than in C_3 plants). Leaf organelles in C_4 plants share the translocators of organelles in leaves of C_3 plants (Heldt and Flügge 1992), but also contain translocators with unique, or considerably altered, kinetic properties. Although transport processes across the envelope of C_4 mesophyll chloroplasts are now adequately characterized, relatively little is known of transport into the bundle sheath chloroplasts, largely because of the difficulty of isolating these intact and in appreciable quantities from any C_4 plants.

Transport in mesophyll chloroplasts

Exchange of phospho*enol*pyruvate (PEP), Pi, glycerate-3-P, and triose-P occurs on a common Pi translocator in the chloroplast envelope of C_4 plants. The C_4 mesophyll Pi translocator (TPT) is similar to the C_3-type phosphate translocator, with between 83% and 94% identity in amino acid residues. Only minor changes in amino acid sequence have occurred to extend the substrate specificity of the C_3 phosphate translocator to recognize PEP in C_4 plants (Fischer et al 1994). A PEP/Pi translocator (PPT) is also present in isolated maize chloroplasts (Huber and Edwards 1977). A gene encoding such a transporter (PPT) from maize is highly expressed in maize endosperm, but the transcript, although reasonably abundant in cauliflower (C_3) leaves, was weakly expressed in maize leaves (Fischer et al 1997). This casts doubt on whether this particular transporter is involved in the high flux of PEP out of the mesophyll chloroplasts of C_4 plants. In C_3 tissues, the PPT is involved in providing plastids with PEP for the shikimate pathway. Interestingly, the *cue*1 mutant of *Arabidopsis*, which is defective in one of two PPT genes, shows a reticulate leaf phenotype that may be due to the lack of generation of phenylpropanoid signals that trigger leaf development in interveinal regions (Voll et al 2003, Knappe et al 2003).

Chloroplasts transport pyruvate on a specific carrier in both C_3 and C_4 plants, but the translocator is more active and is light-dependent in mesophyll chloroplasts of C_4 plants (Flügge et al 1985). Pyruvate transport appears to be driven by an H^+ gradient in NADP (nicotinamide adenine dinucleotide phosphate)-malic enzyme species and a Na^+ gradient, or perhaps Na^+ activation of the transporter (Murata et al 1992) in NAD-malic enzyme and PEP carboxykinase species (Heldt and Flügge 1992, Ohnishi and Kanai 1990, Ohnishi et al 1990, Aoki et al 1992, 1994). However, the pyruvate transporter remains uncharacterized at the molecular level (Hibberd, this volume).

One of the principal fluxes across the chloroplast envelope in C_3 plants is the transport of dicarboxylic acids and glutamate that supports photorespiratory nitrogen cycling (Woo et al 1987). This involves two different translocator proteins with overlapping substrate specificities. DiT1 imports 2-oxoglutarate into the plastid in counter-exchange with export of malate. 2-oxoglutarate is converted to glutamate by GS/GOGAT, and DiT2 subsequently exports glutamate to the cytosol, again in strict counter-exchange with malate. DiT1 is able to catalyze the counter-exchange of malate with fumarate, succinate, 2-oxoglutarate, and, to a lesser extent, with aspartate and oxaloacetate, whereas DiT2 has broader substrate specificity than DiT1 because it also accepts glutamate as a counter-exchange substrate for malate (Renné et al 2003). These translocators have been adapted to suit C_4 photosynthesis.

In C_4 plants, the dicarboxylates (oxaloacetate and malate) are counter-exchanged at the plastid envelope. In isolated mesophyll chloroplasts, the *Km* for uptake of a particular dicarboxylic acid is similar to the *Ki* for the inhibition of uptake by other dicarboxylates. In maize, for example, the *Ki* (0.3 mM) for oxaloacetate inhibition of malate transport is comparable to the *Km* (0.5 mM) for malate uptake (Day and Hatch 1981). Oxaloacetate uptake would clearly not occur when oxaloacetate concentrations are several orders of magnitude less than malate concentrations, as occurs in $NADP^+$-ME plants such as maize, in which OAA concentrations are probably less than 50 μM. There is, therefore, a very active oxaloacetate carrier in maize mesophyll chloroplasts (*Km* 45 μM), which is little affected by malate (*Ki* 7.5 mM) (Hatch et al 1984, Taniguchi et al 2004) and which is likely to be distinct from the other dicarboxylate translocators. A dicarboxylate transporter with the requisite high affinity for OAA, and which possibly functions to import OAA into mesophyll chloroplasts, has been identified in *Arabidopsis* (*AtOMT1*, Taniguchi et al 2002) and in maize, in which transcripts are expressed in the mesophyll cells (*ZmpOMT1*, Taniguchi et al 2004). In sorghum, DiT2 transcripts are confined to bundle sheath cells, whereas DiT1 transcript levels were significantly higher in mesophyll cells than in bundle sheath cells (Renné et al 2003). In maize, four dicarboxylate translocator genes were identified and they largely accumulated in green tissues. These comprise *OMT1*, as discussed above, and three *DCT* genes. *DCT1* (=DiT1) was expressed in the mesophyll and *DCT2* (=DiT2) and *DCT3* in the bundle sheath, where they likely import malate into the bundle sheath chloroplasts (Taniguchi et al 2004).

Transport in bundle sheath chloroplasts

Bundle sheath chloroplasts are unusual because they must export glycerate-3-P at high rates *in vivo*. In C_3 plants, glycerate-3-P is not exported by chloroplasts to any great extent because it is transported by the translocator as the glycerate-3-P^{2-} ion, whereas glycerate 3-P^{3-} is the form that predominates at the pH that occurs in the illuminated stroma (Heldt and Flügge 1992). Investigations of transport in bundle sheath chloroplasts showed that they had a higher *Km* (Pi) (0.33 mM) than the mesophyll chloroplasts (0.14 mM), but the *Ki* (glycerate-3-P) was little different from that of the mesophyll chloroplasts. The bundle sheath chloroplast has a high affinity for PEP (Ohnishi et al 1989).

Little is known about transport of dicarboxylates into bundle sheath chloroplasts, but aspartate has been shown to stimulate malate decarboxylation by bundle sheath chloroplasts of maize. It has therefore been suggested that a carrier specific for malate uptake is present that depends on the presence of aspartate for maximum activity (Boag and Jenkins 1985).

Chloroplast peripheral reticulum

A feature that has received comparatively little attention is the function of the peripheral reticulum in the chloroplasts of C_4 plants. The peripheral reticulum is a membrane system of anastomosing tubules contiguous with the inner membrane of the chloroplast envelope (i.e., the membrane that contains the metabolite transporters). This is present in both bundle sheath and mesophyll chloroplasts, but is generally more highly developed in the mesophyll chloroplasts. Transporters are highly abundant proteins in chloroplasts of C_3 plants. For example, the phosphate translocator in C_3 plants accounts for 10–15% of total envelope protein (Flügge and Heldt 1991), with a transport capacity for Pi of approximately 2.5 $\mu mol\ m^{-2}\ s^{-1}$, assuming 300 mg chlorophyll m^{-2} leaf area (Flügge 1995). Moreover, it is not in great excess, since it exerts considerable control over the photosynthetic rate in tobacco leaves photosynthesizing in elevated CO_2 (Häusler et al 2000). Substantially higher (at least 10–20-fold) fluxes of metabolites are required during C_4 photosynthesis. In maize, a typical rate of photosynthesis might be approximately 50 $\mu mol\ m^{-2}\ s^{-1}$ or more (Leegood and von Caemmerer 1989). A reduction in photosynthetically generated glycerate-3-P in the mesophyll has to occur at a similar rate (plus any overcycling that occurs). This increase in transport capacity might only be met by increasing the surface area of the inner chloroplast envelope to accommodate these transporters (Laetsch 1968, Hatch and Osmond 1976). However, it must be stressed that the peripheral reticulum is also known in C_3 plants, and it can develop particularly under stress conditions, such as chilling (Kratsch and Wise 2000) or ascorbate deficiency (Olmos et al 2006), indicating that there is flexibility in its development.

Altered properties of enzymes of carbohydrate synthesis

Much of our knowledge about sucrose synthesis in plants comes from C_3 species. It is evident, from those studies of C_4 species that are available, that there are many similarities between sucrose synthesis in C_3 and C_4 plants. However, some of the basic regulatory mechanisms common to both C_3 and C_4 plants seem to have been adapted for the specialized photosynthetic metabolism of C_4 plants (Lunn and Furbank 1999). In all C_4 subtypes, the reduction of glycerate-3-P to triose-P is shared between the Benson–Calvin cycle in bundle sheath chloroplasts and mesophyll chloroplasts (Fig. 1). Transport between the mesophyll and bundle sheath is driven by diffusion, dependent upon large concentration gradients (Leegood 1985, Stitt and Heldt 1985b). Metabolite measurements on intact leaves of maize and *Amaranthus edulis* (NAD-ME) show that the amounts of glycerate-3-P and triose-P are extremely high when compared with those of C_3 species (Leegood and Furbank 1984, Usuda 1987a,b, Leegood and von Caemmerer 1988, 1989), with amounts of triose-P typically 20 times higher than in the leaves of C_3 plants. This has an important impact on sucrose synthesis. Interestingly, C_3-C_4 intermediate species also show elevated amounts of triose-P (Leegood and von Caemmerer 1994), suggesting that a triose-P/glycerate 3-P shuttle might have appeared early in the evolution of C_4 photosynthesis (presumably as a means of reducing O_2 evolution in the bundle sheath). A further consequence of elevated glycerate 3-P in these intermediates is elevated PEP (Leegood and von Caemmerer 1994), which could promote PEP carboxylase activity.

Since triose-P is present in the mesophyll of C_4 plants at far higher concentrations than in C_3 plants, the amount of fructose 1,6-bisP_2 (FBP), formed through the action of aldolase, is also higher in the mesophyll cytosol. In maize, the cytosolic fructose 1,6-bisP_2ase (which is the first step in the synthesis of sucrose) shows a much higher *Km* for FBP (20 μM in maize compared with 3 μM in spinach and, in the presence of fructose-2,6-P_2, 3–5 mM in maize and 20 μM in spinach; Stitt and Heldt 1985a). This lowered affinity for substrate makes it possible to reconcile the use of triose-P for sucrose synthesis with the maintenance of a high concentration of triose phosphate in the mesophyll needed for diffusion-driven transport between the mesophyll and bundle sheath. In addition, higher concentrations of triose-P and glycerate-3-P are needed to inhibit the fructose 6-phosphate, 2-kinase from maize, and the enzyme is also inhibited by PEP and OAA. Thus, fructose-2,6-P_2 will build up and inhibit sucrose synthesis when both C_3 and C_4 metabolites are low.

Although C_4 species generally show preferential localization of sucrose-P synthase (and hence sucrose synthesis) in mesophyll cells, the proportion of total leaf sucrose-P synthase activity in mesophyll cells ranges from 65% in *Sorghum bicolor* to 99% in *Atriplex spongiosa* (Lunn and Furbank 1999).

A key enzyme of starch synthesis, ADP-glucose pyrophosphorylase, also shows altered properties in leaves of C_4 plants. ADP-glucose pyrophosphorylase from C_3 spinach chloroplasts is activated by glycerate-3-P and inhibited by Pi, with ratios of glycerate-3-P/Pi for half-maximal activation typically being less than 1.5, whereas the enzyme from maize leaves requires a ratio of glycerate-3-P to Pi of between 7 and 10 for half-maximal activation in the bundle sheath, and even higher ratios in the mesophyll (Spilatro and Preiss 1987).

Mitochondrial specialization

Mitochondria can play major roles during photosynthesis in both C_3 and C_4 plants. In C_3 plants, the flux of carbon through the photorespiratory pathway is about equal to the flux through the photosynthetic Benson–Calvin cycle, thus making it a major metabolic pathway in leaves. Glycine decarboxylase (GDC) has a pivotal role in this cycle as the site of photorespiratory CO_2 and NH_3 release. The rate of NH_3 release is nearly 10 times higher than the rate of primary NH_3 fixation, and the released NH_3 must be recycled efficiently to prevent unacceptable levels of N loss (Keys and Leegood 2002). GDC is a light-induced multienzyme complex that, together with serine hydroxymethyltransferase, can reach 40% of mitochondrial soluble proteins (Oliver et al 1990), leading to an increase in mitochondrial density because of modification of the protein-lipid ratio. The vast abundance of GDC makes it analogous to the similarly abundant Rubisco of C_3 chloroplasts. The occurrence of photorespiration and the photorespiratory pathway thus results in large investments of nitrogen, not only into GDC in the mitochondria but also into other photorespiratory components, such as glutamine synthetase (GS2) in the chloroplasts, for the reassimilation of photorespiratory ammonia, and into catalase to dissimilate the H_2O_2 generated by the oxidation of glycolate, as well as the increased amount of Rubisco that is needed to support both oxygenation and carboxylation. GDC has been lost from mesophyll mitochondria during the evolution of C_4 photosynthesis (Ohnishi and Kanai 1983). Decreasing photorespiration might also affect folate biosynthesis, which occurs in the mitochondria of C_3 leaves (Oliver 1994).

In C_4 plants, the carbon fluxes through the mitochondria in leaves of NAD-malic enzyme plants are equivalent to the rate of photosynthesis and are therefore much higher than the rates of respiration in other tissues. For example, they are several-fold greater than photorespiratory fluxes in C_3 plants and 10- to 20-fold greater than respiratory carbon fluxes in leaves. This might have beneficial effects on the C_4 system because such high rates of respiration serve to lower the O_2 concentration in the bundle sheath and to improve the CO_2/O_2 ratio. Mitochondria are also intimately involved in photosynthesis in PEPCK-type C_4 species.

Mitochondrial transport processes are also specialized. The gene encoding a mitochondrial oxaloacetate/malate transporter (OMT) in *Panicum miliaceum* (NAD-ME type) is highly expressed in leaves, and is up-regulated during greening and cell development in concert with C_4 photosynthetic enzymes (Taniguchi and Sugiyama 1997). This expression profile is unique to the *P. miliaceum* gene because the orthologous genes of *Arabidopsis* and tobacco are expressed nearly constitutively in all tissues (Picault et al 2002). Furthermore, the expression of the *P. miliaceum* mitochondrial OMT gene is restricted to bundle sheath cells, where the mitochondrial OMT presumably accommodates the high rates of exchange of C_4 photosynthetic metabolites across mitochondrial membranes.

Regulation of C_4 photosynthesis

A major deficiency in our understanding of the regulation of C_4 photosynthesis is how carboxylation by PEPC (PEP carboxylase) in the mesophyll and decarboxylation in the bundle sheath by NADP-malic enzyme, NAD-malic enzyme, or PEPCK (PEP carboxykinase) are co-regulated. Enzymes of the C_4 and Benson–Calvin cycle are regulated by similar mechanisms in both C_3 and C_4 plants. These mechanisms include redox regulation by the thioredoxin system (e.g., NADP-malate dehydrogenase in the C_4 cycle and the sedoheptulose bisphosphatase and fructose bisphosphatase, glyceraldehyde 3-P dehydrogenase, and ribulose 5-P kinase in the Benson–Calvin cycle), and by phosphorylation (e.g., PEPC, pyruvate, Pi dikinase, and PEPCK), although this last group of enzymes primarily has nonphotosynthetic functions in C_3 plants. The fact that photosynthetic carbon assimilation in C_4 plants is divided between two cell types requires coordination of regulation between the two cell types. Mechanisms must exist that allow processes in one cell to integrate with those in the other. The most likely candidates for signals between the two cell types are the transported metabolites.

In vivo evidence of coordination between the C_3 and C_4 cycles can be seen in the relationship between the assimilation rate and amounts of metabolites in *Amaranthus edulis* and maize as affected by the intercellular concentration of CO_2. As the rate of photosynthesis decreases with decreasing CO_2, the RuBP pool rises, but the amounts of PEP and metabolites of the C_4 cycle decrease, together with triose-P and glycerate-3-P (Leegood and von Caemmerer 1988, 1989, 1994, Usuda 1987a,b). This behavior indicates the operation of a feedback loop from the Benson–Calvin cycle. Interconversion of glycerate-3-P and PEP, catalyzed by phosphoglycerate mutase and enolase in the mesophyll cytosol, provides metabolic communication between the C_4 cycle and the Benson–Calvin cycle, so that the amount of PEP declines when the assimilation rate falls because the amount of glycerate-3-P exported to the mesophyll declines. This is regulation by substrate availability.

Considerable evidence exists for the regulation of C_4 enzymes by metabolite effectors, particularly for PEPC. Triose-P and hexose-P, which are products of the Benson–Calvin cycle, act as positive effectors of PEPC and relieve inhibition by malate. In leaves of C_4 plants, the amount of triose-P is always closely related to the assimilation rate whether the flux is changed by alterations in irradiance or CO_2 (Leegood and von Caemmerer 1988, 1989, 1994). Increasing concentrations of triose-P and hexose-P (indicating increased output from the Benson–Calvin cycle) would promote the activity of PEPC, thus increasing the rate of CO_2 fixation in the mesophyll. On the other hand, if too much glycerate 3-P were diverted to malate because of an excessive activity of PEPC, this would lead to an accumulation of inhibitors of PEPC (malate and aspartate). The decrease in glycerate 3-P would lead to decreases in the amounts of the activators of PEPC, triose-P, and hexose-P, and thus decrease the rate at which glycerate-3-P is consumed. By such means, rates of initial carboxylation in the C_4 cycle may be linked to the rates of Benson–Calvin cycle turnover and to rates of product synthesis. Photorespiratory intermediates, such as glycine and serine, also activate PEPC (Nishikido and Takanashi 1973, Doncaster and Leegood 1987). For example, glycine accumulates in leaves when the intercellular concentration of CO_2 falls (Leegood and von Caemmerer 1994).

The factors regulating the phosphorylation of PEPC remain enigmatic. PEPC kinase is the smallest known protein kinase with no regulatory domains, and with no obvious mechanism of control other than protein turnover (Nimmo 2003). There is no evidence that PEPC phosphatase activity is regulated by metabolites (Nimmo 2003). Giglioli-Guivarc'h et al (1996) suggested that glycerate-3-P might act as a messenger between the bundle sheath and mesophyll that triggers cytosolic acidification in the mesophyll cells, which leads to a signaling cascade that activates PEPC kinase, presumably by *de novo* synthesis (see also Bakrim et al 2001, Osuna et al 2004).

As far as the regulation of the decarboxylases is concerned, PEPCK is perhaps the best understood (Leegood and Walker 2003). PEPCK is subject to reversible phosphorylation in the leaves of some C_4 plants, such as *Panicum maximum*, although regulation by phosphorylation appears to have been lost in the leaves of others, such as *Urochloa panicoides* (Walker et al 1997) or maize (Leegood and Walker 2003). In *P. maximum*, PEPCK is phosphorylated in darkened leaves and dephosphorylated in illuminated leaves (Walker and Leegood 1996). Changes in PEPCK phosphorylation state lead to diurnal changes in its sensitivity to regulation by adenylates, which lead to its activation in illuminated leaves (Walker et al 2002). Coordination is very precise between the phosphorylation/activation states of PEPC and PEPCK (Bailey et al 2007).

Somewhat surprisingly, light and CO_2 response curves have not been measured for many enzymes of C_4 photosynthesis, although previous measurements of the kinetics of activation of PEPC in the C_4 plants maize, sorghum, and *Salsola soda* suggest similar kinetics of light activation and dark inactivation, with each process taking 1–2 h (Karabourniotis et al 1983, Jiao and Chollet 1988, Nimmo et al 1987, Bakrim et al 1992), and similarly slow kinetics for PEPCK (Leegood and Walker 2003). This prolonged activation of PEPC and PEPCK by light compares with the very short times (5–10 min) required to fully activate NADP-malate dehydrogenase via thioredoxin (Johnson and Hatch 1970) or pyruvate, Pi dikinase by phosphorylation (Burnell and Hatch 1985) in maize leaves, or comparably short times to activate thioredoxin-linked enzymes and Rubisco in the leaves of C_3 plants (for example, Sassenrath-Cole and Pearcy 1994). Prolonged activation/deactivation also demonstrates that light activation is not merely acting as an on-off switch between light and dark.

Of particular importance for engineering C_4 photosynthesis into rice are the properties of NADP-malic enzyme (NADP-ME). Apart from the availability of NADP, which is dependent upon glycerate-3-P reduction in the Calvin cycle, NADP-ME is likely to be activated by the change in stromal conditions occurring upon illumination. This is because its regulatory properties are pH and Mg^{2+}-dependent (Johnson and Hatch 1970, Asami et al 1979). This interaction among malate, pH, and Mg^{2+} is reminiscent of the regulation of certain enzymes of the Benson–Calvin cycle, such as FBPase and ribulose-5-P kinase (Leegood et al 1985). This would achieve coordination with electron transport, but the role of metabolite effectors in regulation remains unclear (Asami et al 1979).

C_4 mutants

Mutants of the C_4 dicot *Amaranthus edulis* have been generated by conventional mutagenesis. Plants were grown in elevated CO_2 and mutant plants were selected by screening in air. Several mutant lines lack PEPC, the P-protein of glycine decarboxylase, or NAD-ME (Dever et al 1995). Interestingly, these mutants show no obvious structural differences at the leaf level from their wild-type counterparts. The interveinal spacing is not affected (Dever et al 1995, Bailey 1998) nor is Rubisco induced in the mesophyll in heterozygous plants (Bailey 1998). Those mutants lacking the C_4 pathway grew extremely poorly in air, but grew as well as the wild types in high CO_2. The poor growth and low rate of photosynthesis are a consequence of the low permeability of the bundle sheath to external CO_2. What is important is that, in these plants grown in high CO_2, C_3 photosynthesis can proceed effectively in a C_4 structural and enzymic background. In other words, it can proceed with a low-affinity C_4-type Rubisco (although few problems would be expected because of the high CO_2 supply), with C_4 NAD-ME–type mitochondria in the bundle sheath, rather than the GDC-rich mitochondria of C_3 plants, and with the enzymatic apparatus for carbohydrate synthesis that is typical of C_4 plants. C_3 photosynthesis can also cope with the potential structural hindrances of the bundle sheath/mesophyll compartmentation. In addition, the mesophyll would be largely redundant as far as carbon fixation is concerned, with only part of the Calvin cycle (reduction of glycerate-3-P to triose-P) in the mesophyll chloroplasts. The extent of acclimation in these plants is not clear. How far, for example, are some C_3 traits, such as GDC-rich mitochondria that support photorespiration, altered in the mutants? However, it is clear that the whole system adjusts without any external intervention. It could be argued that the C_4 system is intrinsically more flexible, as it comprises two cell types with different cytosolic, mitochondrial, and chloroplastic enzymic and transport capacities. However, further study of any of the pleiotropic effects that occur in these plants could be rewarding and might contain lessons for achieving the reverse process of introducing C_4 photosynthesis into rice.

References

Aoki N, Ohnishi J, Kanai R. 1992. Two different mechanisms for transport of pyruvate into mesophyll chloroplasts of C_4 plants—a comparative study. Plant Cell Physiol. 33:805-809.

Aoki N, Ohnishi J, Kanai R. 1994. Proton/pyruvate cotransport into mesophyll chloroplasts of C_4 plants. Plant Cell Physiol. 35:801-806.

Asami S, Inoue K, Akazawa T. 1979. NADP-malic enzyme from maize leaf: regulatory properties. Arch. Biochem. Biophys. 196:581-587.

Bailey KJ. 1998. Control of photosynthesis by PEP carboxylase in leaves of *Amaranthus edulis*. PhD thesis, University of Sheffield.

Bailey KJ, Gray JE, Walker RP, Leegood RC. 2007. Co-ordinate regulation of phospho*enol*pyruvate carboxylase and phospho*enol*pyruvate carboxykinase by light and CO_2 during C_4 photosynthesis. Plant Physiol. 144: doi: 10.1104/pp.106.093013.

Bakrim N, Echevarria C, Crétin C, Arrio-Dupont M, Pierre JN, Vidal J, Chollet R, Gadal P. 1992. Regulatory phosphorylation of *Sorghum* leaf phospho*enol*pyruvate carboxylase: identification of the protein-serine kinase and some elements of the signal transduction cascade. Eur. J. Biochem. 204:821-830.

Bakrim N, Brulfert J, Vidal J, Chollet R. 2001. Phospho*enol*pyruvate carboxylase kinase is controlled by a similar signaling cascade in CAM and C_4 plants. Biochem. Biophys. Res. Commun. 286:1158-1162.

Boag S, Jenkins CLD. 1985. CO_2 assimilation and malate decarboxylation by isolated bundle sheath chloroplasts from *Zea mays*. Plant Physiol. 79:165-170.

Botha CEJ. 1992. Plasmodesmatal distribution, structure and frequency in relation to assimilation in C_3 and C_4 grasses in southern Africa. Planta 187:348-358.

Botha CEJ, Evert RF. 1988. Plasmodesmatal distribution and frequency in vascular bundles and contiguous tissues of the leaf of *Themeda triandra*. Planta 173:433-441.

Burnell JN, Hatch MD. 1985. Light-dark modulation of leaf pyruvate, Pi dikinase. Trends Biochem. Sci. 10:288-291.

Day DA, Hatch MD. 1981. Dicarboxylate transport in maize mesophyll chloroplasts. Arch. Biochem. Biophys. 211:738-742

Dever LV, Blackwell RD, Fullwood NJ, Lacuesta M, Leegood RC, Onek L, Pearson M, Lea PJ. 1995. The isolation and characterisation of mutants of the C_4 photosynthetic pathway. J. Exp. Bot. 46:1363-1376.

Doncaster HD, Leegood RC. 1987. Regulation of phospho*enol*pyruvate carboxylase activity in maize leaves. Plant Physiol. 84:82-87.

Fischer K, Arbinger B, Kammerer B, Busch C, Brink S, Wallmeier H, Sauer N, Eckerskorn C, Flügge U-I. 1994. Cloning and in vivo expression of functional triose phosphate/phosphate translocators from C_3- and C_4-plants: evidence for the putative participation of specific amino acid residues in the recognition of phospho*enol*pyruvate. Plant J. 5:215-226.

Fischer K, Kammerer N, Gutensohn M, Arbinger B, Weber A, Häusler RE, Flügge UI. 1997. A new class of plastidic phosphate translocators: a putative link between primary and secondary metabolism by the phospho*enol*pyruvate/phosphate antiporter. Plant Cell 9:453-462.

Fladung M. 1994. Genetic variants of *Panicum maximum* (Jacq.) in C_4 photosynthetic traits. J. Plant Physiol. 143:165-172.

Flügge U. 1995. Phosphate translocation in the regulation of photosynthesis. J. Exp. Bot. 46:1317-1323.

Flügge UI, Stitt M, Heldt HW. 1985. Light-driven uptake of pyruvate into mesophyll chloroplasts from maize. FEBS Lett. 183:335-339.

Flügge UI, Heldt HW. 1991. Metabolite translocators of the chloroplast envelope. Annu. Rev. Plant Physiol. Plant Mol. Biol. 42:129-144.

Giglioli-Guivarc'h N, Pierre J-N, Brown S, Chollet R, Vidal J, Gadal P. 1996. The light-dependent transduction pathway controlling the regulatory phosphorylation of C_4 phospho*enol*-pyruvate carboxylase in protoplasts from *Digitaria sanguinalis*. Plant Cell 8:573-586.

Haberlandt G. 1884. Physiological plant anatomy (translated by Drummond M). London (UK): MacMillan.

Hatch MD, Osmond CB. 1976. Compartmentation and transport in C_4 photosynthesis. In: Stocking CR, Heber U, editors. Encyclopedia of plant physiology, New Series Vol. 3. Berlin (Germany): Springer-Verlag. p 144-184.

Hatch MD, Dröscher L, Flügge UI, Heldt HW. 1984. A specific translocator for oxaloacetate transport in chloroplasts. FEBS Lett. 178:15-19.

Hattersley PW. 1992. C_4 photosynthetic pathway variation in grasses (Poaceae): its significance for arid and semi arid lands. In: Chapman GP, editor. Desertified grasslands: their biology and managements. Linnean Society Symposium Series 13. London (UK): Academic Press. p 181-212.

Häusler RE, Schlieben NH, Flügge UI. 2000. Control of carbon partitioning and photosynthesis by the triose phosphate/phosphate translocator in transgenic tobacco plants (*Nicotiana tabacum*). II. Assessment of control coefficients of the triose phosphate/phosphate translocator. Planta 210:383-390.

Heldt HW, Flügge UI. 1992. Metabolite transport in plant cells. In: Tobin AK, editor. Plant organelles. Cambridge (UK): Cambridge University Press. p 21-47.

Huber SC, Edwards GE. 1977. Transport in C_4 mesophyll chloroplasts: evidence for an exchange of inorganic phosphate and phospho*enol*pyruvate. Biochim. Biophys. Acta 462:603-612.

Jiao J-A, Chollet R. 1988. Light/dark regulation of maize leaf phospho*enol*pyruvate carboxylase by *in vivo* phosphorylation. Arch. Biochem. Biophys. 261:409-417.

Johnson HS, Hatch MD. 1970. Properties and regulation of leaf nicotinamide-adenine dinucleotide phosphate-malate dehydrogenase and malic enzyme in plants with the C_4-dicarboxylic acid pathway of photosynthesis. Biochem. J. 119:273-280.

Karabourniotis G, Manetas Y, Gavalas NA. 1983. Photoregulation of phospho*enol*pyruvate carboxylase in *Salsola soda* L. and other C_4 plants. Plant Physiol. 73:735-739.

Keys AJ, Leegood RC. 2002. Photorespiratory carbon and nitrogen cycling: evidence from studies of mutant and transgenic plants. In: Foyer CH, Noctor G, editors. Photosynthetic nitrogen assimilation and associated carbon and respiratory metabolism. Dordrecht (Netherlands): Kluwer Academic Publishers. p 115-134.

Knappe S, Löttgert T, Schneider A, Voll L, Flügge U-I, Fischer K. 2003. Characterization of two functional phospho*enol*pyruvate/phosphate translocator (PPT) genes in *Arabidopsis-AtPPT1* may be involved in the provision of signals for correct mesophyll development. Plant J. 36:411-420.

Kratsch HA, Wise RR. 2000. The ultrastructure of chilling stress. Plant Cell Environ. 23:337-350.

Laetsch WM. 1968. Chloroplast specialization in dicotyledons possessing the C_4-dicarboxylic acid pathway of photosynthetic CO_2 fixation. Am. J. Bot. 55:875-883.

Leegood RC. 1985. The intercellular compartmentation of metabolites in leaves of *Zea mays*. Planta 164:163-171.

Leegood RC, Foyer CH, Walker DA. 1985. Regulation of the Benson–Calvin cycle. In: Barber J, Baker NR, editors. Photosynthetic mechanisms and the environment. Amsterdam (Netherlands): Elsevier. p 189-258.

Leegood RC, Furbank RT. 1984. Carbon metabolism and gas exchange in leaves of *Zea mays* L.: changes in CO_2 fixation, chlorophyll fluorescence and metabolite levels during photosynthetic induction. Planta 162:450-456.

Leegood RC, von Caemmerer S. 1988. The relationship between contents of photosynthetic intermediates and the rate of photosynthetic carbon assimilation in leaves of *Amaranthus edulis* L. Planta 174:253-262.

Leegood RC, von Caemmerer S. 1989. Some relationships between contents of photosynthetic metabolites and the rate of photosynthetic carbon assimilation in leaves of *Zea mays*. Planta 178:258-266.

Leegood RC, von Caemmerer S. 1994. Regulation of photosynthetic carbon assimilation in leaves of C_3-C_4 intermediate species of *Moricandia* and *Flaveria*. Planta 192:232-238.

Leegood RC, Walker RP. 2003. Regulation and roles of phospho*enol*pyruvate carboxykinase in plants. Arch. Biochem. Biophys. 414:204-210.
Lunn JE, Furbank RT. 1999. Sucrose biosynthesis in C_4 plants. New Phytol. 143:221-237.
Murata S, Kobayashi M, Matoh T, Sekiya J. 1992. Sodium stimulates regeneration of phospho*enol*pyruvate in mesophyll chloroplasts of *Amaranthus tricolor*. Plant Cell Physiol. 33:1247-1250.
Nelson T, Langdale JA. 1992. Developmental genetics of C_4 photosynthesis. Annu. Rev. Plant Physiol. Plant Mol. Biol. 43:25-47.
Nimmo HG. 2003. Control of the phosphorylation of phospho*enol*pyruvate carboxylase in higher plants. Arch. Biochem. Biophys. 414:189-196.
Nimmo GA, McNaughton GAL, Fewson CA, Wilkins MB, Nimmo HG. 1987. Changes in the kinetic properties and phosphorylation state of phospho*enol*pyruvate carboxylase in *Zea mays* leaves in response to light and dark. FEBS Lett. 213:18-22.
Nishikido T, Takanashi H. 1973. Glycine activation of PEP carboxylase from monocotyledonous C_4 plants. Biochem. Biophys. Res. Commun. 53:126-133.
Ohnishi J, Kanai R.1983. Differentiation of photorespiratory activity between mesophyll and bundle sheath cells of C_4 plants. I. Glycine oxidation by mitochondria. Plant Cell Physiol. 24:1411-1420.
Ohnishi J, Kanai R. 1990. Pyruvate uptake induced by a pH jump in mesophyll chloroplasts of maize and sorghum, NADP-malic enzyme type C_4 species. FEBS Lett. 269:122-124.
Ohnishi J, Flügge U-I, Heldt HW. 1989. Phosphate translocator of mesophyll and bundle sheath chloroplasts of the C_4 plant, *Panicum miliaceum*: identification and kinetic characterization. Plant Physiol. 91:1507-1511.
Ohnishi J, Flügge U-I, Heldt HW, Kanai R. 1990. Involvement of Na^+ in active uptake of pyruvate in mesophyll chloroplasts of some C_4 plants: Na^+/pyruvate transport. Plant Physiol. 94:950-959.
Oliver DJ. 1994. The glycine decarboxylase complex from plant mitochondria. Annu. Rev. Plant Physiol. Plant Mol. Biol. 45:323-337.
Oliver DJ, Neuburger M, Bourguignon J, Douce R. 1990. Glycine metabolism by plant mitochondria. Physiol. Plant. 80:487-491.
Olmos E, Kiddle G, Pellny TK, Kumar S, Foyer CH. 2006. Modulation of plant morphology, root architecture, and cell structure by low vitamin C in *Arabidopsis thaliana*. J. Exp. Bot. 57:1645-1655.
Osuna L, Coursol S, Pierre J-N, Vidal J. 2004. A Ca^{2+}-dependent protein kinase with characteristics of protein kinase C in leaves and mesophyll cell protoplasts from *Digitaria sanguinalis*: possible involvement in the C_4-phospho*enol*pyruvate carboxylase phosphorylation cascade. Biochem. Biophys. Res. Commun. 314:428-433.
Picault N, Palmieri L, Pisano I, Hodges M, Palmieri F. 2002. Identification of a novel transporter for dicarboxylates and tricarboxylates in plant mitochondria: bacterial expression, reconstitution, functional characterization, and tissue distribution. J. Biol. Chem. 277:24204-24211.
Renné P, Dreßen U, Hebbeker U, Hille D, Flügge UI, Westhoff P, Weber APM. 2003. The *Arabidopsis* mutant *dct* is deficient in the plastidic glutamate/malate translocator DiT2. Plant J. 35:316-331.
Sassenrath-Cole G, Pearcy RW. 1994. Regulation of photosynthetic induction state by the magnitude and duration of low light exposure. Plant Physiol. 105:1115-1123.

Spilatro SR, Preiss J. 1987. Regulation of starch synthesis in the bundle-sheath and mesophyll of *Zea mays* L.: intercellular compartmentation of enzymes of starch metabolism and the properties of the ADP glucose pyrophosphorylases. Plant Physiol. 83:621-627.

Stitt M, Heldt HW. 1985a. Control of photosynthetic sucrose synthesis by fructose-2,6-bisphosphate: intercellular metabolite distribution and properties of the cytosolic fructose bisphosphatase in leaves of *Zea mays* L. Planta 164:179-188.

Stitt M, Heldt HW. 1985b. Generation and maintenance of concentration gradients between the mesophyll and bundle-sheath in maize leaves. Biochim. Biophys. Acta 808:400-414.

Taniguchi, M, Sugiyama, T. 1997. The expression of 2-oxoglutarate/malate translocator in the bundle-sheath mitochondria of *Panicum miliaceum*, a NAD-malic enzyme-type C_4 plant, is regulated by light and development. Plant Physiol. 114:285-293.

Taniguchi M, Taniguchi Y, Kawasaki M, Takeda S, Kato T, Sato S, Tabata S, Miyake H, Sugiyama T. 2002. Identifying and characterizing plastidic 2-oxoglutarate/malate and dicarboxylate transporters in *Arabidopsis thaliana*. Plant Cell Physiol. 43:706-717.

Taniguchi Y, Nagasaki J, Kawasaki M, Miyake H, Sugiyama T, Taniguchi M. 2004. Differentiation of dicarboxylate transporters in mesophyll and bundle sheath chloroplasts of maize. Plant Cell Physiol. 45:187-200.

Usuda H. 1987a. Changes in levels of intermediates of the C_4 cycle and reductive pentose phosphate pathway under various concentrations of CO_2 in maize leaves. Plant Physiol. 83:29-32.

Usuda H. 1987b. Changes in levels of intermediates of the C_4 cycle and reductive pentose phosphate pathway under various light intensities in maize leaves. Plant Physiol. 84:549-554.

Voll L, Häusler RE, Hecker R, Weber A, Weissenböck G, Fiene G, Waffenschmidt S, Flügge U-I. 2003. The phenotype of the *Arabidopsis cue1* mutant is not simply caused by a general restriction of the shikimate pathway. Plant J. 36:301-317.

Walker RP, Leegood RC. 1996. Phosphorylation of phospho*enol*pyruvate carboxykinase in plants: studies in plants with C_4 photosynthesis and Crassulacean acid metabolism and in germinating seeds. Biochem. J. 317:653-658.

Walker RP, Acheson RM, Técsi LI, Leegood RC. 1997. Phospho*enol*pyruvate carboxykinase in C_4 plants: its role and regulation. Aust. J. Plant Physiol. 24:459-468.

Walker RP, Chen Z-H, Acheson RM, Leegood RC. 2002. Effects of phosphorylation on phospho*enol*pyruvate carboxykinase from the C_4 plant, Guinea grass. Plant Physiol. 128:165-172.

Weiner H, Burnell JN, Woodrow IE, Heldt HW, Hatch MD. 1988. Metabolite diffusion into bundle sheath cells from C_4 plants: relation to C_4 photosynthesis and plasmodesmatal function. Plant Physiol. 88:815-822.

Woo KC, Flügge U-I, Heldt HW. 1987. A two-translocator model for the transport of 2-oxoglutarate and glutamate in chloroplasts during ammonia assimilation in the light. Plant Physiol. 84:624-632.

Notes

Author's address: Robert Hill Institute and Department of Animal and Plant Sciences, University of Sheffield, Sheffield, S10 2TN, UK, e-mail: r.leegood@sheffield.ac.uk.

C_4 photosynthesis and CO_2 diffusion

S. von Caemmerer, J.R. Evans, A.B. Cousins, M.R. Badger, and R.T. Furbank

The advantages of C_4 photosynthesis for plant productivity, particularly in warmer climates, are well characterized. High rates of biomass accumulation and high water-use efficiency and N-use efficiency make the installation of the C_4 pathway (or some other form of CO_2-concentrating mechanism) into C_3 plants an attractive proposition for biotechnologists. Here, we compare anatomical properties of leaves of C_3 and C_4 species to compare characteristics of CO_2 diffusion. We show that leaves of a wide variety of C_3 species are characterized by high exposed mesophyll and chloroplast surface area to leaf area ratios (S_m and S_c). Combining measurements of the internal conductance to CO_2 diffusion (derived from measurements of carbon isotope discrimination) with measurements of S_c shows that the CO_2 conductance across the cell wall, plasma membranes, and chloroplast membrane interface is on average 0.02 mol m^{-2} chloroplast area s^{-1} bar^{-1} for C_3 annual species (including rice) and 0.01 to 0.02 mol m^{-2} chloroplast area s^{-1} bar^{-1} for deciduous and evergreen trees. Measurements of anatomical properties of a number of C_4 species show that S_m is less in C_4 species than in C_3 species, but that high photosynthetic rates require higher conductances for CO_2 diffusion across the C_4 mesophyll cytosol interface. There is little variation in bundle sheath surface area to leaf area ratio (S_b), with average values of 1.77 ± 0.11, such that S_m is from 6 to 10 times greater than S_b. Bundle sheath conductance to CO_2 diffusion cannot be measured directly; however, the efficiency of the C_4 photosynthetic pathway can be assessed through measurements of carbon isotope discrimination. Using a mathematical model of C_4 photosynthesis, we examine the relationship between bundle sheath conductance (or its inverse, resistance) to CO_2 diffusion and the biochemical capacity of the C_4 photosynthetic pathway and conclude that bundle sheath resistance to CO_2 diffusion must vary with biochemical capacity if the efficiency of the C_4 pump is to be maintained. Finally, we use a mathematical model of single-cell C_4 photosynthesis in a C_3 mesophyll cell and examine the importance of CO_2 diffusion on such a C_4 photosynthetic CO_2 pump.

Keywords: C_4 photosynthesis, CO_2 diffusion, carbon isotope discrimination

The C_4 photosynthetic pathway is a CO_2-concentrating mechanism that combines a mix of biochemical and anatomical specializations to concentrate CO_2 at the site of Rubisco (Hatch 1987, Edwards et al 2004). Most C_4 species have specialized leaf anatomy in which photosynthetic cells are organized in two concentric circles. Mesophyll cells adjacent to intercellular air space radiate from bundle sheath cells. After diffusing into mesophyll cells, CO_2 is initially assimilated into C_4 acids by phospho*enol*pyruvate carboxylase (PEPC) in the mesophyll cytosol. The C_4 acids then diffuse to the bundle sheath, where they are decarboxylated. This concentrates CO_2 in the bundle sheath and enhances Rubisco carboxylation while at the same time inhibiting Rubisco oxygenation (Hatch and Osmond 1976). Diffusion of CO_2 out of the bundle sheath limits the efficiency of the CO_2-concentrating mechanism. Many attempts have been made to quantify the diffusion resistance across the bundle sheath mesophyll interface, but estimates vary widely (Jenkins 1989, Brown and Byrd 1993, Hatch et al 1995, He and Edwards 1996, Kiirats et al 2002).

C_4 species are characterized by high photosynthetic rates, nitrogen-use efficiency, and water-use efficiency relative to plants with the C_3 photosynthetic pathway (Hatch 1987). High photosynthetic rates require high rates of CO_2 diffusion from intercellular air space to mesophyll cytosol. Limitations to CO_2 diffusion across this interface have received less attention so far than the bundle sheath/mesophyll interface (Longstreth et al 1980). Recently, single-cell variants of the pathway have been identified, where CO_2 is fixed by PEPC in one part of the cell and decarboxylated in another part (for a review, see Edwards et al 2004). These species provide us with new challenges for understanding the constraints CO_2 diffusion places on the C_4 pathway. In this review, we compare the CO_2 diffusive properties between C_3 and C_4 leaves and use the results to explore the feasibility of single-cell C_4 photosynthesis. Improved technology has facilitated the ease with which carbon isotope discrimination can be made concurrently with measurements of CO_2 uptake (Cousins et al 2006). We highlight that this technique provides new opportunities for measuring the internal conductance to CO_2 diffusion in C_3 leaves under a variety of environmental conditions and also provides a measure for the efficiency of the C_4 photosynthetic pathway that has yet to be exploited to assess the efficiency of the C_4 pathway under different environmental conditions.

Carbon isotope discrimination and CO_2 diffusion

About 1% of the atmospheric CO_2 contains the heavy isotope ^{13}C. $^{13}CO_2$ diffuses more slowly than $^{12}CO_2$ and Rubisco discriminates against it during carboxylation so that in C_3 species carbon fixed during photosynthesis is depleted in ^{13}C. Carbon fixed during C_4 photosynthesis is much less depleted in ^{13}C compared to C_3 photosynthesis because Rubisco located in the bundle sheath has less opportunity to discriminate. In C_4 species, both the fractionations occurring during hydration and PEP carboxylation of CO_2 as well as the leakiness of the bundle sheath (ϕ) (which is defined as the fraction of CO_2 fixed by PEPC and decarboxylated in the bundle sheath that subsequently leaks back out) affect the overall carbon isotope discrimination (Δ), which is defined as

$$\Delta = R_a/R_p - 1 \tag{1}$$

where R_a and R_p are the molar abundance ratios (^{13}C to ^{12}C) of the carbon in the atmosphere and of the carbon fixed (Farquhar et al 1989). In C_3 species, concurrent measurements of carbon isotope discrimination and CO_2 assimilation rate have become a major tool for studying CO_2 diffusion inside leaves (Evans et al 1986, von Caemmerer and Evans 1991, Evans and von Caemmerer 1996, Hanba et al 2004). In C_4 species, concurrent measurements of carbon isotope discrimination and CO_2 assimilation rate have provided important insights into the leakiness of the C_4 photosynthetic pathway (Farquhar 1983, Henderson et al 1992, von Caemmerer et al 1997). Because of this importance, we give a brief description of the theory below.

Theory developed by Farquhar et al (1982) and Farquhar (1983) showed that carbon isotope discrimination during C_3 and C_4 photosynthesis can be described by an equation having terms dependent on diffusion and on biochemistry:

$$\Delta = a + (b_s + a_l - a)\frac{p_i}{p_a} + (\Delta_{bio} - b_s - a_l)\frac{p_m}{p_a} \tag{2}$$

where $\Delta_{bio} = R_m/R_p - 1$ and R_m is the ratio of ^{13}C to ^{12}C of the CO_2 at the site of leaf mesophyll carboxylation. The symbols p_a, p_i, and p_m denote ambient, intercellular, and mesophyll CO_2 partial pressures (pCO_2). The symbol a denotes the fractionation during diffusion in air (4.4×10^{-3}), b_s is the fractionation as CO_2 enters solution (1.1×10^{-3}), and a_l is the fractionation occurring during diffusion in water (0.7×10^{-3}) (Farquhar et al 1982). The biochemical discrimination factor Δ_{bio} differs for the different biochemical pathways. For C_3 species, $\Delta_{bio} = b_3$, the fractionation by Rubisco (29×10^{-3}) (Roeske and O'Leary 1984). For C_4 photosynthesis,

$$\Delta_{bio} = b_4 - \phi(b_3 - s) \tag{3}$$

where b_4 (-5.7×10^{-3} at 25 °C) is the combined fractionation of PEPC carboxylation and the preceeding dissolution and conversion of CO_2 to bicarbonate and $s = a_l + b_s$ is the fractionation associated with bundle sheath leakiness (Farquhar 1983, Henderson et al 1992). The fractionation factor b_4 is temperature dependent and increases when carbonic anhydrase activity limits the hydration of CO_2 (Farquhar 1983, Henderson et al 1992, Cousins et al 2006).

Using the fact that $A = g_l\,(p_i - p_m)$ where A stands for CO_2 assimilation rate and g_i for the conductance to CO_2 diffusion from intercellular air space to the site of carboxylation, equation (2) can be rewritten to show the influence of internal CO_2 diffusion on Δ:

$$\Delta = a + (\Delta_{bio} - a)\frac{p_i}{p_a} - (\Delta_{bio} - a_l - b_s)\frac{A}{g_i \times p_a} \tag{4}$$

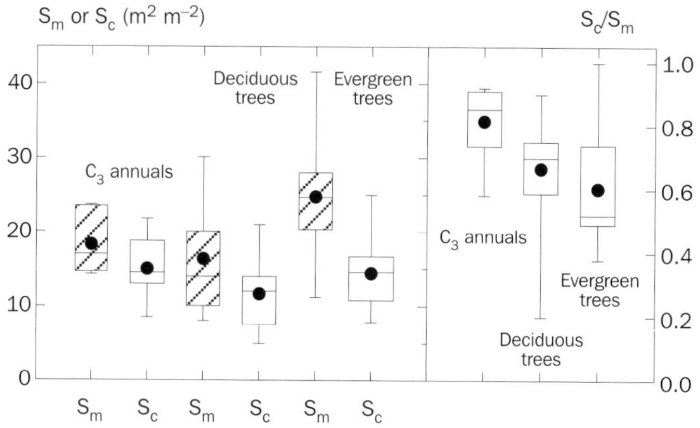

Fig. 1. Box and whisker plots of mesophyll surface area exposed to intercellular air space to leaf area ratio (S_m), chloroplast surface area exposed to intercellular air space to leaf area ratio (S_c), and the ratio of S_c/S_m for various C_3 species. Box and whiskers represent the 25 to 75 percentile and the minimum and maximum distribution. Means are denoted by (●). The graph is derived from data collated in Table 1 of Terashima et al (2006).

The first part of the equation quantifies the dependence on stomatal conductance and the ratio of intercellular to ambient CO_2 (p_i/p_a) and the second part quantifies the effect of internal diffusion on carbon isotope discrimination.

CO_2 diffusion from intercellular air space to chloroplast stroma in C_3 species

In C_3 species, the fractionation factor in the second parentheses of equation (4) is 27‰, which results in a large discrimination that varies with CO_2 assimilation rate. This has been exploited to estimate g_i from concurrent measurements of carbon isotope discrimination and CO_2 assimilation rate under different irradiances, and these measurements have been made on a number of different species (for a review, see Evans and von Caemmerer 1996, Terashima et al 2006). Von Caemmerer and Evans (1991) found that the magnitude of g_i correlated with photosynthetic capacity and suggested that this correlation might be driven by a correlation of photosynthetic capacity and chloroplast surface area appressing intercellular air space, arguing that diffusion across the liquid phase was most likely to be the limiting step. Several studies have now correlated internal diffusion conductance (measured through carbon isotope discrimination) with anatomical measurements of leaves and these have been reviewed by Evans et al (2004) and Terashima et al (2006). In particular, Hanba et al (1999, 2001, 2002) have examined variations in these measurements with environmental perturbations and among different functional types. Figure 1 summarizes measurements made of ratios of mesophyll (S_m) and chloroplast surface area (S_c) appressing intercellular air spaces per unit leaf area. The data are compiled from data presented in Table 1 of Terashima et al (2006). C_3 annuals have an average S_m of 18.3 ± 0.04 and both

Table 1. Leaf properties of several monocot and dicot C_4 species. Anatomical measurements were made on 3 leaf sections per species that had been fixed and embedded after measurements of carbon isotope discrimination and CO_2 assimilation were made. The gas exchange data are from Henderson et al (1992). Fixation and embedding of leaf material and anatomical measurements were made as described by Evans et al (1994) using a curvature correction factor of 1.43.

Species	Type	Suberin	S_m	S_b	Fraction of S_b exposed to intercellular air space	Interveinal distance (μm)	Vascular bundle diameter (μm)	Bundle sheath cell diameter (μm)	Mesophyll cell diameter (μm)	Assimilation rate (μmol m^{-2} s^{-1})	p_i/p_a	Online Δ (‰)
Monocots		+										
Sorghum bicolor (Success 40W)	NADP	+	11.1 ± 1.5	1.49 ± 0.02	0.27 ± 0.21	161 ± 17	61 ± 9.4	21.3 ± 1.6	23.6 ± 1.9	Not determined		
S. bicolor (Tx610SR)	NADP	+	19.5 ± 1.5	1.53 ± 0.09	0.17 ± 0.04	169 ± 21	106 ± 18.7	20.3 ± 0.4	22.1 ± 1.1			
S. bicolor (Trojan II)	NADP	+	12.1 ± 2.0	1.98 ± 0.08	0.15 ± 0.005	127 ± 4.4	71 ± 2.9	19.6 ± 0.5	19.9 ± 2.3	55.9	0.35	3.13
Zea mays (low light)	NADP	+	9.1 ± 0.8	1.58 ± 0.062	0.23 ± 0.02	142 ± 14	62.9 ± 2.6	20 ± 1.1	26.5 ± 2.2			
Zea mays (high light)	NADP	+	10.0 ± 0.6	1.56 ± 0.9	0.22 ± 0.01	101 ± 9.2	63.9	20 ± 0.8	27.5 ± 2.9	58.6	0.34	3.16
Panicum schinzii	NAD/PCK	+	8.35 ± 0.5	2.3 ± 0.07	0.23 ± 0.01	187 ± 13	118 ± 12	35 ± 3.7	25.9 ± 2.9	67.4	0.38	2.98
Urochloa panicoides	PCK	+	9.1 ± 1.5	1.45 ± 0.24	0.26 ± 0.04	161 ± 8.8	89.2 ± 3.7	31 ± 1.9	33 ± 1.3	49.4	0.41	2.97
Chloris gayana	PCK	+	10.8 ± 0.59	2.3 ± 0.02	0.19 ± 0.02	180 ± 11	98.2 ± 6.1	28.4 ± 2.2	20.6 ± 0.7	53.2	0.29	3.64
Eleusine coracana	NAD	–	9.3 ± 0.74	1.82 ± 0.133	0.15 ± 0.02	227 ± 8.8	117 ± 7.0	40.4 ± 2.0	35.6 ± 3.1	48.7	0.38	3.93
Dicots												
Atriplex rosea	NAD	–	13.8 ± 0.4	1.93 ± 0.15	0.19 ± 0.005	165 ± 31	124.2 ± 24	32.8 ± 1.3	23.7 ± 2.5	45.8	0.47	2.64
Amaranthus edulis	NAD	–	12.8 ± 0.75	2.42 ± 0.12	0.12 ± 0.004	102 ± 5.18	71.6 ± 1.12	24.5 ± 0.5	20.7 ± 0.5	45.6	0.43	2.87
Flaveria bidentis	NADP	–	17.0 ± 0.42	1.7 ± 0.13	0.17 ± 0.02	194 ± 24	83.3 ± 4.06	28.4 ± 1.1	22.9 ± 0.7	37	0.43	3.05
Gomphrena globosa	NADP	–	18.9 ± 1.3	0.99 ± 0.9	0.1 ± 0.02	397 ± 54	86.4 ± 3.5	28.2 ± 1.1	27.9 ± 3.8	41.3	0.42	2.97

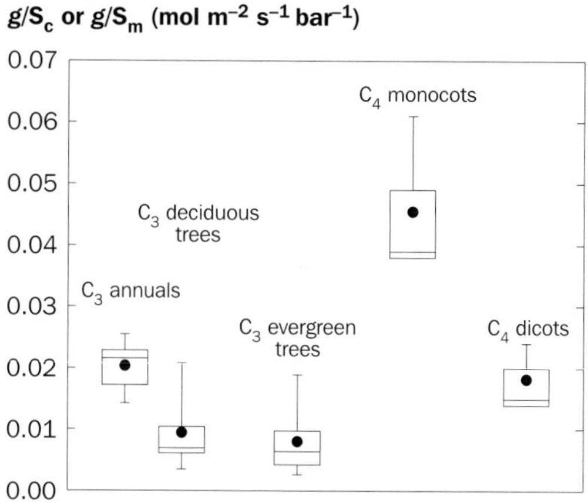

Fig. 2. Box and whisker plots of internal CO_2 diffusion conductance (g_i) expressed per unit chloroplast surface area for C_3 species and per mesophyll surface area for C_4 species. Box and whisker plots are as described in Figure 1. Data for C_3 species were calculated from measurements of S_c and g_i collated in Table 1 of Terashima et al (2006). For C_4 species, g_i was calculated on a leaf area basis from gas exchange measurements as the minimum g_i required to sustain measured CO_2 assimilation rates.

deciduous broadleaf trees and evergreen trees had higher S_m values of 22.5 ± 6 and 24.8 ± 1.7. These greater values of S_m are offset, however, by a lower coverage by chloroplasts such that S_c shows little variation (S_c = 15.0 ± 1.4, 12 ± 1.5, and 14.5 ± 1 for annual, deciduous broadleaf, and evergreen trees, respectively). Rice measured by Hanba et al (2004) had S_m and S_c values of 19 and 14, respectively, typical for annuals. Here, we use these anatomical data together with the measured g_i values to calculate a conductance to CO_2 diffusion on a chloroplast area basis in Figure 2. It is clear that the conductance to CO_2 diffusion across the liquid phase to the chloroplast stroma is greater in annuals than in trees. It is likely that both thickness and composition of cell walls (Kogami et al 2001) and factors relating to membrane permeability (Terashima and Ono 2002, Uehlein et al 2003, Hanba et al 2004) are the cause of this variability. Recent studies have shown that aquaporins modify the CO_2 diffusivity of the plasma membrane, which suggests that g_i may not only be related to anatomy but vary more dynamically with environmental conditions. Bernacchi et al (2002) used combined measurements of chlorophyll fluorescence and CO_2 assimilation rate to demonstrate that g_i varied with temperature in *Nicotiana tabacum*. However, a different temperature response was found for the tropical tree *Eperua* (Pons and Welschen 2002). Recent development of a direct, real-time mass spectrometric system that allows concurrent measurements of CO_2 uptake and carbon isotope discrimination to be made at rapid time intervals without the need for purification of CO_2 should

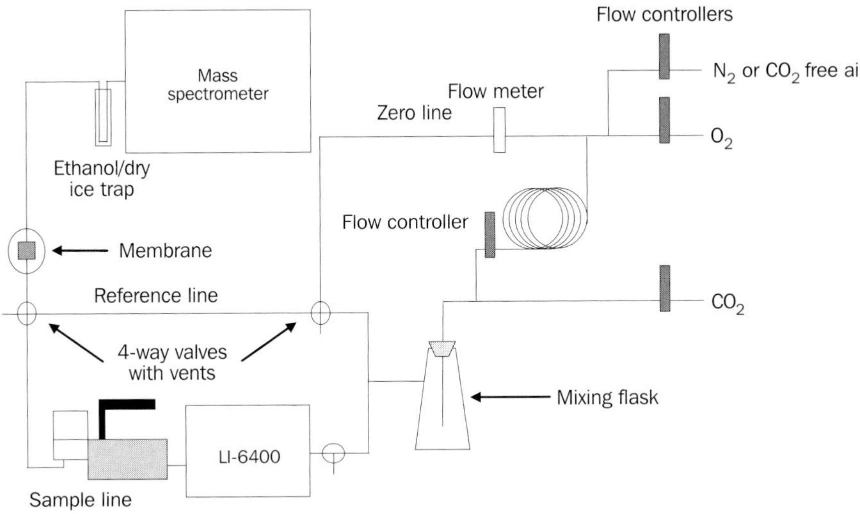

Fig. 3. Arrangement of the gas flow controllers, the LI-6400 gas exchange system, and the mass spectrometer system used for simultaneous measurements of leaf gas exchange and carbon isotope discrimination. Switching between gas samples was controlled by a manual four-way valve. The zero and reference readings were made before and after each leaf measurement and averaged during the calculations.

facilitate more rapid measurements of g_i under different environmental conditions (Fig. 3; Cousins et al 2006).

CO_2 diffusion from intercellular air space to mesophyll cytosol in C_4

A prerequisite for high rates of C_4 photosynthesis is a high rate of CO_2 diffusion from intercellular air spaces to the mesophyll cytosol and thus high values of g_i. The initial CO_2 fixation occurs in the mesophyll cytosol, where CO_2 is converted to bicarbonate and then fixed by PEPC. So, unlike in C_3 species, CO_2 has to diffuse across the cell wall, but only one membrane, the plasmalemma. It must be assumed, however, that PEPC is distributed throughout the cytoplasm, some of which may not appress intercellular air space, thus providing uncertainty about the diffusion path length within the cytosol. Limitations to CO_2 diffusion across this interface have been difficult to assess. Carbon isotope discrimination measurements cannot be used in C_4 species since the absolute value of the biochemical fractionation factor Δ_{bio} for C_4 photosynthesis is less than 1‰, making carbon isotope discrimination insensitive to variations in assimilation rate (A, equation 4). Attempts have been made to estimate g_i from measurements of $C^{18}O^{18}O$ discrimination, but these measurements have a number of confounding factors (Gillon and Yakir 2000).

In Table 1 and Figure 4, we present anatomical data for C_4 leaves that were used for combined measurements of carbon isotope discrimination and CO_2 assimilation by Henderson et al (1992). The mean values of S_m (11.6 ± 1.1 for C_4 monocot species and 15.6

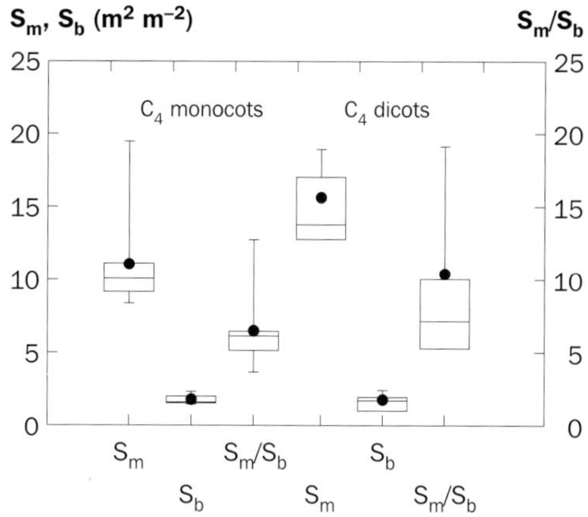

Fig. 4. Box and whisker plots of mesophyll surface area exposed to intercellular air space (S_m), bundle sheath surface area per unit leaf area (S_b), and the ratio of S_m/S_b. Box and whisker plots are as described in Figure 1. Anatomical measurements were made on leaf sections used for gas exchange and carbon isotope discrimination measurements reported by Henderson et al (1992). Species included 9 C_4 monocots and 4 C_4 dicots (see Table 1).

$\pm$ 1.4 for C_4 dicot species, Fig. 4) are slightly lower than those for C_3 species (Fig. 1) but similar to chloroplast surface area appressing intercellular air space in C_3 species, S_c. For both C_3 and C_4 species, one can calculate a minimum g_i required to sustain CO_2 assimilation. This limit is reached when mesophyll CO_2 reaches the compensation point (Γ):

$$g_{i\min} = A/(P_i - \Gamma) \tag{5}$$

In C_4 species, Γ is close to zero. We used the gas exchange data of Henderson et al (1992) to calculate $g_{i\min}$ for the C_4 species listed in Table 1 and these values are expressed on a mesophyll surface area basis in Figure 2. For the C_4 dicots, these minimum values are similar to the values calculated for C_3 annuals. The C_4 monocots, however, require nearly twice the values estimated for the C_3 annuals. Given that these are absolute minimum values, g_i for both C_4 monocots and dicots need to be greater than what has been observed for C_3 species, which raises the question of how this is achieved. It may be that mesophyll cell walls of C_4 species are thinner than those of C_3 species (0.07 μm in *Amaranthus retroflexus*, Longstreth et al 1980, compared with 0.3 μm in *Nicotiana tabacum*, Evans et al 1994), but more measurements of mesophyll cell wall thickness are required. It would also be interesting to know whether the fact that only one membrane needs to be traversed by CO_2 increases the conductance. There is still considerable uncertainty on the CO_2 permeability of the plasmalemma and chloroplast envelopes. Raven (1977) suggested that diffusion of CO_2 and O_2 should be fast across membranes because of the high lipid solubilities and diffusion coefficients of CO_2 and O_2. Direct measurements are difficult because of the unstirred layer effects. At present, only the permeability measured across an artificial lipid bilayer of 3.5×10^{-3} m s^{-1} (Gutknecht et al 1977), which translates into a conductance of 0.14 mol m^{-2} s^{-1} bar^{-1}, is great enough to account for the rapid diffusion that occurs across membranes in C_3 leaves and the mesophyll interface in C_4 species (Fig. 2, Evans et al 1994, 2004).

CO_2 diffusion across bundle sheath/mesophyll interface

It has always been assumed that a low conductance to CO_2 diffusion across the interface of the bundle sheath and mesophyll is an essential feature of the C_4 pathway since CO_2 arrives in the form of dicarboxylic acids and loss of CO_2 (as such) from the bundle sheath is to be minimized. Jenkins et al (1989) demonstrated that inhibition of PEP carboxylase effectively eliminated CO_2 exchange with the normal atmosphere and that high external CO_2 concentrations were required to restore CO_2 fixation. The conductance to CO_2 diffusion across bundle sheath walls has been estimated in intact leaves using this PEP carboxylase inhibitor. Most resistance values are in the range of 400–1,600 m^2 leaf s bar mol^{-1}, which correspond to conductances of 0.6 to 2.5 mmol m^{-2} leaf s^{-1} bar^{-1} (Jenkins et al 1989, Brown and Byrd 1993), but lower bundle sheath resistances ranging from 50 to 180 m^2 leaf s bar mol^{-1} have also been estimated (He and Edwards 1996, Kiirats et al 2002). This raises an important question of how low the bundle sheath conductance can be without compromising the efficiency of the CO_2-concentrating mechanism, which we address below.

Bundle sheath conductance on a leaf area basis is the product of the conductance across the mesophyll/bundle sheath interface and the bundle sheath surface area to leaf area ratio (S_b). Estimates of S_b in the literature range from 0.6 to 3.1 m^2 m^{-2} (Apel and Peisker 1978, Brown and Byrd 1993). Our measured mean value of 1.8 m^2 m^{-2} in Figure 4 and Table 1 is within that range. We found surprisingly little variation in this measurement (Table 1). Mesophyll surface area exposed to intercellular air space per unit leaf area (S_m) was 6.5 to 10 times greater than S_b and on average 18% of the bundle sheath surface area was directly exposed to intercellular air space (Table 1, Fig. 4).

Using conductance values for CO_2 diffusion across the cell wall membrane interface of C_3 species estimated in Figure 2 together with the average S_b, one estimates possible bundle sheath conductances of 18 to 36 mmol m^{-2} s^{-1}, which correspond to resistance values between 28 and 55 m^2 leaf s mol^{-1}. These estimates of resistance are too low compared with most measured values. The conductance to CO_2 diffusion across the bundle sheath/mesophyll interface must therefore be less than the equivalent conductance across the cell wall/chloroplast interface in C_3 species. The diffusion path for CO_2 from the site of C_4 acid decarboxylation to the mesophyll in C_4 plants is complicated by the unique morphological and biochemical specialization of the three biochemical decarboxylation types, and this was discussed in detail by von Caemmerer and Furbank (2003). In the NADP-ME types, decarboxylation occurs within the bundle sheath chloroplast. In this type, the conductance of the chloroplast envelope (or even the stroma) to CO_2 could be an important factor affecting overall CO_2 leakage. In the NAD-ME types, C_4 acid decarboxylation occurs in the mitochondria. The bundle sheath chloroplasts of these plants are arranged centripetally, surrounding the mitochondria, and are located at the vascular bundle side of the bundle sheath cell (Hatch et al 1975). This may be significant in that CO_2 leaking back to the mesophyll must traverse the length of the bundle sheath chloroplast (where it can be refixed by Rubisco), the bundle sheath cytosol, and then the plasma membrane/cell wall barrier. In the PCK types, decarboxylation occurs in the bundle sheath cytosol and mitochondria, but chloroplasts are arranged centrifugally, toward the mesophyll interface. In this case, CO_2 must pass through the chloroplast to leak out from the bundle sheath cell. A further complication is the presence or absence of the suberin lamellae in the bundle sheath cell wall (see Hattersley and Browning 1981). Most C_4 plants (with the exception of C_4 dicotyledons and the NAD-ME types) have one or several layers of suberin laid down in the bundle sheath cell wall adjacent to the mesophyll (Hattersley and Browning 1981). However, the significance of this as a barrier to gaseous diffusion is largely unknown. Surprisingly, measurements of bundle sheath cell wall thickness revealed that it was similar to the thickness of C_3 mesophyll cell walls. Furthermore, NAD-ME-type C_4 species that lack suberin in their cell walls did not appear to compensate with thicker cell walls (von Caemmerer and Furbank 2003). Von Caemmerer and Furbank (2003) concluded that it was difficult to estimate bundle sheath diffusion resistance from anatomical measurements and that cytosolic diffusion and the traversing of a number of membranes may account for much of the final calculated resistance.

Relationship between bundle sheath resistance to CO_2 diffusion and leakiness of the bundle sheath

Farquhar (1983) coined the term "leakiness" to quantify the efficiency of the CO_2 pump of C_4 photosynthesis. Leakiness (ϕ) was defined as the ratio of the CO_2 leak rate out of the bundle sheath to the rate of PEP carboxylation (assumed to be in equilibrium with the rate of C_4 acid decarboxylation in the steady state). This term has been particularly useful in relation to carbon isotope discrimination (see equation 3). The leak rate of CO_2 out of the bundle sheath is dependent on the bundle sheath resistance to CO_2 and the gradient of CO_2 partial pressure between the bundle sheath and mesophyll, which in turn is influenced by the relative biochemical capacity of the C_4 cycle and the C_3 cycle in the bundle sheath (von Caemmerer and Furbank 1999, von Caemmerer 2000). The C_4 cycle consumes energy in the form of ATP during the regeneration of PEP so that the leak rate of CO_2 from the bundle sheath is an energy cost to the leaf.

In Figure 5, the dependency of net CO_2 assimilation rate on bundle sheath resistance is modeled. CO_2 assimilation rates decline steeply in all cases below a resistance value of 100 m^2 leaf s bar mol^{-1} (corresponding to a bundle sheath conductance of 0.01 mol m^{-2} leaf s^{-1} bar^{-1}) and CO_2 assimilation is also sensitive to O_2 partial pressure under these conditions. Most measured resistance values are in the range of

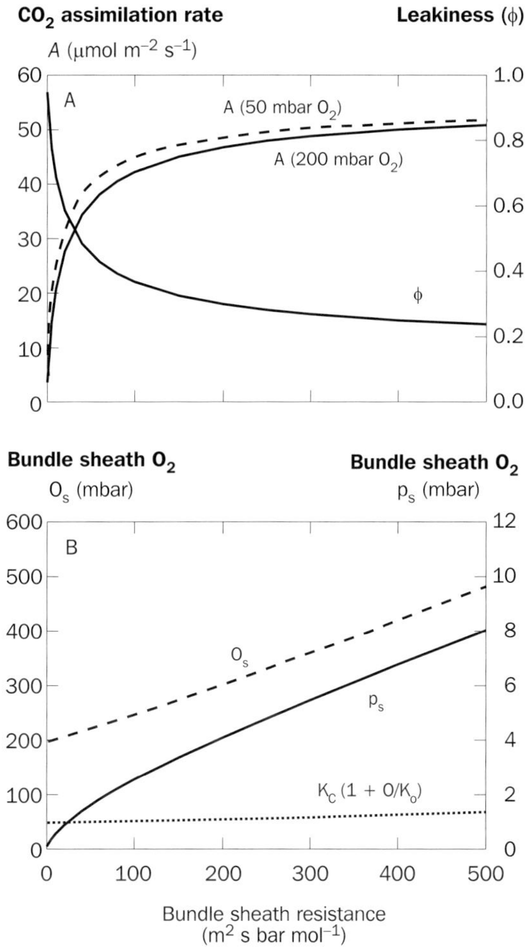

Fig. 5. (A) Modeled C_4 photosynthetic rate as a function of bundle sheath resistance for partial pressures of mesophyll CO_2 of 100 μbar and O_2 of 200 mbar (solid lines) and 50 mbar (dashed lines). Also shown is bundle sheath leakiness (ϕ), the ratio of CO_2 leak rate from the bundle sheath to the rate of PEP carboxylation at 200 mbar O_2. The simulations use the model described by von Caemmerer and Furbank (1999) with kinetic constants given in their Table 2. (B) Predicted bundle sheath O_2 and CO_2 concentrations when ambient O_2 is 200 mbar. Also shown is the effect that increasing bundle sheath O_2 has on the apparent K_m for CO_2 of Rubisco ($K_c(1+O/K_o)$ (right-hand axis).

400–1,600 m^2 leaf s bar mol^{-1} (Jenkins et al 1989, Brown and Byrd 1993), well above this threshold, except for estimates by He and Edwards (1996) and Kiirats et al (2002), who estimated resistances in the range of 50 to 90 m^2 s bar mol^{-1}. However, since bundle sheath resistance cannot be measured directly, uncertainty remains for all of these estimates. Bundle sheath leakiness (ϕ) decreases with increasing bundle sheath resistance. The simulation in Figure 5 assumes that half of the oxygen is evolved in the bundle sheath. Because oxygen diffusivities are such that bundle sheath resistance to O_2 diffusion is 20 times greater than for CO_2 (Raven 1977, Berry and Farquhar 1978), this leads to a buildup of O_2 in the bundle sheath (Fig. 5B). The degree of leakiness of the bundle sheath compartment is thus a compromise between retaining CO_2, allowing O_2 to diffuse out, and permitting metabolites to diffuse at rates fast enough to support the rate of CO_2 fixation (Hatch 1987). It is generally assumed that Rubisco is CO_2 saturated in the bundle sheath, but this may not always be the case (Fig. 5B). The $K_m(CO_2)$ in C_4 species is considerably greater than in C_3 species, probably around 1,000 μbar at ambient O_2 partial pressures (Yeoh et al 1981). There are at present very few measurements of $K_m(O_2)$ so that there is uncertainty on how much the apparent K_m of Rubisco increases *in vivo* with increasing bundle sheath resistance (von Caemmerer and Quick 2000).

The photosynthetic capacities of leaves change with leaf age and environmental growth conditions. The C_4 model predicts that, if leakiness (i.e., efficiency) is to be maintained, bundle sheath conductance has to scale with C_4 and C_3 cycle capacity and thus leaves with low photosynthetic rates, such as very young or old leaves, require lower conductance (higher resistance) to maintain the same efficiency (von Caemmerer and Furbank 1999, 2003). Once the leaf is fully expanded, the geometry of the bundle sheath and properties such as bundle sheath area to leaf surface area are fixed such that there would need to be changes in the CO_2 permeability of cell walls and membranes to achieve a constant efficiency. It is possible that this could be achieved by increasing secondary wall thickening and lignification with leaf age. It is equally intriguing how a balance is achieved in young expanding leaves, where bundle sheath surface area to leaf area is likely to be larger than in mature tissue. Measurements of carbon isotope discrimination made concurrently with measurements of CO_2 assimilation could be used to examine these. It would be especially interesting to couple these with measurements of bundle sheath resistance using a PEPC inhibitor (Jenkins et al 1989).

Estimates of bundle sheath leakiness from measurements of carbon isotope discrimination

The leakiness of the bundle sheath cannot be measured directly. The range of estimates and techniques used to estimate leakiness were recently reviewed by von Caemmerer and Furbank (2003). Concurrent measurements of carbon isotope discrimination and CO_2 assimilation have given the most consistent results. Henderson et al (1992) made short-term measurements of carbon isotope discrimination on several different C_4 species and found that variation in leakiness was conservative and ranged between

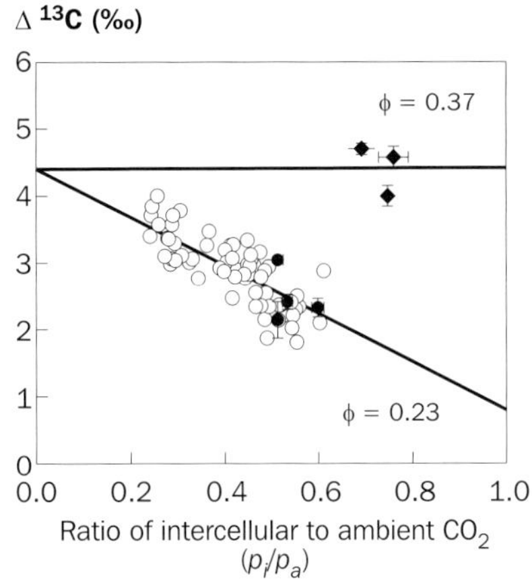

Fig. 6. Carbon isotope discrimination measured concurrently with gas exchange as a function of the ratio of intercellular to ambient CO_2, p_i/p_a, for some dicotyledons (○) and *Flaveria bidentis* wild type (•) and transgenic *F. bidentis* with reduced amounts of Rubisco (♦). Data are taken from Henderson et al (1992) and von Caemmerer et al (1997). The lines depict the theoretical relationship $\Delta = 4.4 + (-5.7 - 4.4 + \phi(30 - 1.8))\ p_i/p_a$ for *F. bidentis* measured at 25 °C. Measurements by Henderson et al (1992) were made at 28 °C and, because the fractionation during the combined hydration and PEP carboxylation of CO_2 has a temperature dependence, −5.7 is replaced by −5.2 and the calculated $\phi = 0.21$.

0.2 and 0.3 for these species (Fig. 6). Modification of photosynthetic enzyme amounts and properties *in vivo*, using transgenic plants of the C_4 dicot *Flaveria bidentis*, has provided a powerful tool to test the link between carbon isotope discrimination and leakiness. It was shown that Rubisco could be progressively reduced to quite low amounts without an effect on PEPC activity (Furbank et al 1996). Intuitively, one would expect the bundle sheath CO_2 concentration to become elevated in these plants. This would result in an increase in leakage of CO_2 from the bundle sheath and a decline in photosynthetic efficiency per net CO_2 fixed. This is, in fact, what is observed when the Rubisco-antisense *Flaveria* transgenic plants are examined by carbon isotope discrimination (Fig. 6, von Caemmerer et al 1997). Measurements of carbon isotope discrimination at different irradiance showed that leakiness increased at low irradiance, making the C_4 pathway less efficient under these conditions (Fig. 7, Henderson et al 1992). Henderson et al (1992) observed a greater increase in the dicot *Amaranthus edulis* than in the moncots *Zea mays* and *Sorghum bicolor*. Recent development of a direct, real-time mass spectrometric system that allows concurrent measurements

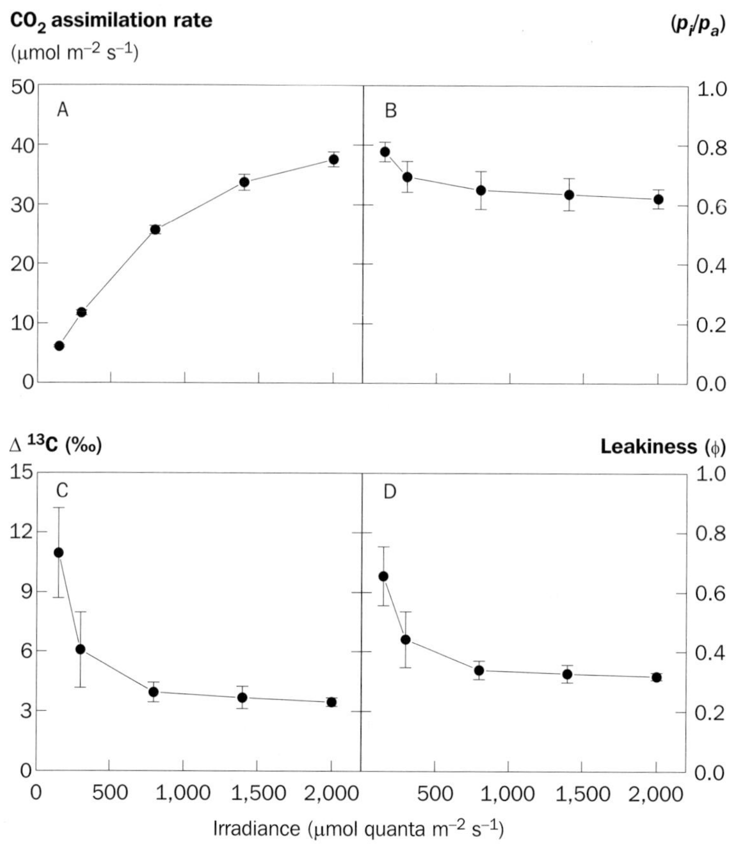

Fig. 7. (A) CO_2 assimilation rate, (B) the ratio of intercellular to ambient CO_2 partial pressures (p_i/p_a), (C) carbon isotope discrimination ($\Delta^{13}C$), and (D) bundle sheath leakiness to CO_2 (ϕ) in *F. bidentis* as a function of irradiance. Measurements were made at a CO_2 partial pressure (pCO_2) of 520 µbar and a leaf temperature of 30 °C. Shown are the means ± SE measurements made on 3 to 5 leaves. The figure is adapted from Cousins et al (2006).

of CO_2 uptake and carbon isotope discrimination to be made at rapid time intervals without the need for purification of CO_2 should facilitate measurements of leakiness under a range of environmental conditions and throughout leaf development and this could provide important new information on the relationship between leaf anatomy and bundle sheath resistance (Fig. 3; Cousins et al 2006).

What are the possibilities for C_4 rice?

The advantages of C_4 photosynthesis for plant productivity, particularly in warmer climates, are well characterized and have been reviewed (Sage and Kubien 2003). High rates of biomass accumulation and high water-use efficiency and N-use efficiency make the installation of the C_4 pathway (or some other form of CO_2-concentrating mechanism) into C_3 plants an attractive proposition for biotechnologists (see Sheehy et al 2000, Edwards et al 2001, Matsuoka et al 2001, Häusler et al 2002, Leegood 2002, Lieman-Hurwitz et al 2003). There are two possible routes: Kranz-type or single-cell C_4 photosynthesis.

Kranz-type C_4 photosynthesis has evolved several times in a large number of both dicot and monocot genera (reviewed in Sage et al 1999) and there is thus the possibility of discovering the molecular switches that can convert rice to a C_4 species where the pathway functions across the interface between the mesophyll and bundle sheath (Brown et al 2005). Anatomically, this would require an increase in vein density and a reduction in cell numbers between vein bundles (Evans and von Caemmerer 2000). The anatomical data presented here suggest that an increase in conductance to CO_2 diffusion from intercellular air space to the mesophyll cytosol would also be required (Fig. 2).

Recently, single-cell variants of the C_4 photosynthetic pathway have been identified in terrestrial plant species, where CO_2 is fixed by PEPC in one part of the cell and decarboxylated in another part (for a review, see Edwards et al 2004). These species provide us with new challenges for understanding the constraints that CO_2 diffusion places on such a C_4 pathway. They raise the interesting question of whether a CO_2-concentrating mechanism that leads to enhanced photosynthetic rates could be installed in rice mesophyll cells. Von Caemmerer and Furbank (2003) adapted their C_4 photosynthetic model (Fig. 5; von Caemmerer and Furbank 1999) to see how such a single-cell C_4 system would operate from a CO_2 diffusion perspective in current C_3 mesophyll cells, where chloroplast surface area is large and the difference between mesophyll surface area and chloroplast surface area exposed to intercellular air space is small (Fig. 1). The model uses C_3 kinetic constants for Rubisco since C_3 Rubisco has a lower K_m CO_2 than C_4 Rubisco (Yeoh et al 1981). In the simulation shown, the internal diffusive conductances for the cell wall/plasmalemma interface and the chloroplast envelope were based on measurements of internal leaf conductance to CO_2 diffusion in rice (0.2 mol m^{-2} s^{-1} bar^{-1}; Hanba et al 2004) using a value of 0.4 mol m^{-2} s^{-1} bar^{-1} for both the cell wall/plasmalemma interface (g_w) and the chloroplast envelope (g_{chlor}) in one simulation and in the second simulation g_{chlor} decreased to 0.25 mol m^{-2} $s^{-1}$$bar^{-1}$ and g_w increased to 1 mol m^{-2} s^{-1} bar^{-1}. This gives the same total conductance of 0.2 mol m^{-2} s^{-1} bar^{-1} for the C_3 comparison. The output from these models shows that CO_2 assimilation rate increases with C_4 cycle activity (quantified in the model by the maximal PEP carboxylase activity, V_{pmax}) and the compensation point decreases (Fig. 8). Thus, the model demonstrates that the chloroplast envelope may enable limited C_4 cycle activity, particularly at low pCO_2 (see von Caemmerer and Furbank 2003). Also shown is the actual PEPC rate, and the difference between this rate and the net CO_2 assimilation rate is the leak rate. It is very clear that, with a high CO_2 conductance across the chloroplast/cytosol interface, the system is energetically inefficient since 2 ATP are required to regenerate each PEP. Of course, reduction in g_{chlor} leads to an increase in CO_2 assimilation rate. In the second simulation, as well as reducing $g_{chlor,}$ g_w was increased, which led to an increase in PEPC rate. It is

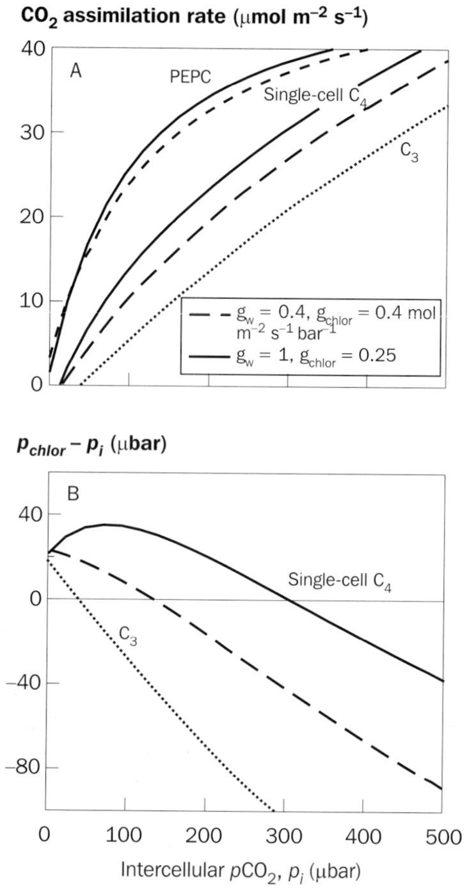

Fig. 8. (A) Hypothetical single-cell C_4 photosynthesis as a function of intercellular CO_2 partial pressure, p_i (dashed and solid lines), for two different partitionings of total internal CO_2 diffusion conductance, g_i (0.2 mol m^{-2} s^{-1} bar^{-1}), between cell wall (g_w) and chloroplast envelope (g_{chlor}). C_4 cycle capacity was V_{pmax}= 50 μmol m^{-2} $s^{-1}$and Rubisco activity was V_{cmax}=100 μmol m^{-2} s^{-1}. The CO_2 assimilation rate is compared to Rubisco-limited C_3 photosynthesis with the same maximal Rubisco activity (smaller broken line). Other values are as described in von Caemmerer and Furbank (2003). (This simulation becomes unrealistic at higher p_i, as it does not consider RuBP or PEP regeneration limitations). (B) The difference between chloroplast and intercellular CO_2 partial pressure (p_{chlor} – p_i) as a function of p_i for the examples given in A.

therefore easily demonstrated that a high conductance to CO_2 diffusion at the cell wall and plasmalemma is required at this interface. On the other hand, a low conductance at the chloroplast/cytosol interface is important since it separates the initial CO_2 fixation from the decarboxylation step. Therefore, by decreasing the chloroplast envelope conductance, the efficiency of the CO_2-concentrating mechanism can be increased. Perhaps the CO_2 conductance of the chloroplast envelope could be manipulated by alteration in aquaporin expression (Hanba et al 2004).

A single-cell C_4 system with current C_3 diffusion characteristics is clearly not an efficient CO_2-concentrating mechanism with its high leakiness and energy requirements. Nevertheless, the model demonstrates that, by introducing a C_4 system (or any other form of CO_2-concentrating mechanism such as an algal bicarbonate pump), it may be possible to reverse the CO_2 draw-down between intercellular air space and chloroplast (Fig. 8B). Introducing a CO_2-concentrating mechanism at the cytosol/chloroplast interface into C_3 cells may therefore be useful in ameliorating this CO_2 draw-down, particularly at low intercellular pCO_2. In C_3 leaves, the largest draw-down in pCO_2 between intercellular air space and chloroplast occurs at high light, where ATP supply is not limiting and considerations of energy efficiency may be of secondary importance. The model shows that this is particularly effective at lower intercellular pCO_2 (von Caemmerer and Furbank 2003).

Conclusions

Leaf internal CO_2 diffusion properties need to be considered in the biotechnological designs of C_4 rice. An efficient CO_2-concentrating mechanism requires first that the CO_2 pump have ready access to CO_2, and second that CO_2 leakage from the high CO_2 compartment be minimized.

1. Any high-capacity C_4 photosynthetic pathway requires a high conductance to CO_2 diffusion from intercellular air space to mesophyll cytosol. To increase CO_2 conductance across the exposed mesophyll interface to values greater than currently present in rice requires either an increase in exposed mesophyll surface per unit leaf area or greater CO_2 conductance per unit exposed mesophyll surface area. Since rice mesophyll is already tightly packed with highly lobed cells, increasing CO_2 conductance per unit exposed mesophyll surface area would appear to be the more favorable option.

2A. Minimizing CO_2 leakage across the mesophyll/bundle sheath interface requires a low bundle sheath conductance to CO_2 diffusion while enabling rapid metabolite shuttling for the C_4 cycle. Bundle sheath conductance is the combination of a cell wall component and diffusion across the bundle sheath cell from the site of decarboxylation, so it is possible to counter high permeability through plasmodesmata for metabolite diffusion by locating chloroplasts in bundle sheath cells adjacent to the vascular tissue. However, many C_4 species have centrifugally arranged chloroplasts. This arrangement is associated with the presence of a suberin layer in the mesophyll/bundle sheath cell wall that has been hypothesized to act as a barrier to CO_2 diffusion.

2B. An alternative to using a two-cell Kranz-type C_4 pathway is to design an efficient single-cell C_4 photosynthetic system that can sustain high rates of photosynthesis. In addition to requirement 1, it requires that CO_2 leakage from the chloroplast (assuming that C_4 decarboxylation is targeted there) is restricted by reducing the permeability of the chloroplast envelope to CO_2.

We have hypothesized that membranes such as the plasmalemma or the chloroplast envelopes are barriers to CO_2 diffusion, which might be amenable to manipulation. Further research is required to learn whether altering the expression of protein channels like the aquaporins can be used to manipulate the CO_2 permeability of these membranes. To manipulate diffusion at the intercellular air space/mesophyll interface and the mesophyll/bundle sheath interface, we need a better understanding of CO_2 diffusivity properties of plant cell walls.

References

Apel P, Peisker M. 1978. Einfluss hoher Sauerstoffkonzentrationen auf den CO_2-Kompensationspunkt von C_4-pflanzen. Kulturpflanze 26:99-103.

Bernacchi CJ, Portis AR, Nakano H, von Caemmerer S, Long SP. 2002. Temperature response of mesophyll conductance: implications for the determination of Rubisco enzyme kinetics and for limitations to photosynthesis in vivo. Plant Physiol. 130:1992-1998.

Berry JA, Farquhar GD. 1978. The CO_2 concentrating function of C_4 photosynthesis a biochemical model. In: Hall D, Coombs J, Goodwin T, editors. The Proceedings of the Fourth International Congress on Photosynthesis. Biochemical Society of London, London. p 119-131.

Brown NJ, Parsley K, Hibberd JM. 2005. The future of C_4 research: maize, *Flaveria* or *Cleome*? Trends Plant Sci. 10:215-221.

Brown RH, Byrd GT. 1993. Estimation of bundle sheath cell conductance in C_4 species and O_2 insensitivity of photosynthesis. Plant Physiol. 103:1183-1188.

Cousins AB, Badger MR, von Caemmerer S. 2006. Carbonic anhydrase and its influence on carbon isotope discrimination during C_4 photosynthesis: insights from antisense RNA in *Flaveria bidentis*. Plant Physiol. 141:232-242.

Edwards GE, Franceschi VR, Voznesenskaya EV. 2004. Single-cell C_4 photosynthesis versus the dual-cell (Kranz) paradigm. [Review]. Annu. Rev. Plant Biol. 55:173-196.

Edwards GE, Furbank RT, Hatch MD, Osmond CB. 2001. What does it take to be C_4? Lessons from the evolution of C_4 photosynthesis. Plant Physiol. 125:46-49.

Evans JR, Sharkey TD, Berry JA, Farquhar GD. 1986. Carbon isotope discrimination measured concurrently with gas-exchange to investigate CO_2 diffusion in leaves of higher-plants. Aust. J. Plant Physiol. 13:281-292.

Evans JR, Terashima I, Hanba YT, Loreto F. 2004. Chloroplast to leaf. In: Smith WK, Vogelmann TC, Chritchley C, editors. Ecological studies of photosynthetic adaptation, chloroplast to landscape. Vol. 178. Berlin (Germany): Springer.

Evans JR, von Caemmerer S. 1996. Carbon dioxide diffusion inside leaves. Plant Physiol. 110:339-346.

Evans JR, von Caemmerer S. 2000. Would C_4 rice produce more biomass than C_3 rice? In: Sheehy JE, Mitchell PL, Hardy B, editors. Redesigning rice photosynthesis to increase yield. Elsevier Science B.V., International Rice Research Institute, Amsterdam, Netherlands. 239 p.
Evans JR, von Caemmerer S, Setchell BA, Hudson GS. 1994. The relationship between CO_2 transfer conductance and leaf anatomy in transgenic tobacco with a reduced content of Rubisco. Aust. J. Plant Physiol. 21:475-495.
Farquhar GD. 1983. On the nature of carbon isotope discrimination in C_4 species. Aust. J. Plant Physiol. 10:205-226.
Farquhar GD, Ehleringer JR, Hubick KT. 1989. Carbon isotope discrimination and photosynthesis. Annu. Rev. Plant Physiol. Plant Mol. Biol. 40:503-537.
Farquhar GD, O'Leary MH, Berry JA. 1982. On the relationship between carbon isotope discrimination and the inter-cellular carbon-dioxide concentration in leaves. Aust. J. Plant Physiol. 9:121-137.
Furbank RT, Chitty JA, von Caemmerer S, Jenkins CLD. 1996. Antisense RNA inhibition of *RbcS* gene expression reduces rubisco level and photosynthesis in the C_4 plant *Flaveria bidentis*. Plant Physiol. 111:725-734.
Gillon JS, Yakir D. 2000. Naturally low carbonic anhydrase activity in C_4 and C_3 plants limits discrimination against (COO)-O-18 during photosynthesis. Plant Cell Environ. 23:903-915.
Gutknecht J, Bisson MA, Tosteson FC. 1977. Diffusion of carbon dioxide through lipid bilayer membranes. J. Gen. Physiol. 69:779-794.
Hanba YT, Kogami H, Terashima I. 2002. The effect of growth irradiance on leaf anatomy and photosynthesis in *Acer* species differing in light demand. Plant Cell Environ. 25:1021-1030.
Hanba YT, Miyazawa SI, Kogami H, Terashima I. 2001. Effects of leaf age on internal CO_2 transfer conductance and photosynthesis in tree species having different types of shoot phenology. Aust. J. Plant Physiol. 28:1075-1084.
Hanba YT, Miyazawa SI, Terashima I. 1999. The influence of leaf thickness on the CO_2 transfer conductance and leaf stable carbon isotope ratio for some evergreen tree species in Japanese warm-temperate forests. Funct. Ecol. 13:632-639.
Hanba YT, Shibasaka M, Hayashi Y, Hayakawa T, Kasamo K, Terashima I, Katsuhara M. 2004. Overexpression of the barley aquaporin HvPIP2;1 increases internal CO_2 conductance and CO_2 assimillation in the leaves of transgenic rice plants. Plant Cell Physiol. 45:521-529.
Hatch MD. 1987. C_4 photosynthesis: a unique blend of modified biochemistry, anatomy and ultrastructure. Biochim. Biophys. Acta 895:81-106.
Hatch MD, Agostino A, Jenkins CLD. 1995. Measurement of the leakage of CO_2 from bundle-sheath cells of leaves during C_4 photosynthesis. Plant Physiol. 108:173-181.
Hatch MD, Kagawa T, Craig S. 1975. Subdivision of C_4-pathway species based on differing C_4 acid decarboxylating systems and ultrastructural features. Aust. J. Plant Physiol. 2:111-128.
Hatch MD, Osmond CB. 1976. Compartmentation and transport in C_4 photosynthesis. In: Stocking CR, Heber U, editors. Transport in plants. Vol. 3. Berlin (Germany): Springer-Verlag.
Hattersley PW, Browining AJ. 1981. Occurrence of the suberised lamella in leaves of grasses of different photosynthetic types. I. In parenchymatous bundle sheath and PCR (Kranz) sheaths. Protoplasma 109:371-401.

Häusler RE, Hirsch HJ, Kreuzaler F, Peterhänsel C. 2002. Overexpression of C_4-cycle enzymes in transgenic C_3 plants: a biotechnological approach to improve C_3-photosynthesis. [Review]. J. Exp. Bot. 53:591-607.

He D, Edwards GE. 1996. Estimation of diffusive resistance of bundle sheath cells to CO_2 from modeling of C_4 photosynthesis. Photosynth. Res. 49:195-208.

Henderson S, von Caemmerer S, Farquhar GD. 1992. Short-term measurements of carbon isotope discrimination in several C_4 species. Aust. J. Plant Physiol. 19:263-285.

Jenkins CLD. 1989. Effects of the phospho*enol*pyruvate carboxylase inhibitor. 3,3-dichloro-2-(dihydroxyphosphinoylmethyl) propenoate on photosynthesis: C_4 selectivity and studies on C_4 photosynthesis. Plant Physiol. 89:1231-1237.

Jenkins CLD, Furbank RT, Hatch MD. 1989. Inorganic carbon diffusion between C_4 mesophyll and bundle sheath cells: direct bundle sheath CO_2 assimilation in intact leaves in the presence of an inhibitor of the C_4 pathway. Plant Physiol. 91:1356-1363.

Kiirats O, Lea PJ, Franceschi VR, Edwards GE. 2002. Bundle sheath diffusive resistance to CO_2 and effectiveness of C_4 photosynthesis and refixation of photorespired CO_2 in a C_4 cycle mutant and wild-type *Amaranthus edulis*. Plant Physiol. 130:964-976.

Kogami H, Hanba YT, Kibe T, Terashima I, Masuzawa T. 2001. CO_2 transfer conductance, leaf structure and carbon isotope composition of *Polygonum cuspidatum* leaves from low and high altitudes. Plant Cell Environ. 24:529-538.

Leegood RC. 2002. C_4 photosynthesis: principles of CO_2 concentration and prospects for its introduction into C_3 plants. [Review]. J. Exp. Bot. 53:581-590.

Lieman-Hurwitz J, Rachmilevitch S, Mittler R, Marcus Y, Kaplan A. 2003. Enhanced photosynthesis and growth of transgenic plants that express *ictB*, a gene involved in HCO_3^- accumulation in cynobacteria. Plant Biotechnol. J. 1:43-50.

Longstreth DJ, Hartsock TL, Nobel PS. 1980. Mesophyll cell properties for some C_3 and C_4 species with high photosynthetic rates. Physiol. Plant. 48:494-498.

Matsuoka M, Furbank RT, Fukayama H, Miyao M. 2001. Molecular engineering of C_4 photosynthesis. [Review]. Annu. Rev. Plant Physiol. Plant Mol. Biol. 52:297-314.

Pons TL, Welschen RAM. 2002. Overestimation of respiration rates in commercially available clamp-on leaf chambers: complications with measurement of net photosynthesis. Plant Cell Environ. 25:1367-1372.

Raven JA. 1977. Ribulose bisphosphate carboxylase activity in terrestrial plants: significance of O_2 and CO_2 diffusion. Curr. Adv. Plant Sci. 9:579-590.

Roeske CA, O'Leary MH. 1984. Carbon isotope effects on the enzyme-catalyzed carboxylation of ribulose bisphosphate. Biochemistry 23:6275-6284.

Sage RF, Kubien DS. 2003. Quo vadis C_4? An ecophysiological perspective on global change and the future of C_4 plants. Photosynth. Res. 77:209-225.

Sage RF, Li M, Monson R. 1999. The taxonomic distribution of C_4 photosynthesis. In: Sage RF, Monson R, editors. C_4 plant biology. San Diego, Calif. (USA): Academic Press. p 551-584.

Sheehy JE, Mitchell PL, Hardy B, editors. 2000. Redesigning rice photosynthesis to increase yield. Elsevier Science B.V., International Rice Research Institute, Amsterdam, Netherlands. 239 p.

Terashima I, Hanba YT, Tazoe Y, Vyas P, Yano S. 2006. Irradiance and phenotype: comparative eco-development of sun and shade leaves in relation to photosynthetic CO_2 diffusion. J. Exp. Bot. 57:343-354.

Terashima I, Ono K. 2002. Effects of HgCl2 on CO_2 dependence of leaf photosynthesis: evidence indicating involvement of aquaporins in CO_2 diffusion across the plasma membrane. Plant Cell Physiol. 43:70-78.

Uehlein N, Lovisolo C, Siefritz F, Kaldenhoff R. 2003. The tobacco aquaporin NtAQP1 is a membrane CO_2 pore with physiological functions. Nature 425:734-737.

von Caemmerer S. 2000. Biochemical models of leaf photosynthesis. Vol 2. CSIRO Publishing, Collingwood, Australia.

von Caemmerer S, Evans JR. 1991. Determination of the average partial-pressure of CO_2 in chloroplasts from leaves of several C_3 plants. Aust. J. Plant Physiol. 18:287-305.

von Caemmerer S, Furbank RT. 1999. Modeling of C_4 photosynthesis. In: Sage RF, editor. The biology of C_4 photosynthesis. San Diego, Calif. (USA): Academic Press. p 169-207.

von Caemmerer S, Furbank RT. 2003. The C_4 pathway: an efficient CO_2 pump. Photosynth. Res. 77:191-207.

von Caemmerer S, Millgate A, Farquhar GD, Furbank RT. 1997. Reduction of ribulose-1,5-bisphosphate carboxylase/oxygenase by antisense RNA in the C_4 plant *Flaveria bidentis* leads to reduced assimilation rates and increased carbon isotope discrimination. Plant Physiol. 113:469-477.

von Caemmerer S, Quick WP. 2000. Rubisco: physiology in vivo. In: Leegood RC, Sharkey TD, von Caemmerer S, editors. Photosynthesis: physiology and metabolism. Dordrecht (Netherlands): Kluwer Academic Publishers. p 85-113.

Yeoh H-H, Badger MR, Watson L. 1981. Variations in kinetic-properties of ribulose-1,5-bisphosphate carboxylases among plants. Plant Physiol. 67:1151-1155.

Notes

Authors' addresses: S. von Caemmerer and A.B. Cousins, Molecular Plant Physiology Group, Research School of Biological Sciences, Australian National University, GPO Box 475, Canberra ACT 2601, Australia; J.R. Evans, Environmental Biology Group, Research School of Biological Sciences, Australian National University, GPO Box 475, Canberra ACT 2601, Australia; M.R. Badger, ARC CoE Plant Energy Biology, Research School of Biological Sciences, Australian National University, GPO Box 475, Canberra ACT 2601, Australia; R.T. Furbank, CSIRO Plant Industry, GPO Box 1600, Canberra ACT 2601, Australia.

Acknowledgments: Anatomical measurements presented in Table 1 were made by Daria Schulze during a vacation studentship at the Research School of Biological Sciences during the summer of 1995.

Nuclear regulation of chloroplast development in C_4 and C_3 plants

J.A. Langdale, M. Waters, E.C. Moylan, and A. Bravo-Garcia

This paper reviews the identification and functional characterization of *GLK* genes in diverse plant species. In so doing, a hypothesis is proposed to suggest how *GLK* gene function differs in C_3 and C_4 species. Finally, suggestions are made about how this research and other genetic strategies could be used to produce C_4 rice.

Keywords: chloroplast, GOLDEN2, maize, rice, development

Land-plant chloroplasts originated from a primary endosymbiotic event in which a photosynthetic cyanobacterium was sequestered by a eukaryotic host (Moreira et al 2000). Following endosymbiosis, genes were lost from the chloroplast and transferred to the host-cell nucleus (Martin et al 2002). As a consequence, most of the multi-subunit protein complexes that accumulate in the chloroplast are encoded by both plastid and nuclear genomes. The resultant interdependence between organelle and nucleus requires complex signaling mechanisms.

Retrograde signaling pathways communicate the functional status of the chloroplast back to the nucleus (Oelmuller 1989, Oelmuller et al 1986, Surpin et al 2002). At least two independent retrograde pathways have so far been defined: one mediated by tetrapyrrole signals (Larkin et al 2003, Strand et al 2003) and the other by redox signals (Escoubas et al 1995).

Anterograde signaling pathways conversely communicate the functional state of the nucleus back to the chloroplast. The "structural" pathway was the first type of anterograde signaling to be described in which the nucleus contributes chloroplast-targeted proteins that have structural, functional, or regulatory roles within the organelle. Many examples of this type have been reported (reviewed in Leon et al 1998). A second type of anterograde regulation uses nuclear-localized proteins that either control the transcription of genes encoding chloroplast-targeted proteins or regulate the process of chloroplast biogenesis per se. Examples of this type are more limited but include AKR (Zhang et al 1992) and PALE CRESS (PAC) (Reiter et al 1994) proteins in *Arabidopsis*, OsHAP3 in rice (Miyoshi et al 2003), and the GOLDEN2-

LIKE (GLK) transcription factors that were first identified in maize (Cribb et al 2001, Hall et al 1998, Langdale and Kidner 1994).

Chloroplast development in the C_4 plant maize

Maize develops three distinct chloroplast types. The most commonly considered are the dimorphic bundle sheath (BS) and mesophyll (M) chloroplasts found in the C_4 leaf blade. M chloroplasts have stacked thylakoid lamellae (grana) and function to shuttle carbon from oxaloacetate (OAA) to malate. In contrast, BS chloroplasts have unstacked thylakoid lamellae and function to decarboxylate malate and re-fix CO_2 in the Calvin cycle. The third chloroplast type is found in the leaf sheath and in leaf-like organs of the plant such as husk leaves. These C_3-type chloroplasts have stacked lamellae and fix CO_2 directly in the Calvin cycle.

The developmental mechanisms that regulate the differentiation of C_4 BS, C_4 M, and C_3-type chloroplasts in maize were first elucidated using gene expression analyses and physiological manipulations. These studies demonstrated that, in the presence of light, cells within a two-cell radius of a vein differentiate C_4 BS and C_4 M chloroplasts (with BS closest to the vein) (Langdale et al 1988). In the dark, and in cells farther than two cells away from a vein, C_3-type chloroplasts develop. These observations indicated that positional signals regulate cell-type chloroplast differentiation. This suggestion was confirmed by a clonal analysis that demonstrated that BS and M cells differentiate according to cell position, not lineage (Langdale et al 1989). Thus, at least in maize, light and cell position determine whether cells develop C_4 BS, C_4 M, or C_3-type chloroplasts.

The developmental mechanism outlined above was first put forward nearly 20 years ago, yet we are no closer to understanding the nature of the proposed signals. However, a potential downstream effector was identified in a genetic screen for maize mutants with disrupted chloroplast development. The strategy adopted was to look for mutations that perturbed the development of only a subset of the three chloroplast types, the logic being that structural defects would perturb all chloroplasts whereas regulatory defects would perturb only one or two types (Langdale et al 1995). The transposon-induced *g2-bsd1-m1* mutant allele was identified in this screen and was used to clone the *G2* gene. Characterization of both the wild-type gene and the mutant phenotype revealed that *G2* encodes a transcription factor that regulates the development of C_4 BS and C_3-type chloroplasts (Hall et al 1998, Langdale and Kidner 1994, Rossini et al 2001). Loss of gene function leads to reduced thylakoid stacking of lamellae but does not directly influence the accumulation of photosynthetic enzymes (Cribb et al 2001). Notably, *G2* is expressed in C_4 BS cells but not in C_4 M cells of the leaf blade, C_4 M chloroplasts are unperturbed in *g2* mutants, and a second *G2*-like gene (*ZmGlk1*) is expressed in C_4 M cells but not in C_4 BS cells. These observations led to the proposal that *G2* is expressed by default in maize and that activity becomes repressed in cells that are at a two-cell distance from a vein in response to the proposed light and positional cues. Suppression of *G2* may then lead to expression of *ZmGlk1* and differentiation of C_4 M cell chloroplasts. Unfortunately, this hypothesis remains untested because a loss of function allele of *ZmGlk1* has yet to be identified. However, *GLK* genes are certainly potential targets of the elusive signals.

GLK genes in C_4 and C_3 plants

Although work in maize suggested that *GLK* gene function differs in C_4 BS and M cells, progress in this system was limited by the lack of genome sequence, insufficient reverse genetics resources to identify mutant alleles, and, most importantly, by the difficulty of generating transgenic plants economically on a large scale. As such, attempts to unravel the GLK pathway moved into the C_3 plants rice (Rossini et al 2001), *Arabidopsis* (Fitter et al 2002), and *Physcomitrella patens* (Yasumura et al 2005), all of which can be transformed.

As in maize, two *GLK* genes were identified in rice, *Arabidopsis*, and *P. patens*. *GLK* genes are members of the GARP superfamily, all of which encode transcription factors with the defining GARP DNA binding domain (Riechmann et al 2000). *GLK* genes are distinguished from the rest of the GARP family (which in *Arabidopsis* and rice comprises 56 and 65 members, respectively) by the presence of a C terminal domain referred to as the GCT box. The eight *GLK* genes from maize, rice, *Arabidopsis*, and *P. patens* form a monophyletic group within the GARP superfamily of transcription factors, with duplications having occurred independently in *Arabidopsis* and *P. patens* and in the monocots before the divergence of rice and maize (Yasumura et al 2005).

If gene duplications are not followed by functional divergence and specialization, one of the genes should be selected against. Therefore, the maintenance of two functional copies following three independent duplication events suggests that individual *GLK* genes have specific functions. In maize, the spatial compartmentalization of *G2* and *ZmGlk1* transcripts in C_4 tissue may represent a specialization required for the development of distinct BS and M chloroplasts (Rossini et al 2001). However, it is not known whether chloroplast dimorphism results from the action of functionally distinct proteins or from the activity of functionally equivalent proteins operating in different developmental contexts. In rice, there is no evidence for specialization of gene function. *OsGlk1* and *OsGlk2,* which are orthologous to *ZmGlk1* and *G2*, respectively, are expressed in overlapping domains in photosynthetic tissues, suggesting genetic redundancy. Unfortunately, however, mutations in *OsGlk* genes are not yet available to confirm the suggestion. The two *P. patens* genes are similarly expressed in overlapping domains but in this case mutational analysis has confirmed genetic redundancy in that mutations in individual *PpGLK* genes do not perturb chloroplast development, whereas double mutants are pale green (Yasumura et al 2005). These observations suggest that specialization of *GLK* gene activity may be a feature of C_4 plants but not C_3 plants.

In the context of genetic redundancy, the most informative work has been carried out in *Arabidopsis*. In vegetative tissue, *AtGLK1* and *AtGLK2* are expressed in overlapping domains and double but not single mutants exhibit perturbed chloroplast development (Fitter et al 2002). However, only *AtGLK2* is expressed in the infloresence, and thus *Atglk2* but not *Atglk1* single mutants have pale green siliques. These observations suggest phase-specific or tissue-specific differences in gene action, but again these differences could result from the temporal and spatial separation of functionally equivalent proteins or could indicate that the two proteins are functionally distinct. Notably, constitutive expression of either *AtGLK1* or *AtGLK2* in double mutant plants leads to normal chloroplast development in both vegetative and inflorescence tissues and expression of *AtGLK1* in *Atglk2* single mutants rescues the pale green silique phenotype (E.C. Moylan and J.A. Langdale, unpublished data). Thus, the two *Arabidopsis* GLK proteins are functionally equivalent.

On the basis of the data discussed above, the retention of two gene copies in maize and *Arabidopsis* can be argued to reflect full and partial separation of *trans*-acting regulators of gene expression patterns, respectively. In *P. patens,* as *GLK* genes function in the haploid gametophyte, two copies may be retained to protect against mutation. In rice, the most pertinent to this conference, we cannot be sure because no genetic analysis has been carried out. Significantly, however, it appears that the GLK proteins are functionally equivalent, not just within *Arabidopsis* but also across species. The *Arabidopsis* double mutant phenotype is at least partially rescued by constitutive expression of a maize gene (*ZmGlk1*) (M. Waters, E.C. Moylan, and J.A. Langdale, unpublished data) and a moss gene (*PpGlk1*) (Yasumura et al 2005). These observations lead us to believe that details of the downstream pathway can be elucidated in the most tractable species and that this information can then be used to assess variations in different developmental contexts.

GLK gene function

The most consistent phenotype of *glk* mutants is the presence of chloroplasts with rudimentary, unstacked thylakoid lamellae. Thus, it can be assumed that GLK function either directly or indirectly promotes thylakoid formation and granal stacking. (A somewhat paradoxical conclusion since *G2*, the first gene identified, acts on C_4 BS chloroplasts that are agranal.) In all chloroplasts, the thylakoid membrane contains the components required for light harvesting and photophosphorylation (reviewed in Staehelin and DeWit 1984). The membrane is composed of a lipid bilayer that is interspersed with the proteins, pigments, and other components of the five functional complexes that are vital for photosynthesis. These complexes are the mobile chlorophyll a/b light-harvesting complex (LHC), photosystem II (PSII), cytochrome f/b_6 (cytf/b_6), photosystem I (PSI), and adenosine triphosphate (ATP) synthase. The photosystems are portioned in a characteristic fashion between the granal and stromal (unstacked) thylakoids. PSI is localized preferentially in the stromal thylakoids and PSII is localized preferentially in the granal thylakoids (reviewed in Allen and Forsberg 2001). To date, the only membrane component found to be affected by loss of *GLK* gene function in *Arabidopsis* is the LHC (Fitter et al 2002).

To understand how GLK proteins control thylakoid formation, it is essential to consider the physiological context in which the process takes place. In the case of chloroplast biogenesis, both environmental and endogenous cues have to be considered. The most obvious environmental influence is light; and the transition from proplastid or etioplast to chloroplast is dependent on the perception, reception, and transmission of red light by the phytochrome pathway (Chory 1997). Cytokinin also promotes chloroplast biogenesis and one of the type A response regulators (ARR4) acts to integrate responses to the hormone and to light (Hwang and Sheen 2001, Sweere et al 2001). *ARR4* expression is induced by both cytokinin and phytochrome B (PhyB) and the ARR4 protein binds to the active form of PhyB to stabilize it. Acting antagonistically to light and cytokinin are the retrograde signaling pathways discussed above and at least one sugar signaling pathway that acts to suppress chloroplast biogenesis when the end-products of photosynthesis are abundant (Rolland et al 2002). Preliminary evidence suggests that *AtGLK* genes act independently of the cytokinin pathway because cytokinin does not rescue the double-mutant phenotype and *ARR4* expression is normal in both double mutants and in plants overexpressing *AtGLK* genes (M. Waters, E.C. Moylan, and J.A. Langdale, unpublished data). However, *GLK* expression is promoted by phytochrome B and repressed by the tetrapyrrole-derived retrograde signal (M. Waters and J.A. Langdale, unpublished data). This suggests that GLK function promotes thylakoid formation in response to red light and that gene function is repressed when high levels of chlorophyll intermediates accumulate.

But how do *GLK* genes regulate chloroplast development? The answer to this question awaits the identification of downstream targets. A number of approaches are being taken to identify both downstream targets and upstream regulators. However, a combination of low protein abundance and high rates of protein turnover *in vivo* and the fact that the protein is very insoluble *ex planta* have contributed to frustratingly slow progress. Another unanswered question arises from the apparent functional equivalence of GLK proteins. If *ZmG2* and *ZmGLK1* are found to be functionally equivalent, presumably their different effects on chloroplast development arise as a consequence of association with different partner proteins in different cellular contexts or through activation of different downstream targets in different cellular contexts. But if this is the case, why are the two genes expressed in distinct cell types? Clearly, many fundamental questions still need to be addressed with respect to the role of *GLK* genes in both C_3 and C_4 plants.

C_4 rice—wishful thinking or potential reality?

If asked whether it would be possible to "invent" C_4 rice, most plant scientists would probably say "not in my lifetime." There is no doubt that it is an extremely ambitious goal. Because plant development is both iterative and flexible in response to environmental and endogenous cues, the "insult" of an introduced transgene is nearly always tolerated, is often accommodated by an unpredicted change in metabolism, and is occasionally accompanied by an unexpected phenotype. Although these three features can caution against attempting to design metabolic pathways, they also argue for having a go, because the outcome is rarely predictable. But where do you start? One of the most encouraging observations is that both the evolution of C_4 and the domestication of rice are polyphyletic (Londo et al 2006, Moore 1982). As such, plant genomes have already told us that there is more than one way to make it happen. The challenge is how to make the two happen together in a reasonable timeframe.

From a geneticist's perspective, one of the first questions to ask would be, "Is it easier to convert a C_3 plant to C_4 or vice versa and, if possible either way, does it require loss of gene function or gain of gene function? These are very fundamental questions that we currently have no answers for. Intuitively, most people would assume that it would be easier to convert a C_4 plant to C_3 because C_4 plants develop both C_3 and C_4 leaf structures. Thus, conversion of C_4 to C_3 would just require repression of the C_4 component. In contrast, conversion of C_3 to C_4 intuitively seems more complicated because it would require the spatial and temporal re-organization of a number of gene expression patterns and the addition of a biochemical pathway. So, in the absence of data, it is likely that designs to invent C_4 rice will focus on adding new functions to existing C_3 varieties. However, just because C_4 plants have gained additional functions, it does not mean that the evolutionary process was necessarily kick-started by a gain of gene function. Indeed, loss of function of a particular gene may have been a prerequisite for the subsequent addition of new function. With this in mind, it seems important to establish whether mutagens can induce a change in physiology and, if they can, what type of mutation needs to occur.

To our mind, the first experiment to set up is a large-scale mutagenesis program in both maize and rice. EMS should be used to induce loss of function mutations and activation tagging to induce gain of function mutations. Mutated populations should be screened as seedlings in controlled environments that are too dry and hot for healthy growth of C_3 plants. Pigment deficiencies can be used as a marker for average mutation rate but otherwise pale plants should be ignored. Instead, healthy rice and fully green but wilty maize should be analyzed to determine whether photosynthetic capability has been directly affected. This experiment will answer the following questions: Can gain or loss of gene function in rice increase drought and heat tolerance? If so, does the increase correlate with changes in photosynthesis? (Does it matter if it doesn't?) How often does loss of drought tolerance in maize correspond to repression of the C_4 syndrome? When it does, is it caused by gain or loss of gene function?

The frequency with which wilty maize and healthy rice occur in the screens described above, as compared to the average mutation rate, should provide an indicator of how easy the genome is able to move from C_4-like to C_3 -like and vice versa. Or at least how easy it is to achieve through random mutation of one or a few genes. This frequency will be an important factor to consider in the design of transgene experiments. In addition to providing a feasibility assessment, this experiment may provide viable drought-tolerant rice plants worthy of further study. Ideally, it would also provide an activation-tagged mutation that changes photosynthetic physiology, enabling the mutated gene to be cloned. Again, intuitively, you would bet against that happening, but the data are not currently available to provide any realistic odds for or against.

The experiment proposed above suggests mutagenizing rice and maize. The genus *Oryza* contains 22 wild species plus two domesticated species, one of which is *Oryza sativa*. Because *O. sativa* has two subspecies (japonica and indica) and over 100,000 named cultivars, the choice of germplasm for any designer experiments will not be straightforward. What is the most important trait to start with—yield? taste? drought tolerance? height? More importantly, how can we work it out? In the absence of any better criteria, the tropical *O. sativa* javanica and the upland drought-tolerant aus cultivars would seem reasonable sources of germplasm to have at hand. With this in mind, any current breeding programs aimed at capitalizing on the genetic diversity in wild rice species could also be directed toward the screen described above. In combination with any existing or proposed quantitative trait loci (QTL) analyses, this could lead to the elucidation of any multigenic contributions to the development of C_4.

Finally, what other experiments can be carried out? When the maize genome sequence is available, bioinformatics will be used to dissect differences between rice and maize in an unprecedented fashion. But will the genetic basis of C_4 be found? Only time will tell. Should effort be made to ensure that a second C_4 genome (sorghum?) is sequenced for comparative purposes? From a "GLK perspective," an obvious and easy experiment to do is to transform rice with both *ZmG2* and *ZmGLK1* under the control of their own promoters, constitutive promoters, and the endogenous rice promoters. Would this induce more substantial chloroplast development in any cell type? Data from transgenic *Arabidopsis* lines that constitutively express *AtGLK* genes say no, because epidermal and root cells do not differentiate chloroplasts. Thus, it appears that GLK function is dependent on the cellular context within which *GLK* genes act. However, unlike *Arabidopsis* epidermal and root cells, rice BS plastids are already on the proplastid to chloroplast trajectory in that they accumulate ribulose bisphosphate carboxylase (Miyake and Maeda 1976). Therefore, the experiment is worth trying because it will assess the degree to which plastid differentiation state is fixed in the rice leaf.

In summary, our view is that experiments to invent C_4 rice should proceed with cautious optimism. Most importantly, designer transgenic experiments should be formulated as far as possible on the basis of information that the rice genome has already provided through domestication and habitat occupation, and will continue to provide through genetic experimentation.

References

Allen JF, Forsberg J. 2001. Molecular recognition in thylakoid structure and function. Trends Plant Sci. 6:317-326.

Chory J. 1997. Light modulation of vegetative development. Plant Cell 9:1225-1234.

Cribb L, Hall LN, Langdale JA. 2001. Four mutant alleles elucidate the role of the G2 protein in the development of C_4 and C_3 photosynthesizing maize tissues. Genetics 159:787-797.

Escoubas J-M, Lomas M, LaRoche J, Falkowski PG. 1995. Light intensity regulation of cab gene transcription is signaled by the redox state of the plastoquinone pool. Proc. Natl. Acad. Sci. USA 92:10237-10241.

Fitter DW, Martin DJ, Copley MJ, Scotland RW, Langdale JA. 2002. *GLK* gene pairs regulate chloroplast development in diverse plant species. Plant J. 31:713-727.

Hall LN, Rossini L, Cribb L, Langdale JA. 1998. GOLDEN2: a novel transcriptional regulator of cellular differentiation in the maize leaf. Plant Cell 10:925-936.

Hwang I, Sheen J. 2001. Two-component circuitry in *Arabidopsis* cytokinin signal transduction. Nature 413:383-389.

Langdale JA, Hall LN, Roth R. 1995. Control of cellular differentiation in maize leaves. Philos. Trans. R. Soc. Lond. (Biol.) 350:53-57.

Langdale JA, Kidner CA. 1994. *bundle sheath defective*, a mutation that disrupts cellular differentiation in maize leaves. Development 120:673-681.

Langdale JA, Lane B, Freeling M, Nelson T. 1989. Cell lineage analysis of maize bundle sheath and mesophyll cells. Dev. Biol. 133:128-139.

Langdale JA, Zelitch I, Miller E, Nelson T. 1988. Cell position and light influence C_4 versus C_3 patterns of photosynthetic gene expression in maize. EMBO J. 7:3643-3651.

Larkin RM, Alonso JM, Ecker JR, Chory J. 2003. GUN4, a regulator of chlorophyll synthesis and intracellular signaling. Science 299:902-906.

Leon P, Arroyo A, Mackenzie S. 1998. Nuclear control of plastid and mitochondrial development in higher plants. Annu. Rev. Plant Physiol. Plant Mol. Biol. 49:453-480.

Londo JP, Chiang Y-C, Hung K-H, Chiang T-Y, Schaal BA. 2006. Phylogeography of Asian wild rice, *Oryza rufipogon*, reveals multiple independent domestications of cultivated rice, *Oryza sativa*. Proc. Natl. Acad. Sci. USA 103:9578-9583.

Martin W, Rujan T, Richly E, Hansen A, Cornelsen S, Lins T, Leister D, Stoebe B, Hasegawa M, Penny D. 2002. Evolutionary analysis of *Arabidopsis*, cyanobacterial, and chloroplast genomes reveals plastid phylogeny and thousands of cyanobacterial genes in the nucleus. Proc. Natl. Acad. Sci. USA 99:12246-12251.

Miyake H, Maeda E. 1976. Development of bundle sheath chloroplasts in rice seedlings. Can. J. Bot. 54:556-565.

Miyoshi K, Ito Y, Serizawa A, Kurata N. 2003. *OsHAP3* genes regulate chloroplast biogenesis in rice. Plant J. 36:532-540.

Moore PD. 1982. Evolution of photosynthetic pathways in flowering plants. Nature 295:647-648.

Moreira D, Le Guyader H, Philippe H. 2000. The origin of red algae and the evolution of chloroplasts. Nature 405:69-72.

Oelmuller R. 1989. Photooxidative destruction of chloroplasts and its effect on nuclear gene expression. Photochem. Photobiol. 49:229-239.

Oelmuller R, Levitan I, Bergfield R, Rajasekhar VK, Mohr H. 1986. Expression of nuclear genes as affected by treatments acting on the plastids. Planta 168:482-492.

Reiter RS, Coomber SA, Bourrett TM, Bartley GE, Scolnik PA. 1994. Control of leaf and chloroplast development by the *Arabidopsis* gene *pale cress*. Plant Cell 6:1253-1264.

Riechmann JL, Heard J, Martin G, Reuber L, Jiang C, Keddie J, Adam L, Pineda O, Ratcliffe OJ, Samaha RR, Creelman R, Pilgrim M, Broun P, Zhang JZ, Ghandehari D, Sherman BK, Yu G. 2000. *Arabidopsis* transcription factors: genome-wide comparative analysis among eukaryotes. Science 290:2105-2110.

Rolland F, Moore B, Sheen J. 2002. Sugar sensing and signaling in plants. Plant Cell 14 Suppl: S185-205.

Rossini L, Cribb L, Martin DJ, Langdale JA. 2001. The maize *golden2* gene defines a novel class of transcriptional regulators in plants. Plant Cell 13:1231-1244.

Staehelin LA, DeWit M. 1984. Correlation of structure and function of chloroplast membranes at the supramolecular level. J. Cell Biochem. 24:261-269.

Strand A, Asami T, Alonso J, Ecker JR, Chory J. 2003. Chloroplast to nucleus communication triggered by accumulation of Mg-protoporphyrinIX. Nature 421:79-83.

Surpin M, Larkin RM, Chory J. 2002. Signal transduction between the chloroplast and the nucleus. Plant Cell 14:S327-338.

Sweere U, Eichenberg K, Lohrmann J, Mira-Rodado V, Baurle I, Kudla J, Nagy F, Schafer E, Harter K. 2001. Interaction of the response regulator ARR4 with phytochrome B in modulating red light signaling. Science 294:1108-1111.

Yasumura Y, Moylan EC, Langdale JA. 2005. A conserved transcription factor mediates nuclear control of organelle biogenesis in anciently diverged land plants. Plant Cell 17:1894-1907.

Zhang H, Scheirer DC, Fowle WH, Goodman HM. 1992. Expression of antisense or sense RNA of an ankyrin repeat-containing gene blocks chloroplast differentiation. Plant Cell 4:1575-1588.

Notes

Authors' addresses: J.A. Langdale, M. Waters, E.C. Moylan, and A. Bravo-Garcia, Department of Plant Sciences, University of Oxford, South Parks Rd., Oxford OX1 3RB, UK; E.C. Moylan, current address: Biomed Central Middlesex House, 32-42 Cleveland Street, London, W1T 4LB.

Acknowledgments: Work in J.A. Langdale's research group is funded by grants from the BBSRC and the Gatsby Charitable Foundation.

Balancing light capture with distributed metabolic demand during C_4 photosynthesis

J.R. Evans, T.C. Vogelmann, and S. von Caemmerer

In C_3 leaves, photosynthetic electron transport is closely coupled to the carbon reduction cycle in each chloroplast due to the small pools of nicotinamide adenine dinucleotide phosphate (NADP) and adenosine triphosphate (ATP). In contrast, in C_4 leaves, photosynthetic electron transport occurring in two different cell layers is buffered via extensive metabolite exchange and larger pools of metabolites. The demand for NADPH and ATP in the mesophyll and bundle sheath cells depends on the decarboxylation type. At one extreme, NADP malic enzyme (ME) species such as *Zea mays* produce almost no NADPH by linear electron flux in the bundle sheath. The NADPH required by the photosynthetic carbon reduction (PCR) cycle is either transferred into the bundle sheath via malate or phosphoglyceric acid (PGA) is cycled into the mesophyll for reduction. At the other extreme, nicotinamide adenine dinucleotide (NAD)-ME species such as *Panicum miliaceum* only need to produce ATP in the mesophyll to supply the C_4 cycle. These two extremes restrict linear electron flux to either the mesophyll (NADP-ME) or the bundle sheath (NAD-ME), with the remaining ATP requirement being generated from cyclic electron flux in either cell type. Biochemical diversity within C_4 species means that intermediate solutions with some linear electron flux in both mesophyll and bundle sheath cells also exist. Additional flexibility is also required for any given decarboxylation type because the requirements change depending on the leakiness of the bundle sheath to CO_2. Leakiness tends to increase at lower irradiance and under fluctuating light.

To maximize quantum yields, the different cellular locations for linear electron flux between the decarboxylation types require different distributions of light absorption between the mesophyll and bundle sheath cells. The majority of chlorophyll is co-located within the cells with linear electron flux. However, light absorption is not simply proportional to chlorophyll distribution because of the complex leaf anatomy. We visualized profiles of light absorption through leaves of *Flaveria bidentis* and *Z. mays* by imaging chlorophyll fluorescence emerging from the transverse face of a cut leaf. Green light was absorbed throughout the leaf. In contrast, blue light was strongly absorbed near the surface, with little light penetrating the bundle sheath. This resulted in the rate of CO_2 assimilation under blue light being half that under green light of the same photon irradiance. The decline in the rate of CO_2 assimilation after switching from green to blue light occurred over 100 s and represented a change in metabolite pool size of 400 μmol m^{-2}. We predict that leakiness is greater under blue light than under green light for a given photon irradiance. Engineering a single-cell CO_2-concentrating mechanism would be simpler than a Kranz-type C_4 system as it would require little cellular-specific adjustment to thylakoid composition and function.

Keywords: Linear electron flux, cyclic electron flux, leakiness, blue light, green light, chlorophyll fluorescence imaging

In 1999, a meeting was held at IRRI to discuss the possibility of redesigning rice photosynthesis (Sheehy et al 2000). The case was advanced that plants using the C_4 photosynthetic pathway produced more biomass for a given input of solar radiation. This arose from the combination of greater quantum yields and greater photosynthetic rates under high irradiance (Evans and von Caemmerer 2000, Mitchell and Sheehy 2000). At the time, no field trial had ever compared the radiation-use efficiency of rice concurrently with that of a C_4 plant at the same location. However, Sheehy et al (this volume) have subsequently conducted such an experiment. Rice and maize were grown side by side at IRRI and the radiation-use efficiency of maize was 50% greater than that of rice. This result establishes the ultimate goal being sought for engineered C_4 rice.

The biochemical complexity of the C_4 cycle is usually simplified to represent the major pathways of carbon. Three major groups of C_4 plants are distinguished by their decarboxylation enzymes, which are located in different organelles within the bundle sheath cells. The requirements for ATP and NADPH in the mesophyll and bundle sheath cells differ between C_4 types (Edwards et al 2000, Hatch 1987). In addition, the requirements change depending on how much of the CO_2 that is pumped into the bundle sheath by the C_4 cycle leaks back out (von Caemmerer 2000). After describing the requirements, we then consider the implications for light absorption between cell types and between linear electron flux and cyclic electron flux within each cell type.

For C_3 leaves, the balance between light capture and carbon reduction occurs within each chloroplast. Consequently, any imbalance results in the loss of potential electron transport. Light absorption through a leaf depends on the distribution of chlorophyll and, in C_3 leaves, the profile of photosynthetic capacity through a leaf tends to match the profile of light absorption (Evans and Vogelmann 2003). For C_4 leaves, the linkage between light capture and carbon reduction is more flexible because of the extensive shuttling of metabolites between cells. However, chloroplasts within bundle sheath cells are surrounded by mesophyll cells containing chloroplasts. Therefore, one would expect that they potentially receive less light.

How does one determine where light is absorbed within a leaf? There have been four approaches. First, ray tracing can be modeled through a virtual leaf. This has been done for a typical C_3 leaf with palisade and spongy cell layers (Ustin et al 2001), but has not yet been applied to leaves with Kranz anatomy. Second, light exiting from the surface of a leaf opposite the side receiving light can be measured for leaves that have been cut to remove different proportions of the mesophyll (Terashima and Saeki 1983), but this technique is best suited to thick leaves. Third, fiber optic probes can be inserted through leaves. As the collection angle of these probes is limited and light is rapidly scattered on entry into a leaf, the profile of light has to be calculated by combining the results from probes inserted in three different directions and with multiple replicates (Cui et al 1991). The resulting profile of space irradiance does not yield the profile of light absorption. The fourth approach is to image chlorophyll fluorescence as the intensity is proportional to the absorbed irradiance (Vogelmann and Evans 2002, Vogelmann and Han 2000). Since the fluorescence is emitted by photosystem II, the application of this method suffers some limitations in C_4 leaves because not all chloroplasts engage in linear electron flux. We applied this method to investigate light absorption through leaves of *Flaveria bidentis* (NADP-ME, dicot) and *Zea mays* (NADP-ME, monocot). We manipulated where light was absorbed within *F. bidentis* leaves by varying the color of actinic light and monitored the rate of CO_2 assimilation both under steady state and following changes in color while keeping the incident photon irradiance constant.

NADPH and ATP requirements

For the following discussion, we consider linear electron flux from water to NADP involving the absorption of quanta by both photosystem II and photosystem I. We assume the operation of the Q cycle such that the absorption of 4 quanta by each photosystem results in the formation of 2 NADPH and the deposition of 12 protons into the lumen, which regenerates 3 ATP, assuming 4 protons per ATP (Berry and Rumberg 1996, von Caemmerer 2000). For cyclic electron flux, 2 quanta absorbed by photosystem I transport 4 protons into the thylakoid lumen and regenerate 1 ATP. There is some evidence that a Mehler reaction can also occur where linear electron flux reduces oxygen rather than NADP and consequently produces 3 ATP from 8 quanta, but no NADPH. Assuming the operation of a Mehler reaction instead of cyclic electron flux would increase the quantum requirements for CO_2 fixation but result in a similar distribution of light absorption between mesophyll and bundle sheath cells.

Initially, we consider the case where there is no leakage of CO_2 from the bundle sheath and there is complete suppression of photorespiration, as we will address these complications later. The simplest C_4 type with respect to energy requirements is the NAD-ME type (Fig. 1A). The C_4 cycle converts ATP and pyruvate to adenosine monophosphate (AMP) and phospho*enol*pyruvate (PEP) and therefore mesophyll cells consume the equivalent of 2 ATP per CO_2 transported into the bundle sheath. The complete C_3 cycle can occur in the bundle sheath cells and it requires 2 NADPH and 3 ATP per CO_2 fixed. The other extreme is evident for NADP-ME species such as *Z. mays* (Fig. 1B). Rather than converting oxaloacetate (OAA) into aspartate to transport CO_2 into the bundle sheath, OAA is reduced to malate. This requires 1 NADPH, which is regenerated within the bundle sheath, effectively reducing the need for linear electron flux within bundle sheath chloroplasts. Instead, linear electron flux and the

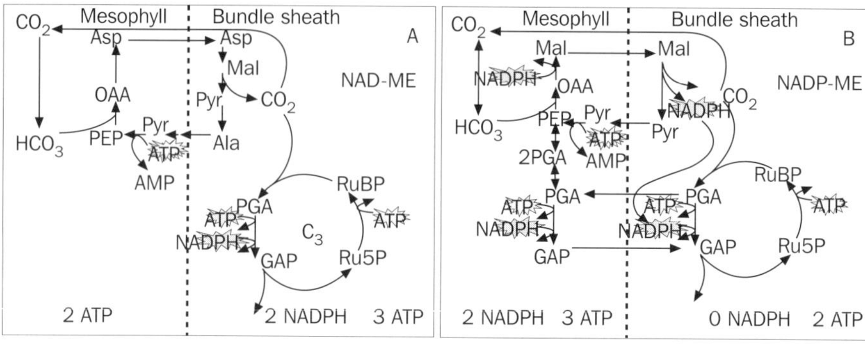

Fig. 1. Carbon flows and cellular location of steps involving ATP and NADPH for NAD-ME (A) and NADP-ME (B) decarboxylation types. In this simplification of the pathways, it is assumed that no photorespiration occurs and that no leakage of CO_2 from the bundle sheath to the mesophyll occurs.

accompanying O_2 production occur in the mesophyll chloroplasts. In the case of *Z. mays*, which generally has little photosystem II activity in the bundle sheath, some of the PGA formed from carboxylation of RuBP then shuttles into the mesophyll for conversion to glyceraldehyde phosphate (GAP). Overall, the ATP and NADPH requirements for the mesophyll and bundle sheath cells are exactly reversed between the NAD-ME and NADP-ME types.

Additional metabolic variation can also result in requirements that are intermediate between these two extremes. For example, the shuttling of some PGA into the mesophyll for reduction to GAP may also occur in NAD-ME species (Hatch and Osmond 1976, Leegood and von Caemmerer 1988). Second, in the NADP-ME dicot *F. bidentis*, a significant part of the C_4 flux into the bundle sheath cells occurs via aspartate, which is converted to malate within the bundle sheath chloroplasts. This reduces the transfer of NADPH reductant from the mesophyll into the bundle sheath and requires linear electron flux to generate NADPH in the bundle sheath. Compared with *Z. mays*, which generally has little photosystem II activity in the bundle sheath, *F. bidentis* has photosystem II activity that is about half that observed in NAD-ME bundle sheath cells (Meister et al 1996).

The requirement for the C_4 cycle can vary with respect to the C_3 cycle because a proportion of the CO_2 released inside the bundle sheath cells leaks back out. One way of assessing the leakiness of the C_4 cycle is to measure the discrimination against $^{13}CO_2$ during photosynthesis (Farquhar 1983). Leakiness, ϕ, has been defined as the proportion of carbon fixed by PEP carboxylation, V_P, which subsequently leaks out of the bundle sheath, where L is the leak rate: $\phi = L/V_P$. The method combines gas exchange measurements with sampling of the inlet and outlet air for determining the isotopic composition of CO_2 (Evans et al 1986). Leakiness has not been extensively measured, but it seems to be similar between monocots and dicots and between decarboxylation types (Henderson et al 1992). The precision of the method declines at low rates of CO_2 assimilation, but it has been possible to examine how leakiness

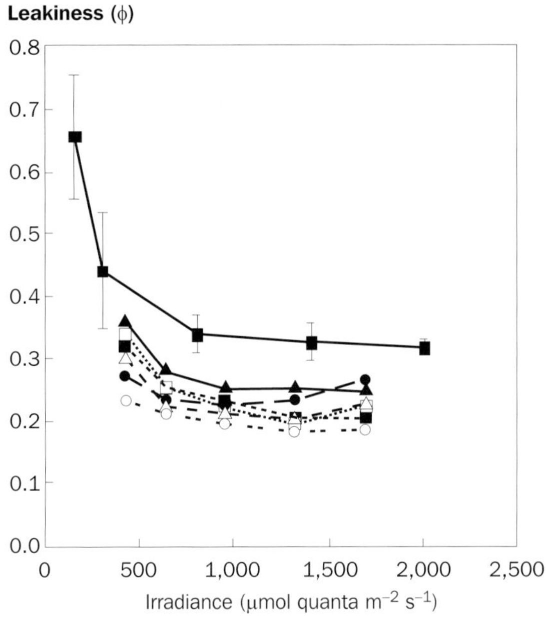

Fig. 2. Relationship between leakiness and irradiance for various C_4 species. Triangles = *Amaranthus edulis*, squares = *Zea mays*, circles = *Sorghum bicolor* (from Henderson et al 1992), solid squares with S.E. bars = *Flaveria bidentis* (Cousins et al 2006).

varies with irradiance (Fig. 2). Leakiness was constant for irradiances greater than 500 μmol quanta m^{-2} s^{-1}, but increased at irradiances less than 500, with leakiness exceeding 0.6 in *F. bidentis* at 150 μmol quanta m^{-2} s^{-1} (Cousins et al 2006).

The variation in leakiness alters the energy requirements for photosynthesis. This can be represented in the following equations.

NAD-ME

The C_4 cycle requires $2/(1 - \phi)$ ATP in mesophyll cells, which can be generated by $4/(1 - \phi)$ quanta absorbed by photosystem I via cyclic electron flux. The C_3 cycle requires 3 ATP and 2 NADPH in bundle sheath cells, which can be generated by 8 quanta divided equally between photosystem II and photosystem I via linear electron flux (Fig. 3A). The fraction of quanta needed by the mesophyll increases as leakiness increases (Fig. 3C). The total quantum requirement is $8 + 4/(1 - \phi)$.

If 1 PGA is shuttled back to the mesophyll for reduction to GAP, then the linear electron flux is split equally between mesophyll and bundle sheath cells, with each requiring 4 quanta divided equally between photosystem II and photosystem I. Since this generates 1.5 ATP in the mesophyll, the quanta required for cyclic electron flux is reduced to $(3 + \phi)/(1 - \phi)$. Since linear electron flow in the bundle sheath cells now generates only 1.5 ATP, an additional 0.5 ATP needs to be made via cyclic electron

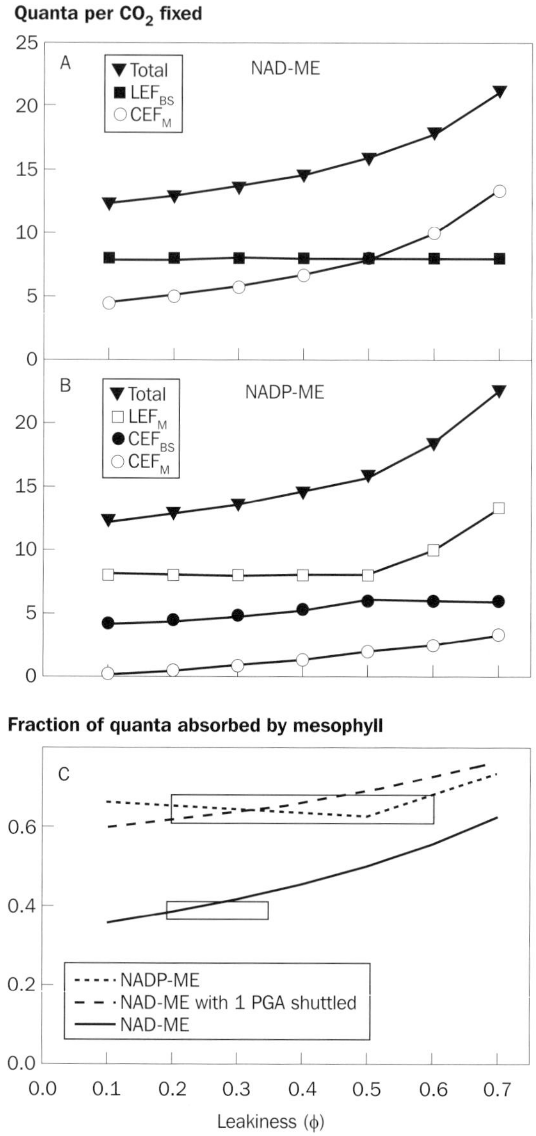

Fig. 3. Quanta required for linear electron flux (LEF) and cyclic electron flux (CEF) in mesophyll (M) and bundle sheath (BS) cells for NAD-ME (A) and NADP-ME (B) C_4 decarboxylation types as a function of leakiness. The fraction of absorbed quanta required by the mesophyll cells varies depending on leakiness (C). Also shown are boxes that represent the observed ranges in leakiness and proportion of chlorophyll present in the mesophyll cells. The NAD-ME type also includes a second case where 1 PGA is shuttled into the mesophyll for reduction to GAP, which shifts 4 quanta for linear electron flux from the bundle sheath into the mesophyll and makes slight changes to cyclic flow required for ATP synthesis. The net result is that a fraction of quanta needs to be absorbed in the mesophyll similar to that required for the NADP-ME type.

flux, which requires 1 quantum. The total quanta required is unchanged by this shuttle, but the fraction of quanta absorbed by the mesophyll needs to be much greater (Fig. 3C).

NADP-ME

For each CO_2 pumped into the bundle sheath, 1 NADPH is required in the mesophyll, which is regenerated in the bundle sheath when malate is decarboxylated. When $\phi > 0.5$, more NADPH is formed in the bundle sheath than can be used in processing PGA (Fig. 1B). Consequently, the equations for energy requirements differ on either side of this threshold.

$0 \leq \phi \leq 0.5$. Two NADPH and 3 ATP are produced by linear electron flux from 8 quanta divided equally between photosystem II and photosystem I in the mesophyll. The C_4 pump requires $2/(1 - \phi)$ ATP. The amount of PGA shuttled back to the mesophyll for reduction to GAP is $(1 - 2\phi)/(1 - \phi)$. Therefore, the ATP required are $(3 - 2\phi)/(1 - \phi)$. Since 3 ATP are produced by linear electron flux, cyclic electron flux must produce $\phi/(1 - \phi)$ ATP, which requires an additional $2\phi/(1 - \phi)$ quanta for photosystem I in the mesophyll. The C_3 cycle requires $1 + 1/(1 - \phi)$ ATP, which requires $2(2 - \phi)/(1 - \phi)$ quanta for photosystem I in the bundle sheath (Fig. 3B). The total quantum requirement is the same as for NAD-ME, being $8 + 4/(1 - \phi)$.

$\phi > 0.5$. The number of quanta required for linear electron flux to supply NADPH for the C_4 cycle is $4/(1 - \phi)$. Since linear electron flux generates 1.5 ATP per NADPH, an additional 0.5 ATP is required for each CO_2 pumped. This requires $1/(1 - \phi)$ quanta for photosystem I in the mesophyll for cyclic electron flux. Three ATP are needed for the C_3 cycle in the bundle sheath cells, which require 6 quanta for photosystem I in the bundle sheath for cyclic electron flux. In contrast to the NAD-ME requirements, about two-thirds of the quanta used in photosynthesis by NADP-ME types need to be absorbed by the mesophyll and this is nearly independent of leakiness. The total quantum requirement, $6 + 5/(1 - \phi)$, is greater than that for NAD-ME because surplus NADPH is produced and transferred to the bundle sheath cells.

In deriving these equations, we have assumed that no photorespiration occurs and have used the most efficient method for generating ATP (cyclic electron flux rather than a Mehler reaction). Therefore, these equations represent the minimum quantum requirement. Oxygen uptake can be distinguished from oxygen evolution during photosynthesis by isotopic methods where $^{16}O_2$ is replaced by $^{18}O_2$. Early work suggested that there was little Mehler reaction in the bundle sheath cells of *Z. mays* (Chapman et al 1980, Furbank and Badger 1982), but that a significant capacity existed in mesophyll chloroplasts (Furbank et al 1983). Siebke et al (2003) found an oxygen uptake equivalent to about 18% of gross oxygen evolution in leaves of C_4 grasses, with the rate increasing as irradiance increased. At the CO_2 compensation point, photorespiration accounted for 70% of the oxygen uptake. They estimated that a Mehler reaction could supply up to half of the ATP required by the C_4 cycle. Significant rates of cyclic electron flux are therefore required to regenerate ATP for the C_4 cycle or where linear electron flux is absent, such as in agranal bundle sheath chloroplasts of *Z. mays* (Chapman et al 1980, Leegood et al 1983). Several mechanisms for cyclic electron flux are possible (Kramer et al 2004). The ferredoxin-plastoquinone oxidoreductase pathway is inhibited by antimycin A while the NAD(P)H-plastoquinone oxidoreductase (NDH) is analogous to complex I in mitochondria. NDH is more highly expressed in bundle sheath than in mesophyll chloroplasts of *S. bicolor* (Kubicki et al 1996), enabling the oxidation of malate or NADPH to produce ATP (Ivanov et al 2005). This possibility was used in the C_4 model of Laisk and Edwards (2000) to deal with any excess NADPH produced in the bundle sheath by NADP-ME, by converting it to ATP. To account for the observed quantum requirement of 14.3, Laisk and Edwards (2000) assumed that 1 ATP was formed for each quantum absorbed for cyclic electron flux in the bundle sheath cells and that extra ATP required in the mesophyll was produced by a Mehler reaction. (We assumed that 0.5 ATP is produced per quantum from cyclic electron flux and 3/8 ATP per quantum from a Mehler reaction.) Using our equations and a leakiness of 0.3, the quantum requirement is 13.7 (or 15.6 if a Mehler reaction rather than cyclic electron flux is used to generate additional ATP). The precision of quantum yield measurements, the uncertainties that still surround the mechanism of cyclic electron flux (Kramer et al 2004), and the difficulty in measuring leakiness currently prevent one from distinguishing between the various possible solutions.

The ability to vary the balance between linear electron flux and cyclic electron flux presumably is associated with changes to the protein composition of the thylakoid membranes. For example, the limited linear electron flux observed for chloroplasts isolated from bundle sheath cells of *Z. mays* has been linked to a lack of the nuclear-encoded oxygen-evolving complex polypeptides of photosystem II because the intrinsic part of the complex is present (Meierhoff and Westhoff 1993). The agranal morphology of bundle sheath chloroplasts of NADP-ME species implies a reduction in the proportion of chlorophyll associated with light-harvesting chlorophyll a/b complexes associated with photosystem II. The increased requirement for cyclic electron flux within bundle sheath chloroplasts is matched to the increased expression of NAD(P)H dehydrogenase (Kubicki et al 1996). Altering the demand for cyclic electron flux through expression of NADP-ME in chloroplasts may not result in an altered supply of NADPH and ATP if it is not accompanied by changes in the thylakoid composition. When maize NADP-ME was expressed in rice chloroplasts, the leaves that developed were pale and susceptible to photoinhibition, and became agranal (Takeuchi et al 2000, Tsuchida et al 2001). The change in ultrastructure could be argued as evidence that changing demand will trigger self-correction in the protein composition of the chloroplasts, but more likely it seems to be the manifestation of damage. Therefore, in addition to the tissue-specific expression of carbon-cycle enzymes, expression of thylakoid proteins will probably need to be modified to adjust the balance of light distribution between mesophyll and bundle sheath chloroplasts and between the photosystems within each chloroplast and between protein complexes enabling linear or cyclic electron flow.

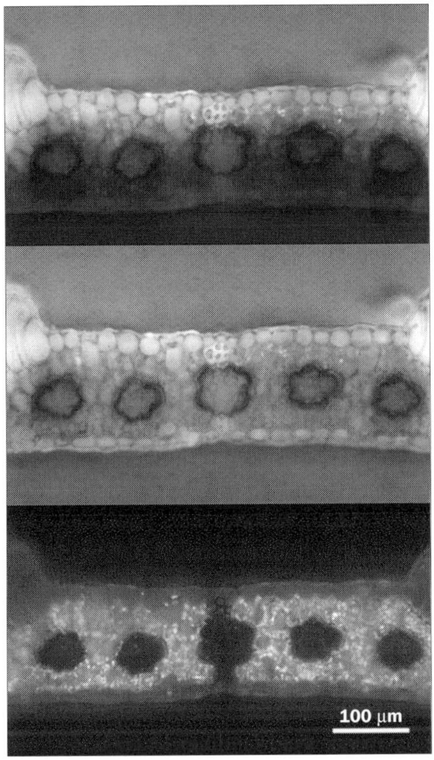

Fig. 4. Fluorescence (680 nm) images of the transverse face of *Zea mays*. The adaxial surface of the leaf was irradiated with monochromatic blue or green light while fluorescence exiting from the transversely cut face of the leaf was imaged. Then, light perpendicular to the transverse face was applied to capture the epi-illumination fluorescence image. The central vein has bundle sheath extensions that contain no chloroplasts.

Distribution of chlorophyll and light absorption

There have been several estimates of the fraction of chlorophyll present in mesophyll cells of C_4 leaves. The NAD-ME species *Panicum miliaceum* and *P. coloratum* had about 40%, while the NADP-ME species *Sorghum bicolor, Cenchrus ciliaris*, and *Z. mays* had 62–67% of chlorophyll in the mesophyll (Ghannoum et al 2005). These two fractions correspond remarkably well to the fraction of quanta that are required (Fig. 3C) given the range of leakiness that has been observed (Fig. 2). However, the Kranz anatomy means that bundle sheath chlorophyll is generally shielded by the surrounding mesophyll cells. Therefore, one would expect that relatively less light would be absorbed per chlorophyll in the bundle sheath compared with the mesophyll.

To investigate this, we imaged chlorophyll fluorescence emitted from the cut transverse face of leaves of *Z. mays* and *F. bidentis* (methodology described in Vogelmann and Evans 2002, Vogelmann and Han 2000). Three images are shown for each species (Figs. 4, 5). Under epi-illumination, light is directed onto the transverse

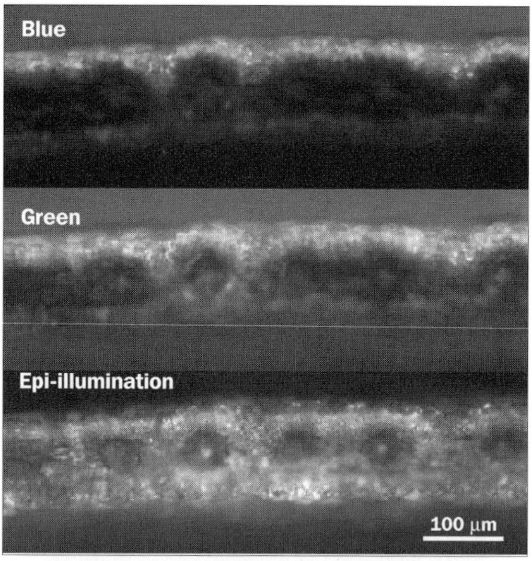

Fig. 5. Fluorescence (680 nm) images of the transverse face of *Flaveria bidentis*. The adaxial surface of the leaf was irradiated with monochromatic blue or green light while fluorescence exiting from the transversely cut face of the leaf was imaged. Then, light perpendicular to the transverse face was applied to capture the epi-illumination fluorescence image.

face through the microscope lens, which also captures the fluorescence. The distribution of fluorescence represents that of chlorophyll. Bright fluorescence can be seen throughout the mesophyll of both species, with chloroplasts clearly visible in *Z. mays*. The lack of functional photosystem II in bundle sheath chloroplasts of *Z. mays* means that, despite containing about one-third of the chlorophyll, no fluorescence is emitted from the bundle sheath (Fig. 4). The bundle sheath extension above and below the central vein in the image lacks chlorophyll and could act as a light guide for the chloroplasts in the bundle sheath. Bundle sheath extensions are commonly found in rice leaves, but their bundle sheath cells contain only a few chloroplasts, which are small in size. The lower intensity of fluorescence within bundle sheath cells is less distinct in *F. bidentis* leaves (Fig. 5), which do have some photosystem II activity in the bundle sheath chloroplasts (Meister et al 1996).

By applying light to the adaxial surface, the fluorescence image reveals the gradient in light absorption through the leaf. Blue light is strongly absorbed by chlorophyll and is rapidly scattered on entry into the leaf. This results in intense fluorescence near the adaxial surface, but little fluorescence from the lower half of the leaf. In contrast, green light penetrates further into the leaf and some fluorescence is still emitted from chloroplasts near the lower surface. The *F. bidentis* images are less distinct, partly because being a dicot leaf, the vascular bundles are not parallel and uniformly spaced as in the *Z. mays* leaf.

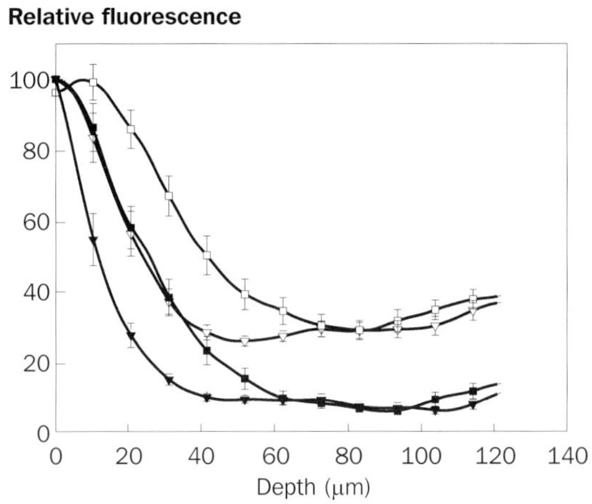

Fig. 6. Quantitative analysis of the fluorescence images in Figure 5. Chlorophyll fluorescence profiles through *Flaveria bidentis* leaves with sampling through veins (triangles) or mesophyll regions between veins (squares) when blue (solid symbols) or green (hollow symbols) light was applied to the adaxial surface. Error bars = S.E. for 15 measurements (total) taken from 4 leaves. Lines join data sampled at 2.6 μm, with symbols shown only every 10.4 μm for clarity.

The profile of fluorescence across the leaf was quantified from *F. bidentis* images by sampling transects through mesophyll or vascular tissue (Fig. 6). Depth is measured from the boundary between the epidermis and mesophyll. Under blue light, fluorescence declines rapidly, falling below 20% within 40 μm through the mesophyll. In contrast, it takes about 300 μm for blue light absorption to decline by a similar amount in leaves of *Spinacia oleracea* (Vogelmann and Evans 2002). Green light penetrated further, declining to a plateau of 30% halfway through the leaf. Clearly, significant amounts of green light reached the bundle sheath chloroplasts compared with relatively little blue light. Fluorescence declined more rapidly through the vascular tissue, but reabsorption by bundle sheath chloroplasts makes interpretation of this problematic.

Effect of color on photosynthesis

The differential penetration of blue and green light into *F. bidentis* leaves led us to investigate the consequent effect on photosynthesis. Initially, we compared steady-state rates of CO_2 assimilation and photochemical efficiency under white, green, or blue light with equivalent incident photon fluxes. Photochemical efficiency was measured with a PAM fluorometer using a blue modulated light rather than the usual red one. As a control, we also measured leaves of *Spinacia oleracea* under the same conditions (Table 1). Despite giving the same incident photon irradiance for each color, the

Table 1. Rate of CO_2 assimilation (μmol m^{-2} s^{-1}), A, and photochemical efficiency, ϕPSII, for leaves of *Spinacia oleracea* and *Flaveria bidentis* measured under different-colored light (mean ± S.E., n = 3). Measurement conditions were 370 μmol CO_2 mol^{-1}, 12–15 mbar leaf to air vapor pressure difference, and a leaf temperature of 25 °C.

Photon irradiance (μmol m^{-2} s^{-1})	Color	*Spinacia oleracea*		*Flaveria bidentis*	
		A	ϕPSII	A	ϕPSII
350	White	9.8 ± 0.1	0.60	12.2 ± 0.5	0.66
	Green	9.2 ± 0.1	0.63	11.7 ± 0.4	0.71
	Blue	7.0 ± 0.2	0.59	5.9 ± 0.06	0.54
170	White	8.1 ± 0.03	0.65	6.8 ± 0.4	0.69
	Green	7.7 ± 0.05	0.66	5.8 ± 0.2	0.73
	Blue	6.1 ± 0.1	0.64	4.0 ± 0.3	0.60

rates of CO_2 assimilation were lower under blue light by about 25% for *Spinacia* and 50% for *Flaveria*. Although leaf absorptance of blue light is greater than green light, quantum yields are lower (Evans 1987, McCree 1971). This was not reflected in the photochemical efficiency signal, which was unchanged for *Spinacia* and decreased by 15% for *Flaveria*. This illustrates the fact that gas exchange integrates the flux through the depth of the leaf over a given area, while fluorescence samples from a layer of chloroplasts near the adaxial surface. For *Flaveria*, the rate of CO_2 assimilation under blue light was approximately the same as under half the photon irradiance of green light. This suggested that the poor penetration of blue light into the bundle sheath cells did not allow sufficient ATP formation to match the rate of CO_2 pumping. One would predict that leakiness should be greater under blue light than under green light. However, there appears to have been some feedback on mesophyll electron transport, which had a lower photochemical efficiency under blue light than under green light.

It was apparent when changing the color of the light that it took much longer for the rates to stabilize after changing between green and blue light than it did when simply changing the photon irradiance of a given color. This was investigated by following the transients (Fig. 7). For *Spinacia*, upon changing either color or photon irradiance, there was a rapid change in rate that approached a new steady state within 20 seconds. This was also evident for *Flaveria* when green photon irradiance was decreased or increased. However, when changing from green to blue light with the same photon irradiance or vice versa, a slow transient was produced that took about 100 seconds to stabilize. This is shown in more detail in Figure 8. Following the change from green to blue light, there is only a slight decline over the first 10 seconds. Integrating the area under the transient yields a pool size of about 400 μmol m^{-2}. This is equivalent to the pool sizes of PGA and GAP in C_4 leaves (Leegood and von Caemmerer 1988, 1989).

The relationship between light and dark reactions of photosynthesis during and following 20-s lightflecks has been studied with *Z. mays* (Krall and Pearcy 1993). Oxygen evolution essentially follows the changes in irradiance, with a small burst at

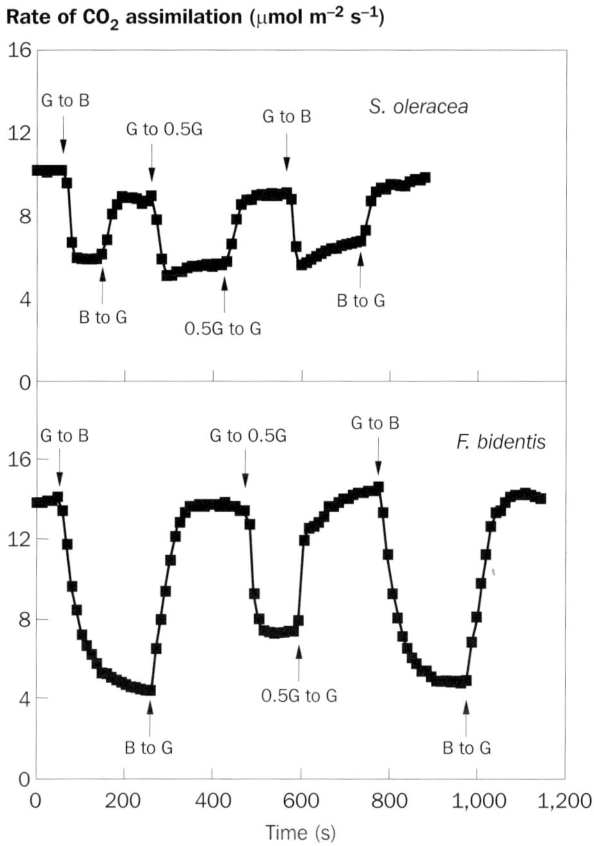

Fig. 7. Transient changes in the rate of CO_2 assimilation following step changes to photon irradiance for leaves of *Spinacia* and *Flaveria*. Incident photon irradiances were 320, 323, and 170 μmol m^{-2} s^{-1} for the green (G), blue (B), and 0.5 green (0.5G) treatments.

the start of bright lightflecks. CO_2 assimilation rate rises more slowly and declines exponentially during the 20 s after the lightfleck. The integrated CO_2 assimilation represented 50–70% of the oxygen evolved. When lightflecks were shorter than 10 s, integrated CO_2 assimilation represented only 10–40% of the oxygen evolved. The difference represents the energy cost involved in establishing the high bundle sheath CO_2 concentrations. Bundle sheath CO_2 leaks back out to the mesophyll following the lightfleck as the Calvin cycle is unable to regenerate ribulose bisphosphate. The pattern for CO_2 assimilation with lightflecks is similar to that of the transients we observed when changing green photon irradiance (Figs. 7, 8). However, the transient following a change in color without altering photon irradiance is much slower, taking over 100 s to approach steady state.

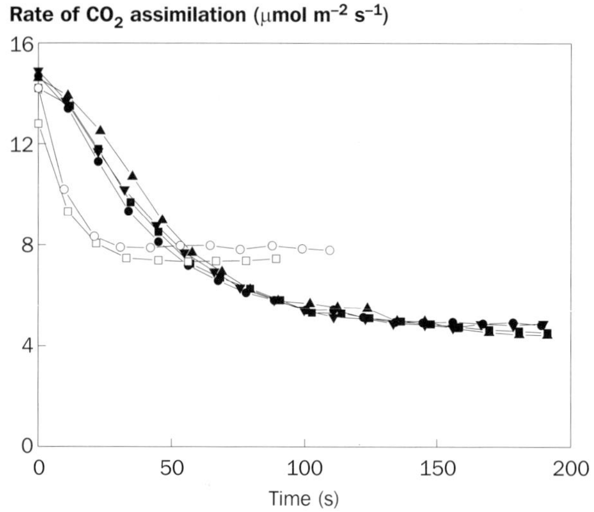

Fig. 8. Transient change in the rate of CO_2 assimilation following the step change from 320 μmol green quanta m^{-2} s^{-1} to either 323 μmol blue quanta m^{-2} s^{-1} (solid symbols) or 170 μmol green quanta m^{-2} s^{-1} (hollow symbols) for *Flaveria bidentis*.

Both lightfleck and color change experiments illustrate the constraints imposed by the C_4 cycle. Unlike in C_3 photosynthesis where light and dark reactions are closely coupled within each chloroplast, in C_4 photosynthesis, large pools of metabolites are necessary to enable CO_2 pumping into the bundle sheath, which results in considerable leeway in the relative photochemical rates between mesophyll and bundle sheath cells. Shuttling of PGA from the bundle sheath into the mesophyll for reduction to triose phosphate also increases the flexibility in coupling between light and dark reactions within a cell.

Conclusions

We take as our starting point that the yield advantage for a C_4 rice requires that the cycle operate with both high efficiency and high capacity, and that the sink will be capable of dealing with the increased supply of photosynthate. Failure to achieve any of these will compromise the potential yield improvement.

Engineering an efficient Kranz-type C_4 photosynthetic pathway into a C_3 leaf such as rice will require cell-specific modifications to the thylakoid membrane composition in addition to the expression of C_4 enzymes. The requirements differ depending on which decarboxylation pathway is chosen. For the NADP-ME pathway, about 65% of quanta need to be absorbed by mesophyll chloroplasts, compared with about 40% for the NAD-ME pathway. In addition, the electron transport chain has to be specialized for cyclic electron flow in the bundle sheath chloroplasts for NADP-ME or mesophyll chloroplasts for NAD-ME and vice versa for linear electron flow. The distribution of light between mesophyll and bundle sheath chloroplasts is not necessarily simply a function of chlorophyll partitioning between these two cell types because bundle sheath chloroplasts are generally shaded by mesophyll chloroplasts. The consequence of this was clearly evident when blue versus green light was used to drive photosynthesis. An efficient Kranz-type C_4 pathway in rice would require the differentiation of additional intermediate veins to achieve the appropriate number of mesophyll to bundle sheath cells. If this were accompanied by enhanced development of bundle sheath chloroplasts, the proportion of light absorbed by mesophyll chloroplasts would also be substantially closer to what is required.

We have illustrated and experimentally demonstrated that considerable complexity of light use is present in C_4 Kranz-type photosynthesis. To realize the potential efficiency gain offered by a C_4 photosynthetic pathway, it will be necessary to balance the energy absorption appropriately between the mesophyll and bundle sheath cells and modify the thylakoid composition of chloroplasts from these two cell types to enable the correct capacity between linear and cyclic electron flow.

A much simpler approach in terms of light energy partitioning would be to engineer a single-cell C_4 system. This would require little modification to chloroplast thylakoid composition apart from an enhanced capacity for cyclic electron flow to supply ATP for PEP regeneration. Instead, the major requirement is to reduce CO_2 leakage from chloroplasts into which a CO_2 pump has been engineered.

References

Berry S, Rumberg B. 1996. H^+/ATP coupling ratio at the unmodulated CF_0CF_1-ATP synthase determined by proton flux measurements. BBA Bioenergetics 1276:51-56.

Chapman KSR, Berry JA, Hatch MD. 1980. Photosynthetic metabolism in bundle sheath-cells of the C_4 species *Zea mays*: sources of ATP and NADPH and the contribution of photosystem II. Arch. Biochem. Biophys. 202:330-341.

Cousins AB, Badger MR, von Caemmerer S. 2006. Carbonic anhydrase and its influence on carbon isotope discrimination during C_4 photosynthesis: insights from antisense RNA in *Flaveria bidentis*. Plant Physiol. 141:232-242.

Cui M, Vogelmann TC, Smith WK. 1991. Chlorophyll and light gradients in sun and shade leaves of *Spinacia oleracea*. Plant Cell Environ. 14:493-500.

Edwards GE, Kiirats O, Laisk A, Okita TW. 2000. Requirements for the CO_2 concentrating mechanism in C_4 plants relative to limitations on carbon assimilation in rice. In: Sheehy JE, Mitchell PL, Hardy B, editors. Redesigning rice photosynthesis to increase yield. Makati City (Philippines): International Rice Research Institute and Amsterdam (The Netherlands): Elsevier Science B.V. p 99-112.

Evans JR. 1987. The dependence of quantum yield on wavelength and growth irradiance. Aust. J. Plant Physiol. 14:69-79.

Evans JR, Sharkey TD, Berry JA, Farquhar GD. 1986. Carbon isotope discrimination measured concurrently with gas exchange to investigate CO_2 diffusion in leaves of higher plants. Aust. J. Plant Physiol. 13:281-292.

Evans JR, Vogelmann TC. 2003. Profiles of ^{14}C fixation through spinach leaves in relation to light absorption and photosynthetic capacity. Plant Cell Environ. 26:547-560.

Evans JR, von Caemmerer S. 2000. Would C_4 rice produce more biomass than C_3 rice? In: Sheehy JE, Mitchell PL, Hardy B, editors. Redesigning rice photosynthesis to increase yield. Makati City (Philippines): International Rice Research Institute and Amsterdam (The Netherlands): Elsevier Science B.V. p 53-71.

Farquhar GD. 1983. On the nature of carbon isotope discrimination in C_4 species. Aust. J. Plant Physiol. 10:205-226.

Furbank RT, Badger MR. 1982. Photosynthetic oxygen exchange in attached leaves of C_4 monocotyledons. Aust. J. Plant Physiol. 9:553-558.

Furbank RT, Badger MR, Osmond CB. 1983. Photoreduction of oxygen in mesophyll chloroplasts of C_4 plants: a model system for studying an *in vivo* Mehler reaction. Plant Physiol. 73:1038-1041.

Ghannoum O, Evans JR, Chow WS, Andrews TJ, Conroy JP, von Caemmerer S. 2005. Faster Rubisco is the key to superior nitrogen-use efficiency in NADP-malic enzyme relative to NAD-malic enzyme C_4 grasses. Plant Physiol. 137:638-650.

Hatch MD. 1987. C_4 photosynthesis: a unique blend of modified biochemistry, anatomy and ultrastructure. Biochim. Biophys. Acta 895:81-106.

Hatch MD, Osmond CB. 1976. Compartmentation and transport in C_4 photosynthesis. In: Stocking CR, Heber U, editors. Transport in plants. III. Intracellular interactions and transport processes. Berlin (Germany): Springer-Verlag. p 144-184.

Henderson SA, von Caemmerer S, Farquhar GD. 1992. Short-term measurements of carbon isotope discrimination in several C_4 species. Aust. J. Plant Physiol. 19:263-285.

Ivanov B, Asada K, Kramer DM, Edwards G. 2005. Characterization of photosynthetic electron transport in bundle sheath cells of maize. I. Ascorbate effectively stimulates cyclic electron flow around PSI. Planta 220:572-581.

Krall JP, Pearcy RW. 1993. Concurrent measurements of oxygen and carbon-dioxide exchange during lightflecks in maize (*Zea mays* L.). Plant Physiol. 103:823-828.

Kramer DM, Avenson TJ, Edwards GE. 2004. Dynamic flexibility in the light reactions of photosynthesis governed by both electron and proton transfer reactions. Trends Plant Sci. 9:349-357.

Kubicki A, Funk E, Westhoff P, Steinmuller K. 1996. Differential expression of plastome-encoded ndh genes in mesophyll and bundle-sheath chloroplasts of the C_4 plant *Sorghum bicolor* indicates that the complex I-homologous NAD(P)H-plastoquinone oxidoreductase is involved in cyclic electron transport. Planta 199:276-281.

Laisk A, Edwards GE. 2000. A mathematical model of C_4 photosynthesis: the mechanism of concentrating CO_2 in NADP-malic enzyme type species. Photosynth. Res. 66:199-224.

Leegood RC, Crowther D, Walker DA, Hind G. 1983. Energetics of photosynthesis in *Zea mays*. I. Studies of the flash-induced electrochromic shift and fluorescence induction in bundle sheath cells. Biochim. Biophys. Acta 722:116-126.

Leegood RC, von Caemmerer S. 1988. The relationship between contents of photosynthetic metabolites and the rate of photosynthetic carbon assimilation in leaves of *Amaranthus edulis* L. Planta 174:253-262.

Leegood RC, von Caemmerer S. 1989. Some relationships between contents of photosynthetic intermediates and the rate of photosynthetic carbon assimilation in leaves of *Zea mays* L. Planta 178:258-266.

McCree KJ. 1971. The action spectrum, absorptance and quantum yield of photosynthesis in crop plants. Agric. Meteorol. 9:191-216.

Meierhoff K, Westhoff P. 1993. Differential biogenesis of photosystem-II in mesophyll and bundle-sheath cells of monocotyledonous NADP-malic enzyme-type C_4 plants: the non-stoichiometric abundance of the subunits of photosystem-II in the bundle-sheath chloroplasts and the translational activity of the plastome-encoded genes. Planta 191:23-33.
Meister M, Agostino A, Hatch M. 1996. The roles of malate and aspartate in C_4 photosynthetic metabolism of *Flaveria bidentis* (L.). Planta 199:262-269.
Mitchell PL, Sheehy JE. 2000. Performance of a potential C_4 rice: overview from quantum yield to grain yield. In: Sheehy JE, Mitchell PL, Hardy B, editors. Redesigning rice photosynthesis to increase yield. Makati City (Philippines): International Rice Research Institute and Amsterdam (The Netherlands): Elsevier Science B.V. p 145-163.
Sheehy JE, Mitchell PL, Hardy B, editors. 2000. Redesigning rice photosynthesis to increase yield. Makati City (Philippines): International Rice Research Institute and Amsterdam (The Netherlands): Elsevier Science B.V. 293 p.
Siebke K, Ghannoum O, Conroy JP, Badger MR, von Caemmerer S. 2003. Photosynthetic oxygen exchange in C_4 grasses: the role of oxygen as electron acceptor. Plant Cell Environ. 26:1963-1972.
Takeuchi K, Akagi H, Kamasawa N, Osumi M, Honda H. 2000. Aberrant chloroplasts in transgenic rice plants expressing a high level of maize NADP-dependent malic enzyme. Planta 211:265-274.
Terashima I, Saeki T. 1983. Light environment within a leaf. 1. Optical properties of paradermal sections of *Camellia* leaves with special reference to differences in the optical properties of palisade and spongy tissues. Plant Cell Physiol. 24:1493-1501.
Tsuchida H, Tamai T, Fukayama H, Agarie S, Nomura M, Onodera H, Ono K, Nishizawa Y, Lee B-H, Hirose S, Toki S, Ku MSB, Matsuoka M, Miyao M. 2001. High level expression of C_4-specific NADP-malic enzyme in leaves and impairment of photoautotrophic growth in a C_3 plant, rice. Plant Cell Physiol. 42:138-145.
Ustin SL, Jacquemoud S, Govaerts Y. 2001. Simulation of photon transport in a three-dimensional leaf: implications for photosynthesis. Plant Cell Environ. 24:1095-1103.
Vogelmann TC, Evans JR. 2002. Profiles of light absorption and chlorophyll within spinach leaves from chlorophyll fluorescence. Plant Cell Environ. 25:1313-1323.
Vogelmann TC, Han T. 2000. Measurement of gradients of absorbed light in spinach leaves from chlorophyll fluorescence profiles. Plant Cell Environ. 23:1303-1311.
von Caemmerer S. 2000. Biochemical models of leaf photosynthesis. Collingwood, Victoria (Australia): CSIRO Publishing. 165 p.

Notes

Authors' addresses: J.R. Evans, Environmental Biology Group, Research School of Biological Sciences, Australian National University, Canberra, ACT 2601, Australia; T.C. Vogelmann, Department of Botany and Agricultural Biochemistry, University of Vermont, Burlington, VT 05405-0086, USA; S. von Caemmerer, Molecular Plant Physiology Group, Research School of Biological Sciences, Australian National University, Canberra, ACT 2601, Australia.
Acknowledgments: We thank Dr. John Sheehy and the International Rice Research Institute for the opportunity to present this work.

Redesigning C_4 rice from limited C_4 photosynthesis

D.M. Jiao

The photosynthetic characteristics of four transgenic rice lines overexpressing maize phospho*enol*pyruvate carboxylase (PEPC; line PC), pyruvate, orthophosphate dikinase (PPDK; line PK), PEPC + PPDK (line CK), and NADP-malic enzyme (NADP-ME; line ME) were investigated using outdoor-grown plants. Relative to untransformed wild-type (WT) rice, PC transgenic rice exhibited high PEPC activity (a 25-fold increase) and enhanced activity of carbonic anhydrase (more than a twofold increase). The PC transgenic plants also showed a higher CO_2 uptake rate and carboxylation efficiency, and slightly reduced CO_2 compensation point. Furthermore, PC transgenic rice produced 22% more grains than WT plants. Labeling with $^{14}CO_2$ for 20 s showed more ^{14}C distributed to C_4 primary photosynthate aspartate and feeding with exogenous C_4 primary products such as oxaloacetate (OAA), malate (MA), or phospho*enol*pyruvate (PEP) showed an increment of photosynthetic rate in PC transgenic rice, suggesting that a limited C_4 cycle exists in leaves of transgenic rice.

Introduction of the maize PEPC gene could activate or induce activities of the key enzymes scavenging active oxygen, such as superoxide dismutase (SOD) and peroxidase (POD).

The line JAAS45 manifested higher photosynthetic rates and photochemical efficiency of PS II (F_v/F_m). The value of $\delta^{13}C$ in PC transgenic rice was similar to that in untransformed rice, demonstrating that transgenic rice is still a C_3 plant.

How can we redesign C_4 rice from the limited C_4 features of photosynthesis reached currently? In future work, introduction of the PEPC gene from a CAM plant into C_4-enzyme transgenic rice could carry out higher photosynthesis day and night. Simultaneously, the enhancement of endogenous ATP in PK transgenic rice through genetic engineering would increase its operation of the C_4 cycle. Most importantly, recent advanced techniques such as laser capture microdissection enable us to study the mechanisms of cellular differentiation, for example, of bundle sheath cells. From the above suggestions, these techniques might shed light on a new green revolution in rice breeding.

Keywords: transgenic rice, C_4 enzymes, C_4 photosynthetic cycle, photoinhibition, photooxidation, biotechnology, rice breeding

Because of the application of the semidwarf gene in the 1960s and heterosis in the 1970s in China, Chinese rice yield leaped twice, rising 20%, respectively, from the previous level. Nowadays, average efficiency of light use in high-yielding varieties is about 1.5%, while theoretically it should reach 3–5%. Thus, photosynthetic production has prodigious potential to be increased. It is obvious that, in the perspective of photosynthesis, yield consists of two components, "source" and "sink." At present, Chinese superhybrid rice achieves high yield mainly because of the increase in sink, for example, by adjusting plant architecture to obtain a maximum number of grains. However, in the major hybrid rice combinations used so far, the panicles are big, but the empty-seed rate is high as well. To further increase yield, the emphasis should logically be shifted to an increase in "source."

In previous years, Ku et al (1999) introduced key enzymes of the maize C_4 pathway to rice and achieved a significant increase in photosynthetic capacity. We developed a new approach to introduce genes for the C_4 enzymes phospho*enol*pyruvate carboxylase (PEPC) and pyruvate, orthophosphate dikinase (PPDK) into sterile and restorer lines, respectively, and enhanced photosynthetic efficiency up to 50% in the F_1 by crossing the two lines (Wang et al 2004a,b). Therefore, we believe that, to increase the source, we can integrate C_4 photosynthetic pathways into conventional C_3 rice on the current basis of more efficient plant architecture.

Photosynthetic characteristics of transgenic rice expressing C_4 photosynthesis enzymes

Expression of enzymes

Transgenic lines of rice were produced containing the genes for PEPC (line called PC), PPDK (called PK), PEPC and PPDK (called CK), and NADP-malic enzyme (NADP-ME; line called ME). The activities of PEPC, PPDK, NADP-ME, and malate dehydrogenase (MDH) were examined by direct assay (Fig. 1) and by Western immunoblot analysis using specific antibodies against maize PEPC and PPDK (Fig. 2). Both the activity and enzyme protein content of these C_4 enzymes were very low in untransformed rice (wild-type, WT). In contrast, the activities of PEPC in PC transgenic rice were 20–25 times higher than those in untransformed rice, reaching 255 μmol $m^{-2}\ s^{-1}$, with a concomitant increase in protein amount. The activity of Rubisco and its kinetic properties were not altered in the transgenic plants (Table 1). Somewhat unexpected is the stimulation of carbonic anhydrase (CA) activity in the PC transgenic rice, which increased by more than twofold, indicating a metabolic adjustment. This increase was observed in leaves exposed to both low and high light prior to the assay. The activities of NADP-ME in ME transgenic rice and the activities of PPDK in PK transgenic rice were about 5-fold higher than those found in untransformed rice plants. The high activities of PEPC in PC and CK transgenic plants and of PPDK in PK and CK transgenic plants are due to increased amounts of enzyme protein.

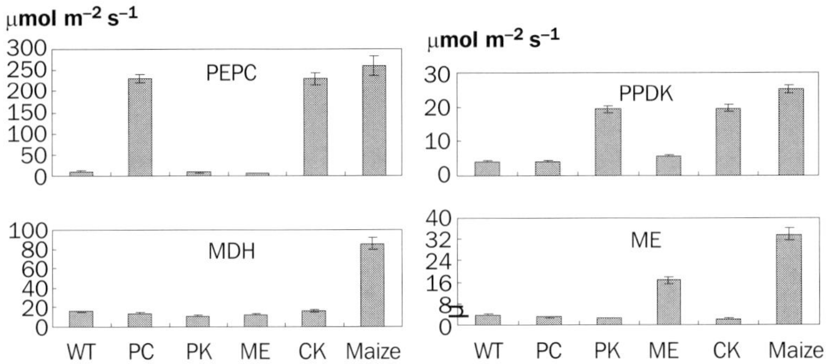

Fig. 1. The activities of four C_4 photosynthetic enzymes, phospho*enol*pyruvate carboxylase (PEPC), pyruvate, orthophosphate dikinase (PPDK), malate dehydrogenase (MDH), and NADP-malic enzyme (ME), in various lines of rice and in maize. The untransformed rice (WT) is Kitaake; PC is transformed with the gene for PEPC, PK with the gene for PPDK, ME with the gene for ME, and CK with the genes for PEPC and PPDK.

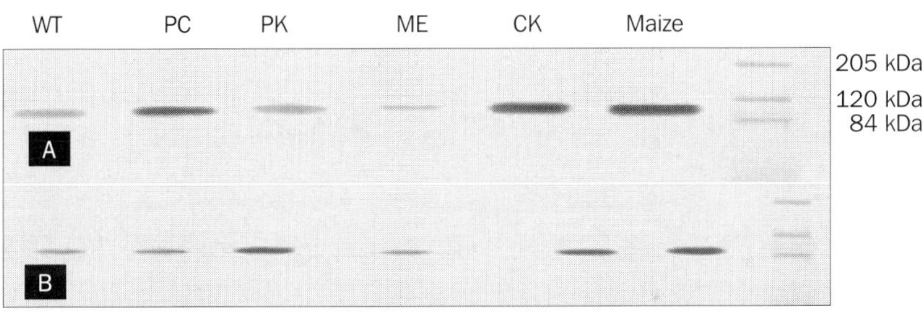

Fig. 2. Western immunoblots of phospho*enol*pyruvate carboxylase (PEPC) in (A) and pyruvate, orthophosphate dikinase (PPDK) in (B), from the leaf protein of untransformed rice (WT), rice transformed with genes for C_4 enzymes (see legend to Fig. 1), and maize.

Photosynthetic characteristics

The photosynthetic characteristics of PC transgenic plants were analyzed in detail. The plants exhibited higher light-saturated photosynthetic rates (55%), higher stomatal conductance (29%) at 1,200 μmol m^{-2} s^{-1}, and higher carboxylation efficiency (50%) than untransformed WT (Fig. 3, Table 2). The PC transgenic plants also had higher (20%) photosynthetic rates at optimal temperature (35 °C). On the other hand, the photosynthetic CO_2 compensation points were slightly lower in the PC transgenic plants, indicating a stronger capability of the plants to assimilate carbon under limited CO_2 conditions. Taken together, these results suggest that the high PEPC coupled with enhanced CA gives the transgenic plants a higher capability to assimilate atmospheric CO_2. However, the exact mechanism for the superior photosynthetic performance of these plants remains to be determined.

Table 1. Changes in activities of phosphoe*nol*pyruvate carboxylase (PEPC), carbonic anhydrase (CA), Rubisco, and K_m for CO_2 in untransformed WT rice and PC transgenic rice under low and high light treatment for 2 hours. Values are mean +/– and SD from three replicates of an assay.

Enzyme	Untransformed rice (WT)		PC transgenic rice	
	Low light[a]	High light[b]	Low light	High light
PEPC (μmol m^{-2} s^{-1})	5.26 ± 0.72	12.00 ± 1.01	90.00 ± 4.70	254.54 ± 11.94
CA (mol mg^{-1} protein min^{-1})	72.95 ± 7.42	203.96 ± 10.75	87.47 ± 8.50	576.92 ± 27.56
Rubisco (μmol m^{-2} s^{-1})	29.95 ± 1.93	59.61 ± 4.18	41.86 ± 1.69	66.73 ± 3.89
Rubisco K_m (CO_2) (μM)	12.0 ± 0.41	11.9 ± 0.31	11.69 ± 0.35	11.53 ± 0.27

[a]Low light, 30 μmol m^{-2} s^{-1}. [b]High light, 1,400 μmol m^{-2} s^{-1}.

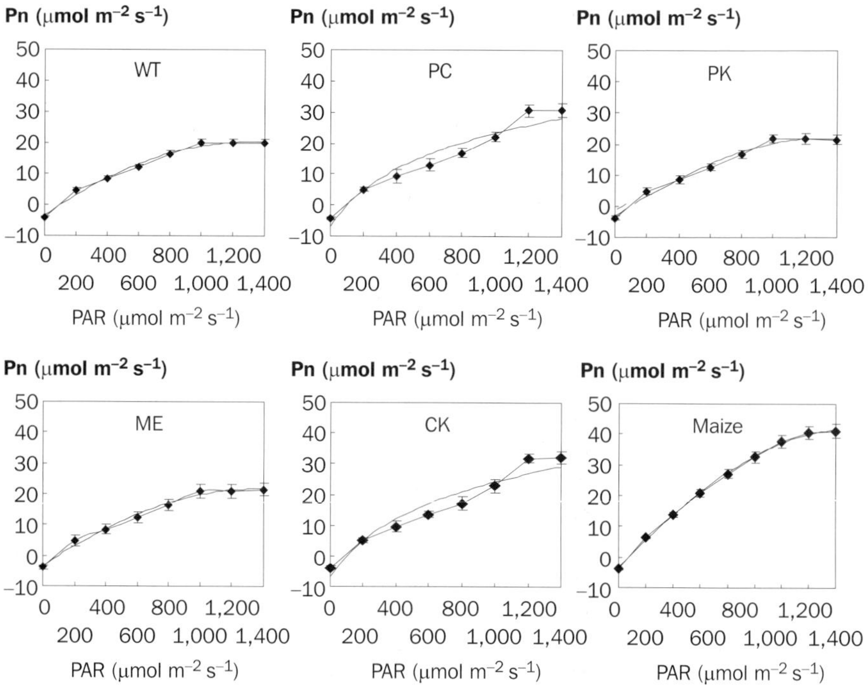

Fig. 3. Photosynthesis–PAR curves measured in the flag leaves of untransformed rice (WT), rice transformed with genes for C_4 enzymes (see legend to Fig. 1), and maize.

Table 2. Activity of phospho*enol*pyruvate carboxylase (PEPC) and some physiological measurements of photosynthesis of untransformed (WT) and PC transgenic rice plants. Values are means and SD from 10–12 measurements.

Genotype	Activity of PC (μmcl m^{-2} s^{-1})	Light-saturated photosynthetic rate (μmol m^{-2} s^{-1})	Photosynthetic rate at optimal temperature (μmol m^{-2} s^{-1})	Carboxylation efficiency (mol mol^{-1})	CO_2 compensation point (mol mol^{-1})	Stomatal conductance (mmol m^{-2} s^{-1})
WT	12.2 ± 1.0	20.34 ± 0.98	23.98 ± 1.07	0.077 ± 0.003	66.3 ± 2.1	282 ± 15
PC	258.4 ± 21.3	31.30 ± 1.38	28.72 ± 1.59	0.115 ± 0.005	52.2 ± 1.3	364 ± 21

Table 3. Grain yield and its components in transgenic rice expressing rice ME and various maize C_4 photosynthesis enzymes. Values are means and SD from 8–12 replicates in pots (5 hills per pot). WT = untransformed rice, ME = ME transgenic rice, PK = PK transgenic rice, PC = PC transgenic rice, CK = transgenic rice simultaneously expressing maize PEPC and PPDK.

Item	WT	ME	PK	PC	CK
Panicles per pot	38.5 ± 1.9	39.8 ± 1.7	41.5 ± 2.1	44.5 ± 1.8	45.1 ± 2.0
Spikelets per panicle	37.6 ± 2.1	37.6 ± 2.6	37.7 ± 1.7	37.8 ± 1.9	38.0 ± 2.2
Filled grains per panicle	28.5 ± 2.0	28.8 ± 2.3	29.0 ± 2.7	29.9 ± 1.9	30.2 ± 2.1
Seed set (%)	75.8	76.6	76.9	79.1	79.5
Seed size (mg $seed^{-1}$)	22.7 ± 0.1	22.7 ± 0.2	22.7 ± 0.1	22.7 ± 0.1	22.7 ± 0.1
Grain yield (g pot^{-1})	24.6 ± 2.1	25.5 ± 2.6	26.2 ± 2.4	30.0 ± 1.8	30.6 ± 2.0

Growth and grain yield

Among the five genotypes tested, PC and CK transgenic rice plants, which have high PEPC activities (Fig. 1), showed significant increases in grain yield on a per pot basis, being 22% and 24%, respectively (Table 3). The increases in grain yield are mainly associated with increased panicle number per plant. No significant changes in seed weight or number of spikelets per panicle were noticed. However, transgenic rice tends to have slightly higher fertility. A more stable photosynthetic capacity under varying climate conditions in the field, such as light intensity, may contribute to the higher productivity in these two transgenic lines.

Photoprotective effects of high-level expression of C_4 phosphoe*nol*pyruvate carboxylase in transgenic rice

Photoinhibition characteristics in PC transgenic rice

As shown in Figure 4A, photosystem II (PS II) photochemical efficiency (F_v/F_m) decreased differently in the two genotypes under high light (1,500 µmol m^{-2} s^{-1}) photoinhibitory conditions. However, photochemical efficiency decreased less in PC transgenic rice, indicating that PC transgenic rice was more tolerant of photoinhibition. This was further shown in the ratio of F_{685}/F_{735} (Fig. 4C), which represents state transfer from PS II to PS I by changes in ratio of fluorescence at 685 nm and 735 nm, respectively. The F_{685}/F_{735} ratio in PC transgenic rice dropped less under strong light, indicating that its PS II was inhibited less. As shown in Figure 4B, photoinhibition by strong light was completely reversed by dark treatment, indicating that this is a dynamic and reversible change. We hypothesized that, in this case, light harvesting complex II (LHC II) might be phosphorylated and separated from PS II, allowing PS II to avoid photodamage from excessive light energy. The disintegrated LHC II could then return to PS II in the dark. To test this conjecture, we treated the experimental system with NaF, a specific inhibitor of phosphorylation. As shown in Figure 4B, NaF treatment blocked both transgenic and untransformed rice from completely recovering

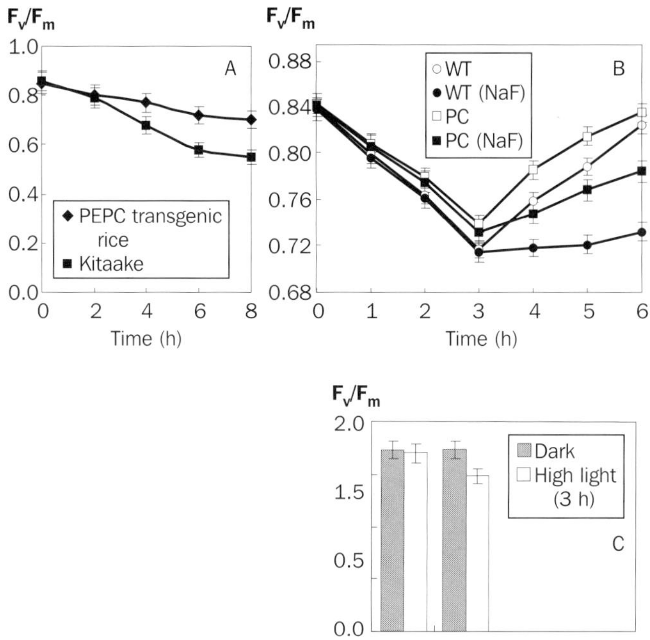

Fig. 4. Time courses of F_v/F_m in (A), F_v/F_m during the light and subsequent dark (with or without NaF treatment) in (B), and the change in F_{685}/F_{735} from the dark to the value after 3 hours of light in (C). The conditions were 1,500 μmol m^{-2} s^{-1} PAR, 26 °C.

PS II, whereas recovery in untransformed rice was less efficient than in PC transgenic rice, indicating that, in untransformed rice, more phosphorylated LHC II moved to PS I to protect PS II from photodamage. For PC transgenic rice, PS II damage was less severe, and LHC II of PS II was phosphorylated less; thus, NaF inhibited less and the change in F_{685}/F_{735} was negligible.

Photooxidative characteristics in PC transgenic rice

As shown in Figure 5A and B, the O_2^- generation rate and membrane peroxidation (malondialdehyde content) in PC transgenic rice were lower than in untransformed rice plants under high light and low CO_2 conditions. These results indicate that photooxidative tolerance was higher in PC transgenic rice, where the introduced maize PEPC gene might enhance the oxygen scavenger system. In both genotypes under photooxidative treatment, the superoxide dismutase (SOD) activity peaked at 4 hours and then dropped gradually, whereas peroxidase (POD) activity increased gradually with treatment time (Fig. 5C, D). These observations suggest that SOD reacted initially to actively scavenge oxygen when O_2^- formed H_2O_2; thereafter, POD acted as the scavenger. The activities of SOD and POD in leaves of PC transgenic rice were higher than in untransformed rice. Similar results were obtained when PC transgenic rice and untransformed rice plants were subjected to other photooxidation treatments,

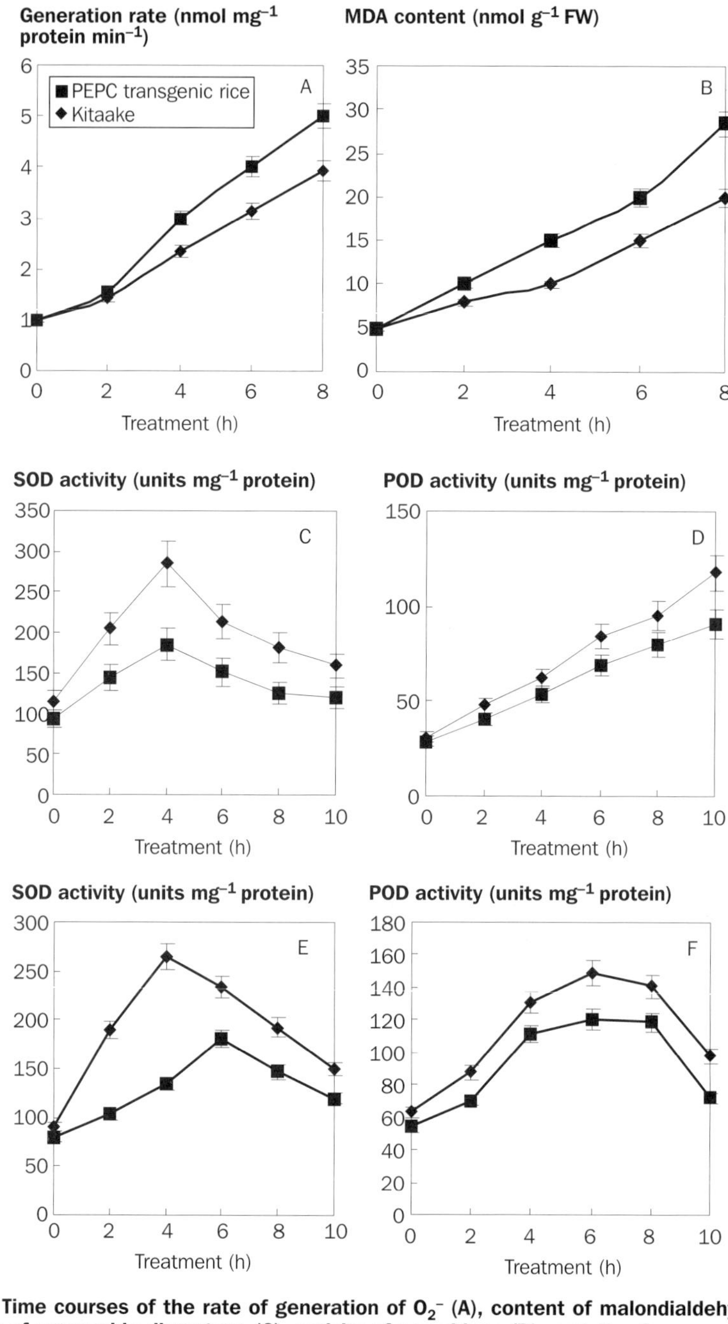

Fig. 5. Time courses of the rate of generation of O_2^- (A), content of malondialdehyde (B), activity of superoxide dismutase (C), activity of peroxidase (D), activity of superoxide dismutase with treatment with methyl viologen (E), and activity of peroxidase with treatment with methyl viologen (F) for Kitaake and PC transgenic rice; MV is methyl viologen.

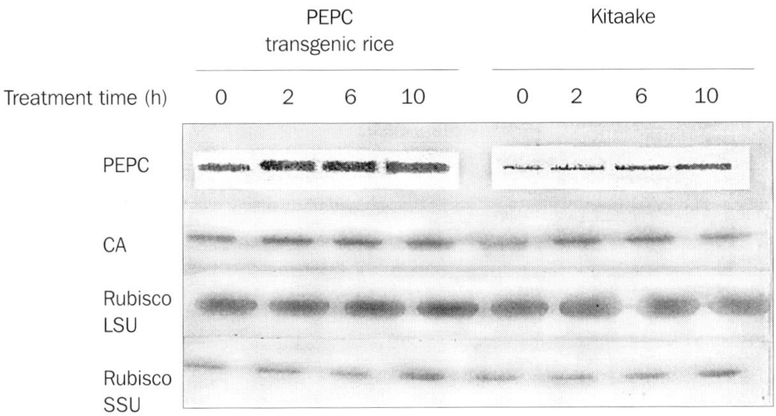

Fig. 6. The occurrence of the proteins phospho*enol*pyruvate carboxylase (PEPC), carbonic anhydrase (CA), Rubisco large subunit, and Rubisco small subunit (Western blot analysis) in PC transgenic rice (left) and untransformed rice (Kitaake, right). Grown in 600 μmol m^{-2} s^{-1} PAR, experimental treatment in 1,400 μmol m^{-2} s^{-1} PAR for 0, 2, 6, or 10 hours.

such as methyl viologen (MV). Taken together, our results show that introduction of the maize PEPC gene into rice increases the activity of not only the gene encoding the photosynthetic enzyme, CA, but also the photooxidative enzymes for SOD and POD.

Physiological characteristics of the primitive CO_2-concentrating mechanism in PC transgenic rice

The expression of main photosynthetic enzymes in PC transgenic rice and untransformed rice

Figure 6 shows that the content of PEPC and CA protein was low in untransformed rice (control) under moderate light (0 h) and increased somewhat under high light for 2–10 h. After the maize gene was introduced into rice, the PC transgenic rice exhibited a higher content of PEPC under moderate light, which was induced to increase further under high light. The CA protein showed a similar trend. The protein expression in both genotypes was similar to the above reported changes in activity.

In addition, Rubisco LSU (a key enzyme in the C_3 pathway) was different, and Rubisco SSU did not change appreciably. These results demonstrated that the enzymes related to CO_2 concentrating in PC transgenic rice were activated to high-level expression under high light.

The relationship between stomatal conductance and photosynthetic rate in PC transgenic rice and untransformed rice

Figure 7 shows that stomatal conductance and photosynthetic rates in both transgenic and control rice were increased as light (PAR, photosynthetically active radiation)

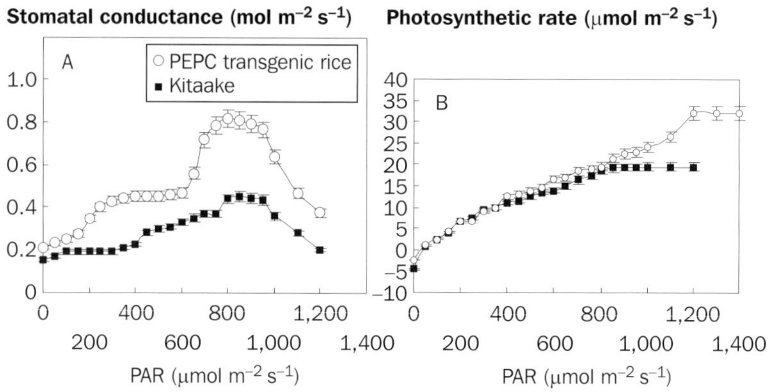

Fig. 7. The change of stomatal conductance (A) and rate of photosynthesis (B) with PAR in untransformed rice (Kitaake) and PC transgenic rice.

increased. Therefore, stomatal conductance correlated with photosynthetic rate in untransformed rice under different light with a correlation coefficient (r) equal to 0.800**. The coefficient in PC transgenic rice was 0.606**, which indicates that a significant positive correlation exists between stomatal conductance and photosynthetic rate. To elucidate whether this relationship was parallel or causal, the correlation coefficient between the increase in stomatal conductance and the increase in photosynthetic rate was tested statistically, and found to have a coefficient of 0.06, which indicates no correlation. Although PC transgenic rice showed 80% more stomatal conductance (Fig. 7A) under PAR of 700–1,000 μmol m^{-2} s^{-1} as compared with untransformed rice, the photosynthetic rate did not increase correspondingly (Fig. 7B). When PAR was higher than 1,000 μmol m^{-2} s^{-1}, stomatal conductance decreased in both genotypes but the photosynthetic rate in PC transgenic rice increased by 50% under PAR of 1,200–1,400 μmol m^{-2} s^{-1}. These results demonstrated that the increment in photosynthetic capacity in PC transgenic rice under high light might not be due to the increment in stomatal conductance supplying more CO_2, but to enhancement of the C_4 metabolism, thus using CO_2 more effectively.

The CO_2 exchange characteristic in PC transgenic rice and untransformed rice

Rice leaves were treated under high light and at different CO_2 concentrations (Fig. 8). Under atmospheric CO_2 (350 μmol mol^{-1}), the carboxylation efficiency of untransformed rice was 0.077, while that of PC transgenic rice was 0.115 (an increase of 50%). This increased carboxylation efficiency may be related to the expression of PEPC and CA, which are key enzymes for concentrating CO_2 (Fig. 6). Figure 8B shows the performance of CO_2 exchange under CO_2-free or low-CO_2 conditions. CO_2 release in untransformed rice Kitaake (control) was 62 μmol mol^{-1}, while that in PC transgenic rice was 50 μmol mol^{-1}. The equilibrium values for C_i were 62 μmol mol^{-1} for the untransformed rice (Kitaake) and 50 μmol mol^{-1} for PC transgenic rice. The

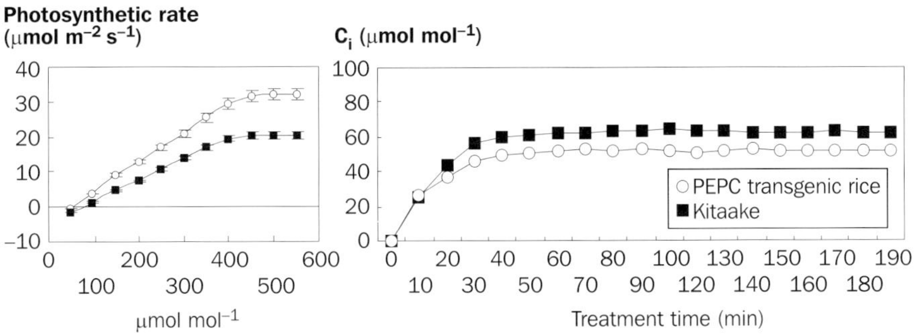

Fig. 8. The photosynthesis–C_i (intercellular concentration of carbon dioxide) curves (A) and the time courses of C_i (B) in untransformed rice (Kitaake) and PC transgenic rice. In (A), PAR was 1,200 μmol m^{-2} s^{-1}; in (B), 1,000 μmol m^{-2} s^{-1}.

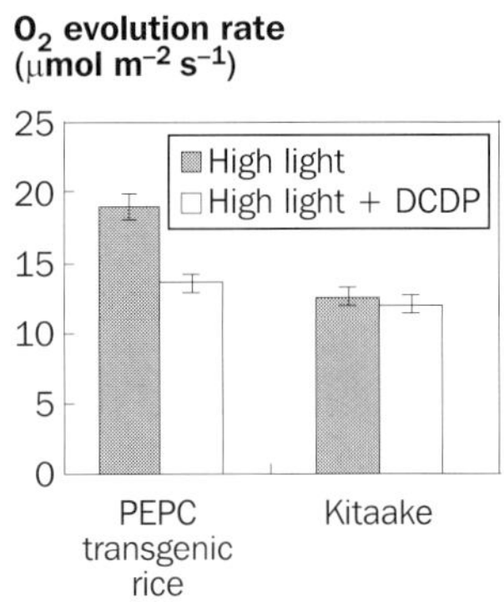

Fig. 9. The effect of DCDP on oxygen evolution rate in untransformed rice (Kitaake) and PC transgenic rice, grown under 1,400 μmol m^{-2} s^{-1} PAR, measured under 1,000 μmol m^{-2} s^{-1} PAR.

results indicate that the introduction of the C_4 photosynthesis enzyme PEPC allows more fixation of CO_2 released in leaves under high light, causing the CO_2 compensation point to decrease by 20%.

To examine whether the enhancement of photosynthesis was related to the introduced PEPC gene, transgenic rice leaves were treated with DCDP, a specific inhibitor of PEPC. The results in Figure 9 show that the photosynthetic rate in untransformed rice did not vary while that in PC transgenic rice decreased until it was close to that of the untransformed rice. This indicated that the increase in photosynthetic capacity in PC transgenic rice was due to the action of the maize PEPC gene. Figure 10 shows the results of the ^{14}C pulse-chase experiment. The proportions of the C_4 photosynthetic primary products and the C_3 photosynthetic primary products

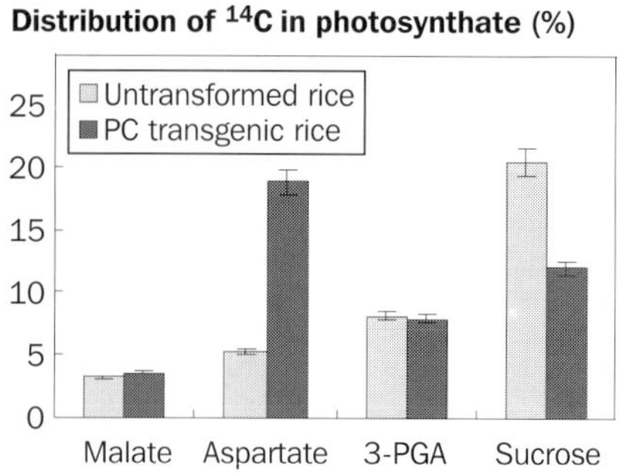

Fig. 10. The distribution of ^{14}C in the products of photosynthesis after labeling for 20 seconds in untransformed rice (Kitaake) and PC transgenic rice.

(3-PGA) in transgenic rice were close to those in untransformed rice. However, in PC transgenic rice, more label was distributed in aspartate, indicating that, although the C_4 pathway of NADP-ME type in maize might not be integrated into transgenic rice, the metabolic capacity of some C_4 photosynthesis is enhanced by the introduction of the maize PEPC gene.

Effects of DCDP, a specific inhibitor of PEPC, on PS II photochemical efficiency in PC transgenic rice

Figure 11A shows that the transgenic and untransformed rice plants exhibited photoinhibition and reduced PS II photochemical efficiency (F_v/F_m) under high light intensity similar to that of noon in summer. However, PC transgenic rice showed less photoinhibition under these conditions. After the treatment with DCDP, the decrease in F_v/F_m in PC transgenic rice was close to that observed in untransformed rice, indicating that PEPC overexpression accelerates CO_2 assimilation and maintains stable efficiency of the energy of the conversion in PS II.

Characteristics of CO_2 exchange and chlorophyll fluorescence for rice pollen lines transferred with the PEPC gene

In China, stable PC transgenic rice was used as a male parent to cross with restorer lines in Sichuan Province or sterile lines in Anhui Province, and a PC transgenic hybrid rice combination has been selected. In Jiangsu Province, the stable PC transgenic rice was used as a male parent to cross with common japonica rice cultivar 9516, and the F_1 hybrids exhibiting high PEPC activity underwent anther culture. Then, through the identification and selection of generations, new rice lines, namely, JAAS45 pollen lines expressing the PEPC gene stably, have been obtained. Furthermore, physiological indices were measured for leaves of JAAS45 pollen lines and their parents.

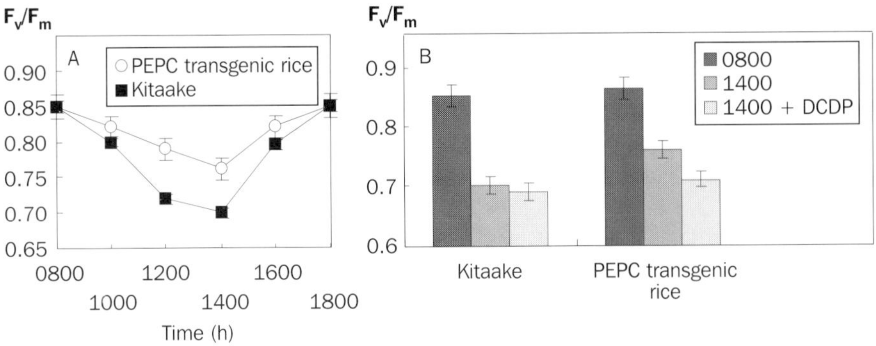

Fig. 11. The diurnal change in F_v/F_m (A) and the effect of DCDP on F_v/F_m (B) in untransformed rice (Kitaake) and PC transgenic rice.

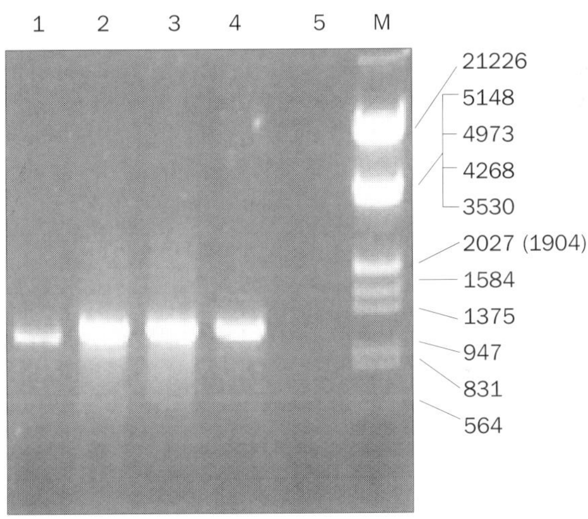

Fig. 12. The results of PCR amplification. Lane 1 is maize, lane 2 is PC transgenic rice, lanes 3 and 4 are JAAS45, lane 5 is 9516, and lane M is markers.

Expression of JAAS45, 9516, and PC transgenic rice

With the maize genome as a positive control, all materials were amplified by PCR according to the prime characteristic of the maize-specific C_4-type PEPC genome. As shown in Figure 12, 1,190 base pairs were amplified in the DNA of three materials, including maize, PC, and JAAS45 pollen lines having the PEPC gene, while they were not in 9516 (lane 5). The results indicated that the application of the prime PEPC gene characteristic could precisely select transgenic rice expressing the maize C_4-type specific PEPC gene. Simultaneously, it had been shown that the maize C_4-type specific PEPC gene was introduced into the JAAS45 pollen line, which expressed high amounts of maize PEPC.

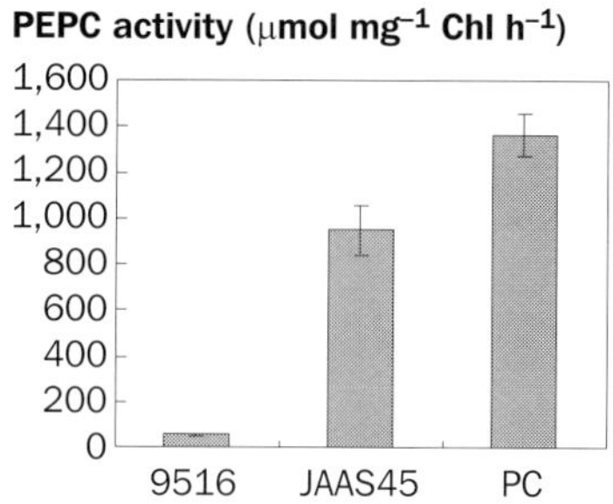

Fig. 13. Activity of phosphoe*nol*pyruvate carboxylase (PEPC) in rice 9516, JAAS45, and PC transgenic rice.

The maize PEPC gene was introduced into a C_3 crop, rice, to increase PEPC activity. Therefore, PEPC activity was further measured in JAAS45 and its parents (Fig. 13). The PEPC activity in PC transgenic rice as the male parent (1,364.8 ± 95.1 μmol mg^{-1} chlorophyll h^{-1}) exceeded by 24.9 times that in 9516 as the female parent (54.8 ± 7.0 μmol mg^{-1} chlorophyll h^{-1}), whereas that in JAAS45 (947.33 ± 106.42 μmol mg^{-1} $chlorophyll^{-1}$ h^{-1}) was significantly higher than that of their female parent 9516. These results revealed that the PEPC gene of the male parent could be inherited stably into the JAAS45 pollen line.

The light curves of photosynthesis of leaves in JAAS45, 9516, and PC

The photosynthetic rates were measured in the fourth generation of JAAS45, 9516, and PC. As shown in Figure 14, under 0–200 μmol m^{-2} s^{-1} PAR, the photosynthetic rate in JAAS45 and PC was slightly higher than in 9516. However, with the increase in PAR, the photosynthetic rate in JAAS45 and in PC was clearly higher than that in 9516. The light-saturated photosynthetic rate was 28.56 ± 1.01 μmol m^{-2} s^{-1} for JAAS45 and 30.84 ± 0.74 μmol m^{-2} s^{-1} for PC when PAR was about 1,200 μmol m^{-2} s^{-1}. For 9516, the light-saturated photosynthetic rate was 20.48 ± 0.81 μmol m^{-2} s^{-1} at about 1,000 μmol m^{-2} s^{-1}. The apparent quantum yield in JAAS45 was calculated to be 0.0577, slightly higher than that in its female parent 9516 of 0.0536, which was close to that of its male parent PC (0.0591). These results showed that high light could induce overexpression of the C_4 photosynthetic PEPC enzyme in JAAS45 to intensify some C_4 photosynthetic metabolic activity or accelerate the operation of a limited C_4 pathway to distinctively modify the photosynthetic characteristics of JAAS45.

Effect of OAA, MA, or PEPC on CO_2 exchange in rice leaves

To investigate whether there is a C_4 photosynthetic microcycle in JAAS45, 9516, and PC, the detached leaves from rice plants were immediately immersed into oxaloacetate (OAA, 200 μmol L^{-1}), malate (MA, 200 μmol L^{-1}), or phospho*enol*pyruvate (PEP, 100 μmol L^{-1}) solution, and distilled water as a control for 30 min under saturating PAR of 1,200 μmol m^{-2} s^{-1} for light induction. Then, net photosynthetic rates of rice

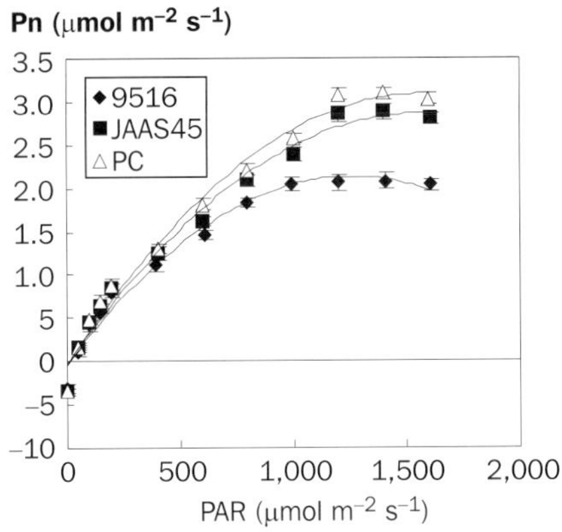

Fig. 14. Photosynthesis–PAR curves measured in the leaves of rice 9516, JAAS45, and PC transgenic rice. Means with error bars for SD are shown, n = 5.

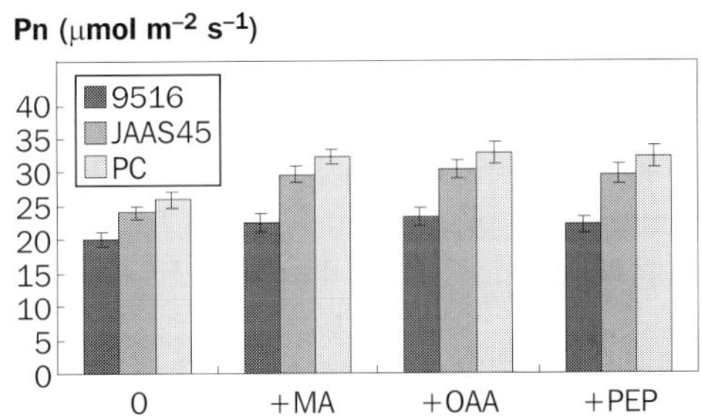

Fig. 15. The effect on net rate of photosynthesis in the leaves of rice 9516, JAAS45, and PC transgenic rice of incubation with distilled water (control, labeled 0), malate (MA, 200 μmol L^{-1}), oxaloacetate (OAA, 200 μmol L^{-1}), or phospho*enol*pyruvate (PEP, 100 μmol L^{-1}) solution. Incubation was for 30 minutes under 1,200 μmol m^{-2} s^{-1} PAR. Means with error bars for SD are shown, n = 5.

leaves were measured with a portable photosynthetic gas analysis system. Figure 15 showed that OAA, MA, or PEP could promote photosynthesis of JAAS45, 9516, and PC. Net photosynthetic rates of the leaves treated with OAA, MA, or PEP increased by 17%, 12%, and 11% in 9516, by 26%, 23%, and 23% in JAAS45, and by 26%, 25%, and 24% in PC, respectively, as compared with the respective control leaves. Thus, it appears that OAA, MA, and PEP are important for raising the net photosynthetic rates of JAAS45, 9516, and PC.

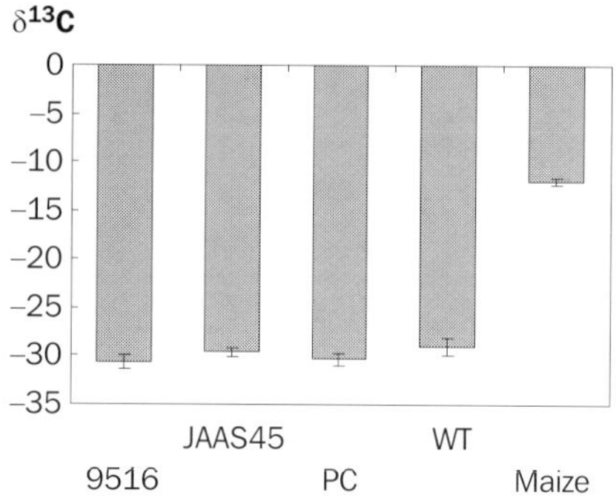

Fig. 16. Values of $\delta^{13}C$ for rice 9516, JAAS45, PC transgenic rice, untransformed rice (Kitaake, WT), and maize.

Transgenic rice lines were still a C_3 plant

The value of $\delta^{13}C$ can reflect the capability of fixing CO_2 by PEPC in different plants. To examine whether expression of the PEPC gene increased C_4 photosynthesis and photosynthetic capacity in rice, we measured the value of $\delta^{13}C$ in JAAS45, 9516, PC, and maize. As shown in Figure 16, the $\delta^{13}C$ value in JAAS45 was similar to that in 9516, PC transgenic rice, and untransformed rice, which was –29.62‰, –30.64‰, –30.33‰, and –29.05‰, respectively, whereas it was –12‰ in maize. Clearly, the rice plants were carrying out C_3 photosynthesis.

Future work

Integration of high efficiencies of photosynthetic productivity and plant architecture

Since scientists have successfully approached the goal of increasing "source" by crossing the genetically engineered PEPC enzyme contained in sterile and restorer lines, we can apply such a strategy to superhybrid rice with high efficiency and good architecture. This would introduce C_4 enzymes into parental lines of super-hybrid rice, and integrate the two improved traits together.

Further modification of photosynthetic productivity

In our previous work, we found that the photosynthetic rate of PK transgenic rice is limited by the amount of light. But this limitation can be overcome by applying extra ATP. Therefore, we believe that if we can increase the production of ATP through genetic engineering, photosynthetic productivity should be further increased.

In addition, to further increase photosynthetic productivity, we can also try to re-fix the CO_2 released by respiration by introducing PEPC of CAM plants, with a dark-activated enzyme, into available C_4-enzyme transgenic rice. In this way, the transgenic plants could have higher photosynthesis during day and night.

Genetic modification of leaf anatomy

Almost all C_4 plants found in nature have a specific Kranz anatomy adapted for their metabolic characteristics. So, it is reasonable to hypothesize that genetic modification of leaf anatomy could also be a useful approach to increase photosynthetic productivity. It was found that stems of tobacco, a C_3 plant, as well as the veins of celery stalks had photosynthetic cells with C_4 characteristics, just like the bundle sheath cells in maize leaves. These cells could be involved in this aspect of genetic research on constructing C_4 rice. It is worth noting that the true C_4 structure of leaves is induced at the five-leaf stage in the leaves of maize.

Recent advanced techniques such as laser capture microdissection enable us to study the regulatory mechanism of cellular differentiation related not only to leaf anatomy but also to metabolic pathways. We believe that such study will lead us to finally build up an anatomic base for highly efficient photosynthetic productivity of transgenic rice with C_4 pathways.

Bibliography

Chi W, Jiao DM, Huang XQ, Li X, Kuang TY, Ku MSB. 2001. Photosynthetic characteristics of transgenic rice plants overexpressing maize phospho*enol*pyruvate carboxylase. Acta Bot. Sin. 43:650-660.

Huang XQ, Jiao DM, Chi W, Ku MSB. 2002. Characteristics of CO_2 exchange and chlorophyll fluorescence of transgenic rice with C_4 genes. Acta Bot. Sin. 44:405-412.

Huang XQ, Jiao DM. 2001. The characteristics of resistance to photooxidation of transgenic rice (*Oryza sativa* L.) plants with maize genes coding for C_4 photosynthesis enzyme. Acta. Phytophysiol. Sin. 27:393-400.

Ji BH, Tan HH, Zhou R, Jiao DM, Shen YG. 2005. Promotive effect of low concentrations of $NaHSO_3$ on photophosphorylation and photosynthesis in phospho*enol*pyruvate carboxylase transgenic rice leaves. Acta Bot. Sin. 47:178-186.

Ji BH, Zhu SQ, Jiao DM. 2004. A limited photosynthetic C_4-microcycle and its physiological function in transgenic rice plant expressing the maize *PEPC* gene. Acta Bot. Sin. 46:542-551.

Jiao DM, Huang XQ, Li X, Chi W, Kuang TY, Zhang QD, Ku MSB. 2002. Photosynthetic characteristics and tolerance to photooxidation of transgenic rice expressing C_4 photosynthesis enzymes. Photosynth. Res. 72:85-93.

Jiao DM, Kuang TY, Li X, Ge QY, Huang XQ, Hao NB, Bai KZ. 2003. Physiological characteristics of the primitive CO_2 concentrating mechanism in PEPC transgenic rice. Sci. China Ser. C 33:33-39.

Jiao DM, Li X, Huang XQ, Chi W, Kuang TY, Ku MSB. 2001. The characteristics of CO_2 assimilation of photosynthesis and chlorophyll fluorescence in transgenic PEPC rice. Chin. Sci. Bull. 46:414-418.

Jiao DM, Li X, Ji BH. 2005. Photoprotective effects of high level expression of C_4 phospho*enol*pyruvate carboxylase in transgenic rice during photoinhibition. Photosynthetica 43:501-508.

Ku MSB, Agarie S, Nomura M, Fukayama H, Tsuchida H, Ono K, Hirose S, Toki S, Miyao M, Matsuoka M. 1999. High-level expression of maize phospho*enol*pyruvate carboxylase in transgenic rice plants. Nature Biotechnol. 17:76-80.

Wang DZ, Chi W, Wang SH, Jiao DM, Wu S, Li X, Li CQ, Wang XF, Luo YC. 2004a. Characteristics of transgenic rice overexpressing maize photosynthetic enzymes for breeding two-line hybrid rice. Acta Agron. Sin. 30:248-252.

Wang DZ, Jiao DM, Wu S, Li X, Li L, Chi W, Wang SH, Li CQ, Luo YC, Wang XF. 2002. Breeding for parents of hybrid rice with maize *pepc* gene. Agric. Sci. Chin. 35:1165-1170.

Wang DZ, Wang SH, Wu S, Li CQ, Jiao DM, Luo YC, Wang XF, Du SY. 2004b. Inheritance and expression of the maize *pepc* gene in progenies of transgenic rice bred by crossing. Acta Gen. Sin. 31:195-201.

Zhang Q, Jiao DM, Ling LL, Zhang YH, Huang XQ. 2005. Study of the protective effects in PEPC transgenic rice. Agric. Sci. Chin. 4(2):94-100.

Notes

Author's address: 84-40148, Zhongling Street, Nanjing, 210014, China; e-mail: jiaodm_123@yahoo.com.cn.

Overexpression of C_4 pathway genes in the C_3 dicots potato, tobacco, and *Arabidopsis*: experiences and future challenges

C. Peterhänsel, H.-J. Hirsch, and F. Kreuzaler

C_3 plants lose a significant part of previously fixed CO_2 in the process of photorespiration. Reduction in photorespiration is expected to increase the productivity of crop plants and reduce the requirements for irrigation and fertilization. For more than ten years, research at our institute has focused on the genetic engineering of dicotyledonous crop plants toward improved CO_2 fixation. In this paper, we summarize results form our work vis-à-vis reports from other laboratories and define future challenges. Furthermore, we introduce an alternative approach based on the installation of a bypass of photorespiration in the chloroplast.

Keywords: carbonic anhydrase, NADP-malic enzyme, phospho*enol*pyruvate carboxylase, phospho*enol*pyruvate/Pi translocator, photorespiration

Most crop plants are classified as C_3 plants because the first product of CO_2 fixation by ribulose 1,5-bisphosphate carboxylase/oxygenase (Rubisco) is the 3-carbon compound 3-phosphoglycerate (3-PGA). Molecular oxygen competes with CO_2 for the binding of CO_2 to the active site of Rubisco. Under standard conditions, roughly 20–30% of all catalytic events are oxygenase reactions (Ogren 1984). The products of the oxygenase reaction are one molecule of 3-PGA and one molecule of 2-phosphoglycolate (2-PG). The latter is toxic to the cell and is recycled to 3-PGA in the complex process of photorespiration (Tolbert 1997, cf. Fig. 3). In the course of these reactions, CO_2 is lost from reduced carbon compounds. Moreover, NH_3 is released in the same reaction and has to be refixed in energy-consuming processes. The relative oxygenase activity of Rubisco increases with temperature (Brooks and Farquhar 1985). Furthermore, plants close their stomata in hot and arid environments to reduce water evaporation from the intercellular space. CO_2 inside the leaf is rapidly used up by the carboxylase activity of Rubisco and additional O_2 is formed in the light reactions of photosynthesis. This progressively favors the oxygenase activity of Rubisco and ultimately results in growth arrest or even net CO_2 release from already fixed compounds. Conversely, it has been repeatedly shown that an artificial increase in

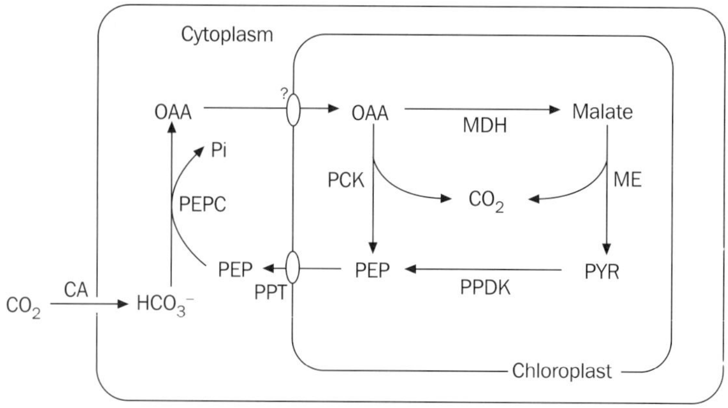

Fig. 1. Possible enzymatic activities required for the installation of C_4-like metabolism in C_3 plants. OAA = oxaloacetic acid, PEP = phospho*enol*pyruvate, PYR = pyruvate, CA = carbonic anhydrase, PEPC = PEP carboxylase, PCK = PEP carboxykinase, MDH = malate dehydrogenase, ME = malic enzyme, PPDK = PYR/Pi dikinase, PPT = PEP/Pi translocator.

atmospheric CO_2 concentration can improve crop growth and yield under otherwise optimal conditions (Kimball 1983, Arp et al 1998).

Adaptations to unfavorable growth conditions with low water supply and high temperature developed independently many times in evolution. C_4 plants separated primary and secondary CO_2 fixation in two different tissues (von Caemmerer and Furbank 2003). Primary fixation takes place in the mesophyll and is catalyzed by the oxygen-insensitive enzyme phospho*enol*pyruvate carboxylase (PEPC). The resulting C_4 acid diffuses to the bundle sheath, where CO_2 is released by different decarboxylases dependent on the species. The bundle sheath is separated from the air space by reduction of the intercellular space, thick cell walls, and often by additional reinforcements such as a suberin lamella (Nelson and Langdale 1992). Through these features, CO_2 cannot easily diffuse from this tissue but is efficiently concentrated. Rubisco is limited to the bundle sheath in C_4 plants and consequently the oxygenase activity of Rubisco and photorespiration are suppressed. The cycle is closed by regeneration of the primary acceptor molecule in the mesophyll.

There is evidence from submerged aquatic C_3 plants that a C_4-like cycle can also function within a single cell. Under low CO_2 concentrations, these plants induce the accumulation of C_4 enzymes in the cytosol and chloroplast, resulting in a significant reduction in the CO_2 compensation point (Reiskind et al 1997, Casati et al 2000, Rao et al 2002). This mechanism may serve as a blueprint for the installation of C_4-like photosynthesis in C_3 crops (Fig. 1), although evidence exists that functioning of the pathway might depend on low CO_2 diffusion coefficients under water (Leegood 2002).

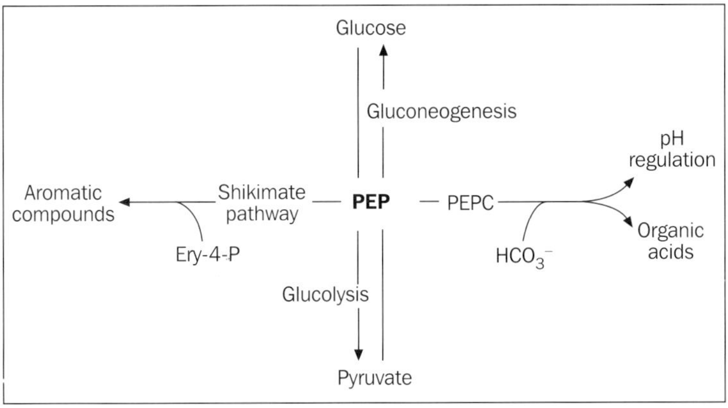

Fig. 2. The central position of phospho*enol*pyruvate in metabolism. PEP is an intermediate of glycolysis and gluconeogenesis and a precursor of the shikimate pathway in the chloroplast. PEP carboxylation participates in anaplerotic reaction pathways for the replenishment of citric acid cycle intermediates and in the fine regulation of cytosolic pH.

The carboxylation reaction

All but one approach to install a C_4-like CO_2 fixation pathway in C_3 plants started with overexpression of PEPC, the primary carboxylase of C_4 plants. An overview of the major results of our and other groups during the last two decades is given in Table 1. We started our work with a bacterial gene (*cppc*) under constitutive promoter control and transformed potato plants because of the ease of vegetative propagation (Gehlen et al 1996). The activity of PEPC in vitro was 4-fold higher in transgenic lines than in wild types. Antisense lines generated in parallel showed only half the wild-type activity. Phospho*enol*pyruvate (PEP) is an important intermediate in plant metabolism (Fig. 2). Besides its role in glycolysis and gluconeogenesis, PEP is a precursor of the shikimate pathway. The PEPC of C_3 plants participates in anaplerotic reaction pathways important for amino acid metabolism. Furthermore, in concert with malic enzyme (Davies 1986), it is involved in the fine regulation of cytoplasmic pH and compensates for the alkalinization of the cytoplasm during nitrate reduction (Manh et al 1993). In stomatal guard cells, PEPC counteracts the alkalinization of the cytosol via the plasma-membrane proton pump and is therefore important for stomatal movement (Asai et al 2000). A significant overexpression of PEPC was therefore expected to interfere with the basal metabolism of the plant. However, only a few very high expressing lines showed slight growth retardation. PEPC overexpression enhanced dark respiration of the plants, but, on the other hand, improved electron use for CO_2 fixation and the CO_2/O_2 ratio in the vicinity of Rubisco as deduced from the CO_2 compensation point independent of dark respiration in the light (Γ^*, Brooks and Farquhar 1985). We interpreted these data as an increased CO_2 release from the products of PEP carboxylation resulting in enhanced intracellular CO_2 concentration.

Table 1. Overexpression of PEPC genes in C_3 plants.

Plant	Source of gene	Promoter	Major effect	Reference
Tobacco	Maize C_4 gene or cDNA	Various	Higher malate contents with maize cDNA and chlorophyll a/b binding protein promoter	Hudspeth et al (1992)
Tobacco	Maize cDNA	35S	Chlorosis, retarded growth	Kogami et al (1994)
Potato	*Corynebacterium*	35S	Reduced CO_2 compensation point (Γ^*) Enhanced dark respiration Induction of cytosolic malic enzyme	Gehlen et al (1996), Häusler et al (1999), Häusler et al (2001)
Rice	Maize C_4-gene	Maize C_4 gene	>100-fold overexpression Reduced O_2 inhibition of photosynthesis Improper phosphorylation	Ku et al (1999), Fukayama et al (2003)
Potato	*Sorghum* C_4 cDNA	35S	No major effects	Beaujean et al (2001)
Potato	Potato (engineered for reduced product inhibition)	35S	Redirection of carbon flow from sugars to organic acids More primary fixation into malate Induction of cytosolic malic enzyme Reduction in phosphorylated intermediates Stunted growth	Rademacher et al (2002)
Arabidopsis	*Synechococcus* (little sensitivity to product inhibition)	35S	Less aromatic amino acids More Gln, Asn, Arg Chlorosis, retarded growth Can be rescued by supplementation with aromatic amino acids	Chen et al (2004)
Arabidopsis	Maize C_4 cDNA	35S	No major effects	Chen et al (2004)

For instance, the complete oxidation of the reaction product malate would produce four CO_2 molecules in the citric acid cycle for each HCO_3^- fixed. Interestingly, later analyses revealed that the endogenous cytosolic NADP malic enzyme was strongly induced on the transcriptional level in PEPC-overexpressing potato plants, supporting the idea of CO_2 release from the products of PEPC (Häusler et al 2001). Consequently, only slight changes were observed in the steady-state metabolite concentrations, with the exception of a clear increase in the sucrose content of transgenic lines (Häusler et al 1999).

We hypothesized that the *in vivo* activity of the transgenic bacterial PEPC might be limited by its affinity for PEP or its sensitivity to allosteric inhibition by malate. In general, the C_3 isoforms of PEPC display a high substrate affinity, but also a high sensitivity for product inhibition, whereas both properties are clearly lower for the photosynthetic isoforms of C_4 plants (Svensson et al 2003). The sensitivity of C_4-PEPC for product inhibition is further diminished by phosphorylation of an N-terminal serine residue in the light (Bakrim et al 1993, Ueno et al 1997). The overexpression of the complete maize C_4-PEPC gene in rice revealed the importance of this modification for *in vivo* activity. Although the maize gene was highly expressed in rice and in vitro activity was more than 100-fold higher than in the wild type, no major changes in carbon metabolism could be observed (Ku et al 1999, Fukayama et al 2003). This indicated a low *in vivo* activity and was attributed to inadequate posttranslational regulation (Fukayama et al 2003). A suitable enyzme for the installation of a C_4-like pathway in C_3 plants should combine high substrate affinity with low product inhibition and we managed to create such an enzyme by genetic engineering of the potato PEPC. The N-terminal phosphorylation site was modified in a way mimicking constitutive phosphorylation. Additionally, a central part was exchanged for the homologous region of PEPC from the C_4 plant *Flaveria trinervia* (Rademacher et al 2002). The resulting engineered PEPC enzyme redirected carbon flow in potato leaves from sugars to organic acids. The primary CO_2 fixation into malate increased compared to the wild type and plants with a more than 3-fold increase in PEPC expression were strongly impaired in growth although photosynthetic performance was not compromised. Similar growth retardation is also observed when the engineered potato PEPC is overexpressed in *Arabidopsis*, whereas tobacco lines overexpressing the gene to a similar extent show strong chlorosis (unpublished results). A recent publication on the overexpression of a cyanobacterial PEPC that naturally shows low sensitivity to product inhibition indicates that growth retardation can be complemented by an exogenous supply of aromatic amino acids, suggesting that a decreased partitioning of PEP for the shikimate pathway is the major cause of growth inhibition (Chen et al 2004). This is in line with a reduction in flavonoid contents in potato plants overexpressing PEPC (Häusler et al 2001). Alternatively, the growth effects might be attributed to a reduction in phosphorylated intermediates by the overexpression of PEPC (Rademacher et al 2002). Both effects could be potentially rescued by completion of a C_4-like pathway. Therefore, enyzmes with low malate sensitivity constitute an optimized starting point for establishment of the metabolic cycle.

The decarboxylation reaction

Less information is available about the installation of a decarboxylating activity in the chloroplasts of C_3 plants. Our research focused on the overexpression of NADP-malic enzyme (ME) in the chloroplasts of potato and tobacco (Lipka et al 1999, Häusler et al 2001). Earlier, only the bacterial PEPC (*cppc*) was available and therefore combined with ME. These enzymes should form a minimal CO_2 pump with HCO_3^- uptake in the cytosol and CO_2 release in the chloroplast (if the metabolite transport and the regeneration of the acceptor are not taken into consideration, cf. Fig. 1). Indeed, the transgenic lines overexpressing both enzymes showed photosynthetic characteristics indicative of improved CO_2 fixation. The electron requirement for CO_2 fixation was again reduced compared to the wild type and to a line overexpressing only NADP-ME. This effect appeared to depend on high light intensities, elevated temperatures (Lipka et al 1999), and atmospheric O_2 concentration (Häusler et al 2001) and might therefore be attributed to a reduction in photorespiration.

As desired, the simultaneous overexpression of ME and *cppc* also relieved some of the pleiotropic effects of single *cppc* expression. The enzymatic activity of the endogenous cytoplasmic NADP-ME gene was again reduced to background and the amount of flavonoids increased to wild-type levels. These data suggest that cytoplasmic carboxylase (PEPC) and plastidal decarboxylase (ME) cooperate to a certain extent and that a fully functional C_4-like cycle with well-balanced enzymatic acitivities could ultimately work decoupled from the residual metabolism (Häusler et al 2002).

The combination of both enzymatic activities in tobacco unexpectedly resulted in less clear effects (Häusler et al 2001). Although enzymatic activities were comparable in both species, neither the effects on endogenous enzymes nor the improvement of photosynthetic parameters could be reproduced. Our current attempts to construct identical *Arabidopsis* lines will indicate whether this result is an exception.

The ectopic expression of a PEP carboxykinase (PCK) from *Urochloa panicoides* in rice chloroplasts was sufficient to increase the fixation of radiolabeled CO_2 into organic acids. Suzuki et al (2000) suggest that the endogenous cytosolic PEPC activity made use of additional substrate produced by PCK in the chloroplast. These data provide independent evidence that cytosolic PEP consumption and plastidal PEP production can be coordinated in C_3 plants (see also below). We are currently attempting to combine the *Urochloa* PCK with our engineered potato PEPC enzyme in several species and thereby hope to install an optimized minimal C_4-like cycle. A recent report about rice plants overexpressing *Urochloa* PCK in combination with maize PEPC did not support this idea because PEPC added little to the effects already obtained with PCK alone (Suzuki et al 2006). However, this might again be attributed to low *in vivo* activity of maize PEPC in rice (Fukayama et al 2003).

More C_4 proteins and future challenges

Two enzymatic activities were combined in one potato line and this resulted in improved photosynthetic properties and relief of pleiotropic effects. However, a true C_4 cycle is composed of more proteins (Fig. 1). Most importantly, when using malic enzyme as the decarboxlyase, an enzyme for the regeneration of the primary acceptor molecule PEP has to be installed in C_3 chloroplasts. We tested the bacterial PEP synthase enzyme in potato. Some changes in the contents of individual amino acids were found, but no impact of PEP synthase (PEPS) overexpression on photosynthetic properties could be observed. The most significant effect was a clear delay in stomatal closure in the dark (Panstruga et al 1997, unpublished results). As an alternative, the C_4-pyruvate-Pi-dikinase (PPDK) from *Flaveria trinervia* (Rosche and Westhoff 1990) was overexpressed in tobacco chloroplasts and appreciable activities were detected (unpublished data). These plants await further physiological characterization. We focused on the by then less successful tobacco system because this species can be genetically crossed and therefore allows the efficient combination of multiple transgenes in one plant.

Additionally, we attempted to study transport from the chloroplast to the cytosol. C_3 plants contain a PEP/Pi translocator (PPT) in the inner chloroplast membrane that normally transports PEP into the chloroplast as a precursor of the shikimate pathway and aromatic amino acid biosynthesis (Voll et al 2003, Weber et al 2005). Transport is directed by the concentration gradient of PEP and, therefore, strong production of this compound in the chloroplast would allow export into the cytosol. Data from the overexpression of ME or PCK in the chloroplast of C_3 plants support this idea (see above). To analyze whether transport capacities limit the exchange of PEP from the chloroplast to the cytosol, we overexpressed the PPT from *Brassica oleracea* in tobacco in combination with additional C_4 enzymes. The current analyses do not provide evidence that this factor influences the efficiency of C_4-like metabolite flow in C_3 plants (unpublished results). However, this question should be re-examined in future approaches with optimized expression systems. Even less is known about the import of oxaloacetate into the chloroplast. Operation of a C_4-like pathway in C_3 plants would require a net exchange of oxaloacetate or malate for pyruvate or PEP, although this connection is not necessarily direct. However, all dicarboxylate transporters from C_3 plants characterized thus far exchange only a C_4 compound for another C_4 compound (Weber and Flügge 2002). It will be a major assignment of future projects to analyze how C_4 bundle sheath chloroplasts perform this task (Taniguchi et al 2004).

The initial step of C_4 photosynthesis is the conversion of CO_2 to HCO_3^- catalyzed by carbonic anhydrase (CA) to provide the substrate for CO_2 fixation by PEPC. Antisense *Flaveria bidentis* plants for C_4-CA do not show clear effects when the CA amounts decrease to 20% of those of the wild type, but further reduction strongly diminishes photosynthesis (von Caemmerer et al 2004), indicating that cytosolic CA activity is essential for C_4 photosynthesis. The situation is complicated by the fact that CAs are also abundant proteins in C_3 plants, but here the main part of the activity is associated with the chloroplast stroma and assumed to facilitate diffusion of inorganic carbon across the chloroplast envelope. At least 14 putative CA genes have been identified in the *Arabidopsis* genome and it remains to be shown which of these enzymes are active to what extent in which compartment of the cell (Moroney et al 2001). C_3 plants overexpressing cytosolic CA in combination with other C_4 enzymes have not been analyzed to date.

Taken together, the data available so far do not allow a decision as to whether a C_4-like unicellular cycle might be functional in a C_3 plant or not. Apparently, the installation of such a pathway will require the overexpression of multiple chimeric genes in one plant. Moreover, the amount of expression will have to be precisely co-ordinated to avoid pleiotropic effects. We have currently combined PEPC, ME, PEPS, and PPT activities in tobacco without obtaining clear effects on growth or biomass production at ambient conditions. A further physiological analysis will reveal whether photosynthetic properties are improved compared to plants expressing only some of these genes.

An alternative approach

The C_4 cycle evolved to concentrate CO_2 at the site of fixation by Rubisco, which suppresses the alternative fixation of atmospheric oxygen. This always requires transport between organelles and perhaps also anatomical adaptations as specified above. We designed an alternative approach that accepts the oxygenase activity of Rubisco and instead aims to metabolize the resulting glycolate inside the chloroplast (Kebeish et al 2007). The system is based on the glycerate pathway from *Escherichia coli* that enables the bacterium to grow on glycolate as the sole carbon source (Lord 1972). Figure 3 shows this metabolic pathway in the background of the higher plant photorespiratory pathway. Glycolate is first converted to glyoxylate by glycolate dehydrogenase (GDH). The bacterial enzyme differs from the plant peroxisomal glycolate oxidase in using organic co-factors instead of molecular oxygen as a co-factor and therefore no reactive oxygen species are produced. However, it is composed of three subunits and is therefore difficult to transfer to plant chloroplasts. In the next reaction step, two molecules of glyoxylate are converted to one molecule of tartronic semialdehyde (TS) and CO_2 is released. Since this reaction takes place in the chloroplast, we expect that the CO_2 can be more efficiently refixed by Rubisco compared to mitochondrial CO_2 release during photorespiration. Moreover, this reaction does not include any release of reduced nitrogen. Ultimately, TS is reduced to glycerate by TS reductase. Hence, this metabolic pathway is capable of creating a bypass of photorespiration in the chloroplast.

We installed the complete pathway in *Arabidopsis* by sequential transformation and crossing. The resulting plants show clearly enhanced growth and biomass production at ambient conditions. This can be correlated with a reduction in photorespiratory flow, enhanced photosynthetic properties, and an increase in leaf sugar contents (unpublished data). These results provide strong evidence that an improvement in CO_2 fixation can improve the productivity of C_3 crop plants.

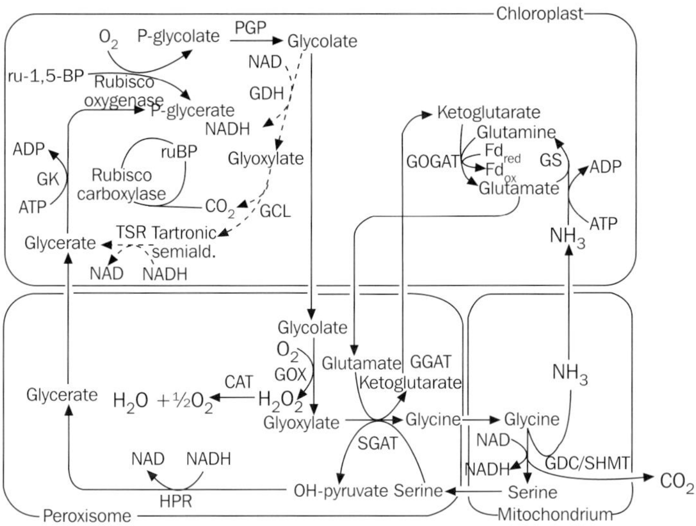

Fig. 3. The photorespiratory pathway (solid lines) and the bypass reaction (broken lines) aiming to reduce photorespiratory losses. Rubisco = ribulose-1,5-bisphosphate carboxylase/oxygenase, PGP = phosphogycolate phosphatase, GOX = glycolate oxidase, CAT = catalase, GGAT = glyoxylate/glutamate aminotransferase, GDC/SHMT = glycine decarboxylase/serine hydroxymethyl transferase, SGAT = serine/glutamate aminotransferase, HPR = hydroxypyruvate reductase, GK = glycerate kinase, GS = glutamine synthetase, GOGAT = glutamate/oxoglutarate aminotransferase, GDH = glycolate dehydrogenase, GCL = glyoxylate carboxyligase, TSR = tartronic semialdehyde reductase.

References

Arp WJ, Van Mierlo JEM, Berendse F, Snijders W. 1998. Interactions between elevated CO_2 concentration, nitrogen and water: effects on growth and water use of six perennial plant species. Plant Cell Environ. 21:1-11.

Asai N, Nakajima N, Tamaoki M, Kamada H, Kondo N. 2000. Role of malate synthesis mediated by phospho*enol*pyruvate carboxylase in guard cells in the regulation of stomatal movement. Plant Cell Physiol. 41:10-15.

Bakrim N, Prioul JL, Deleens E, Rocher JP, Arrio-Dupont M, Vidal J, Gadal P, Chollet R. 1993. Regulatory phosphorylation of C_4 phospho*enol*pyruvate carboxylase (a cardinal event influencing the photosynthesis rate in *Sorghum* and maize). Plant Physiol. 101:891-897.

Beaujean A, Issakidis-Bourguet E, Catterou M, Dubois F, Sangwan R, Sangwan-Norreel B. 2001. Integration and expression of *Sorghum* C_4 phospho*enol*pyruvate carboxylase and chloroplastic NADP(+)-malate dehydrogenase separately or together in C_3 potato plants. Plant Sci. 160:1199-1210.

Brooks A, Farquhar GD. 1985. Effect of the temperature on the CO_2/O_2 specificity of ribulose-1,5-bisphosphate carboxylase/oxygenase and the rate of respiration in the light. Planta 165:397-406.

Casati P, Lara MV, Andreo CS. 2000. Induction of a C_4-like mechanism of CO_2 fixation in *Egeria densa*, a submersed aquatic species. Plant Physiol. 123:1611-1622.

Chen LM, Li KZ, Miwa T, Izui K. 2004. Overexpression of a cyanobacterial phospho*enol*-pyruvate carboxylase with diminished sensitivity to feedback inhibition in *Arabidopsis* changes amino acid metabolism. Planta 219:440-449.

Davies DD. 1986. The fine control of cytosolic pH. Physiol. Plant. 67:702-706.

Fukayama H, Hatch MD, Tamai T, Tsuchida H, Sudoh S, Furbank RT, Miyao M. 2003. Activity regulation and physiological impacts of maize C_4-specific phospho*enol*pyruvate carboxylase overproduced in transgenic rice plants. Photosynth. Res. 77:227-239.

Gehlen J, Panstruga R, Smets H, Merkelbach S, Kleines M, Porsch P, Fladung M, Becker I, Rademacher T, Hausler RE, Hirsch HJ. 1996. Effects of altered phospho*enol*pyruvate carboxylase activities on transgenic C_3 plant *Solanum tuberosum*. Plant Mol. Biol. 32:831-848.

Häusler RE, Hirsch HJ, Kreuzaler F, Peterhänsel C. 2002. Overexpression of C_4-cycle enzymes in transgenic C_3 plants: a biotechnological approach to improve C_3-photosynthesis. J. Exp. Bot. 53:591-607.

Häusler RE, Kleines M, Uhrig H, Hirsch HJ, Smets H. 1999. Overexpression of phospho*enol*-pyruvate carboxylase from *Corynebacterium glutamicum* lowers the CO_2 compensation point (Γ*) and enhances dark and light respiration in transgenic potato. J. Exp. Bot. 50:1231-1242.

Häusler RE, Rademacher T, Li J, Lipka V, Fischer KL, Schubert S, Kreuzaler F, Hirsch HJ. 2001. Single and double overexpression of C_4-cycle genes had differential effects on the pattern of endogenous enzymes, attenuation of photorespiration and contents of UV protectants in transgenic potato and tobacco plants. J. Exp. Bot. 52:1785-1803.

Hudspeth RL, Grula JW, Dai Z, Edwards GE, Ku MSB. 1992. Expression of maize phospho-*enol*pyruvate carboxylase in transgenic tobacco. Plant Physiol. 98:458-464.

Kebeish R, Thiruveedhi K, Niessen M, Bari R, Hirsch HJ, Stäbler N, Schönfeld B, Rosenkranz R, Kreuzaler F, Peterhänsel C. 2007. Bypassing photorespiration in the chloroplast improves photosynthesis and increases biomass production in transgenic *Arabidopsis* plants. Nature Biotechnol. (In press.)

Kimball BA. 1983. Carbon dioxide and agricultural yield: an assemblage and analysis of 430 prior observations. Agron. J. 75:779-788.

Kogami H, Shono M, Koike T, Yanagisawa S, Izui K, Sentoku N, Tanifuji S, Uchimiya H, Toki S. 1994. Molecular and physiological evaluation of transgenic tobacco plants expressing a maize phospho*enol*pyruvate carboxylase gene under the control of the cauliflower mosaic virus 35S promoter. Transgenic Res. 3:287-296.

Ku MSB, Agarie S, Nomura M, Fukayama H, Tsuchida H, Ono K, Hirose S, Toki S, Miyao M, Matsuoka M. 1999. High-level expression of maize phospho*enol*pyruvate carboxylase in transgenic rice plants. Nature Biotechnol. 17:76-80.

Leegood RC. 2002. C_4 photosynthesis: principles of CO_2 concentration and prospects for its introduction into C_3 plants. J. Exp. Bot. 53:581-590.

Lipka V, Häusler RE, Rademacher T, Li J, Hirsch HJ, Kreuzaler F. 1999. *Solanum tuberosum* double transgenic expressing phospho*enol*pyruvate carboxylase and NADP-malic enzyme display reduced electron requirement for CO_2 fixation. Plant Sci. 144:93-105.

Lord JM. 1972. Glycolate oxidoreductase in *Escherichia coli*. Biochim. Biophys. Acta 267:227-237.

Manh CT, Bismuth E, Boutin JP, Provot M, Champigny ML. 1993. Metabolite effectors for short-term nitrogen-dependent enhancement of phospho*enol*pyruvate carboxylase activity and decrease of net sucrose synthesis in wheat leaves. Physiol. Plant. 89:460-466.

Moroney JV, Bartlett SG, Samuelsson G. 2001. Carbonic anhydrases in plants and algae. Plant Cell Environ. 24:141-153.

Nelson T, Langdale JA. 1992. Developmental genetics of C_4 photosynthesis. Annu. Rev. Plant Physiol. Plant Mol. Biol. 43:25-47.

Ogren WL. 1984. Photorespiration: pathways, regulation and modification. Annu. Rev. Plant Physiol. 35:415-442.

Panstruga R, Hippe-Sanwald S, Lee Y-K, Lataster M, Lipka V, Fischer R, Cai Liao Y, Häusler RE, Kreuzaler F, Hirsch H-J. 1997. Expression and chloroplast-targeting of active phospho*enol*pyruvate synthetase from *Escherichia coli* in *Solanum tuberosum*. Plant Sci. 127:191-205.

Rademacher T, Häusler RE, Hirsch HJ, Zhang L, Lipka V, Weier D, Kreuzaler F, Peterhänsel C. 2002. An engineered phospho*enol*pyruvate carboxylase redirects carbon and nitrogen flow in transgenic potato plants. Plant J. 32:25-39.

Rao SK, Magnin NC, Reiskind JB, Bowes G. 2002. Photosynthetic and other phospho*enol*pyruvate carboxylase isoforms in the single-cell, facultative C_4 system of *Hydrilla verticillata*. Plant Physiol. 130:876-886.

Reiskind JB, Madsen TV, Van Ginkel LC, Bowes G. 1997. Evidence that inducible C_4-type photosynthesis is a chloroplastic CO_2-concentrating mechanism in *Hydrilla*, a submersed monocot. Plant Cell Environ. 20:211-220.

Rosche E, Westhoff P. 1990. Primary structure of pyruvate, orthophosphate dikinase in the dicotyledonous C_4 plant *Flaveria trinervia*. FEBS Lett. 273:116-121.

Suzuki S, Murai N, Burnell JN, Arai M. 2000. Changes in photosynthetic carbon flow in transgenic rice plants that express C_4-type phospho*enol*pyruvate carboxykinase from *Urochloa panicoides*. Plant Physiol. 124:163-172.

Suzuki S, Murai N, Kasaoka K, Hiyoshi T, Imaseki H, Burnell JN, Arai M. 2006. Carbon metabolism in transgenic rice plants that express phospho*enol*pyruvate carboxylase and/or phospho*enol*pyruvate carboxykinase. Plant Sci. 170:1010.

Svensson P, Blasing OE, Westhoff P. 2003. Evolution of C_4 phospho*enol*pyruvate carboxylase. Arch. Biochem. Biophys. 414:180-188.

Taniguchi Y, Nagasaki J, Kawasaki M, Miyake H, Sugiyama T, Taniguchi M. 2004. Differentiation of dicarboxylate transporters in mesophyll and bundle sheath chloroplasts of maize. Plant Cell Physiol. 45:187-200.

Tolbert NE. 1997. The C_2 oxidative photosynthetic carbon cycle. Annu. Rev. Plant Physiol. Plant Mol. Biol. 48:1-25.

Ueno Y, Hata S, Izui K. 1997. Regulatory phosphorylation of plant phospho*enol*pyruvate carboxylase: role of a conserved basic residue upstream of the phosphorylation site. FEBS Lett. 417:57-60.

Voll L, Häusler RE, Hecker R, Weber A, Weissenbock G, Fiene G, Waffenschmidt S, Flügge UI. 2003. The phenotype of the *Arabidopsis cue1* mutant is not simply caused by a general restriction of the shikimate pathway. Plant J. 36:301-317.

von Caemmerer S, Furbank RT. 2003. The C_4 pathway: an efficient CO_2 pump. Photosynth. Res. 77:191-207.

von Caemmerer S, Quinn V, Hancock NC, Price GD, Furbank RT, Ludwig M. 2004. Carbonic anhydrase and C_4 photosynthesis: a transgenic analysis. Plant Cell Environ. 27:697-703.

Weber A, Flügge U-I. 2002. Interaction of cytosolic and plastidic nitrogen metabolism in plants. J. Exp. Bot. 53:865-874.
Weber AP, Schwacke R, Flügge UI. 2005. Solute transporters of the plastid envelope membrane. Annu. Rev. Plant Biol. 56:133-164.

Notes

Authors' address: RWTH Aachen University, Institute for Biology I, Worringer Weg 1, 52056 Aachen, Germany; phone +49-241-8026632, fax +49-241-8022637, email: cp@bio1.rwth-aachen.de.

Acknowledgments: This work was financially supported by grants from the Federal Ministry of Research and Education, the Deutsche Forschungsgemeinschaft, and Bayer Cropscience. We are indebted to Rainer Häusler (Cologne University) for his continuous support during the physiological evaluation of our plants and to numerous PhD students for generating and characterizing transgenic lines.

Molecular evolution of C_4 photosynthesis in the dicot genus *Flaveria*: implications for the design of a C_4 plant

U. Gowik and P. Westhoff

C_4 photosynthesis is characterized by a division of labor between two different photosynthetic cell types, mesophyll and bundle sheath cells. Relying on phosphoe*nol*pyruvate carboxylase (PEPC) as the primary carboxylase in the mesophyll cells, a CO_2 pump is established in C_4 plants that concentrates CO_2 at the site of ribulose 1,5-bisphosphate carboxylase/oxygenase in the bundle sheath cells. The C_4 photosynthetic pathway evolved polyphyletically, implying that the genes encoding the C_4 genes originated from nonphotosynthetic progenitor genes that were already present in the C_3 ancestral species. To establish a C_4 cycle in a C_3 plant, detailed knowledge about the components of C_4 photosynthesis and the differences of these components in C_3 and C_4 plants is needed.

Keywords: *Flaveria*, C_4 photosynthesis, evolution, phospho*enol*pyruvate carboxylase

The C_4-photosynthetic carbon cycle is an elaborated addition to the C_3 photosynthetic pathway. It is an adaptation to high light, high temperatures, and dryness. Therefore, C_4 plants dominate grassland floras and biomass production in the warmer climates of tropical and subtropical regions (Brown 1999, Sage et al 1999). Consequently, the transfer of C_4 traits into C_3 plants is one strategy that could be adopted for improving the photosynthetic performance of C_3 crop plants such as rice, as one possibility to improve yields.

The high photosynthetic capacity of C_4 plants is achieved by their unique mode of carbon assimilation, which involves two different cell types, mesophyll and bundle sheath cells. In the mesophyll cells, CO_2 is initially fixed by phospho*enol*pyruvate carboxylase (PEPC) into the C_4 acids malate or aspartate or both, which are then transported to the bundle sheath. There, the C_4 acids are decarboxylated, and the CO_2 is refixed by ribulose 1,5-bisphosphate carboxylase/oxygenase (Rubisco) (Hatch 1987). C_4 plants show drastically reduced rates of photorespiration because CO_2 is concentrated at the site of Rubisco. This largely excludes the competitive inhibition of this enzyme by oxygen, which becomes prominent at higher temperatures. In C_3

plants, photorespiration may reduce the efficiency of photosynthesis by up to 40% (Ehleringer and Monson 1993).

The CO_2 pump ensures high rates of photosynthesis even when CO_2 concentrations are low in the intercellular air spaces of the leaf. Therefore, C_4 plants are able to limit the opening of their stomata and minimize water loss due to transpiration. As the CO_2 pump delivers saturating concentrations of CO_2 to the site of Rubisco, high photosynthetic rates are maintained with less enzyme than is required in C_3 species. This is reflected in a higher nitrogen-use efficiency (Long 1999).

C_4 plants differ from C_3 plants in several properties. Because close contact between mesophyll and bundle sheath cells is vital for C_4 photosynthesis, the leaf structure of C_4 plants is altered compared with that of most C_3 plants. The interveinal distance is usually smaller and the leaf thickness is limited to maximize the contact of mesophyll and bundle sheath cells. leading to the typical Kranz anatomy (Leegood 2002). To guarantee the high flux of metabolites between the two cell types, they are connected via numerous plasmodesmata (Botha 1992). Finally, the expression of the genes encoding C_4 pathway components, which are evolutionarily derived from C_3 ancestral genes, had to be modulated for the needs of this photosynthetic pathway as did the kinetic properties of at least some enzymes encoded by these genes (Sheen 1999, Drincovich et al 2000, Svensson et al 2003).

C_4 plants occur in at least 18 families of monocotyledonous and dicotyledonous plants, which are phylogenetically quite separate from each other (Sage et al 1999). This indicates that C_4 plants must have evolved several times independently from C_3 ancestors during the evolution of angiosperms. There is strong evidence that, even within a single taxon, for instance, the Gramineae, this transition from C_3 to C_4 may have occurred more than once (Kellogg 1999, Monson 1999). The multiple independent origin of C_4 photosynthesis suggests that the evolution of a C_3 into a C_4 species must have been relatively easy in genetic terms and that just a few master genes, for example, genes that are responsible for creating a certain leaf anatomy, would have been involved. The available molecular data on the C_4-cycle enzymes support this point of view. None of the C_4 enzymes are unique to C_4 plants. Nonphotosynthetic isoforms of these enzymes are also present in C_3 species and in the nonphotosynthetic tissues of C_4 species. The ubiquity of these nonphotosynthetic isoforms of the C_4-cycle enzymes in C_3 plants strongly indicates that these "C_3 isoforms" served as the starting point for the evolution of C_4 genes (Monson 1999, Westhoff and Gowik 2004).

To adapt a C_4 progenitor gene for its function in C_4 photosynthesis, at least three major changes were necessary. C_4 isoform genes are highly expressed (Harpster and Taylor 1986, Hermans and Westhoff 1990, Crétin et al 1991), but C_3 isoform genes are only moderately transcribed (Crétin et al 1991, Kawamura et al 1992, Ernst and Westhoff 1996). The effectiveness of gene expression had therefore to be increased. Because strict compartmentation of C_4 enzymes is imperative for proper functioning of the C_4 cycle, the C_4 isoform genes additionally had to evolve expression patterns that are specific to organs or cells (Hatch 1987). Finally, it is known, at least for phospho*enol*pyruvate carboxylase, that the C_4-cycle enzymes differ from their C_3 counterparts in kinetic and regulatory characteristics (Ting and Osmond 1973a,

Parameter	*F. pringlei*	*F. linearis*	*F. pubescens*	*F. brownii*	*F. trinervia*
Kranz anatomy	No	Poorly developed		Well developed	
CO_2 compensation point (µbar)	62	27	21	6	3
PEPC activity (µmol mg^{-1} Chl × h)	24	123	207	460	900
C_4 cycle	–	+	++	+++	++++
Photosynthesis	C_3	C_3-C_4	C_3-C_4	C_4-like	C_4

→ C_4 photosynthesis

Fig. 1. C_4 photosynthesis in the genus *Flaveria*: a stepwise evolution of a quantitative trait (Edwards and Ku 1987).

Svensson et al 2003). Therefore, the coding regions had to be changed to achieve the required adaptations of the enzymatic properties.

To engineer C_4 photosynthesis into a C_3 plant, it is important to know precisely the properties of C_4 plants that are necessary for the proper function of the C_4 cycle and how these properties are realized at the molecular level. It is also important to know how these properties can be implemented into a C_3 plant without fatally disturbing the metabolism of this plant. One way to get this information is to follow how C_4 photosynthesis was introduced into C_3 plants by nature during evolution.

To obtain insight into an evolutionary process, it is ideal to have closely related species where one species has acquired the complete character of the new trait while the other species has not. This means that their common ancestor existed recently, in evolutionary time scales. Hence, the morphological, biochemical, and genetic differences observed would mostly be due to selection for the new character and would arise only to a minor extent from random genetic changes. A series of intermediate species with progressively more advanced traits for the new character would allow deciphering of the discrete evolutionary steps and even the order of these steps.

To gain insight into the evolution of C_4 genes, the genus *Flaveria* (Asteraceae; Powell 1978) is being used as our experimental system because it comes very close to the ideal scenario described above. *Flaveria* contains C_3 and C_4 species and, in addition, a large number of C_3-C_4 intermediates (Edwards and Ku 1987). These intermediates differ in the expression of C_4 photosynthetic traits (see Fig. 1), and there is convincing evidence that at least some of them are true evolutionary intermediates (Monson and Moore 1989). Since suitable phylogenetic data are available (Kopriva et al 1996, Westhoff and Gowik 2004, McKown et al 2005), it is possible to directly compare orthologous genes from closely related species with different types of photosynthesis and to identify the evolutionary changes in these genes that were necessary to establish C_4 photosynthesis.

C_4 *Flaveria* species, like the vast majority of C_4 plants, use the two-cell mode, that is, Kranz anatomy, to concentrate CO_2 at the site of Rubisco. However, some plants have inducible or permanent CO_2 pumps established within a single cell (Holaday and Bowes 1980, Freitag and Stichler 2000, 2002, Voznesenskaya et al 2001, 2002). These one-cell types of C_4 photosynthesis are rare exceptions, which apparently evolved only under very special environmental conditions.

One point of discussion is which of these CO_2-concentrating mechanisms should be introduced into a C_3 plant to improve its photosynthetic capacity, the complex two-cell C_4 photosynthesis, which might be very difficult to realize because of the required changes in leaf anatomy, or the less complex but also less efficient one-cell type (Leegood 2002, von Caemmerer 2003, Long et al 2006). This raises the question whether the investigation of the evolution of C_4 photosynthesis in *Flaveria* can help to fulfill one of these approaches. In our opinion, this is the case.

The comparative investigation of orthologous genes from closely related species performing different types of photosynthesis allows the identification of C_4-related alterations at the molecular level. If the genes involved in establishing C_4 traits are known, it is possible to identify the variant nucleotides that are required for the C_4 functioning of these genes. To identify "new" genes involved in the establishment of C_4 photosynthesis, for example, genes responsible for the Kranz anatomy, *Flaveria* might not be the ideal system because of the lack of genetic resources. But it might be possible to identify such genes by searching for mutants showing an altered leaf anatomy using C_3 model plants like *Arabidopsis thaliana* and benefit from the genetic resources existing for this plant. *Flaveria* could then serve as a model system to investigate the function of such genes in a C_4 context.

C_4-specific protein properties

The enzymes involved in the C_4 cycle also occur in C_3 plants and the nonphotosynthetic tissues of C_4 plants. The C_4 enymes differ in their kinetic and regulatory properties from the nonphotosynthetic isoforms. When establishing C_4 metabolism in a C_3 plant, the question arises as to which isoforms of the enzymes should be used for overexpression in the C_3 plant.

Some attempts have been made to overexpress the C_4 isoform enzymes using the full-length genes or cDNAs of the C_4 plant maize in the C_3 plant rice. The introduction of the intact gene of maize phospho*enol*pyruvate carboxylase led to a high expression of the maize protein associated with a high PEPC activity in the leaves of transgenic rice plants. Physiologically, the transgenic plants exhibited reduced O_2 inhibition of photosynthesis and photosynthetic rates comparable with those of untransformed plants (Ku et al 1999). However, later investigations of transgenic rice plants could not confirm reduced O_2 inhibition (Fukayama et al 2003). Here, a reduced CO_2 assimilation rate, caused by the stimulation of respiration in light, was found. Additionally, it was shown that the maize enzyme was regulated in a manner similar to the endogenous rice PEPC, which is contrary to the regulation of the maize enzyme in maize leaves (Fukayama et al 2003).

When a cDNA of the C_4 NADP-dependent malic enzyme (NADP-ME) of maize was introduced into rice and expressed under control of the rice *Cab* promoter, high amounts of the maize protein and very high NADP-ME activity could be detected in the leaves of transgenic plants. However, transgenic plants showed leaf bleaching and growth hindrance under natural light. These effects resulted from enhanced photoinhibition of photosynthesis due to an increase in the amount of NADPH inside the chloroplast (Tsuchida et al 2001). Another group that performed a similar experiment additionally found aberrant chloroplasts in rice plants highly expressing maize NADP-ME. The chloroplasts were agranal without thylakoid stacking, and chlorophyll content and photosystem II activity were inversely correlated with the activity of NADP-ME (Takeuchi et al 2000).

In contrast, the high expression of maize C_4 pyruvate orthophosphate dikinase (PPDK) by introducing the complete maize *Pdk* gene into the C_3 plant rice did not lead to abnormalities in growth behavior or fertility (Fukayama et al 2001).

Bacterial or genetically modified enzymes have also been overexpressed in the leaves of C_3 plants using their cDNAs and strong promoters such as the 35S promoter. To maximize PEPC activity in the leaves of potato plants, Rademacher et al (2002) expressed different modified potato PEPCs under control of the 35S promoter. With the modifications introduced, the affinity toward PEP increased and sensitivity to malate decreased. Plants expressing these modified enzymes did not grow properly and showed drastically reduced tuber yield. These effects were due to the redirection of carbon and nitrogen fluxes in the transgenic plants. When the unmodified potato enzyme was overexpressed, no disadvantageous effects on the transgenic plants were observed, but they also showed no increased PEPC activities when compared with wild-type plants (Rademacher et al 2002).

Chen et al (2004) obtained a similar result. They overexpressed a PEP-carboxylase from the cyanobacterium *Synechococcus vulcanus* under control of the 35S promoter in *Arabidopsis*. The cyanobacterial enzyme is almost insensitive to feedback inhibition by malate or aspartate. Strong expression of this enzyme led to severe visible phenotypes such as leaf bleaching and infertility, which were caused by changes in the amino acid metabolism of the transgenic plants (Chen et al 2004).

These examples show that a nonregulated overexpression of enzymes involved in the C_4 pathway can cause severe problems for transgenic plants. The kinetic and regulatory properties of the C_4 enzymes differ from the properties of their nonphotosynthetic counterparts. These differences represent necessary adaptations to assure the proper function of the enzymes under the conditions of C_4 metabolism.

To gain insight into the evolution of C_4 enzymes, we are using the entry enzyme of the C_4 cycle, PEPC, as our model C_4 enzyme and gene. By comparison of orthologous enzymes from closely related C_3 and C_4 species of the genera *Flaveria* and *Alternanthera*, we are trying to unravel the alterations in kinetic and regulatory properties at the molecular level that were required to create a proper C_4 PEPC from a nonphotosynthetic enzyme.

C_4 PEPCs exhibit substrate saturation constants (*Km*) for PEP that are usually about ten times larger than those of their C_3 counterparts (Ting and Osmond 1973b). On the other hand, the saturation constant for bicarbonate, the second substrate, is lower in C_4 PEPCs than in C_3 PEPCs (Bauwe 1986). Finally, C_4 PEPCs are more tolerant of the inhibitor malate and more sensitive to the activator glucose-6-phosphate when compared with the C_3 PEPC isozymes (Dong et al 1998, Bläsing et al 2002). This was also found for the orthologous *ppcA* enzymes of *Flaveria trinervia* (C_4) and *F. pringlei* (C_3) or *Alternanthera pungens* (C_4) and *A. sessilis* (C_3) (Svensson et al 1997, Gowik et al 2006). The differences in substrate affinity and inhibition by malate suggest that C_4 PEPCs harbor specific C_4 determinants that were acquired during the evolution of C_4 photosynthesis. To investigate how PEPC enzyme characteristics changed during evolution toward C_4 photosynthesis, *ppcA* PEPCs from the C_3-C_4 intermediate plant *F. pubescens* and the C_4-like C_3-C_4 intermediate *F. brownii* were investigated. Both the *Km* (PEP) values and the malate inhibition constants, *Ki*, were found to be intermediate between the C_3 and C_4 *ppcA* PEPCs (Fig. 2). This indicates that the C_3 PEPC evolved step by step into a C_4 enzyme (Engelmann et al 2003).

Since the C_3 and C_4 PEPC isoforms share 96% identical amino acid positions, it should be feasible to pinpoint changes in the amino acid sequence responsible for the C_4 characteristics (Svensson et al 1997). Therefore, reciprocal domain swapping experiments combined with site-specific mutagenesis were conducted with the two *ppcA* PEPCs of *F. trinervia* and *F. pringlei* to locate regions and amino acid residues in the enzyme that influence *Km* (PEP) (Bläsing et al 2000). With this approach, two regions, from amino acids 296 to 437 (region 2) and from amino acids 645 to 966 (region 5), were identified that contain the major C_4 determinants for the saturation kinetics of the substrate PEP, whereas the C_4-specific properties in region 5 were confined to a single amino acid, serine 774 (Fig. 2; Bläsing et al 2000). This was confirmed by inserting region 2 of the C_4 enzyme and the C_4-specific serine into an otherwise C_3 background. The resulting chimerical enzyme possessed about two-thirds of C_4 PEPC characteristics with respect to *Km* (PEP) (Engelmann et al 2002).

In region 2, 16 differences were detected between the C_3 and C_4 *ppcA* PEPCs of *Flaveria* (Fig. 2). There is only one amino acid residue, a lysine at position 347, which both *F. trinervia* and *F. brownii* have in common and which differs from the arginine in this position in *F. pubescens* and *F. pringlei* (Fig. 2; Engelmann et al 2003). This lysine is also conserved in the C_4 PEPC of maize, where it is located between helices 12 and 13 (Matsumura et al 2002). In the corresponding region of the *Alternanthera* enzymes (amino acids 297–438), no corresponding amino acid exchanges could be detected. Nevertheless, in *Alternanthera* as in *Flaveria*, the C_4 enzyme has a lower affinity to the substrate PEP than the C_3 enzyme, indicating that some alterations of the enzyme kinetic properties were realized by different modifications at the molecular level in both genera (Gowik et al 2006).

The distinct serine residue in the carboxy terminus (serine 774 in the *F. trinervia* enzyme and serine 775 in the *A. pungens* enzyme), which is the main determinant for the lower PEP affinity, is very well conserved in C_4 PEPCs. All C_4 enzymes studied to date contain a serine at this position, whereas in all nonphotosynthetic and CAM PEPCs this site is occupied by an alanine (Svensson et al 2003). It has to be concluded that serine 774 is of central importance for the evolution of C_4 characteristics, at least

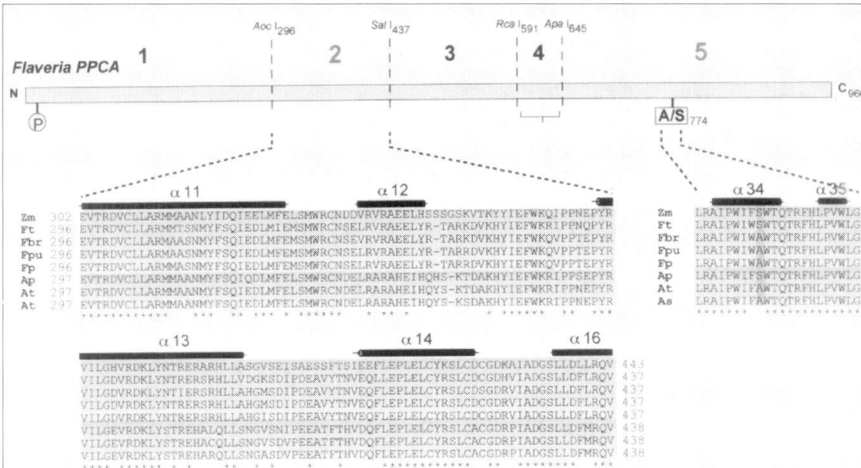

$K_{0,5}$ (PEP) (μM)

Species	F. pringlei	F. pubescens	F. brownii	F. trinervia	Species	A. sessilis	A. tenella	A. pungens
Photosynthesis	C_3	C_3-C_4	C_4-like	C_4	Photosynthesis	C_3	C_3-C_4	C_4
–Glc6P	29	53	108	269	–Glc6P	36	42	157
+Glc6P	17	20	17	53	+Glc6P	13	25	20

C_4 photosynthesis → C_4 photosynthesis →

Fig. 2. C_4-specific molecular and kinetic properties of PEPC proteins in *Flaveria* and *Alternanthera*. From five investigated enzyme domains (in *Flaveria*), region 2 (positions 296–437) and region 5 (amino acids 645–966) contain the major C_4 determinants for the saturation kinetics of PEP. P indicates the target phosphorylation site at position 11. The secondary structures indicated on top of the sequence alignments (black bars) were obtained from the 3D structure of the C_4 PEPC of *Zea mays* (Matsumura et al 2002). Sequence positions, which are identical in all shown PEPCs, are marked by stars below the strings of sequences. At position 774 (gray column), serine occurs only in C_4 PEPCs, whereas PEPCs from C_3 and C_3-C_4 intermediate plants contain an alanine at this position. The amino acid numbering follows that of the *F. trinervia* protein.

with regard to the *Km* (PEP). All investigated C_3-C_4 intermediate PEPCs, even from the C_4-like species *F. brownii*, still show an alanine at this position (Engelmann et al 2003). This suggests that the change from alanine to serine occurred only recently during evolution from C_3 to C_4 photosynthesis. One wonders why this change occurred so late during evolution and, more importantly, why it apparently had to occur. One consequence of the alanine to serine substitution is an increase in *Km* (PEP) (Bläsing et al 2000). When, in addition, residues 296–437 are swapped from C_3 to C_4, the *Km* (PEP) value increases further, almost reaching that of the C_4 enzyme (Engelmann et al 2002).

So far, we have not investigated the influence of the phosphorylation of a conserved serine residue in the N-terminus on the kinetic properties of the *Flaveria* enzymes. It is known that phosphorylated PEPCs are less sensitive to the inhibitor malate and more sensitive to the activator glucose-6-phosphate (Vidal and Chollet 1997). The *Flaveria* system provides the opportunity to investigate whether C_4 and C_3 enzymes respond differently to phosphorylation and, if so, the chimeric enzymes will give us the chance to map the molecular determinants responsible for the effects of phosphorylation.

The physiological significance of the alanine to serine exchange during C_4 evolution is not clear yet. Is this increase in *Km* (PEP) important for an efficient C_4 PEPC or are these changes required for another important characteristic of the C_4 isoform? Is the rise in *Km* (PEP) therefore an unavoidable side effect of creating a lower *Km* (bicarbonate) or a higher malate tolerance? Both possibilities are reasonable. An increase in *Km* (PEP) may have been necessary to adequately regulate C_4 PEPC, since the PEP concentration *in vivo* is significantly higher around C_4 PEPCs than around nonphotosynthetic PEPCs (Leegood and Walker 1999). This assumption is supported by the experiments of Rademacher et al (2002) and Chen et al (2004) indicating that high amounts of PEPC enzyme with high affinity to PEP or high malate tolerance, or both, disturb carbon and nitrogen metabolism.

Experiments with recombinant chimerical enzymes demonstrated that differences in malate tolerance are associated with region 5 of the *Flaveria* enzymes but not with the serine or alanine 774 residue in that region (Jacobs and Westhoff, unpublished data). So, the importance of lower *Km* (bicarbonate) may be more vital than the apparent disadvantage of a higher *Km* (PEP). Consequently, if serine 774 is important for that characteristic, a higher *Km* (PEP) might be "the price to pay."

To distinguish between these various possibilities, the available recombinant enzymes from C_3, C_3-C_4, and C_4 *Flaveria* and *Alternanthera* will be crucial. Thorough investigation of the kinetic and regulatory properties of these enzymes, with and without phosphorylation combined with activators and inhibitors (Tovar-Méndez et al 2000), will provide detailed information about the evolutionary steps during C_3 to C_4 PEPC evolution.

However, in vitro studies suffer from inherent limitations. Therefore, in the end, *in vivo* analyses will be necessary to test critically the predictions inferred from in vitro enzyme studies. Such studies would probably involve a knockout of the C_4 PEPC gene in a C_4 plant combined with its replacement by a PEPC gene whose properties will be assessed. A transformation system for the C_4 plant *F. bidentis* is currently available (Chitty et al 1994); therefore, *Flaveria* would be the study system of choice in which to pursue this *in vivo* approach.

C_4-specific gene regulation

In recent years, more and more evidence has been collected that changes in the spatio-temporal expression of genes were the starting points for the development of novel biochemical or morphological traits (Doebley and Lukens 1998, Carroll 2000). Since C_4 photosynthesis was developed several times during the evolution of angiosperms and as C_4 photosynthesis largely depends on the correct compartmentation of the enzymes involved in this metabolic pathway, it can be expected that such changes in gene expression have played a key role during the design of C_4 plants by evolution. For the proper and efficient function of the two-cell variant of C_4 photosynthesis, tissue-specific expression of the C_4-cycle genes is essential (Hatch 1987). This is also true for other genes, for example, genes for proteins of the photorespiration pathway or genes encoding proteins participating in intercellular and intracellular transport of metabolites relevant for the C_4 pathway. Examples are the gene encoding the P subunit of glycine decarboxylase (GdcP) of the C_4 plant *Flaveria trinervia*, which is involved in photorespiration and exclusively expressed in bundle sheath cells (Burscheidt 1998, Bauwe and Kolukisaoglu 2003) or the gene encoding the plastidic glutamate/malate translocator (DiT2) of the C_4 plant *Sorghum bicolor*, which is also exclusively expressed in bundle sheath cells (Renné et al 2003).

The differential expression of C_4 genes is largely controlled at the transcriptional level (Sheen 1999). The transcription of a gene is determined by *cis*-regulatory elements, which are part of the gene and serve as transcription factor binding sites, and by proteins that bind to these *cis*-regulatory elements—the *trans*-regulatory factors. During the evolution of the C_4 genes from C_3 nonphotosynthetic ancestors, the *cis*-regulatory elements of these genes were altered to provide the required tissue specificity of gene expression. In some cases, this was possible without changes in the *trans*-regulatory network in the plant; this is indicated by C_4 genes that are expressed in a C_4-specific manner when introduced into a C_3 plant (Matsuoka et al 1994). Other C_4 genes do not behave as C_4 genes when introduced into C_3 plants, indicating also that the *trans*-regulatory factors, involved in the expression of these genes in the C_4 plant or in their expression pattern, were altered during C_4 evolution (Matsuoka et al 1994, Stockhaus et al 1994, Nomura et al 2000a, 2005a,b).

To induce a functional two-cell C_4 cycle in a C_3 plant, one must be able to correctly express the C_4 genes in this C_3 plant in time and space. In some cases, this is not challenging and the promoters of C_4 genes can be used to drive C_4-like gene expression also in C_3 plants. The promoter of the photosynthetic *Ppc* gene of the C_4 plant maize, for example, causes a mesophyll-specific and light-dependent expression when introduced into the C_3 plant rice, what is essentially the same behavior as in maize (Matsuoka et al 1994). The same is true for the promoter of the chloroplastic pyruvate, orthophosphate dikinase gene (*cPdk*) of maize. When this promoter is fused to a GUS reporter gene and introduced into rice, the reporter gene is expressed specifically in the mesophyll cells in a light-dependent manner, which reflects the function of this promoter in maize (Matsuoka et al 1993, Nomura et al 2000b). These experiments demonstrate also that the *trans*-regulatory proteins, which are necessary to interpret the *cis*-regulatory signals of these two mesophyll-specific promoters from C_4 genes of maize correctly, are present in the C_3 plant rice. There are also bundle sheath-specific promoters from C_4 plants known, which show bundle sheath-specific expression in the C_3 plant rice. The promoter of the phospho*enol*pyruvate carboxykinase gene (*Pck*) from *Zoysia japonica*, a PCK-type C_4 plant, showed bundle sheath specificity in rice as in *Z. japonica* but lacked inducibility by light, which was present in *Z. japonica* (Nomura et al 2005a).

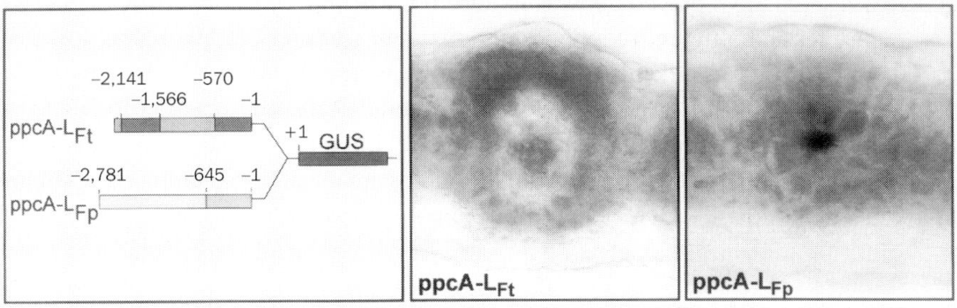

Fig. 3. Histochemical analysis of the activities of the *ppcA1* promoters of *F. trinervia* (C_4) and *F. pringlei* (C_3) in transgenic *F. bidentis* (C_4) (cf. Stockhaus et al 1997).

However, the cell-specific expression of C_4 genes is not necessarily maintained in a C_3 background. Especially, many genes with bundle sheath-specific expression do not show this tissue specificity in C_3 plants. Examples are the genes for mitochondrial aspartate aminotransferase of *Panicum miliaceum* (Nomura et al 2005b), for NADP-malic enzyme of maize (Nomura et al 2005a), or for the small subunit of Rubisco (Matsuoka et al 1994, Nomura et al 2000a). In C_4 plants, these genes are exclusively expressed in bundle sheath cells. In the C_3 plant rice, the promoters of all these genes are also active in mesophyll cells.

To design a functional two-cell C_4 cycle in a C_3 plant, it is important to have more detailed information about the *cis*-regulatory and *trans*-regulatory components that are important for C_4-specific gene expression, for example, to elucidate precisely the involved *cis*-regulatory sequences and to identify the transcription factors binding to these *cis*-elements. This knowledge could be used to trigger the expression of C_4-cycle genes in a C_4-specific manner in C_3 plants.

We performed a detailed analysis of the promoter of the photosynthetic phospho*enol*pyruvate carboxylase gene (*ppcA*) of the C_4 plant *Flaveria trinervia*. To identify relevant *cis*-regulatory elements, we performed reporter gene experiments and sequence comparisons with *ppcA* promoter sequences from closely related *Flaveria* species with a C_3, C_4, or C_3-C_4 intermediate type of photosynthesis.

The mesophyll-specific expression of the *ppcA1* gene of *F. trinervia* is controlled at the transcriptional level. About 2,200 base pairs of 5′ flanking sequences (with reference to the AUG translational start codon) are sufficient to cause high β-glucuronidase (GUS) expression exclusively in the mesophyll cells of the closely related C_4 plant *F. bidentis* (Stockhaus et al 1997; Fig. 3). In contrast, the 2,538 base pairs (with reference to the AUG start codon) of the 5′ flanking sequences of the *ppcA1* gene of *F. pringlei* were found to be a weak promoter and did not direct any organ-specific or cell-specific expression (Stockhaus et al 1997; Fig. 3). Both promoters thus exhibited the attributes expected from the accumulation patterns of their corresponding RNAs and proteins. The increase in gene expression, but exclusively in the leaves, and the confinement of expression to the mesophyll cells must be caused by differences between these two promoter sequences.

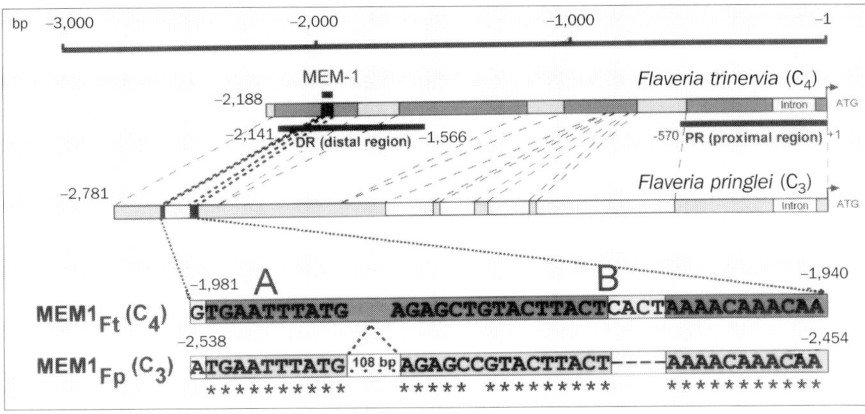

Fig. 4. Structures of the *ppcA1* promoters from *F. trinervia* (C_4) and *F. pringlei* (C_3) and the nucleotide composition of the mesophyll expression module MEM1. The numbers of nucleotides refer to the translation initiation codon. A darker color marks regions with high similarity (>60% identical nucleotides). The positions of MEM1 and its homolog in *F. pringlei* are marked by black boxes. Asterisks label identical nucleotides in the A or B segments of MEM1. The C/T difference in the B segment is not correlated with C_3-C_4 photosynthesis, because all C_4 *Flaveria* except *F. trinervia* contain a C at that position (Gowik et al 2004).

Promoter deletion and recombination studies revealed that a 41-base-pair segment, named MEM1 (mesophyll expression module 1), located in the distal segment of the *F. trinervia* promoter in combination with the proximal segment of this promoter, was sufficient to confer specificity for expression of the GUS reporter gene in the mesophyll. The proximal promoter part alone is expressed only weakly in mesophyll and bundle sheath cells (Fig. 4; Gowik et al 2004).

MEM1 homologous sequences were also detected in the *ppcA1* promoter of *F. pringlei* and in other C_3, C_4, and C_4-like *Flaveria* species (Gowik et al 2004). Their comparison revealed that MEM1 sequences consist of two parts, A and B, that are contiguous in *F. trinervia*, but are separated by 97–108 base pairs in the C_3 species *F. pringlei* and *F. cronquistii*, the C_4 plant *F. bidentis*, and the C_4-like species *F. palmeri* and *F. vaginata* (Gowik et al 2004). The A parts of all C_4 and C_4-like species show a guanine at their first nucleotide position, while an adenine is present in the A-homologs of the two C_3 species. A similar C_4 to C_3 associated difference is also found for the tetranucleotide CACT. This assemblage is present in the B parts of all C_4 and C_4-like species but absent in both C_3 promoters. These C_4 to C_3 correlated differences in MEM1 composition are candidates for *cis*-regulatory elements governing mesophyll-specific gene expression. Indeed, the *F. trinervia* MEM1 loses its ability to direct mesophyll-specific expression when the C_4-specific sequence motives are converted in their C_3 counterparts (Akyildiz and Westhoff, unpublished data).

In a DNA protein interaction screen with the yeast one-hybrid system (Li and Herskowitz 1993) using MEM1 as a bait, basic leucine zipper proteins were isolated that interact with MEM1 of *F. trinervia* but not with the MEM1 homolog of *F. pringlei*

(Akyildiz and Westhoff, unpublished data). The CACT tetranucleotide is embedded in a sequence context (TTACTCACTAA) that can form an imperfect palindrome. The palindrome resembles a binding site for a GCN4-like basic leucine zipper transcription factor (Arndt and Fink 1986, Oñate et al 1999, Matys et al 2003). The precise function of this DNA–protein interaction must be further investigated by gene knockout and/or overexpression experiments and biochemical approaches.

The full-length *ppcA* promoter of *F. trinervia* does not act properly in the C_3 plants tobacco and *Arabidopsis thaliana*. In *Arabidopsis*, it is active in mesophyll and bundle sheath cells (Engelmann and Westhoff, unpublished), whereas its activity is confined to the palisade parenchyma in tobacco (Stockhaus et al 1994). It might be very interesting to investigate the function of the orthologs of the basic leucine zipper proteins, identified as interactors of the C_4 MEM1, in these C_3 plants. That could provide information about the primary function of these transcription factors before they were recruited to control the expression of C_4 genes. Detailed knowledge of these interrelationships would help to better understand the evolution of C_4 photosynthesis as well as manipulate C_3 plants to express C_4 genes in a C_4-specific manner.

An example of a C_4-related promoter of *F. trinervia* with C_4-like cell-specific activity in C_3 plants is the promoter of the *GdcPA* gene. This gene encodes the P subunit of the glycine decarboxylase protein complex, which catalyzes the release of CO_2 during photorespiration. In the C_4 plant *F. bidentis*, this promoter is active only in the bundle sheath cells and vascular bundle (Burscheidt 1998). This is plausible because one of the major steps in the evolution of the C_4 pathway was to dislocate photorespiration, especially the CO_2-releasing reaction of this pathway into the bundle sheath cells (Monson 1989, Rawsthorne et al 1998). In *Arabidopsis*, the cell specificity of this promoter is the same as in the C_4 plant *F. bidentis*. Activity is confined to the bundle sheath cells and vascular bundle (Engelmann, unpublished). A detailed analysis of the *cis*-regulatory elements that are responsible for this promoter activity in *Arabidopsis* and *F. bidentis* is in progress.

These examples demonstrate that it is possible to imitate a C_4-specific gene expression in C_3 plants. But they also demonstrate that further detailed knowledge about the regulatory networks controlling the tissue-specific gene expression in the leaves of C_3 and C_4 plants is necessary before attempting the creation of a C_4 photosynthetic pathway in a C_3 plant.

Function of bundle sheath cells in C_3 plants

In C_4 plants, bundle sheath cells play an important role in the fixation of CO_2 as they represent the compartment in which CO_2 is concentrated and refixed by Rubisco.

Bundle sheath cells can also be found in C_3 plants. In *Arabidopsis thaliana* leaves, they constitute a distinct leaf cell type, as defined by their elongate morphology, their position adjacent to the vein, and differences in their chloroplast development compared with mesophyll cells (Kinsman and Pyke 1998). Developmental differences between the mesophyll and bundle sheath cells of C_3 plants are indicated by, for example, the reticulate mutants of *Arabidopsis* (Kinsman and Pyke 1998). One of these mutants, *cueI*, is defective in the gene encoding the phospho*enol*pyruvate/phosphate translocator (*AtPPT1*) (Streatfield et al 1999). A defect in this gene, which is expressed in the vascular tissues of roots and leaves, leads to a disturbance of chloroplast development only in mesophyll cells but not in bundle sheath cells (Knappe et al 2003, Voll et al 2003). Different physiological properties of mesophyll and bundle sheath cells of *Arabidopsis* can be inferred from, for instance, the specific expression of the gene encoding cytosolic ascorbate peroxidase (*APX2*) in bundle sheath cells under high-light conditions. This regulation of *APX2* may reflect a functional organization of the leaf to resolve two conflicting physiological requirements of protecting the sites of primary photosynthesis from reactive oxygen species and, at the same time, stimulating reactive oxygen species accumulation to signal responses to changes in the light environment (Fryer et al 2003).

The bundle sheath compartment is also defined by the activity of different promoters that are specifically active in the bundle sheath cells of the C_3 plants *Arabidopsis* or rice, for example, the promoter of the *Pck* gene from *Zoysia japonica* in rice (Nomura et al 2005a) or the promoter of the *Oshox1* gene of rice in *Arabidopsis* (Scarpella et al 2005), which is active in leaves in the vascular bundle and bundle sheath cells. The same is also true for the promoter of the *GdcPA* gene of *Flaveria trinervia* (Engelmann and Westhoff, unpublished data). The function of the bundle sheath cells in C_3 plants is largely unknown and we can only speculate about their physiological role. The high proportion of bundle sheath cell surface area in contact with adjacent cells suggests a central role for these cells in the transport of water and solutes into the mesophyll (Esau 1953). The position of bundle sheath cells makes them strong candidates for a role in the transfer of photoassimilate from mesophyll to phloem in source leaves. The bundle sheath may contribute to the mechanical stability of the leaf, as the tight attachment of the bundle sheath to the vascular strand appears to be greater than would be expected on the basis of the proportion of cell-cell contact. Furthermore, the arrangement of the vascular bundle as a cylinder composed of a ring of smaller cylinders enclosing the vascular strand may increase mechanical strength (Kinsman and Pyke 1998).

Bundle sheath cells appear photosynthetically competent since they contain a significant chloroplast population with normal internal morphology, and 20% of the bundle sheath is exposed to intercellular airspace. They constitute 15% of chloroplast-containing cells within the leaf. Consequently, these cells could contribute significantly to overall leaf photosynthesis (Kinsman and Pyke 1998).

It was reported that tobacco, a typical C_3 plant, shows characteristics of C_4 photosynthesis in cells of stems and petioles that surround the xylem and phloem. These photosynthetic cells possess high activities of enzymes characteristic of C_4 photosynthesis, which allow the decarboxylation of four-carbon organic acids from the xylem and phloem (Hibberd and Quick 2002). The existence of C_4 bundle sheath-like cells in C_3 plants, even if they can be found in stems and petioles and not in the leaves, might explain why C_4 photosynthesis could evolve independently many times. The genetic information for C_4 bundle sheath cells is already present in C_3 plants and only the place where this genetic program is expressed must have been changed.

Nevertheless, even if it is possible to induce a C_4 bundle sheath metabolism in the leaf bundle sheath cells by manipulation of one or only a few genes, the problem of disturbing the original function of these cells, which is not exactly known, remains. To create a two-cell type of C_4 photosynthesis in a C_3 plant, the metabolism of bundle sheath cells would have to be altered. Without detailed knowledge of the physiology and development of these cells, this could cause big problems if the original function of the bundle sheath cells is disturbed by manipulations. To avoid this, it is imperative to identify the original functions of this cell type in C_3 plants.

One way for better characterization of bundle sheath function would be a detailed transcriptome and proteome analysis of these cells. Because of the huge genetic resources available, *Arabidopsis thaliana* would be best suited for this kind of analysis. Unfortunately, it is challenging, especially when using dicot plants, to obtain clean preparations of bundle sheath cells, which are necessary for this kind of analysis. One way to overcome this problem might be to use known bundle sheath-specific promoters to tag these cells with *green fluorescent protein (gfp)* expression and to use fluorescence-assisted cell sorting for isolation of the cells. This technique worked well for the analysis of gene expression in different tissues of *Arabidopsis* root (Birnbaum et al 2003). The bundle sheath-specific promoters could also serve as a starting point for the isolation of DNA binding proteins, which are involved in cell-specific gene expression, for example, by using the yeast one-hybrid system. That would provide information about the transcription machinery responsible for bundle sheath-specific transcription in C_3 plants. *Arabidopsis* plants expressing a reporter gene like *gfp* or *gus* under the control of such a bundle sheath-specific promoter can also be used to search for mutants, which are affected in bundle sheath development or gene expression. Thereby, one can think of screening for both loss-of-function mutants and dominant gain-of-function mutants using an activation tagging approach (Walden et al 1994). With such a genetic approach, it should also be possible to identify genes responsible for the different leaf anatomies of C_4 and C_3 plants.

Outlook

The design of an efficient C_4 photosynthetic carbon cycle in a C_3 plant faces many difficulties. These can be overcome by choosing adequate experimental systems. The leaf anatomy of the C_3 plant must be changed and intercellular and intracellular transport systems for metabolites must be inserted. The high expression of the C_4 pathway enzymes must be introduced in a cell-specific manner. Thereby, one has to consider the kinetic and regulatory properties of these enzymes to avoid disadvantageous disturbances of other metabolic pathways. Almost no information is available about the function and physiological properties of bundle sheath cells in C_3 plants. So, for bundle sheath cells, it would not be clear what metabolic pathways one could disturb by overexpressing C_4 pathway enzymes in this compartment.

In our opinion, *Flaveria* is the ideal system to investigate genes known to be involved in C_4 photosynthesis. The availability of orthologous genes from closely related C_4 and C_3 species allows the identification of the molecular signatures that are characteristic for C_4 genes. An alternative to *Flaveria* could be the genus *Cleome*. *Cleome* also contains closely related C_3 and C_4 plants and, moreover, it is closely related to *Arabidopsis*. It might be easier, therefore, to transfer findings between these species than between *Arabidopsis* and *Flaveria* (Brown et al 2005). Unfortunately, the information about the physiology and biochemistry of *Cleome* is limited and so far no *Cleome* species can be genetically modified. The possibility to transform the C_4 plant *F. bidentis* allows us to investigate the physiological properties of C_4 enzymes by knockout or overexpression experiments. This information will be important when it has to be decided which C_4 pathway genes should be used for expression in C_3 plants and what their kinetic and regulatory properties should be. *Flaveria* could also be used to identify so-far-unknown genes important for C_4 metabolism by expression profiling of species with different types of photosynthesis. To identify genes responsible for the typical C_4 leaf anatomy, *Flaveria*, like other C_4 species, is not well suited because of the lack of genetic resources. It might be a better strategy to identify such genes in a model plant like *Arabidopsis* and then use the *Flaveria* system to verify the functions of these genes and investigate differences between the C_3 and C_4 orthologs.

References

Arndt K, Fink GR. 1986. GCN4 protein, a positive transcription factor in yeast, binds general promoters at all 5′ TGACTC 3′ sequences. Proc. Natl. Acad. Sci. USA 83:8516-8520.

Bauwe H. 1986. An efficient method for the determination of Km values for HCO_3^- of phospho*enol*pyruvate carboxylase. Planta 169:356-360.

Bauwe H, Kolukisaoglu Ü. 2003. Genetic manipulation of glycine decarboxylation. J. Exp. Bot. 54:1523-1535.

Birnbaum K, Shasha DE, Wang JY, Jung JW, Lambert GM, Galbraith DW, Benfey PN. 2003. A gene expression map of the *Arabidopsis* root. Science 302(5652):1956-1960.

Bläsing OE, Ernst K, Streubel M, Westhoff P, Svensson P. 2002. The non-photosynthetic phospho*enol*pyruvate carboxylases of the C_4 dicot *Flaveria trinervia*: implications for the evolution of C_4 photosynthesis. Planta 215:448-456.

Bläsing OE, Westhoff P, Svensson P. 2000. Evolution of C_4 phospho*enol*pyruvate carboxylase in *Flaveria*, a conserved serine residue in the carboxyl-terminal part of the enzyme is a major determinant for C_4-specific characteristics. J. Biol. Chem. 275:27917-27923.

Botha CEJ. 1992. Plasmodesmatal distribution, structure and frequency in relation to assimilation in C_3 and C_4 grasses in southern Africa. Planta 187:348-358.

Brown NJ, Parsley K, Hibberd JM. 2005. The future of C_4 research—maize, *Flaveria* or *Cleome*? Trends Plant Sci. 10(5):215-221.

Brown RH. 1999. Agronomic implications of C_4 photosynthesis. In: Sage RF, Monson RK, editors. C_4 plant biology. San Diego, Calif. (USA): Academic Press. p 473-507.

Burscheidt J. 1998. *Cis*-regulatorische Determinanten für mesophyll- und bündelscheidenspezifische gene expression in C_4-Spezies der Gattung *Flaveria*: Die Promotoren der Phospho*enol*pyruvat-Carboxylase- und der Glycin-Decarboxylasegene. Math.-Nat. Fakultät. Düsseldorf, Heinrich-Heine-Universität.

Carroll SB. 2000. Endless forms: the evolution of gene regulation and morphological diversity. Cell 101:577-580.

Chen L-M, Li K-Z, Miwa T, Izui K. 2004. Overexpression of a cyanobacterial phospho*enol*pyruvate carboxylase with diminished sensitivity to feedback inhibition in *Arabidopsis* changes amino acid metabolism. Planta 219(3):440-449.

Chitty JA, Furbank RT, Marshall JS, Chen Z, Taylor WC. 1994. Genetic transformation of the C_4 plant, *Flaveria bidentis*. Plant J. 6:949-956.

Crétin C, Santi S, Keryer E, Lepiniec L, Tagu D, Vidal J, Gadal P. 1991. The phospho*enol*pyruvate carboxylase gene family of *Sorghum*: promoter structures, amino acid sequences and expression of genes. Gene 99:87-94.

Doebley J, Lukens L. 1998. Transcriptional regulators and the evolution of plant form. Plant Cell 10:1075-1082.

Dong LY, Masuda T, Kawamura T, Hata S, Izui K. 1998. Cloning, expression, and characterization of a root-form phospho*enol*pyruvate carboxylase from *Zea mays*: comparison with the C_4-form enzyme. Plant Cell Physiol. 39:865-873.

Drincovich MF, Casati P, Andreo CS. 2000. NADP-malic enzyme from plants: a ubiquitous enzyme involved in different metabolic pathways. FEBS Lett. 490:1-6.

Edwards GE, Ku MSB. 1987. Biochemistry of C_3-C_4 intermediates. In: Hatch MD, Boardman NK, editors. The biochemistry of plants. Vol. 10. New York, N.Y. (USA): Academic Press, Inc. p 275-325.

Ehleringer JR, Monson RK. 1993. Evolutionary and ecological aspects of photosynthetic pathway variation. Annu. Rev. Ecol. Syst. 24:411-439.

Engelmann S, Bläsing OE, Gowik U, Svensson P, Westhoff P. 2003. Molecular evolution of C_4 phospho*enol*pyruvate carboxylase in the genus *Flaveria*: a gradual increase from C_3 to C_4 characteristics. Planta 217:717-725.

Engelmann S, Bläsing OE, Westhoff P, Svensson P. 2002. Serine 774 and amino acids 296 to 437 comprise the major C_4 determinants of the C_4 phospho*enol*pyruvate carboxylase of *Flaveria trinervia*. FEBS Lett. 524:11-14.

Ernst K, Westhoff P. 1996. The phospho*enol*pyruvate carboxylase (*ppc*) gene family of *Flaveria trinervia* (C_4) and *F. pringlei* (C_3): molecular characterization and expression analysis of the *ppcB* and *ppcC* genes. Plant Mol. Biol. 34:427-443.

Esau K. 1953. Plant anatomy. New York, N.Y. (USA): John Wiley.

Freitag H, Stichler W. 2000. A remarkable new leaf type with unusual photosynthetic tissue in a Central Asiatic genus of Chenopodiaceae. Plant Biol. 2:154-160.

Freitag H, Stichler W. 2002. *Bienertia cycloptera* Bunge ex Boiss., Chenopodiaceae, another C_4 plant without Kranz tissues. Plant Biol. 4:121-134.

Fryer MJ, Ball L, Oxborough K, Karpinski S, Mullineaux PM, Baker NR. 2003. Control of Ascorbate Peroxidase 2 expression by hydrogen peroxide and leaf water status during excess light stress reveals a functional organisation of *Arabidopsis* leaves. Plant J. 33(4):691-705.

Fukayama H, Hatch M, Tamai T, Tsuchida H, Sudoh S, Furbank R, Miyao M. 2003. Activity regulation and physiological impacts of maize C_4-specific phospho*enol*pyruvate carboxylase overproduced in transgenic rice plants. Photosynth. Res. 77(2-3):227-239.

Fukayama H, Tsuchida H, Agarie S, Nomura M, Onodera H, Ono K, Lee B-H, Hirose S, Toki S, Ku MSB, Makino A, Matsuoka M, Miyao M. 2001. Significant accumulation of C_4-specific pyruvate, orthophosphate dikinase in a C_3 plant, rice. Plant Physiol. 127(3):1136-1146.

Gowik U, Burscheidt J, Akyildiz M, Schlue U, Koczor M, Streubel M, Westhoff P. 2004. *cis*-regulatory elements for mesophyll-specific gene expression in the C_4 plant *Flaveria trinervia*, the promoter of the C_4 phospho*enol*pyruvate carboxylase gene. Plant Cell 16:1077-1090.

Gowik U, Engelmann S, Bläsing O, Raghavendra A, Westhoff P. 2006. Evolution of C_4 phosphoenolpyruvate carboxylase in the genus *Alternanthera*: gene families and the enzymatic characteristics of the C_4 isozyme and its orthologues in C_3 and C_3/C_4 *Alternantheras*. Planta 223(2):359-368.

Harpster MH, Taylor WC. 1986. Maize phospho*enol*pyruvate carboxylase: cloning and characterization of mRNAs encoding isozymic forms. J. Biol. Chem. 261:6132-6136.

Hatch MD. 1987. C_4 photosynthesis: a unique blend of modified biochemistry, anatomy and ultrastructure. Biochim. Biophys. Acta 895:81-106.

Hermans J, Westhoff P. 1990. Analysis of expression and evolutionary relationships of phospho*enol*pyruvate carboxylase genes in *Flaveria trinervia* (C_4) and *F. pringlei* (C_3). Mol. Gen. Genet. 224:459-468.

Hibberd JM, Quick WP. 2002. Characteristics of C_4 photosynthesis in stems and petioles of C_3 flowering plants. Nature 415:451-454.

Holaday AS, Bowes G. 1980. C_4 acid metabolism and dark CO_2 fixation in a submersed aquatic macrophyte (*Hydrilla verticillata*). Plant Physiol. 65:331-335.

Kawamura T, Shigesada K, Toh H, Okumura S, Yanagisawa S, Izui K. 1992. Molecular evolution of phospho*enol*pyruvate carboxylase for C_4 photosynthesis in maize: comparison of its cDNA sequence with a newly isolated cDNA encoding an isozyme involved in anaplerotic function. J. Biochem. 112:147-154.

Kellogg EA. 1999. Phylogenetic aspects of the evolution of C_4 photosynthesis. In: Sage RF, Monson RK, editors. C_4 plant biology. San Diego, Calif. (USA): Academic Press. p 411-444.

Kinsman EA, Pyke KA. 1998. Bundle sheath cells and cell-specific plastid development in *Arabidopsis* leaves. Development 125:1815-1822.

Knappe S, Lottgert T, Schneider A, Voll L, Flugge UI, Fischer K. 2003. Characterization of two functional phospho*enol*pyruvate/phosphate translocator (PPT) genes in *Arabidopsis*: AtPPT1 may be involved in the provision of signals for correct mesophyll development. Plant J. 36:411-420.

Kopriva S, Chu CC, Bauwe H. 1996. Molecular phylogeny of *Flaveria* as deduced from the analysis of nucleotide sequences encoding the H-protein of the glycine cleavage system. Plant Cell Environ. 19:1028-1036.

Ku MSB, Agarie S, Nomura M, Fukayama H, Tsuchida H, Ono K, Hirose S, Toki S, Miyao M, Matsuoka M. 1999. High-level expression of maize phospho*enol*pyruvate carboxylase in transgenic rice plants. Nat. Biotechnol. 17:76-80.

Leegood RC. 2002. C_4 photosynthesis: principles of CO_2 concentration and prospects for its introduction into C_3 plants. J. Exp. Bot. 53:581-590.

Leegood RC, Walker RP. 1999. Regulation of the C_4 pathway. In: Sage RF, Monson RK, editors. C_4 plant biology. San Diego, Calif. (USA): Academic Press. p 89-131.

Li JJ, Herskowitz I. 1993. Isolation of ORC6, a component of the yeast origin recognition complex by a one-hybrid system. Science 262:1870-1874.

Long SP. 1999. Environmental responses. In: Sage RF, Monson RK, editors. C_4 plant biology. San Diego, Calif. (USA): Academic Press. p 215-249.

Long SP, Zhu X-G, Naidu SL, Ort DR. 2006. Can improvement in photosynthesis increase crop yields? Plant Cell Environ. 29(3):315-330.

Matsumura H, Xie Y, Shirakata S, Inoue T, Yoshinaga T, Ueno Y, Izui K, Kai Y. 2002. Crystal structures of C_4 form maize and quaternary complex of *E. coli* phospho*enol*pyruvate carboxylases. Structure 10:1721-1730.

Matsuoka M, Kyozuka J, Shimamoto K, Kano-Murakami Y. 1994. The promoters of two carboxylases in a C_4 plant (maize) direct cell-specific, light-regulated expression in a C_3 plant (rice). Plant J. 6:311-319.

Matsuoka M, Tada Y, Fujimura T, Kano-Murakami Y. 1993. Tissue-specific light-regulated expression directed by the promoter of a C_4 gene, maize pyruvate, orthophosphate dikinase, in a C_3 plant, rice. Proc. Natl. Acad. Sci. USA 90:9586-9590.

Matys V, Fricke E, Geffers R, Gossling E, Haubrock M, Hehl R, Hornischer K, Karas D, Kel AE, Kel-Margoulis OV, Kloos DU, Land S, Lewicki-Potapov B, Michael H, Munch R, Reuter I, Rotert S, Saxel H, Scheer M, Thiele S, Wingender E. 2003. TRANSFAC: transcriptional regulation, from patterns to profiles. Nucl. Acids Res. 31:374-378.

McKown AD, Moncalvo J-M, Dengler NG. 2005. Phylogeny of *Flaveria* (Asteraceae) and inference of C_4 photosynthesis evolution. Am. J. Bot. 92(11):1911-1928.

Monson RK. 1989. On the evolutionary pathways resulting in C_4 photosynthesis and crassulacean acid metabolism (Cam). Adv. Ecol. Res. 19:57-110.

Monson RK. 1999. The origins of C_4 genes and evolutionary pattern in the C_4 metabolic phenotype. In: Sage RF, Monson RK, editors. C_4 plant biology. San Diego, Calif. (USA): Academic Press. p 377-410.

Monson RK, Moore B. 1989. On the significance of C_3-C_4 intermediate photosynthesis to the evolution of C_4 photosynthesis. Plant Cell Environ. 12:689-699.

Nomura M, Higuchi T, Ishida Y, Ohta S, Komari T, Imaizumi N, Miyao-Tokutomi M, Matsuoka M, Tajima S. 2005a. Differential expression pattern of C_4 bundle sheath expression genes in rice, a C_3 plant. Plant Cell Physiol. 46(5):754-761.

Nomura M, Higuchi T, Katayama K, Taniguchi M, Miyao-Tokutomi M, Matsuoka M, Tajima S. 2005b. The promoter for C_4-type mitochondrial aspartate aminotransferase does not direct bundle sheath-specific expression in transgenic rice plants. Plant Cell Physiol. 46(5):743-753.

Nomura M, Katayama K, Nishimura A, Ishida Y, Ohta S, Komari T, Miyao-Tokutomi M, Tajima S, Matsuoka M. 2000a. The promoter of *rbcS* in a C_3 plant (rice) directs organ-specific, light-dependent expression in a C_4 plant (maize), but does not confer bundle sheath cell-specific expression. Plant Mol. Biol. 44:99-106.

Nomura M, Sentoku N, Nishimura A, Lin JH, Honda C, Taniguchi M, Ishida Y, Ohta S, Komari T, Miyao-Tokutomi M, Kano-Murakami Y, Tajima S, Ku MSB, Matsuoka M. 2000b. The evolution of C_4 plants: acquisition of *cis*-regulatory sequences in the promoter of C_4-type pyruvate, orthophosphate dikinase gene. Plant J. 22:211-221.

Oñate L, Vicente-Carbajosa J, Lara P, Díaz I, Carbonero P. 1999. Barley BLZ2, a seed-specific bZIP protein that interacts with BLZ1 in vivo and activates transcription from the GCN4-like motif of B-hordein promoters in barley endosperm. J. Biol. Chem. 274:9175-9182.

Powell AM. 1978. Systematics of *Flaveria* (Flaveriinae-Asteraceae). Ann. Mo. Bot. Gard. 65:590-636.

Rademacher T, Hausler RE, Hirsch HJ, Zhang L, Lipka V, Weier D, Kreuzaler F, Peterhansel C. 2002. An engineered phospho*enol*pyruvate carboxylase redirects carbon and nitrogen flow in transgenic potato plants. Plant J. 32(1):25-39.

Rawsthorne S, Morgan CL, O'Neill CM, Hylton CM, Jones DA, Frean ML. 1998. Cellular expression pattern of the glycine decarboxylase P protein in leaves of an intergeneric hybrid between the C_3-C_4 intermediate species *Moricandia nitens* and the C_3 species *Brassica napus*. Theor. Appl. Genet. 96:922-927.

Renné P, Dreßen U, Hebbeker U, Hille D, Flügge U-I, Westhoff P, Weber A. 2003. The *Arabidopsis* mutant *dct* is deficient in the plastidic glutamate/malate translocator DiT2. Plant J. 35:316-331.

Sage RF, Li M, Monson RK. 1999. The taxonomic distribution of C_4 photosynthesis. In: Sage RF, Monson RK, editors. C_4 plant biology. San Diego, Calif. (USA): Academic Press. p 551-584.

Scarpella E, Simons EJ, Meijer AH. 2005. Multiple regulatory elements contribute to the vascular-specific expression of the rice HD-zip gene *Oshox1* in *Arabidopsis*. Plant Cell Physiol. 46(8):1400-1410.

Sheen J. 1999. C_4 gene expression. Annu. Rev. Plant. Physiol. Plant Mol. Biol. 50:187-217.

Stockhaus J, Poetsch W, Steinmüller K, Westhoff P. 1994. Evolution of the C_4 phospho*enol*pyruvate carboxylase promoter of the C_4 dicot *Flaveria trinervia*: an expression analysis in the C_3 plant tobacco. Mol. Gen. Genet. 245:286-293.

Stockhaus J, Schlue U, Koczor M, Chitty JA, Taylor WC, Westhoff P. 1997. The promoter of the gene encoding the C_4 form of phospho*enol*pyruvate carboxylase directs mesophyll-specific expression in transgenic C_4 *Flaveria* spp. Plant Cell 9:479-489.

Streatfield SJ, Weber A, Kinsman EA, Häusler RE, Li JM, Post-Beittenmiller D, Kaiser WM, Pyke KA, Flügge UI, Chory J. 1999. The phospho*enol*pyruvate/phosphate translocator is required for phenolic metabolism, palisade cell development, and plastid-dependent nuclear gene expression. Plant Cell 11:1609-1621.

Svensson P, Bläsing O, Westhoff P. 1997. Evolution of the enzymatic characteristics of C_4 phospho*enol*pyruvate carboxylase: a comparison of the orthologous ppcA phospho*enol*pyruvate carboxylases of *Flaveria trinervia* (C_4) and *F. pringlei* (C_3). Eur. J. Biochem. 246:452-460.

Svensson P, Bläsing OE, Westhoff P. 2003. Evolution of C_4 phospho*enol*pyruvate carboxylase. Arch. Biochem. Biophys. 414:180-188.

Takeuchi K, Akagi H, Kamasawa N, Osumi M, Honda H. 2000. Aberrant chloroplasts in transgenic rice plants expressing a high level of maize NADP-dependent malic enzyme. Planta 211:265-274.

Ting IP, Osmond CB. 1973a. Multiple forms of plant phospho*enol*pyruvate carboxylase associated with different metabolic pathways. Plant Physiol. 51:448-453.

Ting IP, Osmond CB. 1973b. Photosynthetic phospho*enol*pyruvate carboxylase: characteristics of allozymes from leaves of C_3 and C_4 plants. Plant Physiol. 51:439-447.

Tovar-Méndez A, Mújica-Jiménez C, Muñoz-Clares RA. 2000. Physiological implications of the kinetics of maize leaf phospho*enol*pyruvate carboxylase. Plant Physiol. 123:149-160.

Tsuchida H, Tamai T, Fukayama H, Agarie S, Nomura M, Onodera H, Ono K, Nishizawa Y, Lee BH, Hirose S, Toki S, Ku MSB, Matsuoka M, Miyao M. 2001. High-level expression of C_4-specific NADP-malic enzyme in leaves and impairment of photoautotrophic growth in a C_3 plant, rice. Plant Cell Physiol. 42:138-145.

Vidal J, Chollet R. 1997. Regulatory phosphorylation of C_4 PEP carboxylase. Trends Plant Sci. 2(6):230-237.

Voll L, Hausler RE, Hecker R, Weber A, Weissenbock G, Fiene G, Waffenschmidt S, Flugge U-I. 2003. The phenotype of the *Arabidopsis cue1* mutant is not simply caused by a general restriction of the shikimate pathway. Plant J. 36(3):301-317.

von Caemmerer S. 2003. C_4 photosynthesis in a single C_3 cell is theoretically inefficient but may ameliorate internal CO_2 diffusion limitations of C_3 leaves. Plant Cell Environ. 26(8):1191-1197.

Voznesenskaya EV, Franceschi VR, Kiirats O, Artyusheva EG, Freitag H, Edwards GE. 2002. Proof of C_4 photosynthesis without Kranz anatomy in *Bienertia cycloptera* (Chenopodiaceae). Plant J. 31:649-662.

Voznesenskaya EV, Franceschi VR, Kiirats O, Freitag H, Edwards GE. 2001. Kranz anatomy is not essential for terrestrial C_4 plant photosynthesis. Nature 414:543-546.

Walden R, Fritze K, Hayashi H, Miklashevichs E, Harling H, Schell J. 1994. Activation tagging: a means of isolating genes implicated as playing a role in plant growth and development. Plant Mol. Biol. 26:1521-1528.

Westhoff P, Gowik U. 2004. Evolution of C_4 phospho*enol*pyruvate carboxylase – genes and proteins: a case study with the genus *Flaveria*. Ann. Bot. 93:1-11.

Notes

Authors' address: Institut für Entwicklungs- und Molekularbiologie der Pflanzen, Heinrich-Heine-Universität, Universitätsstraße 1, D-40225 Düsseldorf, Germany; email: gowik@uni-duesseldorf.de, west@uni-duesseldorf.de.

Learning from nature to develop strategies for the directed evolution of C_4 rice

R. Sage and T.L. Sage

C_4 photosynthesis has evolved naturally over 50 times in 19 families of flowering plants. This repeated evolution of a complex trait indicates that it is either relatively easy or was under an intense directional selection pressure. Either way, the repeated evolution of C_4 photosynthesis indicates that it should be feasible to create C_4 rice plants by engineering C_4 genes into C_3 rice and replicating strong selection pressures for C_4 traits that we think exist in nature. Studies of the natural lineages, particularly those using species with intermediate characteristics of C_3 and C_4 photosynthesis, reveal the probable phases and selection pressures in the evolution of C_4 photosynthesis. A key early step is the formation of leaves with close vein spacing and slightly enlarged bundle sheath cells. Following this, the photorespiratory enzyme glycine decarboxylase is localized to the bundle sheath tissue, which allows for CO_2 to be concentrated into the bundle sheath, thereby improving photosynthetic efficiency at low atmospheric CO_2 concentration. Localization of glycine decarboxylase to the bundle sheath tissues facilitates the creation of Kranz-like anatomy and elaborate transportation networks between the mesophyll and bundle sheath tissues. This prepares the lineage to evolve the C_4 cycle, and is thus considered a key link in the evolutionary bridge to C_4 photosynthesis. Natural populations of C_3-C_4 intermediate species indicate that photorespiratory conditions (high temperature and low CO_2) were the main selection agent favoring the evolution of C_4 photosynthesis. By establishing screens based on high rates of photorespiration, genotypes transformed by natural selection and mutagenized populations could be bred for increasing expression of C_4-like characters. Biotechnology approaches could accelerate the breeding process by introducing critical genes; however, a screening approach will likely be needed to improve the many unknown traits involved in the evolution of C_4 plants.

Keywords: C_3-C_4 intermediates, directed evolution, glycine decarboxylase, phenotype screening, photorespiration, vein spacing

In warm climates of the current atmosphere, C_4 photosynthesis has a greater ability than C_3 photosynthesis to convert radiant energy into biologically useful forms of chemical energy that humans can exploit (Brown 1999, Sheehy 2000). More food, fiber, and fodder can thus be produced on a plot of land than is possible with C_3 plants, and, where the climate is hot and arid, C_4 plants often constitute the only major source of agricultural productivity (Sage 2001). C_4 photosynthesis is also more efficient than C_3 photosynthesis in terms of using water, nitrogen, and other mineral nutrients to produce valuable biomass (Long 1999). Thus, where water and mineral nutrients are scarce, or their exploitation causes large-scale environmental damage, the use of C_4 photosynthesis can increase the profitability and sustainability of agricultural production.

Unfortunately, relatively few C_4 species have been exploited for human use and most domesticated species are cereals and forage grasses (Brown 1999). The main C_4 crops are maize, sugar cane, sorghum, and millets. After these species, the most important C_4 crops are the suite of C_4 grasses that produce forage for grazing animals in low latitudes. Sorghum and millets remain unpopular in much of the world, and thus are not leading contributors to the human food supply on a global basis (Brown 1999). C_3 cereals remain the principal suppliers of food for human consumption, with wheat serving as the dominant grain in the temperate zones while rice dominates food production at lower latitudes (Maclean et al 2002). Wheat is largely a crop grown where the climate is cooler—at high latitudes, high elevation, or during the winter and spring in warmer climate zones (Sage and Pearcy 2000). Because C_4 photosynthesis is less effective than C_3 photosynthesis in cooler climates (defined here as having a mean growing-season temperature of less than 20 °C), it is not perceived as a priority to engineer the C_4 pathway into wheat and related C_3 cereals such as barley, oats, and rye—particularly with atmospheric CO_2 concentration increasing. In contrast, rice is a warm-climate crop that routinely experiences high temperatures where photorespiration is a major limitation and the advantages of C_4 photosynthesis are pronounced (Sage 2000). Indeed, rice is typically grown in flooded situations where the high water level suppresses competition from C_4 weeds (Galinato et al 1999, Sage 2000). Rice production in nonflooded soils is difficult because of severe yield losses from weed competition (Moody 1996). For these reasons, introducing C_4 photosynthesis into rice has considerable merit. Not only will the potential yield increase, but farmers may be able to grow rice effectively in nonflooded dryland soils, thus reducing the huge demand for water in rice-producing areas (Sheehy et al, this volume). C_4 rice should also be more competitive against C_4 weeds, and, with the increase in water-use efficiency (WUE), more land in drier climates could become available for rice production. Even in CO_2-enriched atmospheres, C_4 rice could have great value because of its potential to increase WUE and nitrogen-use efficiency over C_3 rice (Sage 2000). More people could be fed, at reduced production and environmental costs than currently possible with high-yielding varieties of C_3 rice.

Although the potential benefits of C_4 rice are clear, the means to obtain them are not. Initial efforts at importing genes from C_4 plants into rice failed to produce anything resembling C_4 photosynthesis, although some modest yield increases may

Table 1. A list of traits modified during the evolution of C_4 photosynthesis.

- C_4 cycle enzymes: PEP carboxylase (and associated regulatory proteins), pyruvate phosphate dikinase, aspartate aminotransferases, malate dehydrogenase, NAD(P) malic enzymes
- C_3 cycle enzymes: Rubisco and all Calvin cycle enzymes
- Other enzymes: carbonic anhydrase, glycine decarboxylase, all photorespiratory enzymes, organic acid transporters, triose phosphate transporters
- Stomatal regulation
- Structural: mesophyll to bundle sheath size and number, plasmodesmatal frequency, leaf thickness, intercellular space
- Ultrastructural: aquaporin distribution, bundle sheath wall properties, organelle localization, organelle number in mesophyll and bundle sheath, photosystem I and II distribution
- Stem xylem anatomy

have resulted from metabolic alteration of organic acid metabolism (Ku et al 2000, Häusler et al 2002, Matsuoka et al 2001). The failure to produce anything resembling efficient C_4 photosynthesis is not surprising, since it is well recognized that a fully functional C_4 pathway requires a coordinated change in tissue structure and metabolic biochemistry. Leegood (2002) nicely summarizes the requirements for a functional C_4 pathway as follows.

A fully-functional C_4 plant must have
(1) an active, photosynthetically driven CO_2 capture system,
(2) a supply of photosynthetic energy,
(3) an intermediate pool for captured CO_2,
(4) a mechanism to release CO_2 from the intermediate pool,
(5) a compartment in which to concentrate CO_2 around Rubisco, and
(6) a means to reduce CO_2 leakage.

In short, for efficient C_4 photosynthesis, the biochemical machinery has to be embedded in a structure that properly packages the enzymatic machinery, channels metabolite flow, and prevents the occurrence of futile cycles. How this is accomplished is only partially understood because few of the genetic changes converting a C_3 plant into a C_4 plant have been described. Most of the enzymes examined in detail code for metabolic enzymes; few studies have identified genes coding for anatomical and developmental changes in the genome during the evolution of C_4 plants. The total number of biochemical and structural trait changes required to convert a C_3 species into a C_4 species is large, encompassing dozens, if not hundreds, of genes (Table 1). How these changes occurred, and in what order, remains uncertain. To give a sense of the evolutionary challenge to produce C_4 photosynthesis from C_3 ancestors, consider two islands in a hostile sea. One "adaptive" island is C_3 photosynthesis, where the area occupied by the island represents the biochemical and structural configurations that provide a functional photosynthetic system that confers evolutionary fitness. The second island is C_4 photosynthesis, which, like the C_3 island, has a defined space where a specific biochemical and structural configuration confers fitness. Because C_4 photosynthesis has substantially different adaptive configurations than C_3 photosynthesis, the C_4 island is some distance away from the C_3 island, creating a major challenge for evolution to bridge the gap between the two pathways.

Evolution is not directional, in the sense that there is no road map to C_4 photosynthesis that evolution can follow. However, the evolution of complex traits often follows a specific sequence, with initial stages setting the stage for subsequent steps. It is now clear that the evolution of C_4 from C_3 photosynthesis also required a series of stages in a specific order (Monson 1999), such that the bridging of the evolutionary gap between a C_3 mode and a C_4 mode could be likened to building a pontoon bridge between the two adaptive islands. The bridge is laid down in sequence, with earlier stages facilitating the appearance of later stages. The challenge to biologists interested in engineering C_4 photosynthesis into C_3 species is to identify the sequential steps in C_4 evolution, and to develop ways to bring them about rapidly or, alternatively, engineer jumps around the more problematic stages. For this understanding, the best source of material is the many natural examples of C_4 evolution in the plant kingdom.

At least 50 independent evolutionary lineages of C_4 photosynthesis have been postulated, including a number where C_3-C_4 intermediates are present (Sage 2004, Muhaidat et al 2007). Phylogenies for these lineages are becoming available (e.g., McKown et al 2005, Sanchez-Acebo 2005), and the proper positioning of the intermediates in the evolutionary sequence is becoming a reality. With this information, we are in a much better position to evaluate hypotheses of how C_4 photosynthesis evolved, and how the evolutionary sequence might be used to facilitate the engineering of C_4 photosynthesis into many of our most valued species, most notably rice.

In this paper, we will summarize our understanding of the evolution of C_4 photosynthesis in natural populations, and, from this, propose a strategy we might follow to convert a C_3 species to C_4 within the funding and time limits that constrain agricultural research. We will briefly review the age and taxonomic distributions of C_4 photosynthesis, after which we will review the most recent models for the evolution of C_4 photosynthesis in natural populations. Identification of the taxonomic distribution of C_4 photosynthesis allows us to characterize the probable environments in which C_4 photosynthesis arose. Indentification of the habitat characteristics where C_4 species appeared would indicate which environmental conditions most likely selected for C_4 traits. This understanding of the selection environment for C_4 photosynthesis would provide conditions that breeders could replicate in a program to breed for C_4 rice.

The age and taxonomic distribution of C_4 photosynthesis

Current estimates indicate that C_4 photosynthesis first evolved in the grasses about 25 to 35 million years ago, during a time after the Oligocene climate deterioration (Kellogg 1999). This is a period when atmospheric CO_2 concentration dropped from over 500 ppm to below the current concentration, and the climate became drier over

Table 2. The distribution in angiosperm families of the estimated 50 lineages of C_4 plants. "+" indicates that there are likely additional genera remaining to be discovered.

Monocots 15+	Dicots 33+
Poaceae 10+	Acanthaceae 1
Cyperaceae 4	Aizoaceae 3
Hydrocharitaceae 1	Amaranthaceae 3
	Asteraceae 4
Total origins, approx. 50	Boraginaceae 1
	Capparidaceae 1
	Caryophyllaceae 1
	Chenopodiaceae 10
	Euphorbiaceae 1
	Molluginaceae 1
	Nyctaginaceae 1
	Polygonaceae 1
	Portulacaceae 2
	Scrophulariaceae 1
	Zygophyllaceae 2

much of the world (Zachos et al 2001). Since the end of the Oligocene, at least 50 independent evolutionary lines of C_4 plants have appeared in 19 higher plant families. Ten or more lineages are estimated for the grasses, four for the sedges, and some 32 in the eudicots (Table 2). The eudicot family with the most C_4 species is the Chenopodiaceae, with at least ten estimated lineages and some 500 species (Sage et al 1999b, Sage 2004). C_4 chenopods are estimated to have arisen in the early Miocene, some 20 million years ago (Kaderheit et al 2003). Many of the C_4 lineages in the eudicots appear to be recently evolved in geological time, based on relatively low numbers of C_4 species, the presence of C_3-C_4 intermediates, and close phylogenetic affinity to C_3 species. For example, in the genus *Anticharis* (Scrophulariaceae), there are six C_4 species and three C_3 species, while in the genus *Cleome*, there are three known C_4 species (Sage and Hibberd, unpublished). Based on similar reasoning, Ehleringer et al (1997) suggest that there was a burst of C_4 evolution in the past 5 million years, forming many of the current lineages, particularly in the eudicots. These younger lineages may hold the key to understanding how C_4 photosynthesis is assembled during the evolutionary process. To date, the most studied evolutionary lineage is in the genus *Flaveria* (Asteraceae). This group has become the model for C_4 evolution, due to the presence of many species with a range of intermediate traits between C_3 and C_4 forms. Intermediates in the genera *Heliotropium* (Boraginaceae), *Panicum* and *Neurachne* (Poaceae), *Mollugo* (Molluginaceae), and *Alternanthera* (Amaranthaceae) have also contributed to our understanding of C_4 origins, as have species in *Moricandia* and *Parthenium* (Asteraceae). *Moricandia* and *Parthenium* lack C_4 species but exhibit

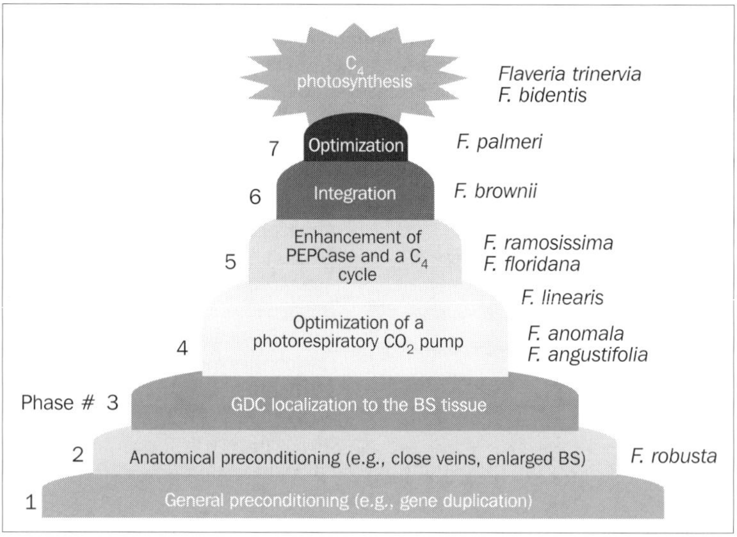

Fig. 1. A conceptual model of the main phases of C_4 photosynthesis, as derived from studies of species in the genus *Flaveria* and other species with C_3-C_4 intermediates. Individual *Flaveria* species are listed beside the phase whose characteristics they best demonstrate. Species between *F. bidentis* and *F. robusta* are considered C_3-C_4 intermediates. Adapted from Sage (2004) and Monson and Rawsthorne (2000).

many traits that may have been critical in the early phases of C_4 evolution (Monson 1999, Sage et al 1999b).

Flaveria and our current understanding of the pathway for C_4 evolution

Flaveria contains 23 known species, with four C_3 species, seven C_4 species, and a dozen or so intermediates ranging from C_3-like to C_4-like intermediates (McKown et al 2005). Two C_4 lineages are apparent in *Flaveria,* one leading to full C_4 photosynthesis in six species and a second leading to one C_4-like species, *F. brownii*. Assuming that the degree of intermediacy reflects the relative evolutionary position between the C_3 ancestors and C_4 progeny, then a step-wise model for C_4 evolution is possible (Monson 1989b, 1999). A recent version of this model is presented in Figure 1 (adapted from Sage 2004). In this model, seven distinct phases are delineated, although in reality the various steps would overlap to a considerable degree.

The initial phase is a preconditioning phase, which represents a group of traits that in some way predispose a lineage to evolve toward C_4 photosynthesis, often multiple times. Suggested traits that would predispose a species to begin evolving toward C_4 photosynthesis are a rapid life cycle, flexible leaf venation, and a high degree of gene duplication (Monson 2003). For example, in grasses, the parallel venation system may be better able to reduce vein spacing than the reticulate venation of the eudicots, allowing for the initiation of C_4 photosynthesis in less extreme selection environ-

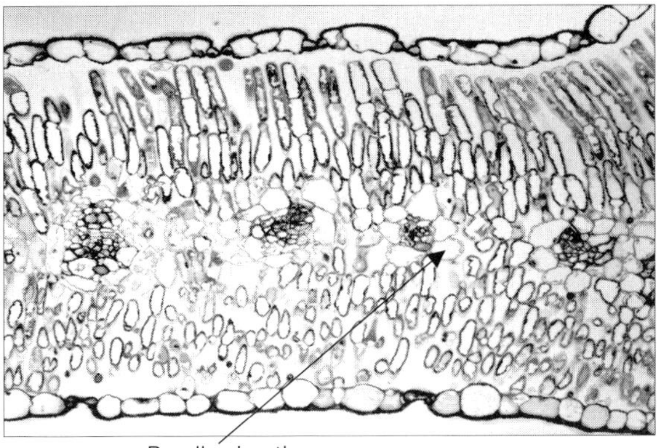

Fig. 2. A cross-section of a leaf from *Cleome sparsifolia*, a C_3 species closely related to the C_4 species *Cleome gynandra* (Sanchez-Acebo 2005). The bundle sheath tissue is indicated by the arrow.

ments (Ehleringer et al 1997). Gene duplication would produce copies of genes that could be altered without becoming lethal, while a rapid life cycle would accelerate the evolutionary process and allow for greater frequency of genetic recombination (Monson 2003).

The second phase of the evolutionary model is changes in the leaf anatomy, represented by a reduction in vein spacing possibly accompanied by an enlargement of bundle sheath size. This phase occurs within the context of C_3 photosynthesis, but is critical for the initial phases in C_4 evolution because it facilitates rapid metabolite flux between mesophyll and bundle sheath cells. Reduction in vein spacing and enhancement of the bundle sheath from diminutive cells to noticeable cells in cross-section are often observed in drought-adapted species, and have been specifically observed in C_3 species that are closely related to C_4 species (Sage 2001). For example, *Flaveria robusta*, the most closely related C_3 species to C_4 *Flaveria* species, is noted to have reduced vein spacing, and *Cleome sparsifolia*, a C_3 species that is a close relative to the C_4 plant *Cleome gynandra* (Sanchez-Acebo 2005), has close veins and enlarged bundle sheath cells relative to the norm in C_3 plants (Fig. 2). The reasons for these anatomical changes in C_3 plants remain unclear, but they tend to occur in species of hot, low-humidity environments where evapotranspiration potential is very high (Sage 2001). Hence, it has been suggested that close vein spacing may serve to increase the delivery rate of water to mesophyll cells when transpiration is high, thereby precluding the need to close stomates, and minimizing the risk of injury in the event of a sudden transpirational surge that can collapse cells and rupture their walls (Sage 2001, 2004).

Following an increase in vein density and bundle sheath size, the diffusion distance between mesophyll cells and the bundle sheath decreases, and the metabolic capacity of the bundle sheath tissue relative to the mesophyll cell increases (von

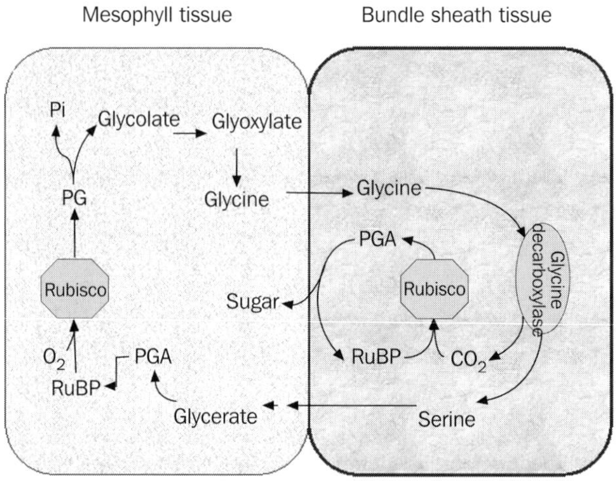

Fig. 3. A schematic diagram of the photorespiratory CO_2 pump that occurs upon localization of glycine decarboxylase to the bundle sheath tissue. When this occurs, the photorespiratory cycle that is shown in the diagram shifts from being contained in one cell to operating between the two distinct cell types of the mesophyll and bundle sheath.

Caemmerer 1989, 2000). Enlargement of the bundle sheath tissue is critical, because more organelles can then form inside, allowing the metabolic capacity of the bundle sheath to increase. These series of changes allow for the first critical metabolic event in the C_4 evolutionary sequence, which is delineated here as phase 3, the localization of the photorespiratory enzyme glycine decarboxylase (GDC) to the bundle sheath tissue. This occurs following a mutation that knocks out either the promoter or coding region of the mesophyll GDC gene (Morgan et al 1993, Monson and Rawsthorne 2000). In *Flaveria*, the mutation is apparently in the mesophyll-specific promoter of GDC, whereas, in *Moricandia*, the gene coding for mesophyll-specific GDC is impaired (Monson and Rawsthorne 2000, Gowik and Westhoff, this volume). In C_3-C_4 intermediates of *Flaveria* and *Moricandia,* bundle sheath localization of GDC is present, and this is apparently sufficient to metabolize all the glycine formed in the leaf via Rubisco oxygenase activity (Fig. 3; Hylton et al 1988, Morgan et al 1993, Monson and Rawsthorne 2000). The result is that all of the photorespiratory CO_2 is released in the bundle sheath compartment, and the serine produced by GDC activity diffuses back to the mesophyll cells, where it is processed to phosphoglyceric acid (PGA) by the remainder of the photorespiratory pathway (Fig. 3). With GDC localization to the bundle sheath, the normal, single-celled photorespiratory pathway instead becomes a two-tissue pathway requiring substantial trafficking of photosynthetic metabolites between mesophyll (MS) and bundle sheath (BS) tissues.

The evolutionary phase following GDC localization to the BS appears to be a series of modifications that optimize the ability of the BS cell to recover CO_2 released by GDC (Fig. 1). This involves further enhancement of the bundle sheath tissue up to sizes that can approach the size seen in the Kranz anatomy of C_4 plants (Brown and Hattersley 1989, Muhaidat et al 2007). In addition, organelles become more numerous and centrally located along the inner wall of the BS tissue (toward the vascular bundle), and the number of plasmodesmata may increase to facilitate rapid flux of photorespiratory metabolites between cells (Monson and Rawsthorne 2000). By the end of this phase, the leaf may have an optimized photorespiratory CO_2-concentrating mechanism, which reduces the photosynthetic CO_2 compensation point to below half the values seen in C_3 plants, and can enhance CO_2 assimilation at high temperature in extreme photorespiratory conditions (Schuster and Monson 1990, von Caemmerer 1989, 2000).

Once Kranz-like anatomy and a fully developed photorespiratory CO_2 pump are in place, the next major stage in C_4 evolution is the enhancement of the C_4 cycle. This initially involves a modest (< 25%) increase in the activities of PEPCase, NAD-ME, or NADP-ME, and other C_4 cycle enzymes (Monson and Rawsthorne 2000). Presumably, enhancement of PEPCase is accompanied by increased expression of the other C_4 cycle enzymes to prevent PEPCase from becoming substrate limited. The initial function of the C_4 cycle enzymes in C_3-C_4 intermediates is thought to be to capture photorespiratory CO_2 leaking out of the bundle sheath and send it back for refixation by Rubisco (Bauwe et al 1987); however, CO_2 newly entering the leaf would also be fixed by PEPCase. Once a limited C_4 cycle is engaged, it could create a selective advantage for further enhancement of bundle sheath activity of Rubisco and other enzymes for photosynthetic carbon reduction (PCR), which, in turn, could favor selection for further enhancements in C_4 cycle enzymes. This positive selection cycle could in relatively short order lead to C_4-like intermediates, where a strong C_4 cycle is operating alongside an attenuated C_3 cycle in the mesophyll (Sage 2004).

The next phase in C_4 evolution is the integration phase, in which the C_3 and C_4 cycles become fully integrated to allow for close coordination and high efficiency. The establishment of a C_4 cycle in the Kranz-like tissue of a C_3-C_4 intermediate would initially be inefficient because the component parts of the photosynthesis apparatus in these leaves would not be closely coordinated. In addition, there are probably many inefficiencies, such as competition between a weak mesophyll C_3 cycle and the C_4 cycle. Key steps during this evolutionary phase include loss of expression of Rubisco and some Calvin cycle enzymes from the mesophyll cells, and adjustment of thylakoid patterns to establish mesophyll to bundle sheath energy distributions that are appropriate for C_4 photosynthesis (Evans et al, this volume). Carbonic anhydrase would also have to be enhanced to provide bicarbonate for high-PEPCase activity (Burnell and Hatch 1988). At the end of the integration phase, a fully functional C_4 pathway is present; however, it may not be fully optimized, as many of the metabolic enzymes may require modification to work optimally in the specialized environment of the C_4 leaf.

The final phase in C_4 evolution is termed optimization. This involves changing the properties of the photosynthetic enzymes to optimize them for the differences in the mesophyll and bundle sheath environments, and adjustments in stomatal control and tissue hydraulics (Huxman and Monson 2003, Kocacinar and Sage 2003, Sage 2004). For example, the substitution of a serine for alanine at position 774 in PEP carboxylase comes late in the evolution of *Flaveria* species (Gowik and Westhoff, this volume). The C_4-like *F. brownii*, for example, lacks this substitution in its mesophyll PEPCase. This alanine to serine substitution alters the kinetics of PEPCase in such a manner that the enzyme operates more efficiently in the mesophyll environment where metabolite concentrations are altered relative to the C_3 situation. Malate pools have to be high in the mesophyll to favor diffusion to the bundle sheath, whereas high PEP pools drive PEP carboxylation in the forward direction. Both metabolites are known effectors of PEPCase (Leegood and Walker 1999). Rubisco is another major enzyme altered in the optimization phase. Typically, its K_{cat} is enhanced in C_4 plants to allow it to operate more rapidly in the high CO_2 conditions of the bundle sheath environment (Seemann et al 1984, Sage 2002). This allows for a reduction in Rubisco content in C_4 plants, and contributes to the high nitrogen-use efficiency of C_4 with respect to C_3 photosynthesis (Sage et al 1987). Numerous C_4 species, however, express forms of Rubisco that are similar to those of C_3 plants in terms of the catalytic efficiency and relative specificity for CO_2 (Sage 2002). The species that express a C_4 type of Rubisco, such as maize, are some of the most productive and efficient plants on the planet. Hence, to fully realize the yield enhancements foreseen in C_4 rice, it would be important to engineer plants through the optimization phase.

The critical significance of photorespiration in C_4 evolution

Much has been made about C_4 plants being favored in environments where photorespiration is a major inhibition to C_3 photosynthesis, and the impression is typically given that a high photorespiration rate would be a negative trait that natural selection should reduce. The irony is, however, that photorespiration also appears to be a positive trait that favors the initial stages of C_4 evolution by creating an internal resource (photorespired CO_2) upon which selection can act. Changes that allow for the compartmentalization and maximum refixation of the CO_2 resource would be beneficial such that selection would favor species that use photorespiration to enhance the efficiency of Rubisco and, in doing so, establish the anatomy, ultrastructure, and intertissue trafficking networks required by C_4 photosynthesis. Because of this, photorespiration is regarded to be the evolutionary bridge to C_4 photosynthesis (Bauwe and Kolukisaoglu 2003).

The bridging potential of photorespiration can be fully appreciated by considering the large number of complex changes that have to occur to establish a C_4 pathway. In addition to changes in expression and localization of PEPCase, Rubisco, the decarboxylating enzymes, and PPDK, many changes are needed in leaf development to establish Kranz anatomy, and in trafficking networks to create rapid diffusion pathways between mesophyll and bundle sheath cells (see Leegood, this volume, and Nelson et al, this volume). Also of significance are cellular modifications to establish a diffusive barrier in the bundle sheath tissue, which restricts CO_2 escape. Establishment of the diffusion barrier involves changes to wall properties and aquaporin characteristics, and the positioning of the chloroplasts and other organelles along the inner wall of the bundle sheath (Dengler and Nelson 1999, von Caemmerer et al, this volume). By establishing the conditions that favor the recovery of CO_2 released in the bundle sheath, a photorespiratory CO_2 pump facilitates the creation of the mesophyll to bundle sheath relationships that are needed for C_4 photosynthesis. In plants evolving a photorespiratory CO_2 pump, the movement of photorespiratory metabolites between the mesophyll and bundle sheath regions establishes the transport networks that can be later co-opted by the C_4 cycle. Moreover, the increase in the capacity of bundle sheath cells to refix photorespired CO_2 provides the enhanced C_3 metabolic capacity in the bundle sheath that is needed once a C_4 cycle has been established. This capacity of photorespiration to facilitate C_4 evolution may be a critically important tool that we could exploit in the directed evolution of C_4 rice. Artificial selection of rice plants in photorespiratory environments may allow for many of the more complicated structural changes to be established in rice, setting the stage for introducing a functional C_4 cycle into rice via molecular engineering.

What are the selection forces favoring C_4 photosynthesis?

In the various models of C_4 evolution, an enhanced potential for photorespiration is the primary selective agent; hence, factors that enhance photorespiration would be the critical environmental variables. Heat and low CO_2 are the two most important factors promoting photorespiration, with drought, salinity, and high light playing accessory roles because they contribute to situations where photorespiration is extreme (Sharkey 1988, Sage 2004). Drought and salinity promote stomatal closure, reducing intercellular CO_2 levels and thereby enhancing photorespiration. High light promotes high surface temperatures, thereby warming plants and promoting photorespiration.

Although these ideas are sound, they remain untested due to the difficulty of experimentally examining C_4 evolution in progress. A way to build support for these hypotheses is to examine the distribution of the postulated C_4 lineages in the plant kingdom. This is possible in the dicots, where the relatively small size of most C_4 groups and the relatively recent evolution allow us to examine the geographic distribution of the various species in the lineages. Centers of origin of C_4 photosynthesis in these lineages would be apparent where closely related C_3 and C_4 species occur, where the maximum diversity of the C_4 species occurs, and where intermediate forms are present. When these features are evaluated for the C_4 dicot lineages, the centers of origin can be predicted for most of the lineages (Fig. 4). All of the predicted centers of origin occur at low latitudes, and in the arid zones of their respective continents. These areas are characterized as being affected by monsoonal summer rains, and are very hot in the summer (peak temperatures of >40 °C), with frequent clear skies and drought between rain events (Sage 2004). The geography of the lineages is thus

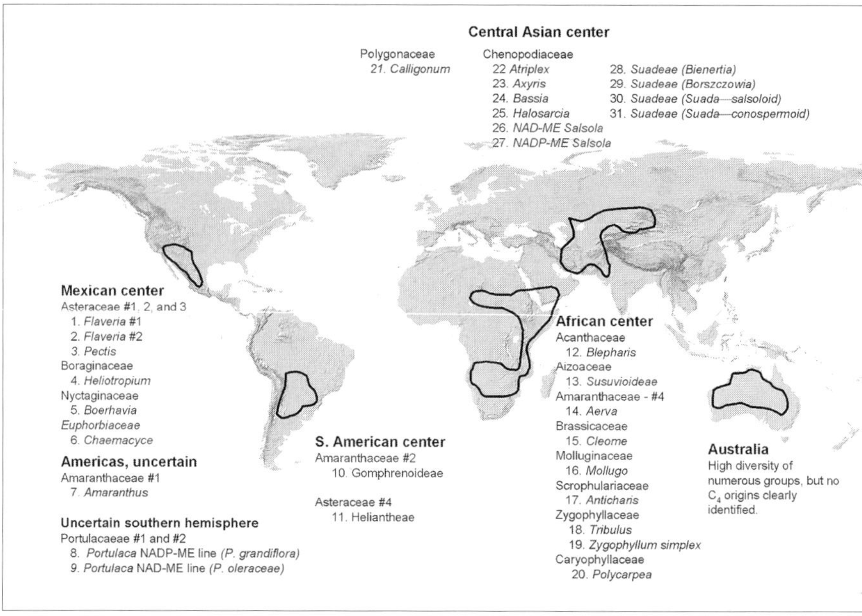

Fig. 4. The location of the postulated centers of origin for the estimated 34 lineages of C_4 photosynthesis in the eudicots. Adapted from Sage (2004).

consistent with the model that high rates of photorespiration promote the evolution of the C_4 pathway.

Further refinement of the potential selection pressures for C_4 evolution can be identified by examining the microhabitat distribution of the C_3 species known to be closely related to C_4 species, and the distribution of known C_3-C_4 intermediate species, notably intermediates in *Flaveria* and *Heliotropium*. A compilation of the habitats of these species reveals a series of common characteristics. These sites are often characterized by harsh substrates, notably sand, gypsum, or saline flats (Fig. 5). These soils are notoriously unproductive, and patches of bare ground are common. Disturbance plays a role, notably from waves, wind, fire, or flooding. This in combination with the low productive potential contributes to openness of the sites and many opportunities for establishment. Competition is likely low at these sites due to sparse vegetation cover. The species are active during hot periods of the year, typically when summer rains are available. A particularly notable feature is that the surfaces the species occur on are highly radiative—emitting high amounts of long-wave radiation and reflecting much short-wave radiation from a very bright sun. As a consequence, herbaceous plants growing on the surface receive an unusually high amount of energy from both the sun and the surrounding open surface. This would promote high leaf temperatures (>40 °C) and extreme rates of photorespiration.

Where the potential for photorespiration would be most extreme is in the surface boundary layer, where solar energy and long-wave radiation from the hot surface can

Fig. 5. The habitat of *Cleome sparsifolia*, on dark sand in the Nevada desert, USA. Plants of *C. sparsifolia* are scattered in the foreground. Note the open nature of the habitat. The insert shows a seedling in the boundary layer of the dark sand. The location of the photo is the junction of Highways 6 and 95, in west-central Nevada. (Photos by R. Sage.)

be trapped. In high-radiation environments, near-surface temperatures can be well above bulk air temperatures, by 10 to 20 °C in low wind conditions (Sage and Sage 2002). Seedlings of summer-active plants experience these intense boundary layer temperatures right after germination (Fig. 5, inset) and, given their lack of deep roots, have a high likelihood of dying of drought or high temperature unless they can rapidly grow above the soil boundary layer. To do this, they require either significant carbon from seed reserves or a robust rate of photosynthesis. The high photorespiration rates expected in the boundary layer would hinder photosynthesis and thereby slow growth out of the boundary layer. Under these dangerous conditions, any traits that can conserve carbon and accelerate growth and photosynthesis would presumably be highly favored by natural selection acting upon seedlings.

To consider fully the factors promoting C_4 evolution, we need to place these considerations in a geological context, and consider the photorespiratory situation in atmospheres of the past 20 million years. Atmospheric CO_2 concentrations were generally 30% to 50% below the current concentration since the early Miocene period (Zachos et al 2001). Although the global climate was cooler on average, surface temperatures in the subtropics and tropics were still very high because the radiation load was high and summer air temperatures at low latitude were still high, even during glacial episodes (Kutzbach et al 1993, Farrera et al 1999). As shown in Figure 6, seedlings in the hot boundary layer of the soil would have experienced very high rates

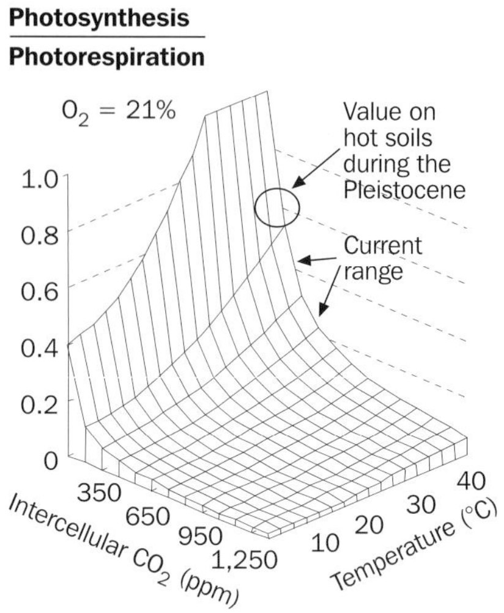

Fig. 6. The theoretical ratio of photorespiration to photosynthesis for a C_3 plant as a function of intercellular CO_2 concentration and leaf temperature. Values were modeled according to Sharkey (1988) using spinach Rubisco kinetic properties. (Adapted from Ehleringer et al 1991.) Arrows indicate the range of intercellular CO_2 values present in the current atmosphere, and the predominant low CO_2 atmosphere of the Pleistocene epoch (18,000 to 2.5 million years ago).

of photorespiration during low CO_2 episodes, probably equivalent to 60% to 80% of the photosynthesis rate. For seedlings at least, these high rates of photorespiration would have been a severe drag on their growth potential and ability to establish before the seedling was killed by drought, infection, or herbivory. In contrast, seedlings able to exploit the photorespired pool of CO_2 to enhance photosynthetic efficiency would have been at a selective advantage, as they would have been more capable of putting a leaf canopy above the boundary layer and establishing an adult plant. Based on this reasoning, it is logical to propose that selection for increased recovery of photorespiratory CO_2 in seedlings growing on hot, barren surfaces was how the process of C_4 evolution began.

Campbell et al (2005) recently examined seedling performance in response to a combination of low CO_2 and elevated temperature in tobacco seedlings (Fig. 7). Over a 3-week exposure period beginning at emergence, seedlings grown at 19 °C days and 15 °C nights were able to establish, as indicated by sustained growth in height to above 2 cm, and the production of enlarged leaves over 4 cm^2 in area. Seedlings at 25/19 °C and 100 ppm failed to establish, and remained tiny. At 30 °C days and 25 °C nights, seedlings at both 100 and 150 ppm failed to establish, remaining in the boundary layer throughout the experiment. The treatment of 150 ppm CO_2 approaches the minimum

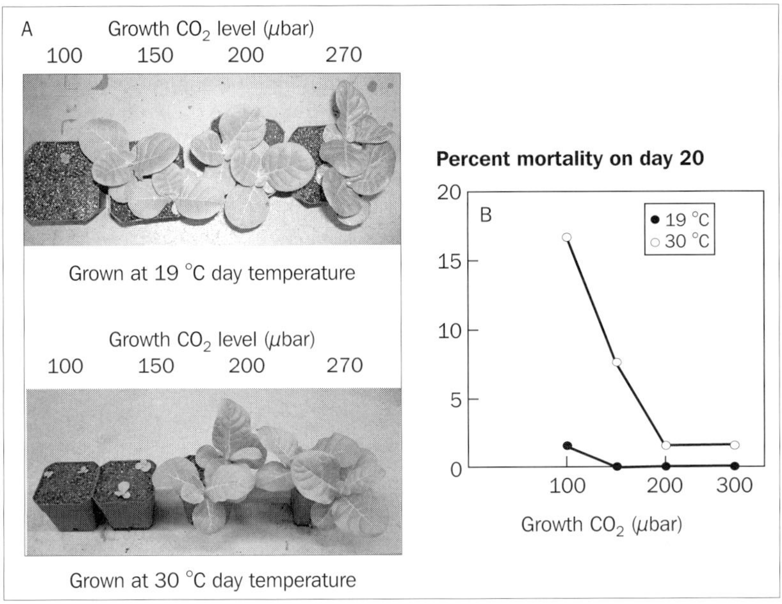

Fig. 7. (A) Photos showing the size of tobacco seedlings grown for 3 weeks at 100, 150, 200, or 270 ppm CO_2 and either 19 °C day and 15 °C night or 30/25 °C day/night temperatures. Adapted from Campbell (2004). (B) Mortality data for plants in the growth conditions described for panel A at 3 weeks after emergence. (Adapted from Campbell et al 2005.)

CO_2 concentration (180 ppm) in Earth's atmosphere during the recent ice ages of the Pleistocene. Presumably, at 180 ppm, temperatures near 35 °C would also prevent establishment, as has been shown in beans (Cowling and Sage 1998). Notably, at 30 °C, tobacco seedlings at 100 ppm CO_2 exhibited enhanced mortality (Fig. 7), which supports the possibility of increased selection pressure on genotypes lacking any ability to recapture photorespiratory CO_2. These experiments support the hypothesis that selection on seedlings would have been an important part of the C_4 evolutionary process. In the context of creating a C_4 rice plant, these results indicate that exposing seedlings to elevated temperatures and low CO_2 may be an excellent screen to filter rice genotypes for traits that could be useful in creating a C_4 rice plant.

Learning from nature: designing systems to select for C_4 traits in an accelerated manner

In considering the many traits and associated genes that have to be modified to create an efficient C_4 pathway, it is apparent that transformation of a series of single genes will not produce the high efficiency and productivity gains noted in comparisons between rice and maize (Sheehy et al, this volume). A functional C_4 cycle might be introduced into rice, and it may yield some benefits, but without the effective compartmentalization that allows for CO_2 concentration, it is difficult to see how such a plant could avoid severe CO_2 leakage, with the consequent loss of efficiency. Therefore, it is envisioned that creating productive C_4 plants will require the establishment of Kranz anatomy, with the associated high transport capacity between mesophyll and bundle sheath cells. This may be the most difficult aspect of C_4 evolution because we do not understand the genes controlling the shift from C_3 to C_4 anatomy. It seems certain that the genetic controls are complex and may not be amenable to a mutant approach or single-gene transformation (see Nelson et al, this volume, and Langdale et al, this volume). Therefore, a directed breeding protocol that emphasizes the establishment of Kranz anatomy may be the most promising initial approach to creating C_4 rice. Establishing high rates of photorespiration may be the best screen for C_4-like anatomical traits that will then facilitate the successful introduction of C_4 metabolism via molecular engineering.

A directed breeding approach for C_4 rice should exploit two salient observations from the studies of natural C_4 evolution. First, the primary selection force for C_4 evolution was most likely extreme rates of photorespiration brought on by low CO_2, heat, drought, and salinity. Second, the selection pressure would have been most intense during establishment, when photosynthesis is most limiting and seedlings are most vulnerable to premature death. Screening seedlings in hot, low-CO_2 conditions may thus be the best approach to select for C_4-like traits in a reasonable amount of time. Photorespiration appears to have been a bridge in natural C_4 evolution. It may also be the bridge that we can exploit to breed for the more complex characteristics arising during C_4 evolution. The following presents a scheme to facilitate the directed evolution of C_4 rice.

Large production greenhouses such as the one shown in Figure 8 could be closed off to the atmosphere except for vents that allow for resupply of atmospheric air at a controlled rate. Maintaining low CO_2 is potentially problematic, as an efficient CO_2 scrubbing system is needed. Chemical scrubbers are expensive at large scales, and a plant-based scrubbing system may be a better approach. Johnson et al (2000) used C_4 plants to deplete CO_2 to low values in a greenhouse setting, indicating that a similar system could maintain low CO_2 levels in a large greenhouse. Precise control of CO_2 could be maintained by allowing air from outside to enter at a controlled rate, using an infrared gas analyzer to open and close the vents as needed. The inside CO_2 concentration in the greenhouse would then be a balance between the leak rate and the scrubbing by productive C_4 plants such as maize or sugar cane. Temperature and humidity would be controlled using standard, compressor-based air-conditioning systems widely used in research greenhouses around the world.

Seedlings of rice or any other species of interest would be sown in flats, small pots, or planting tubes as routinely used in nursery production. Upon emergence, the seedlings would be exposed to a photorespiratory regime of low CO_2 and elevated growth temperature that has been predetermined to prevent growth. The exact conditions used for the screen would depend upon the species used, but, based on the work of Cowling and Sage (1998) and Campbell et al (2005), 150 ppm and 30 to 35 °C may be ideal. Photorespiratory conditions should be maintained until clear differences in seedling size emerge. Fluorescence imaging could also be used to screen high numbers of seedlings; where photosynthetic capacity is improved, the seedlings would show a different fluorescence signal from the bulk of the population. The more productive seedlings would be examined for traits of interest, and, if such traits were present, the plants would be removed and grown to maturity in normal atmospheric conditions, allowing for relatively rapid maturation in a low maintenance environment. Valuable traits would be those that contribute to C_4 evolution; for example, higher vein density, enlarged bundle sheath size, greater organelle number in the bundle sheath, more plasmodesmata between the mesophyll and bundle sheath, increased expression of C_4 enzymes, and the localization of glycine decarboxylase to the bundle sheath. Upon flowering, the plants would be crossed with others selected from the screen. Crossbreeding the offspring would be essential to limit loss of genetic diversity and to create novel genetic combinations. The offspring would then be re-screened as seedlings in the low-CO_2 greenhouse, after which the more productive individuals with improved traits would be segregated and grown to maturity outside. The process would then be repeated many times until the desired results were obtained.

Screening seedlings is advantageous for several reasons. The seedling is the most sensitive stage of the life cycle to high photorespiratory conditions (Campbell et al 2005). Seedlings also require the least amount of space, and only a few weeks are needed to grow seedlings to a size at which differences in performance would arise. At a minimum, it is anticipated that dozens of screening cycles would be required, so compressing the screening periods is necessary.

Ideally, the photorespiratory screen would select for genotypes in a manner analogous to the natural evolutionary pathway postulated for C_4 photosynthesis (Fig. 1). In nature, the time to produce a C_4 plant from a C_3 ancestor is not known, but thousands of years seem like a realistic possibility. Obviously, to convert a C_3 line to a C_4 line, the evolutionary process would have to be compressed to one or two decades. For this to occur, humans would have to use all tools available. Strategic use of mutants, natural diversity, gene transformations, and activation tagging could provide the genotypes upon which the selection process could act in an expeditious manner. Ideally, by creating the right combination of traits through strategic transformations and crossbreeding, we could set up a system in which the screen would filter out inefficient combinations while favoring an optimal combination of traits.

Conclusions: the next steps forward to prove the concept

Studies of the natural process of C_4 evolution have allowed for robust models describing the phases required to produce C_4 progeny from C_3 ancestors. In all cases, high rates of photorespiration play a key role, and the early stages in C_4 evolution appear to be associated with scavenging of the carbon in photorespiratory metabolites. Humans have already demonstrated an excellent ability to introduce genes for C_4 metabolic enzymes into rice; however, we do not have sufficient knowledge of the genetic controls over leaf development to engineer Kranz anatomy. Here, we propose to combine molecular engineering with artificial selection for Kranz-like traits in warm, low-CO_2

Fig. 8. A commercial greenhouse, demonstrating the design of a trial that might be used in a selection protocol for C_4-like traits. Photorespiratory conditions would be established in the greenhouse by reducing CO_2 to 100 to 150 ppm and raising temperatures to 30 to 35 °C. (Photo by R. Sage.)

environments. Rather than trying to directly engineer Kranz anatomy, which may require too many steps for which we now have no knowledge, we suggest it may be better to let selection in a low-CO_2 environment establish the many changes needed for Kranz anatomy. Molecular engineering of key traits would be critical, as this could help establish an efficient selection system. For example, knocking out the mesophyll GDC may be necessary to allow selection to favor traits leading to a photorespiratory CO_2 pump and enlarged bundle sheath cells.

At the current time, insufficient knowledge is available to use photorespiratory screens effectively. Before embarking on a major effort to create C_4 rice, we need to identify many of the genes and associated promoters required for strategic transformation. We need to get a sense of the variation in the rice genome for traits of importance to C_4 photosynthesis, and we need to evaluate in more detail key genetic changes in the natural evolution of C_4 photosynthesis. Furthermore, we need to prove the concept of a low-CO_2 screen, which might be done using *Arabidopsis* transgenics. If the screens were able to select for enhanced Kranz anatomy in *Arabidopsis*, especially in genotypes transformed with a C_4 cycle or GDC knockouts, then we could be reasonably confident that we could use the approach in other species such as rice. To complete this preliminary work in short order, however, would require a consortium of labs working together to attract funding and skilled labor. As photorespiration may be the initial bridge to C_4 photosynthesis, the initial bridge to C_4 rice appears to rest in the early organizational steps, and preliminary experiments, needed to justify the concerted effort to create a C_4 rice plant.

References

Bauwe H, Keerberg O, Bassüner R, Pärnik T, Bassüner B. 1987. Reassimilation of carbon dioxide by *Flaveria* (Asteraceae) species representing different types of photosynthesis. Planta 172:214-218.

Bauwe H, Kolukisaoglu U. 2003. Genetic manipulation of glycine decarboxylation. J. Exp. Bot. 54:1523-1525.

Brown RH. 1999. Agronomic implications of C_4 photosynthesis. In: Sage RF, Monson RK, editors. C_4 plant biology. San Diego, Calif. (USA): Academic Press. p 473-507.

Brown RH, Hattersley PW. 1989. Leaf anatomy of C_3-C_4 species as related to evolution of C_4 photosynthesis. Plant Physiol. 91:1543-1550.

Burnell JN, Hatch MD. 1988. Low bundle sheath carbonic anhydrase is apparently essential for effective C_4 pathway operation. Plant Physiol. 86:1252-1256.

Campbell CD. 2004. Plant performance at low atmospheric CO_2: interactions with phosphorous supply and growth temperature. Ph.D. dissertation. University of Toronto, Toronto, Ontario, Canada. 158 p.

Campbell CD, Sage RF, Kocacinar F, Way DA. 2005. Estimation of the whole-plant CO_2 compensation point of tobacco (*Nicotiana tabacum* L.). Global Change Biol. 11:1956-1967.

Cowling SA, Sage RF. 1998. Interactive effects of low atmospheric CO_2 and elevated temperature on growth, photosynthesis and respiration in *Phaseolus vulgaris*. Plant Cell Environ. 21:427-435.

Dengler NG, Nelson T. 1999. Leaf structure and development in C_4 plants. In: Sage RF, Monson RK, editors. C_4 plant biology. San Diego, Calif. (USA): Academic Press. p 133-172.

Ehleringer JR, Sage RF, Flanagan LB, Pearcy RW. 1991. Climate change and the evolution of C_4 photosynthesis. Trends Ecol. Evol. 6:95-99.

Ehleringer JR, Cerling TE, Helliker BR. 1997. C_4 photosynthesis, atmospheric CO_2 and climate. Oecologia 112:285-299.

Farrera I, Harrison SP, Prentice IC, Ramstein G, Guiot J, Bartlein PJ, Bonnefille R, Bush M, Cramer W, von Grafenstein U, Holmgren K, Hooghiemstra H, Hope G, Jolly D, Lauritzen SE, Ono Y, Pinot S, Stute M, Yu G. 1999. Tropical climates at the last glacial maximum: a new synthesis of terrestrial paleoclimate data. I. Vegetation, lake-levels and geochemistry. Clim. Dyn. 15:823-856.

Galinato MI, Moody K, Piggin CM. 1999. Upland rice weeds of South and Southeast Asia. Manila (Philippines): International Rice Research Institute. 156 p.

Häusler RE, Hirsch HJ, Kreuzaler F, Peterhänsel C. 2002. Overexpression of C_4-cycle enzymes in transgenic C_3 plants: a biotechnological approach to improve C_3 photosynthesis. J. Exp. Bot. 53:591-607.

Huxman TE, Monson RK. 2003. Stomatal responses of C_3, C_3-C_4 and C_4 *Flaveria* species to light and intercellular CO_2 concentration: implications for the evolution of stomatal behaviour. Plant Cell Environ. 26:313-322.

Hylton CM, Rawsthorne S, Smith AM, Jones DA. 1988. Glycine decarboxylase is confined to the bundle-sheath cells of leaves of C_3-C_4 intermediate species. Planta 175:452-459.

Johnson HB, Polley HW, Whitis RP. 2000. Elongated chambers for field studies across atmospheric CO_2 gradients. Funct. Ecol. 14:388-396.

Kaderheit G, Borsch T, Weising K, Freitag H. 2003. Phylogeny of Amaranthaceae and Chenopodiaceae and the evolution of C_4 photosynthesis. Int. J. Plant Sci. 164:959-986.

Kellogg EA. 1999. Phylogenetic aspects of the evolution of C_4 photosynthesis. In: Sage RF, Monson RK, editors. C_4 plant biology. San Diego, Calif. (USA): Academic Press. p 411-444.

Kocacinar F, Sage RF. 2003. Photosynthetic pathway alters xylem structure and hydraulic function in herbaceous plants. Plant Cell Environ. 26:2015-2026.

Kutzbach JE, Guetter PJ, Behling PJ, Stein R. 1993. Simulated climatic changes: results of the COHMAP climate-model experiments. In: Wright HE Jr, Kutzbach JE, Webb T, Ruddiman WF, Street-Perrott FA, Bartlein PJ, editors. Global climates since the last glacial maximum. Minneapolis, Minn. (USA): University of Minnesota Press. p 24-93.

Ku MSB, Cho D, Ranade U, Hsu T-P, Li X, Jiao D-M, Ehleringer J, Miyao M, Matsouka M. 2000. Photosynthetic performances of transgenic rice plants overexpressing maize C_4 photosynthesis enzymes. In: Sheehy JE, Mitchell PL, Hardy B, editors. Redesigning rice photosynthesis to increase yields. Manila (Philippines): International Rice Research Institute, and Amsterdam (Netherlands): Elsevier Science. p 193-204.

Leegood RC. 2002. C_4 photosynthesis: principles of CO_2 concentration and prospects for its introduction into C_3 plants. J. Exp. Bot. 53:581-590.

Leegood RC, Walker RP. 1999. Regulation of the C_4 pathway. In: Sage RF, Monson RK, editors. C_4 plant biology. San Diego, Calif. (USA): Academic Press. p 89-132.

Long SP. 1999. Environmental responses. In: Sage RF, Monson RK, editors. C_4 plant biology. San Diego, Calif. (USA): Academic Press. p 215-249.

Maclean JL, Dawe DC, Hardy B, Hettel GP, editors. 2002. Rice almanac. Manila (Phillipines): International Rice Research Institute. 253 p.

Matsouka M, Furbank RT, Fukayama H, Miyao M. 2001. Molecular engineering of C_4 photosynthesis. Annu. Rev. Plant Physiol. Mol. Biol. 52:297-314.

McKown AD, Moncalco J-M, Dengler NG. 2005. Phylogeny of *Flaveria* (Asteraceae) and inference of C_4 photosynthetic evolution. Am. J. Bot. 92:1911-1928.

Monson RK. 1989a. On the evolutionary pathways resulting in C_4 photosynthesis and crassulacean acid metabolism (CAM). Adv. Ecol. Res. 19:57-101.

Monson RK. 1989b. The relative contributions of reduced photorespiration, and improved water- and nitrogen-use efficiencies, to the advantages of C_3-C_4 intermediate photosynthesis in *Flaveria*. Oecologia 80:215-221.

Monson RK. 1999. The origins of C_4 genes and evolutionary pattern in the C_4 metabolic phenotype. In: Sage RF, Monson, RK, editors. C_4 plant biology. San Diego, Calif. (USA): Academic Press. p 377-410.

Monson RK. 2003. Gene duplication, neofunctionalization, and the evolution of C_4 photosynthesis. Int. J. Plant Sci. 164:S43-S54.

Monson RK, Rawsthorne S. 2000. CO_2 assimilation in C_3-C_4 intermediate plants. In: Leegood RC, Sharkey TD, von Caemmerer SC, editors. Photosynthesis: physiology and metabolism. Dordrecht (Netherlands): Kluwer Academic. p 533-550.

Moody K. 1996. Priorities for weed science research. In: Evenson RE, Herdt RW, Hossain M, editors. Rice research in Asia: progress and priorities. Wallingford (UK): CAB International. p 277-290.

Morgan CL, Turner SR, Rawsthorne S. 1993. Coordination of the cell-specific distribution of the four subunits of glycine decarboxylase and of serine hydroxymethyltransferase in leaves of C_3-C_4 intermediate species from different genera. Planta 190:468-473.

Muhaidat RM, Dengler NG, Sage RF. 2007. Identification of new C_3-C_4 intermediates in the genus *Heliotropium* (Boraginaceae). Abstract #14007. American Society of Plant Biology Conference Proceedings. http://abstracts.aspb.org/pb2006/public/P14/P14007.html.

Sage RF. 2000. C_3 versus C_4 photosynthesis in rice: ecophysiological perspectives. In: Sheehy JE, Mitchell PL, Hardy B, editors. Redesigning rice photosynthesis to increase yields. Manila (Philippines): International Rice Research Institute, and Amsterdam (Netherlands): Elsevier Science. p 13-38.

Sage RF. 2001. Environmental and evolutionary preconditions for the origin and diversification of the C_4 photosynthetic syndrome. Plant Biol. 3:202-213.

Sage RF. 2002. Variation in the k_{cat} of Rubisco in C_3 and C_4 plants and some implications for photosynthetic performance at high and low temperature. J. Exp. Bot. 53:609-620.

Sage RF. 2004. The evolution of C_4 photosynthesis. New Phytol. 161:341-370.

Sage RF, Li MR, Monson RK. 1999b. The taxonomic distribution of C_4 photosynthesis. In: Sage RF, Monson RK, editors. C_4 plant biology. San Diego, Calif. (USA): Academic Press. p 551-584.

Sage RF, Pearcy RW. 2000. The physiological ecology of C_4 photosynthesis. In: Leegood RC, Sharkey TD, von Caemmerer S, editors. Photosynthesis: physiology and metabolism. Dordrecht (Netherlands): Kluwer Academic. p 497-532.

Sage RF, Pearcy RW, Seemann JR. 1987. The nitrogen use efficiency of C_3 and C_4 plants. III. Leaf nitrogen effects on the activity of carboxylating enzymes in *Chenopodium album* L. and *Amaranthus retroflexus* L. Plant Physiol. 85:355-359.

Sage RF, Sage TL. 2002. Microsite characteristics of *Muhlenbergia richardsonis* (Trin.) Rydb., an alpine C_4 grass from the White Mountains, California. Oecologia 132:501-508.

Sage RF, Wedin DA, Li MR. 1999a. The biogeography of C_4 photosynthesis: patterns and controlling factors. In: Sage RF, Monson RK, editors. C_4 plant biology. San Diego, Calif. (USA): Academic Press. p 313-373.

Sanchez-Acebo L. 2005. A phylogenetic study of the new world *Cleome* (Brassicaceae, Cleomoideae). Ann. Mo. Bot. Gard. 92:179-201.

Schuster WS, Monson RK. 1990. An examination of the advantages of C_3-C_4 intermediate photosynthesis in warm environments. Plant Cell Environ. 13:903-912.

Seemann JR, Badger MR, Berry JA. 1984. Variations in specific activity of ribulose-1,5-bisphosphate carboxylase between species utilizing differing photosynthetic pathways. Plant Physiol. 74:791-794.

Sharkey TD. 1988. Estimating the rate of photorespiration in leaves. Physiol. Plant. 73:147-152.

Sheehy JE. 2000. Limits to yield for C_3 and C_4 rice: an agronomist's view. In: Sheehy JE, Mitchell PL, Hardy B, editors. Redesigning rice photosynthesis to increase yields. Manila (Philippines): International Rice Research Institute, and Amsterdam (Netherlands): Elsevier Science. p 39-52.

Svensson P, Bläsing OE, Westhoff P. 2003. Evolution of C_4 phospho*enol*pyruvate carboxylase. Arch. Biochem. Biosci. 414:180-188.

von Caemmerer S. 1989. A model of photosynthetic CO_2 assimilation and carbon-isotope discrimination in leaves of certain C_3-C_4 intermediates. Planta 178:463-474.

von Caemmerer S. 2000. Biochemical models of leaf photosynthesis. Collingwood (Australia): CSIRO Publishing.

Wessinger ME, Edwards GE, Ku MSB. 1989. Quantity and kinetic properties of ribulose 1,5-bisphosphate carboxylase in C_3, C_4, and C_3-C_4 intermediate species of *Flaveria* (Asteraceae). Plant Cell Physiol. 30:665-671.

Zachos J, Pagani M, Sloan L, Thomas E, Billups K. 2001. Trends, rhythms and aberrations in global climate 65 Ma to present. Science 292:686-693.

Notes

Authors' address: Department of Ecology and Evolutionary Biology, University of Toronto, 25 Willcocks Street, Toronto, Ontario M5S3B2, Canada, email: R.sage@utoronto.ca.

The regulation of genes in C_3 plants that have been co-opted into C_4 photosynthesis, and implications for making a C_4 rice

J.M. Hibberd

Placing the efficient C_4 pathway into rice appears ambitious because it would involve modifications to biochemistry, leaf development, and cell biology. The biochemical modifications need to include high expression of genes encoding carbonic anhydrase, phospho*enol*pyruvate carboxylase, malate dehydrogenase, and pyruvate orthophosphate dikinase in the mesophyll, while a decarboxylase and Rubisco are specifically needed in the bundle sheath. Alterations in leaf development required are increased venation, larger bundle sheath cells, and fewer mesophyll cells. Changes in cell biology include chloroplast proliferation and expansion in the bundle sheath, and increased plasmodesmatal connectivity between mesophyll and bundle sheath cells.

Although these modifications appear complex, C_3 species have the ability to accumulate proteins needed for C_4 photosynthesis in defined cell types, and it also appears that they possess *trans*-factors needed for the expression of genes specifically in mesophyll or bundle sheath cells. When intact genes from a C_4 species are placed in a closely related C_3 plant, they are expressed in the correct cell type for C_4 photosynthesis, but in more distantly related species this is less likely. It should therefore be possible to integrate enzymes needed for C_4 photosynthesis into rice if genes are sourced from a closely related C_4 plant.

I propose a dual-track approach to the challenge of integrating C_4 traits into rice. First, studies of rice leaf development are needed. Second, fundamental work is needed on C_4 photosynthesis itself, and the species used should depend on the particular question being asked. The hypothesis that introducing the biochemistry of C_4 photosynthesis into a C_3 plant leads to leaf development associated with the C_4 pathway should be tested. It would be fastest to do this by placing genes from *Cleome gynandra* into *Arabidopsis thaliana*. If this hypothesis is supported, a shortcut to the whole process of generating a C_4 rice could be found. If the hypothesis is not supported, many phenotypes of C_4 plants are shared, and so loci controlling these could be identified in systems other than rice and maize. For example, bundle sheath enlargement and increased plasmodesmatal connectivity should be investigated with *A. thaliana* and *Cleome* because resources and generation times are favorable.

Keywords: C_4 photosynthesis, rice, *Arabidopsis*, *Cleome*, leaf biochemistry, leaf development

The aim of the work being conducted in my laboratory is twofold. First, we wish to understand how, in C_3 plants, genes encoding enzymes that have been recruited into C_4 photosynthesis are regulated. Second, we aim to identify the functions of the proteins encoded by these genes in C_3 species. Comparative analysis of both gene regulation and protein function in C_3 and C_4 plants should then provide insights into the evolution of C_4 photosynthesis.

We have shown that C_3 plants have photosynthetic cells associated with the veins of stems and petioles. Cells containing chlorophyll around veins are particularly easy to see in stems of species such as onion (Fig. 1A), but can also easily be seen in transverse sections of celery, tobacco (Hibberd and Quick 2002), and *Arabidopsis* (Fig. 1B). We set out to understand the role of photosynthesis in cells around the veins of C_3 plants.

Initially, the source of CO_2 for cells containing chlorophyll associated with veins in stems and petioles of C_3 species was unclear for two reasons: first, these cells are distant from stomata on the stem surface; second, there tend to be few intercellular airspaces between the cortical cells (Esau 1955). Both of these factors decrease conductivity for CO_2. The xylem stream is a potential alternative source of CO_2 for these cells. To test whether these chlorophyll-containing cells around veins receive CO_2 from the xylem, radiolabeled carbon was fed to the xylem stream, and the sites of its fixation determined. Insoluble radiolabeled material accumulated in a light-dependent manner in cells associated with veins (Hibberd and Quick 2002). This was true when HCO_3^-, glucose, or malate was supplied. We proposed that HCO_3^- enters the xylem stream, and, as it moves toward the leaves, it exits the xylem and is fixed by Rubisco in cells around the xylem. Incorporation of radiolabeled material around the veins after glucose was supplied to roots is consistent with glucose being respired in roots, with the CO_2 generated entering the xylem stream and then being fixed in photosynthesis. Last, incorporation of isotopically labeled carbon into insoluble material around the veins after radiolabeled malate was supplied to the transpiration stream implied that there was an ability to remove CO_2 from malate, and use that CO_2 in photosynthesis (Hibberd and Quick 2002).

The three decarboxylase enzymes known to release CO_2 from four carbon compounds are NADP-dependent malic enzyme (NADP-ME), NAD-dependent malic enzyme (NAD-ME), and phospho*enol*pyruvate carboxykinase (PEPCK) (Kanai and Edwards 1999). Significant activities of each of these decarboxylation enzymes were detectable in cells surrounding the veins of tobacco petioles (Hibberd and Quick 2002). This implies that any or all of these four-carbon decarboxylases could be used to provide CO_2 to Rubisco in cells around the veins of C_3 species.

When malate is decarboxylated via NADP-ME and NAD-ME, the three-carbon intermediate pyruvate is generated. In C_4 species, pyruvate orthophosphate dikinase (PPDK) phosphorylates pyruvate in chloroplasts to generate PEP and so allows the

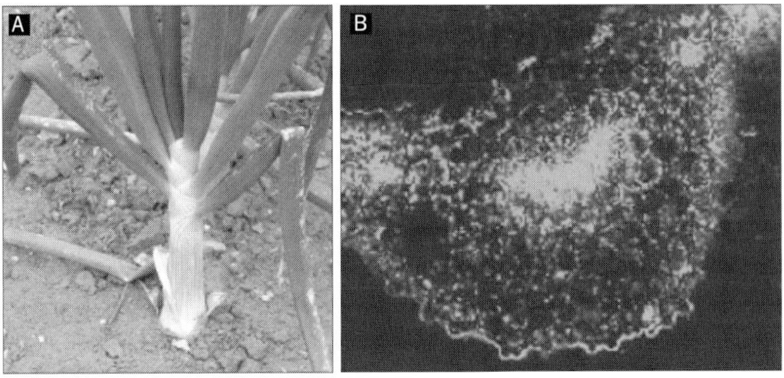

C

```
At2g47490              1 ~~~~~~~~~~~~~~~~~~~~~~~~~~~~~~~~~~~~~~~~~~~~~~~~~~~~~~~~~~~~
Yeast transporter      1 MTQTDNPVPNCGLLPEQQYCSADHEEPLLLHEEQLIFPDHSSQLSSADIIEPIKMNSSTE

At2g47490              1 ~~~~~~MSANSHPPNSKNVLCNAAAGAAAGVVAATFVCPLDVIKTRFQVHGFTLPKLGDA
Yeast transporter     61 SIIGTTLRKKWVPLSSTQI..TALSGAFAGFLSGVAVCPLDVAKTRLQAQGLQ.TRFENP

At2g47490             55 NIKGSLIVGSLEQIFKREGMRGLYRGLSPTVMALLSNWAIYFTMYDQLKSFLFTCSNDHK
Yeast transporter    118 YYRG..IMGTLSTIVRDEGPRGLYKGLVPIVLGYFPTWMIYFSVYEFSKKFFHGIF..PQ

At2g47490            115 LSVGANVLAASGAGAATTIATNPLWVVKTRLQTQGMRVGIVP..YKSTFSALRRIFTAYE
Yeast transporter    174 FDFVAQSCAAITAGAASTTLTNPIWVVKTRLMLQS.NLGEHPTHYKGTFDAFRKLF..YQ

At2g47490            173 EGIRGLYSGLVPALAGISHVAIQFPTYEMIKV.YLAKKGDKSVDNLNARDVAVASSIFTA
Yeast transporter    231 EGFKALYAGLVPSLLGLFHVAIHFPIYEDLKVRFHCYSRENNTNSINLQRLIMASSV..S

At2g47490            232 KIFASTLTYPHEVVRARLQEQGHHSEKRYSGVRDCIKKVFEKDGFPGFYRGCATNLLRFT
Yeast transporter    289 KMIASAVTYPHEILRTRMQLKSDIPDSIQRRLFPLIKATYAQEGLKGFYSGFTTNLVR..

At2g47490            292 TTPAAVITFTSFEMVHRFLVTHIPSEQSSIL~
Yeast transporter    347 TIPASAITLVSFE....YFRNRLENISTMVI*
```

Fig. 1. **(A) Cells containing chlorophyll around the veins are visible in onion (image taken in Cottenham, Cambridge, UK). (B) Transverse section of *Arabidopsis thaliana* petiole viewed with confocal laser scanning microscopy, red represents chlorophyll fluorescence, and the chlorophyll in the center of the section surrounds the central vascular bundle of the petiole. (C) Alignment of an *Arabidopsis* protein with significant similarity to the yeast pyruvate transporter. Yeast transporter is YIL006w, and At2g47490 is the putative *Arabidopsis* transporter.**

C_4 cycle to continue. High activities of PPDK could also be measured around veins of tobacco (Hibberd and Quick 2002). In order to define the regulation of the gene for PPDK (*PPDK*), and function of the protein in C_3 plants, we have used *Arabidopsis thaliana*, which possesses a single gene but with two promoters giving rise to two types of transcript. The longer transcripts are generated from a promoter upstream of the first exon, whereas the shorter transcripts are derived from a promoter found within the first intron of the gene (Parsley and Hibberd 2006). A similar gene structure is found in C_4 dicotyledons (Rosche and Westhoff 1995) as well as in C_3 and C_4 monocotyledons (Sheen 1991, Imaizumi et al 1997). As the same gene structure for *PPDK* is found in maize, rice, *Arabidopsis*, and *Flaveria*, it appears that the chloroplastic version of PPDK has been recruited from the same gene organization in each lineage of C_4 plants. When each cDNA for PPDK from *Arabidopsis* was translationally fused to a green fluorescent protein (GFP) reporter, it was possible to show that the longer transcript encodes a protein targeted to the chloroplast, that its first exon acts as a transit peptide, and that the smaller protein is cytosolic. Transcripts for both cytosolic and plastidic PPDK proteins are detectable in veins of *Arabidopsis*, indicating that both isoforms are likely to perform important roles in these cells of C_3 species (Parsley and Hibberd 2006).

Overall, these results imply the following: (1) that chlorophyll accumulation in cells around the veins of tobacco provides light-harvesting capacity to allow carbon fixation in those cells; (2) that CO_2 from the xylem stream is supplied to these cells as CO_2, HCO_3^-, or malate; (3) that, if malate leaves the xylem, it will be decarboxylated to generate CO_2 (for photosynthesis) and PEP; and (4) that PPDK generates a chloroplastic pool of PEP in cells around veins of C_3 plants. Currently, it is thought that PEP for the chorismate pathway is supplied by glycolysis (Herrmann 1995), and so PPDK may provide an alternative route to deliver PEP for chorismate synthesis, allowing the provision of carbon skeletons important for secondary metabolism, including the biosynthesis of lignin in cells around veins of C_3 plants.

Regulation of genes that have been co-opted into C_4 photosynthesis

The presence of C_4 biochemistry in cells around veins of tobacco indicates that regulatory mechanisms exist in C_3 species to allow genes encoding these enzymes to be expressed at high levels relative to the photosynthetic apparatus. This may be one of the reasons that C_4 has evolved multiple times within the angiosperms. To gain further insight into the evolution of C_4 photosynthesis, we need to determine how the regulation of these genes differs in closely related C_3 and C_4 species.

In C_4 plants, the mechanisms by which enzymes of C_4 photosynthesis accumulate preferentially in mesophyll or bundle sheath cells vary. For example, elements in the promoter of *PEPC* from *Flaveria trinervia* are needed for mesophyll expression (Gowik et al 2005), and, in maize DNA, demethylation of the *PEPC* promoter is associated with transcript accumulation in mesophyll cells (Langdale et al 1991). In contrast, for high expression of NADP-ME in the bundle sheath of *Flaveria*, elements in the promoter, the start of the coding region, and the terminator appear to interact (Ali and Taylor 2001), whereas in *Amaranthus hypochondriacus*, the 5′ and 3′ untranslated regions (UTRs) stabilize transcripts preferentially in the bundle sheath (Patel et al 2006).

Although the mechanisms generating cell-specific accumulation of proteins differ between genes and C_4 lineages, when intact genes from a C_4 plant are placed in a closely related C_3 plant, they are expressed in the correct cell type for C_4 photosynthesis. Intact *PPDK* and *PEPC* genes from maize are highly expressed in rice in the cell types needed for C_4 photosynthesis to operate (Ku et al 1999, Fukayama et al 2001). This indicates that *trans*-factors needed for C_4-like expression of these genes are already present in rice, and, importantly, that they operate in the correct cell types. Analysis of the PEPC promoter from the C_4 *Flaveria trinervia* placed in the C_3 model tobacco also supports this conclusion. Elements within the *FtPEPC* promoter drive expression of *uidA* in mesophyll cells of tobacco (Stockhaus et al 1994, Gowik et al

2005). There are cases where mechanisms regulating protein accumulation in bundle sheath and mesophyll cells are conserved between more distantly related C_3 and C_4 plants. For example, the 5′ and 3′ UTRs from *Amaranthus hypochondriacus* enhance translation in C_3 tobacco as well as in C_4 *Flaveria bidentis* (Patel et al 2004), and these UTRs stabilize transcripts in bundle sheath cells of *F. bidentis* (Patel et al 2006). But there are also examples where mechanisms generating cell specificity in gene expression fail in more distantly related species, for example, when the *FtPEPC* promoter is placed in *A. thaliana*, it is no longer faithful to the mesophyll cells (Westhoff, personal communication). Overall, this indicates that factors that recognize *cis*-elements within the *FtPEPC* promoter are conserved enough in closely related species to generate mesophyll-specific expression, but not in more distantly related plants.

To gain further insight into how gene regulation has altered as C_4 photosynthesis evolved, comparative analysis of orthologous genes in C_3 and C_4 plants is needed. There are more resources associated with *A. thaliana* than with any other C_3 plant, and so progress in understanding the regulation of genes should be fastest with *Arabidopsis*. We therefore set out to identify how genes encoding the four-carbon decarboxylases are regulated in *Arabidopsis*. This analysis includes assessing whether promoters already direct expression in specific cells, whether terminator regions enhance expression, and whether UTRs are regulatory (Brown and Hibberd, unpublished). If analysis of *Arabidopsis* is to provide real insight into how regulation of genes has altered as C_4 photosynthesis evolved, the availability of a model C_4 plant that is closely related to *Arabidopsis* is imperative.

Phylogenetically informed approaches to understand C_4 photosynthesis

We are advocating the use of a genus containing C_4 plants (*Cleome*) that has largely been ignored to date (Brown et al 2005). *Cleome* contains the most closely related C_4 species to *A. thaliana* and so we can use knowledge of *Arabidopsis* to inform our understanding of C_4 photosynthesis. *A. thaliana* itself can also be used as a biological test-tube to define the genetic components necessary for C_4 photosynthesis. Because integrating C_4 photosynthesis into rice is ambitious, it may need several approaches to be successful. It is therefore pertinent to ask whether *Arabidopsis* and *Cleome* are relevant to generating a C_4 rice.

Although some aspects of C_4 photosynthesis are restricted to specific subtypes, many phenotypes are shared by all the subtypes, and a system such as *Arabidopsis* and *Cleome* should provide insights into these features more quickly than restricting work to maize and rice. For example, I argue that *Arabidopsis* is likely to be useful in identifying genes that control bundle sheath cell expansion, proliferation of bundle sheath chloroplasts, and formation of plasmodesmata between the bundle sheath and mesophyll (see later).

Choosing the C_4 system and the sub-type

If integrating C_4 photosynthesis into rice is to be worthwhile, the full suite of adaptations that are found in productive C_4 species is needed. This implies that using a single-cell version of C_4 photosynthesis is unlikely to be useful. One of the first decisions to be made if a C_4 rice is to be engineered relates to which sub-type of C_4 photosynthesis is chosen. Choosing the sub-type is important, as the underlying biology of each sub-type differs with respect to both the enzymes that have been recruited into C_4 photosynthesis and leaf and cell development. For example, sub-types vary in the transporters used to move metabolites between cytosol and chloroplasts or mitochondria, positioning of organelles in the bundle sheath, thylakoid stacking, and suberization of cell walls (Hatch 1987).

There are several reasons to use maize as the source of genetic information to introduce the C_4 pathway into rice: (1) maize and rice are relatively closely related; (2) maize happens to be one of the most productive C_4 crops we have, and the agronomic reasoning for integrating C_4 traits into rice is based on a comparison of yield potentials of rice and maize; (3) maize is probably the best-studied C_4 monocot, and empirical data so far indicate that when intact genes from maize are placed in rice they are expressed in a manner faithful to the C_4 pathway (Ku et al 1999, Matsuoka et al 2001, Fukayama et al 2001); and (4) there are improving genomic resources (genome and EST sequences, mutant collections, and genome annotation) associated with maize. As maize uses NADP-ME, it appears that this is the likely sub-type to be placed into rice.

Two extreme alternatives to engineer a C_4 rice

Once the sub-type of C_4 photosynthesis that is to be integrated into rice has been decided, the sequence of events to allow this genetic engineering should be considered. Two extreme approaches could probably be used. First, research could focus solely on the large number of *Oryza* accessions, and, together with placing genes from maize into rice, an attempt to engineer a C_4 rice could be made. Because the life cycles of maize and rice are long compared to *Arabidopsis*, this approach is likely to take a considerable amount of time. Second, knowledge and strategies being developed with alternative models could be used to convert the most tractable model C_3 species (*Arabidopsis*) into a C_4 plant, and, when this is done, the knowledge gained from this venture could be used to convert rice to C_4 photosynthesis. This would be analogous to manufacturing engineers producing a prototype before the working model. It is likely that the correct way forward would combine these two extremes. For a C_4 rice to be generated, I therefore advocate the adoption of two parallel strategies.

Strategy one

Analysis of *Oryza* accessions

First, as has already been initiated at IRRI, a concerted effort to define developmental flexibility within *Oryza* needs to be made. If it is considered possible that any *Oryza* species uses C_4 photosynthesis, then this should be addressed by screening all accessions for their $\delta^{13}C$ isotopic signatures. Otherwise, the screens being conducted for vein spacing, number of interveinal cells, and chloroplast abundance need to continue, and methods streamlined to allow high throughput. This sort of screen should include wild relatives, cultivars, as well as mutated rice collections, and lines in which *trans*-activation constructs have been placed. The screen is relatively simple; leaves can be cleared in ethanol in multiwell format, and then images digitally recorded. If lines are found in rice with increased venation, these could then be used to accept loci known to control other traits needed for C_4 photosynthesis.

It is possible that accessions of rice possessing all the developmental alterations needed for an efficient C_4 pathway do not exist. This scenario needs to be catered for, and so a parallel program of fundamental research is needed. Within this alternative track of research, should leaf development be altered initially, or should leaf biochemistry be completed first? I suggest that genes known to encode proteins necessary for leaf biochemistry should be placed in a C_3 plant first, or at least in association with work to identify loci important in generating the anatomy of C_4 leaves. Because C_4 photosynthesis appears to have evolved many times (Sage 2004), the most parsimonious explanation for this is that a relatively small number of changes lead to many more of the phenotypes needed for the C_4 pathway to operate efficiently. A simple but potentially enlightening hypothesis can be proposed.

Strategy two

Does the metabolic compartmentation of C_4 photosynthesis alter leaf development?

The hypothesis is that the spatial segregation of metabolites generated by the biochemistry of C_4 photosynthesis induces many of the phenotypes needed for the C_4 pathway to work efficiently, including changes in leaf and cell development. I subsequently refer to this as the "metabolite hypothesis." It would be an explanation for the polyphyletic evolution of the C_4 pathway in angiosperms. If key metabolites need to accumulate *in defined cell types* in order to induce developmental alterations, the existence of C_4 species that operate a C_4 cycle within a single-cell type (Reiskind et al 1997, Reinfelder et al 2000, Voznesenskaya et al 2001) does not invalidate this extended hypothesis. The existence of large bundle sheath cells and increased venation in species that are otherwise C_3 also does not invalidate this hypothesis, as there are likely to be factors downstream of the ones that elicit the large-scale changes associated with the C_4 pathway, and these downstream factors could have altered in isolation.

I suggest that placing the C_4 biochemistry into *Arabidopsis* leaves should be a major aim. The reason for this is that it will allow us to test relatively rapidly the hypothesis that the spatial compartmentation of metabolites found in C_4 leaves leads to the alterations in leaf structure associated with the C_4 pathway. Placing *Cleome gynandra* genes known to encode enzymes of C_4 photosynthesis into *Arabidopsis* in one or two rounds of transformation should be feasible with a recombineering approach (Zhang et al 1998, 2000, Roden et al 2005). If the metabolite hypothesis is supported, the recombineering process can be used to place maize genes into rice. This approach is attractive from a pragmatic viewpoint as the genes encoding the enzymes needed for C_4 photosynthesis have been cloned. Although a number of the enzymes are regulated posttranslationally in C_4 species, genes encoding these regulatory proteins have also been isolated. For example, PEPC and PPDK are both regulated by reversible phosphorylation (Vidal and Chollet 1997, Burnell and Hatch 1983). These regulatory mechanisms are also found in C_3 species (Chastain and Chollet 2003, Gousset-Dupont et al 2005), and it appears likely that, if PEPC and PPDK from maize are placed into rice, they should be regulated by the existing kinases. This needs to be determined empirically. If it proves not to be the case, the regulatory proteins would have to be cloned from maize, and, in addition to the enzymes, placed into rice. The bifunctional regulatory protein from maize (Burnell and Chastain 2006) and *Arabidopsis* (Chastain, Xu, Parsley, Hibberd, and Chollet, unpublished) has now been identified. This implies that the proteins known to be posttranslationally modified should be integrated into rice early so that their regulatory proteins can be added soon after if necessary.

I advocate careful analysis of leaf development and cell ultrastructure of lines after each construct is integrated. Although this "metabolite hypothesis" may be considered unlikely and naïve, there is evidence that manipulating primary metabolism in leaves of C_3 species leads to changes in leaf development. In addition, some of these changes in leaf development are relevant to the C_4 pathway.

Examples where manipulating proteins of primary metabolism leads to changes in leaf development are described below. First, there are at least three examples where manipulating components of the Calvin cycle causes alterations to leaf anatomy. Antisense repression of the small subunit of Rubisco increases specific leaf area (i.e., leaf thickness is reduced) (Fichtner et al 1993). The same is true when sedoheptulose bisphosphatase accumulation is repressed via an antisense approach (Lawson et al 2006), and antisense repression of CP12, which is a regulator of phosphoribulokinase and glyceraldehyde 2-phosphate dehydrogenase (both part of the Calvin cycle), generates abnormal development of tobacco leaves (Raines and Paul 2006).

Second, specific metabolites can act as signaling molecules and alter leaf development. Increasing the amount of trehalose-6-phosphate in tobacco leads to reduced specific leaf area (i.e., thicker leaves) (Pellny et al 2004), while manipulating fructose-6-phosphate levels in tobacco alters leaf shape and induces the development of necrotic regions (Raines and Paul 2006).

Third, it has been reported that introducing high levels of NADP-ME from maize into rice, via the constitutive *cab* promoter, leads to reduced stacking of thylakoids in chloroplasts (Takeuchi et al 2000). Agranal chloroplasts are of course specifically associated with maize bundle sheath cells, which contain high activities of NADP-ME.

Fourth, specific manipulation of translocators that have been co-opted into key roles in C_4 plants can also alter leaf and cell development. When the oxoglutarate malate translocator is repressed constitutively via an antisense approach, leaf development is compromised so that little leaf blade is produced (Schneidereit et al 2006). Moving malate across the chloroplast is one of the key alterations needed for C_4 photosynthesis. Last, plants lacking a chloroplastic translocator for phospho*enol*pyruvate and phosphate (*cue1* mutants) show reduced development of the palisade mesophyll, and the size of the bundle sheath can be increased (Streatfield et al 1999).

Although we lack a mechanistic understanding of each of these changes in leaf development after primary metabolism has been manipulated, it can be argued that altered mesophyll development when a key transporter needed for the C_4 pathway is antisensed is more than a coincidence. It is also worthwhile noting that these manipulations in enzyme activity or transport capacity have either been via constitutive repression or complete gene knockout. This contrasts with the situation in C_4 plants, where alterations occur in defined cell types. Thus, manipulating carbon flow through primary metabolism in the specific cells needed for C_4 photosynthesis may well induce the more defined alterations in leaf development that are associated with C_4 plants.

This metabolite hypothesis could be tested relatively rapidly, and if it is supported would provide a fast way to integrate C_4 traits into rice. Because the biochemistry of C_4 plants is well characterized, only a few genes remain to be identified. These include transporters, and so fundamental research needs to be pursued to isolate these genes. An oxoglutarate/malate translocator was isolated from spinach (Weber et al 1995). Dicarboxylic transporters have since been identified in *Arabidopsis* and maize (Taniguchi et al 2002, 2004). In both species, these transporters belong to multigene families, and, in maize, specific members of the family have been proposed to be preferentially expressed in the mesophyll or bundle sheath cells (Taniguchi et al 2004) to allow the C_4 cycle to function. This needs to be confirmed.

Transport of pyruvate into chloroplasts of the mesophyll is also needed in the NADP-ME sub-type. Pyruvate carriers in the chloroplast membrane were characterized in the C_4 species *Digitaria sanguinalis* and *Panicum miliaceum* as well as C_3 spinach (Huber and Edwards 1977, Ohnishi and Kanai 1987). High rates of pyruvate transport were found in *Digitaria* compared with spinach, and compounds that are effective against other eukaryotic pyruvate carriers inhibited this process (Huber and Edwards 1977). To my knowledge, a plant gene encoding a pyruvate tranporter has yet to be cloned; however, yeast mutants unable to transport pyruvate into mitochondria have been isolated, and the gene identified (Hildeyard and Halestrop 2003). Searching the *Arabidopsis* genome with this yeast gene identifies clear candidates for pyruvate transporters in *Arabidopsis* (e.g., Fig. 1C). The ability of these proteins to transport pyruvate needs to be confirmed, and it should be fastest to do this by cloning these genes from *A. thaliana* and *C. gynandra*, and confirming their transport abilities in yeast.

If the hypothesis that compartmentation of C_4 biochemistry induces C_4 leaf development is not supported, or is supported only partially, then alternative strategies are needed, and these should begin before the metabolite hypothesis is rejected. In the following sections, I therefore propose a range of approaches to identify mechanisms controlling the phenotypes of C_4 photosynthesis that currently are relatively poorly understood.

Alternative approaches to isolating regulators of C_4 leaf development

Maize leaves have increased venation, larger bundle sheath cells, and reduced mesophyll development compared with rice leaves. Our understanding of vein production in leaves is increasing, particularly for *Arabidopsis* (Casson et al 2002, Carland et al 2002, Sieburth et al 2006). This knowledge could be used to inform a candidate gene approach in the maize and rice systems. For example, microarrays comparing developing C_3 husk leaves versus C_4 leaves from maize could be analyzed for alterations in transcripts that are known to be involved in vein production in *Arabidopsis*. We have generated a library of genomic fragments of *C. gynandra* in a binary vector-based bacterial artificial chromosome (BIBAC), and are using this to integrate large regions of *C. gynandra* into *Arabidopsis* (Parsley and Hibberd, unpublished). These lines of *Arabidopsis* could be screened for increased venation in leaves, and then candidate genes identified from the BAC. An advantage is that the line of *Arabidopsis* used to accept these BACs could already have veins labeled with GFP to facilitate the screen, and, although vein development differs between dicots and monocots, it is possible that the signaling molecules controlling venation are shared. This approach may inform the transfer of regions of genomic DNA or individual loci from maize to rice.

Alternative approaches to isolating regulators of cell development in C_4 leaves

Increased expansion of bundle sheath cells is an important C_4 phenotype. *Cleome* has a clear preconditioning to possess large bundle sheath cells (Marshall and Hibberd, unpublished). It is possible that large bundle sheath cells are widespread in plants, and so *Oryza* collections that IRRI has available should be screened for enlarged bundle sheaths. In addition, it would probably be wise to screen the *Arabidopsis* lines, into which BACs containing fragments of the *C. gynandra* genome have been integrated, for increased size of bundle sheath cells. In *Arabidopsis*, this sort of screen should be less time-consuming because the bundle sheath cells of the line accepting the BACs could be marked with green fluorescent protein, which negates the need to clear, fix, or stain sections. As bundle sheath cells border the veins, this sort of screen could also be used to provide information on vein development.

Alternative approaches to isolating regulators of C_4 ultrastructure

In association with altered bundle sheath and mesophyll development, critical aspects in a species such as maize, which uses NADP-ME, include (1) increased proliferation of bundle sheath chloroplasts, (2) reduced photosystem II and thylakoid stacking in the bundle sheath, (3) increased plasmodesmatal frequency linkage between bundle sheath and mesophyll, and (4) suberization of the cell wall between the mesophyll and bundle sheath. I suggest approaches to understand each of these phenomena in order.

Our understanding of chloroplast division and expansion in *Arabidopsis* is improving. Increasing numbers of genes important in these processes have been identified (Maple and Moller 2005, Haswell and Meyerowitz 2006). More chloroplasts in the bundle sheath of C_4 compared with C_3 species must be due to an additional round (or rounds) of chloroplast division. The challenge is to identify the mechanism by which this takes place. Chloroplast division is probably not different in dicotyledons and monocotyledons, and so it would seem sensible to identify the likely mechanisms underlying increased division of chloroplasts in bundle sheath cells using *Arabidopsis*. Several approaches can be envisaged: first, analysis of transcript abundance during stages of leaf development while bundle sheath chloroplast division is occurring may identify candidates: second, placing BACs from *C. gynandra* into an *Arabidopsis* line that possesses chloroplasts in the bundle sheath labeled with GFP should also isolate candidates. Information gained could then be used to increase chloroplast division and expansion in the bundle sheath of rice.

Golden Like (GLK) genes (Hall et al 1998, Rossini et al 2001, Fitter et al 2002) are known to control chloroplast ultrastructure (for further details, see the Chapter by J. Langdale in this volume). Placing maize *Golden* genes into rice would be a sensible first step, if changes in thylakoid structure are not induced when C_4 biochemistry is placed in rice.

Little is known about the components making up plasmodesmata. However, proteomic analysis of cell walls derived from *Arabidopsis* cell suspension cultures is likely to provide advances here (Bayer et al 2004, 2006). It is also possible to label *Arabidopsis* with a translational fusion between the tobacco mosaic virus movement protein and GFP (Oparka et al 1997). Plasmodesmata are then visible with epifluorescence microscopy and could be transformed with BACs harboring genomic fragments of *C. gynandra*. A visual screen for increases in plasomodesmatal number could be carried out at the same time as the one for increased bundle sheath cell size. Once again, this approach would inform loci to select from maize, and integrate into rice. In maize, between mesophyll and bundle sheath cells, the cell wall is impregnated with suberin. There are few clues in the literature as to how this process is controlled. However, suberization of cell walls is not uncommon in plants, for example, the Casparian strip of roots is defined as such because of suberization of walls of the endodermis. It therefore appears that maize leaves have co-opted a process that is occurring in roots, and that identifying regulators of this process and switching them on in the rice leaf would get us over this (final?) hurdle. Suberin deposition could be investigated by analysis of transcript profiles in developing bundle sheath cells of maize, and the developing endodermis of rice or *Arabidopsis*, for example. Candidates could then be misexpressed in *Arabidopsis* and rice to identify whether they are useful. An alternative is to include analysis of suberin in the screen of *Arabidopsis* lines harboring genomic fragments of *C. gynandra* or screening the *Oryza* accessions at IRRI.

Conclusions

Maize appears to be the system of choice to source genes that are placed into rice to make it C_4. A dual-pronged attack to achieve this aim is suggested. First, existing accessions of *Oryza* should be screened for characteristics of the C_4 pathway. Second, fundamental research needs to be conducted, and this includes testing the metabolite hypothesis as well as considering screens for the developmental characteristics needed for efficient C_4 photosynthesis in the most tractable models for each trait. I advocate the use of *Arabidopsis* and *Cleome* for much of the fundamental work.

References

Ali S, Taylor WC. 2001. Quantitative regulation of the *Flaveria Me1* gene is controlled by the 3′-untranslated region and sequences near the amino terminus. Plant Mol. Biol. 46:251-261.

Bayer E, Thomas CL, Maule AJ. 2004. Plasmodesmata in *Arabidopsis thaliana* suspension cells. Protoplasma 223:93-102.

Bayer EM, Bottrill AR, Walshaw J, Vigouroux M, Naldrett MJ, Thomas CL, Maule AJ. 2006. *Arabidopsis* cell wall proteome defined using multidimensional protein identification technology. Proteomics 6:301-311

Brown NJ, Parlsey K, Hibberd JM. 2005. The future of C_4 research: maize, *Flaveria* or *Cleome*? Trends Plant Sci. 10:215-221.

Burnell JN, Chastain CJ. 2006. Cloning and expression of maize-leaf pyruvate, Pi dikinase regulatory protein gene. Biochem. Biophys. Res. Comm. 345:675-680.

Burnell JN, Hatch MD. 1983. Dark/light regulation of pyruvate, Pi dikinase in C_4 plants: evidence that the same protein catalyses activation and deactivation. Biochem. Biophys. Res. Comm. 111:288-293.

Carland FM, Nelson T. 2004. COTYLEDON VASCULAR PATTERN2–mediated inositol (1,4,5) triphosphate signal transduction is essential for closed venation patterns of *Arabidopsis* foliar organs. Plant Cell 16:1263-1275.

Casson SA, Chilley PM, Topping JF, Evans IM, Souter MA, Lindsey K. 2002. The POLARIS gene of *Arabidopsis* encodes a predicted peptide required for correct root growth and leaf vascular patterning. Plant Cell 14:1705-1721.

Chastain CJ, Chollet R. 2003. Regulation of pyruvate, orthophosphate dikinase by ADP-/Pi-dependent reversible phosphorylation in C_3 and C_4 plants. Plant Physiol. Biochem. 41:523-532.

Esau K. 1955. Plant anatomy. New York, N.Y. (USA): Wiley & Sons.

Fichtner K, Quick WP, Schulze ED, Mooney HA, Rodermel SR, Bogorad L, Stitt M. 1993. Decreased ribulose-1,5-bisphosphate carboxylase-oxygenase in transgenic tobacco transformed with antisense *rbc*S. 5. Relationship between photosynthetic rate, storage strategy, biomass allocation and vegetative plant-growth at 3 different nitrogen supplies. Planta 190:1-9.

Fitter DW, Martin DJ, Copley MJ, Scotland RW, Langdale JA. 2002. *GLK* gene pairs regulate chloroplast development in diverse plant species. Plant J. 31:713-727.

Fukayama H, Tsuchida H, Agarie S, Nomura M, Onodera H, Ono K, Lee BH, Hirose S, Toki S, Ku MSB, Makino A, Matsuoka M, Miyao M. 2001. Significant accumulation of C_4 specific pyruvate orthophosphate dikinase in a C_3 plant, rice. Plant Physiol. 127:1136-1146.

Gousset-Dupont A, Lebouteiller B, Monreal J, Echevarria C, Pierre JN, Hodges M, Vidal J. 2005. Metabolite and post-translational control of phospho*enol*pyruvate carboxylase from leaves and mesophyll cell protoplasts of *Arabidopsis thaliana*. Plant Sci. 169:1096-1101.

Gowik U, Burscheidt J, Akyildiz M, Schlue U, Koczor M, Streubel M, Westhoff P. 2005. *cis*-regulatory elements for mesophyll specific gene expression in the C_4 plant *Flaveria trinervia*, the promoter of the C_4 phospho*enol*pyruvate carboxylase gene. Plant Cell 16:1077-1090.

Hall LN, Rossini L, Cribb L, Langdale JA. 1998. GOLDEN2: a novel transcriptional regulator of cellular differentiation in the maize leaf. Plant Cell 10:925-936.

Haswell ES, Meyerowitz EM. 2006. MscS-like proteins control plastid size and shape in *Arabidopsis thaliana*. Curr. Biol. 16:1-11.

Hatch MD. 1987. C_4 photosynthesis: a unique blend of modified biochemistry, anatomy and ultrastructure. Biochem. Biophys. Acta 895:81-106.

Herrmann KM. 1995. The shikimate pathway: early steps in the biosynthesis of aromatic compounds. Plant Cell 7:907-910.

Hibberd JM, Quick WP. 2002. Characteristics of C_4 photosynthesis in stems and petioles of C_3 flowering plants. Nature 415:451-454.

Hildeyard JCW, Halestrap AP. 2003. Identification of the mitochondrial pyruvate carrier in *Saccharomyces cerevisiae*. Biochem. J. 374:607-611.

Huber SC, Edwards GE. 1977. Transport in C_4 mesophyll chloroplast characterisation of the pyruvate carrier. Biochem. Biophys. Acta 4:583-602.

Imaizumi N, Ku MSB, Ishihara K, Samejima M, Kaneko S, Matsuoka M. 1997. Characterisation of the gene for pyruvate orthophosphate dikinase from rice, a C_3 plant, and a comparison of structure and expression between C_3 and C_4 genes for this protein. Plant Mol. Biol. 34:701-716.

Kanai R, Edwards GE. 1999. Structure-function of the C_4 syndrome. In: Sage RF, Monson RK, editors. C_4 plant biology. San Diego, Calif. (USA): Academic Press.

Ku MSB, Agarie S, Nomura M, Fukayama H, Tsuchida H, Ono K, Hirose S, Toki S, Miyao M, Matsuoka M. 1999. High-level expression of maize phospho*enol*pyruvate carboxylase in transgenic rice plants. Nature Biotechnol. 17:76-80.

Langdale JA, Taylor WC, Nelson T. 1991. Cell-specific accumulation of maize phospho*enol*-pyruvate carboxylase is correlated with demethylation at a specific site greater than 3 kb upstream of the gene. Mol. Gen. Genet. 225:49-55.

Lawson T, Bryant B, Lefebvre S, Lloyd JC, Raines CA. 2006. Decreased SBPase activity alters growth and development in transgenic tobacco plants. Plant Cell Environ. 29:48-58.

Maple J, Moller SG. 2005. An emerging picture of plastid division in higher plants. Planta 223:1-4.

Matsuoka M, Furbank RT, Fukayama H, Miyao M. 2001. Molecular engineering of C_4 photosynthesis. Annu. Rev. Plant Physiol. Mol. Biol. 52:297-314.

Ohnishi JI, Kanai R. 1987. Pyruvate uptake by mesophyll and bundle sheath chloroplasts of a C_4 plant, *Panicum miliaceum*. Plant Cell Physiol. 28:1-10.

Oparka KJ, Prior DAM, SantaCruz S, Padgett HS, Beachy RN. 1997. Gating of epidermal plasmodesmata is restricted to the leading edge of expanding infection sites of tobacco mosaic virus (TMV). Plant J. 12:781-789.

Parsley K, Hibberd JM. 2006. The *Arabidopsis PPDK* gene is transcribed from two promoters to produce differentially expressed transcripts responsible for cytosolic and plastidic proteins. Plant Mol. Biol. (In press.)

Patel M, Corey AC, Yin LP, Ali SJ, Taylor WC, Berry JO. 2004. Untranslated regions from C-4 amaranth *AhRbcS1* mRNAs confer translational enhancement and preferential bundle sheath cell expression in transgenic C_4 *Flaveria bidentis*. Plant Physiol. 136:3550-3561.

Patel M, Siegel, AJ, Berry JO. 2006. Untranslated regions of *FbRbcS1* mRNA mediate bundle sheath cell-specific gene expression in leaves of a C_4 plant. J. Biol. Chem. doi:10.1074/jbc.M604162200.

Pellny TK, Ghannoum O, Conroy JP, Schluepmann H, Smeekens S, Andralojc J, Krause KP, Goddijn O, Paul MJ. 2004. Genetic modification of photosynthesis with *E. coli* genes for trehalose synthesis. Plant Biotechnol. J. 2:71-82.

Raines CA, Paul MJ. 2006. Products of leaf primary carbon metabolism modulate the development programme determining plant morphology. J. Exp. Bot. 57:1857-1862.

Reinfelder JR, Kraepiel AML, Morel FMM. 2000. Unicellular C_4 photosynthesis in a marine diatom. Nature 407:996-999.

Reiskind JB, Madsen TV, van Ginkel LC, Bowes G. 1997. Evidence that inducible C_4-type photosynthesis is a chloroplastic CO_2-concentrating mechanism in *Hydrilla*, a submersed monocot. Plant Cell Environ. 20:211-220.

Roden LC, Göttgens B, Mutasa-Göttgens ES. 2005. Precision engineering of plant gene loci by homologous recombination cloning in *Escherichia coli*. Plant Methods 1:6.

Rosche E, Westhoff P. 1995. Genomic structure and expression of the pyruvate, orthophosphate dikinase gene of the dicotyledonous C_4 plant *Flaveria trinervia* (Asteraceae). Plant Mol. Biol. 29:663-678.

Rossini L, Cribb L, Martin DJ, Langdale JA. 2001. The maize *Golden2* gene defines a novel class of transcriptional regulators in plants. Plant Cell 13:1231-1244.

Sage RF. 2004. The evolution of C_4 photosynthesis. New Phytol. 161:341-370.

Schneidereit J, Hausler RE, Fiene G, Kaiser WM, Weber APM. 2006. Antisense repression reveals a crucial role of the plastidic 2-oxoglutarate/malate translocator DiT1 at the interface between carbon and nitrogen metabolism. Plant J. 45:206-224.

Sheen J. 1991. Molecular mechanisms underlying the differential expression of maize pyruvate, orthophosphate dikinase genes. Plant Cell 3:225-245.

Sieburth LE, Muday GK, King EJ, Benton G, Kim S, Metcalf KE, Meyers L, Seamen E, Van Norman JM. 2006. SCARFACE encodes an arf-gap that is required for normal auxin efflux and vein patterning in *Arabidopsis*. Plant Cell 18:1396-1411.

Stockhaus J, Poetsch W, Steinmuller K, Westhoff P. 1994. Evolution of the C_4 phospho*enol*-pyruvate carboxylase promoter of the C_4 dicot *Flaveria trinervia*: an expression analysis in the C_3 plant tobacco. Mol. Gen. Genet. 245:286-293

Streatfield SJ, Weber A, Kinsman EA, Hausler RE, Li JM, Post-Beittenmiller D, Kaiser WM, Pyke KA, Flugge UI, Chory J. 1999. The phospho*enol*pyruvate/phosphate translocator is required for phenolic metabolism, palisade cell development and plastid-dependent nuclear gene expression. Plant Cell 11:1609-1621.

Takeuchi Y, Akagi H, Kamasawa N, Osumi M, Honda H. 2000. Aberrant chloroplasts in transgenic rice plants expressing a high level of maize NADP-dependent malic enzyme. Planta 211:265-274.

Taniguchi M, Taniguchi Y, Kawasaki M, Takeda S, Kato T, Sato S, Tahata S, Miyake H, Sugiyama T. 2002. Identifying and characterizing plastidic 2-oxoglutarate/malate and dicarboxylate transporters in *Arabidopsis thaliana*. Plant Cell Physiol. 43:706-717.

Taniguchi Y, Nagasaki J, Kawasaki M, Miyake H, Sugiyama T, Taniguchi M. 2004. Differentiation of dicarboxylate transporters in mesophyll and bundle sheath chloroplasts of maize. Plant Cell Physiol. 45:187-200.

Vidal J, Chollet R. 1997. Regulatory phosphorylation of C_4 PEP carboxylase. Trends Plant Sci. 2:230-237.

Voznesenskaya EV, Franceschi VR, Kiirats O, Freitag H, Edwards GE. 2001. Kranz anatomy is not essential for terrestrial C_4 plant photosynthesis. Nature 414:543-546.

Weber A, Menzlaff E, Arbinger B, Gutensohn M, Eckerskorn C, Flugge UI. 1995. The 2-oxoglutarate malate translocator of chloroplast envelope membranes: molecular-cloning of a transporter containing a 12-helix motif and expression of the functional protein in yeast-cells. Biochemistry 34:2621-2627.

Zhang YM, Buchholz F, Muyrers JPP, Stewart AF. 1998. A new logic for DNA engineering using recombination in *Escherichia coli*. Nature Genet. 20:123-128.

Zhang YM, Muyrers JPP, Testa G, Stewart AF. 2000. DNA cloning by homologous recombination in *Escherichia coli*. Nature Biotechnol. 18:1314-1317.

Notes

Author's address: Department of Plant Sciences, Downing Street, University of Cambridge, Cambridge, CB2 3EA, UK, email: julian.hibberd@plantsci.cam.ac.uk.

Acknowledgments: I thank Alison G. Smith for discussions and the Biology and Biotechnology Sciences Research Council, The Leverhulme Trust, and The Isaac Newton Trust for funding.

Section 3:

Single-cell C_4 systems

C_4 rice: early endeavors and models tested

J. Burnell

Plans to make a C_4 rice plant date back to a document in 1987 and the first patent application for C_4 rice submitted in 1991. In addition, an attempt to make a C_4 rice plant was made in collaboration with Japan Tobacco Inc. during the 1990s. This collaboration recognized the importance of two compartments in C_4 photosynthesis, normally provided by mesophyll and bundle sheath cells. However, a single-cell system was devised in which the endogenous compartments of the cytosol and the chloroplast of C_3 plants were used to mimic the two C_4 compartments. Phospho*enol*pyruvate carboxykinase (PEPCK) was used as the C_4 acid decarboxylating enzyme and was synthesized with a transit peptide to ensure location in the chloroplasts. The PEPCK gene from *Urochloa panicoides* was transferred to rice and was expressed successfully: carbon flow was altered toward a C_4 pathway but without appreciable increases in photosynthesis or growth. The properties and location of enzymes postulated to be required to convert a C_3 plant to a C_4 plant (carbonic anhydrase, phosphoenolpyruvate carboxylase, PEPCK, and pyruvate, orthophosphate dikinase) are reviewed. Further modifications to maximize the efficiency of a C_4 pathway in C_3 plants are discussed.

Keywords: C_4 rice, C_4 acid decarboxylation, carbonic anhydrase, compartmentation, phospho*enol*pyruvate carboxykinase, phospho*enol*pyruvate carboxylase, pyruvate, orthophosphate dikinase, PPDK regulatory protein, *Urochloa panicoides*

My interest in C_4 photosynthesis began in 1982 following a move to the Commonwealth Scientific and Industrial Research Organization (CSIRO) Division of Plant Industry in Canberra to take up a research scientist position in the Hal Hatch laboratory, where I was immediately given the task of extracting and assaying the activity of pyruvate kinase in phospho*enol*pyruvate carboxykinase-type (PEPCK) C_4 plants. Unbeknown to me at the time, this was a task that had been given to a number of my predecessors. The task was associated with trying to obtain an activity of pyruvate kinase that would be sufficient to convert PEP to pyruvate at a rate commensurate with the rate of inorganic carbon fixation according to the prevailing model for PEPCK-type C_4 photosynthesis. Not surprisingly (in hindsight), activities about one-third of that required to run the photosynthetic pathway were detected and my interests were drawn toward exploring the mechanism of the pyruvate, orthophosphate dikinase (PPDK) regulatory protein (PDRP) in maize. After two years of working on PDRP, I returned to exploring the photosynthetic pathway present in PEPCK-type C_4 plants and this involved isolating and working with bundle sheath cells. Adopting a different approach, which involved extended hours with an oxygen electrode, led to the development of a revised model of PEPCK-type photosynthesis which required the operation of two C_4 acid decarboxylation mechanisms; this model also explained the source of adenosine triphosphate (ATP) for use in the PEPCK-catalyzed reaction. On completion of this work, my life took a completely different path, this time along the corridor to a new laboratory with the expressed instructions to "learn some molecular biology." Under the experienced guidance of Dr. Paul Whitfeld and Dr. John Mason, the first plant carbonic anhydrase (CA) was cloned and sequenced (albeit a spinach CA), and my interest in the possibilities of using molecular biology techniques to improve photosynthetic rates in plants began to develop.

In March 1988, a conference titled *PI2000* was held in Canberra "to explore the challenges facing research in plant science and rural industries into the 21st century." At this conference, I presented a paper titled *Molecular genetic approaches to understanding and manipulating C_4 photosynthesis* co-authored with William Taylor. Two alternative approaches were raised as a possible means of increasing agricultural productivity: either by modifying a C_4 plant to produce an edible product (sugar cane, maize, sorghum, and millet are identified as the only agriculturally beneficial C_4 plants) or increasing the photosynthetic rate of an agriculturally important C_3 crop (such as rice). The possibility of developing cold-tolerant maize was also put forward. So, the possibility of improving plant productivity by genetic manipulation was proposed. At the end of the conference, I discussed (with the then chief of the CSIRO Division of Plant Industry) my strong interest in developing both cold-tolerant maize and transforming a plant to convert it from a C_3 plant to a C_4 plant and was quietly informed that I would have to seek funding independently of CSIRO if I wished to pursue these goals. As circumstances would have it, a new university, Bond University (Australia's first private university), was being set up on the Gold Coast in Queensland, Australia, and I responded to the opportunity to further my research interests in a new environment. (Although the Bond University Graduate School of Science and Technology operated for only two years, the collaborations and friendships engendered in such an environment have lasted to today.)

While at Bond University, I, like my Bond University colleagues, took the opportunity to apply to a cross-section of funding agencies, in my case, to fund projects for developing cold-tolerant maize and C_4 rice. All these applications were rejected. In response to these rejections, I looked toward overseas funding agencies and applied for a number of research opportunities in Japan. In 1990, I was awarded a senior fellowship by the Japanese Society for the Promotion of Science and I spent almost 6 months working in Professor Tatsuo Sugiyama's laboratory at the University of

Nagoya. (I had worked with Prof. Sugiyama in Hal Hatch's laboratory in Canberra a number of years earlier and he generously supported my fellowship application.) Before departing Bond University for Nagoya, I put together a document titled *Opportunities for improved rates of plant growth by genetic manipulation of the C_4 genome.* This document provided a background of C_4 photosynthesis and contained descriptions of five independent research projects:

1. Increased cold tolerance of C_4 plants by introduction of a gene coding for the cold-tolerant form of PPDK.
2. Molecular biology of the PPDK regulatory protein: its location, expression, and regulation.
3. Improvement of photosynthetic rates in C_3 plants by expression of genes coding for C_4-enzymes in specific intracellular compartments.
4. Development of C_4-specific herbicides.
5. A biochemical and molecular biological examination of the plasmodesmata of bundle sheath cells of C_4 plants (with a view to increasing resistance to viral infection).

While in Japan, where I initially cloned maize leaf carbonic anhydrase (Burnell et al 1990, Sugiharto et al 1992a,b), I visited several private companies that might potentially fund plant research and eventually entered into preliminary discussions with representatives of Japan Tobacco Inc. (JTI). These discussions continued after my return to Australia, where I was greeted with the announcement (Melbourne Cup Day 1990—Australians will relate to this date) that Bond University was closing its Graduate School of Science and Technology and that I should seek a position elsewhere. Following my retrenchment from Bond University, I continued to have discussions with representatives of JTI and these discussions were further interrupted by the outbreak of war in Kuwait and Iraq, which prevented international travel by JTI executives. Finally, in February 1991, after seeking the services of a business manager with previous experience in dealing with private Japanese companies, I drew up a Heads of Agreement outlining a research and development plan covering the two projects that JTI representatives had elected to fund (*C_4 rice* and *Cold-tolerant maize*) and submitted Provisional Patent Applications *(Method of enhancing photosynthetic activity* and *Method for developing cold tolerance in plants*) to the Australian Patents Office. Following further discussions in both Australia and Japan, formal Research and Development and Materials Transfer Agreements were signed. Having elected to accept access to laboratory facilities at the Queensland University of Technology in Brisbane, I recruited a small team of researchers that worked on these and related projects following success in gaining funding from the Australian Research Council. Together with Shoichi Suzuki (a member of the JTI plant technology group with whom I had worked at the University of Nagoya), Pat Finnegan (PEP carboxykinase in *Urochloa panicoides*), Martha Ludwig (carbonic anhydrase in C_3, C_4, and C_3/C_4 intermediate plant species), and Peter Cooke (maize PPDK regulatory protein), I collaborated closely with a team of researchers in the Iwata Laboratories of JTI. The results of this collaboration are reported below and have been published (Burnell 2000, Burnell and Ludwig 1996, 1997, Cavallaro et al 1994, Finnegan and Burnell 1994,

Finnegan et al 1999, Ludwig and Burnell 1995, Ohta et al 1996, Suzuki and Burnell 1995, 2003, Suzuki et al 2000, 2006, Usami et al 1995).

(From time to time, I have been asked why do I think I was successful in gaining funding from JTI. On reflection, I believe it was due to a number of contributing factors, but the most important was probably timing. When I initially approached JTI (mid-1990), scientists in the JTI Iwata laboratories were successfully transforming rice and were seeking applications of this newly developed technology; in other words, JTI had the new plant gene transformation technology and I had a means of using the new technology. I also believe that my understanding of matters of confidentiality and intellectual property was also important in gaining the confidence of JTI's senior scientists and senior executives. And, most important of all, I think, was the part I had played in elucidating some of the material that was to be used in the two JTI-funded projects (the discovery of a cold-stable PPDK in *Flaveria brownii*, Burnell 1990a; the first to clone a plant carbonic anhydrase, Burnell et al 1990; and elucidation of the photosynthetic pathway in PEPCK-type C_4 plants, Burnell and Hatch 1988a,b).) For the collaboration with JTI, I experienced a huge amount of enjoyment, good science, and personal satisfaction. Both projects progressed simultaneously. Regarding the C_4 rice project, I believe that scientists at JTI were more accepting of the idea than many of my colleagues that, although at the time all known C_4 plants possessed Kranz anatomy, it might be possible to mimic C_4 photosynthesis in C_3 plants by adopting the intrinsic intracellular compartmentation (i.e., the cytosol and the chloroplast stroma). This might be achieved by expressing specific C_4 genes in specific intracellular compartments. In fact, the diagram included in the initial provisional patent application (11 February 1991) summarized the requirements as expression of carbonic anhydrase and PEP carboxylase in the cytosol and expression of PEP carboxykinase (or NADP-malic enzyme) and pyruvate, Pi dikinase in the chloroplasts. To me, this represented the simplest model for converting a C_3 plant to a C_4 plant. And, my reasoning regarding initially choosing rice as the plant to convert from C_3 to C_4 was twofold. First, rice is the staple food of a large proportion of the world's population and, second, it is usually grown in the tropics and would allow efficient functioning of PPDK, an enzyme known to be cold-sensitive.

In Section 3 of my original 1990 document *Improvement of photosynthetic rates in C_3 plants by expression of genes coding for C_4-enzymes in specific intracellular compartments*, I noted that for many years scientists had been attempting to improve the photosynthetic growth rates of C_3 plants by crossing them with closely related C_4 relatives. Although the activity of some of the enzymes involved in C_4 photosynthesis increased in the F_1 offspring resulting from these crosses, the photosynthetic rates of these plants did not increase. So, a direct approach involving the expression of specific genes in the leaves of C_3 plants was required.

C_4 rice

Following the elucidation of the PEP carboxykinase-type C_4 photosynthetic pathway in *Urochloa panicoides* and the revelation of the involvement of the NAD-malic enzyme pathway in generating ATP required for the reaction catalyzed by PEP carboxykinase, all three types of C_4 photosynthesis were examined with regard to the potential to transfer a minimum number of genes into a C_3 plant to increase the rate of inorganic carbon fixation. From the outset, recognition was given to the fact that compartmentation was critical to modifying photosynthesis in C_3 plants. In the late 1980s, it was believed that Kranz anatomy was critical for C_4 photosynthesis; however, intracellular compartmentation of the C_3 mesophyll cells was recognized and a pathway was designed based on the cytosol representing the mesophyll cell and the chloroplast representing the bundle sheath cell. With this basic architecture, the biochemistry was subdivided between the two locations. More recent research has shown the existence of plants lacking Kranz anatomy yet capable of fixing inorganic carbon via a C_4 pathway (Bowes et al 2002, Voznesenskaya et al 2001).

In selecting the most suitable species for transformation, rice was chosen for four main reasons:

1. The C_4 pathway is most successful in warm, sunny climates and rice is widely grown in countries possessing these climates.
2. PPDK, one of the enzymes required to be expressed in the transformed plant, is cold-sensitive and therefore there is a need to choose a plant that is commonly grown in the tropics.
3. A large population, much of which is located in equatorial countries, consumes rice and therefore rice would be a suitable target.
4. Rice transformation technology had recently been developed at Japan Tobacco.

The enzymes

Carbonic anhydrase

Following recognition of carbonic anhydrase (CA) as the first enzyme in the C_4 photosynthetic pathway (Hatch and Burnell 1990), it was deemed important to locate CA in the cytosol of transgenic plants to maximize the rate of conversion of carbon dioxide to bicarbonate and therefore maximize the rate of diffusion of carbon dioxide into the mesophyll cells. In addition, we used the fact that the CA from monocot and dicot plants differs immunologically, with monocot CA binding to antibodies raised against monocot CA and dicot CA reacting with antibodies raised against CA isolated from dicots (Burnell 1990b). Therefore, the expression of a monocot CA in a dicot plant and, conversely, a dicot expressed in a monocot could be easily assessed. A number of CAs had been cloned by the mid-1990s (Burnell et al 1990, Fawcett et al 1990, Majeau and Coleman 1991, 1992, Roeske and Ogren 1990).

In C_3 plants, CA can represent as much as 2% of the total protein in leaf tissue (Okabe et al 1984) and two CA isozymes are present in C_3 plants, one in the cytosol and one in the chloroplast (Utsunomiya and Muto 1993). The chloroplastic form constitutes more than 95% of the total CA activity present in the leaves. In C_3 plants, the role of CA has been investigated using antisense modification of CA amounts, although results have differed in different species. Decreased CA amounts in *Flaveria* and *Arabidopsis* caused death but plants could be rescued by growth on sucrose or by growing the plants in high CO_2 concentrations (von Caemmerer et al 2004); both these treatments are consistent with CA facilitating its diffusion into the chloroplasts of C_3 plants.

In C_4 plants, the picture is not so clear, especially in NADP-malic enzyme-type C_4 plants. More CA isozymes appear to be present in C_4 plants, as exemplified by the presence of three distinct isozymes in *Flaveria bidentis* (an NADP-malic enzyme-type dicot C_4 plant) (Ludwig and Burnell 1995), and at least four isozymes present in both maize and sorghum (NADP-ME-type monocot C_4 plants) (Burnell and Ludwig 1997, Burnell 2000). The location and role of the CA isozymes in C_4 plants are yet to be identified. It is interesting to note that, of the three isozymes present in *F. bidentis*, one is clearly identified as being located in the cytosol (not possessing a transit peptide sequence) and one in the chloroplast stroma (inferred due to similarity with the C_3 homolog). The location and function of the third isozyme are yet to be determined but the third isozyme also appears to have a transit peptide.

The story in maize and sorghum is a little more complex, with three genes identified and four proteins detected by western blotting (Burnell and Ludwig 1997). All three CA genes have repeat sequences present (A, B, and C) and translation of the three genes does not coincide with the proteins detected by western blotting. Bacterial expression studies are currently in progress to characterize the proteins produced by translation of full-length cDNA in an effort to correlate molecular data with proteins isolated from maize and sorghum. Interestingly, expression of a single repeat can produce a protein with CA activity and with the same subunit size as detected by western blotting of leaf extracts (Tems and Burnell, unpublished results). The native form of maize CA is yet to be determined; however, it is possible that in maize a single repeat region may be active as a monomer, making expression of the enzyme in a transgenic plant simpler. A clearer understanding of the roles of CA isozymes in C_4 plants may be critical to increasing photosynthetic rates in rice.

As mentioned above, C_3 plants have a majority of their CA activity located in the chloroplast stroma, where it functions to catalyze the conversion of carbon dioxide to bicarbonate, lowering the stromal concentration of carbon dioxide, increasing the CO_2 gradient across the chloroplast membrane, and thus increasing the rate of CO_2 diffusion into the stroma (see review by Badger and Price 1994). In close proximity, the CA would be expected to catalyze the conversion of bicarbonate to CO_2, the inorganic carbon substrate of Rubisco.

In the case of C_4 transgenic rice, it was initially thought that the presence of an endogenous chloroplast stromal CA would be deleterious to the efficient operation of the C_4 photosynthetic pathway as it would decrease the availability of carbon dioxide to Rubisco after being released from the C_4 acid transported across the chloroplast membrane. In addition, it has been reported that low amounts of CA in the bundle sheath cells of C_4 plants may be essential for the effective functioning of the C_4 photosynthetic pathway (Burnell and Hatch 1988c). Therefore, it was suggested that the endogenous stromal CA in C_3 plants might have to be down-regulated (by antisense technology) to maximize the rates of inorganic carbon fixation. To this end, rice chloroplastic CA was isolated, purified, its N-terminal sequence determined, and the gene sequence determined (Suzuki and Burnell 1995).

PEP carboxylase

PEP carboxylase catalyzes the carboxylation of PEP to form oxaloacetate and uses bicarbonate as the inorganic carbon substrate. The importance of bicarbonate being the substrate for the primary inorganic carbon-fixing reaction is twofold. First, the stereochemistry of bicarbonate is quite different from that of oxygen and therefore the two compounds do not compete for the active site of the protein (compared with the situation for Rubisco), and the conversion of carbon dioxide to bicarbonate catalyzed by CA serves to increase the rate of diffusion of inorganic carbon into the mesophyll cells. PEPC is not affected by the presence of oxygen and is located in the cytosol of all C_4 plants. The PEPC gene had been thoroughly characterized and was available for use in the transformation of rice (see Chollet et al 1996 for a review).

It is probably relevant to raise the issue of the regulation of PEP carboxylase as the mechanism of the light-dark regulation of C_4 PEP carboxylase has been elucidated (see Chollet et al 1996 for a review). The C_4 isozyme has been shown to be reversibly light activated *in vivo* by a mechanism involving phosphorylation of a single serine residue near the N-terminal end of the protein. Phosphorylation of PEP carboxylase (up-regulation) is catalyzed by a highly regulated protein kinase and dephosphorylation (down-regulation) is catalyzed by a protein phosphatase 2A that up-regulates the protein. Phosphorylation of the protein makes PEPC considerably less sensitive to malate-dependent inhibition and both more active and more sensitive to glucose-6-phosphate-dependent activation. This reversible regulatory mechanism, coupled with the allosteric properties of PEPC, appears to be unique to the plant enzyme and may be indicative of the requirement for regulation in a C_4 pathway. The efficient operation of a C_4 pathway in C_3 plants may be dependent on the coordinated activity of both PEP carboxylase and its regulatory protein.

PEP carboxykinase

The choice of which C_4 acid decarboxylation mechanism to adopt for use in the transformation of rice was narrowed to one of two mechanisms; the NAD-malic enzyme-type mechanism was eliminated from consideration due to its reliance on NAD/NADH and mitochondria. This left the choice between NADP-ME and the PEPCK-type C_4 mechanisms. The biochemical pathways of both mechanisms were examined and the strengths and weaknesses of adopting either of the two alternatives were identified. In the end, it was decided that C_4 rice would be transformed with the PEPCK gene as it liberated PEP (the substrate of PEP carboxylase) as its end product and this would eliminate the requirement for high PPDK activity in the chloroplast. The PEPCK from *Urochloa panicoides*, a C_4 monocot species, was chosen due to the ease of preparing bundle sheath cells and therefore aiding in the purification of PEPCK. The enzyme was purified, the N-terminal amino acid sequence determined, and four PEPCK genes identified (Finnegan and Burnell 1995). Of the four genes identified, two were found to be expressed in the leaves and two in the roots (Finnegan et al 1999). A cDNA encoding *U. panicoides* PEPCK was successfully expressed in rice chloroplasts. Constructs contained the PEPCK cDNA and the transit peptide of the Rubisco small subunit under the control of the promoter of the maize PEPC or PPDK genes. In experiments with excised leaves of transgenic plants, up to 20% of the carbon dioxide fixed was incorporated into C_4 acids (malate, aspartate, and oxaloacetate) compared with about 1% in excised leaves of control nontransgenic plants (Suzuki et al 2000). Significantly, there was a positive correlation between PEPCK activity and the amount of labeling of 4C compounds. Finally, when the L-[4-^{14}C]malate was fed to excised leaves, the amount of radioactivity incorporated into sucrose was three times higher in transgenic plants than in control plants. These results indicated that the expression of PEPCK in rice chloroplasts was able to partially change the flow of carbon in mesophyll cells into a C_4-like photosynthetic pathway (Suzuki et al 2000).

More recently, experiments were conducted to determine whether the *pck1* promoter from *U. panicoides* could be used to control gene expression in transgenic plants, and the promoter was tested in both rice and maize. Results indicated that a 1.3-kb 5′-flanking region of *U. panicoides pck1* contains *cis*-acting elements for preferential and abundant expression in bundle sheath cells of the leaf blade with light dependence in maize, but rice lacks some *trans*-acting elements required for expression controlled by *pck1* (Suzuki and Burnell 2003).

The *U. panicoides* PEPCK is a homohexameric protein containing an N-terminal region important for light/dark regulation. The PEPCK from *Panicum maximum* (a PEPCK-type C_4 plant) is phosphorylated in the dark; however, the PEPCK from *U. panicoides* (also a PEPCK-type C_4 plant) and maize (a NADP-ME type C_4 plant) is not phosphorylated in the dark. If the PEPCK mechanism is adopted for use in transgenic C_4 rice, it may be important to regulate the activity of transgenic PEPCK.

Pyruvate, orthophosphate dikinase

Pyruvate, orthophosphate dikinase (PPDK) is often cited as the rate-limiting enzyme in C_4 photosynthesis and this claim is supported by the fact that PPDK activity in leaves of C_4 plants is high enough to support the rate of photosynthetic inorganic carbon fixation (expressed on a chlorophyll basis) (Edwards et al 1985). Two PPDK isoforms are located in most C_4 plants, with the major form located in the chloroplast stroma and the minor form located in the cytosol; both forms are derived from differential expression of a single gene (Taniguchi et al 2000).

PPDK catalyzes the ATP-dependent phosphorylation of pyruvate to PEP in a reaction that essentially uses two high-energy bonds (two ATPs) and, unless closely regulated, has the potential to consume energy. As such, PPDK is a highly regulated enzyme. Like most photosynthetic enzymes, its expression is light-dependent but, more significantly, the enzyme is subject to a variety of regulatory mechanisms, including a light/dark phosphorylation/dephosphorylation mechanism. The regulation of PPDK differs from the phosphorylation/dephosphorylation mechanisms that regulate both PEPC and PEPCK in that phosphorylation completely inactivates PPDK in contrast to the phosphorylation of PEPC and PEPCK that renders them more or less sensitive to their substrates or allosteric effectors. In addition, the dark-dependent phosphorylation uses ADP (adenosine diphosphate) as the phosphate donor in contrast to ATP that is used to regulate both PEPC and PECK. In addition, both the phosphorylation and dephosphorylation of PPDK are catalyzed by a single enzyme (Burnell and Hatch 1985, 1986).

PPDK and its regulatory protein (RP) are found in the chloroplasts of C_3 plants (Chastain et al 2002). RP-like activity has also been detected in crude extracts of immature rice seeds (Chastain et al 2002) and it has been suggested that RP-like activity may be ubiquitous throughout C_3 plants, occurring in both leaves and other organs such as developing seeds (Chastain et al 2002, Fukayama et al 2001).

When considering which genes were to be used to generate a transgenic C_4 rice plant, consideration was given to maximizing the efficiency of the enzymes to be introduced. For PPDK, this involved considering the cold sensitivity of the maize leaf PPDK; a cold-resistant PPDK had been identified in *Flaveria brownii*, a C_3-C_4 intermediate plant species (Burnell 1990a). Subsequent research identified the amino acids responsible for the cold sensitivity of maize PPDK (Usami et al 1995, Ohta et al 1996), and a form of maize PPDK that, when expressed in bacteria, exhibits the same cold tolerance as *F. brownii* PPDK has now been constructed and expressed in maize (Ohta et al 2006). Expression of cold-tolerant PPDK in maize plants transformed with a chimeric cDNA made from *Flaveria bidentis* and *F. brownii* PPDK produced cold-tolerant homotetramers and heterotetramers of intermediate cold sensitivity formed with endogenous PPDK (Ohta et al 2004). Introduction of an antisense gene for maize PPDK significantly increased the ratio of heterologous PPDK to endogenous PPDK. More recent research has demonstrated that the cold tolerance of PPDK in crude leaf extracts was greatly improved in plants that expressed a large amount of a modified PPDK gene. This was achieved by introducing modified genomic sequences of maize PPDK. It was suggested that the high expression achieved may have been due to the exon-intron structure of the gene (Ohta et al 2006). Similar effects have also been reported when maize PPDK and PEPC gene constructs that included all introns produced much higher amounts of enzymes in transgenic rice than those that contained the cDNAs with the same promoter and terminator sequences (Ku et al 1999, Fukayama et al 2001). The photosynthetic rate in the transgenic maize plants containing the cold-tolerant PPDK at 8 °C was significantly increased with no deleterious effect at higher temperatures, indicating that PPDK is one of the limiting factors in the C_4 photosynthetic pathway of maize under cold conditions (Ohta et al 2006).

I still think that it is wise to introduce a cold-tolerant form of PPDK when designing the next generation of C_4 rice.

Transporters

The carboxylation of PEP to form oxaloacetate (OAA) was designed to occur in the cytosol, whereas the PEPCK-dependent decarboxylation of OAA was designed to occur in the chloroplast stroma. Significantly, this requires the presence and efficient operation of an OAA transporter that operates to transport OAA into the stroma from the cytosol. Similarly, a PEP transporter is required to transport PEP from the chloroplast stroma to the cytosol, thus facilitating the efficient completion of the C_4 cycle in the photosynthetic cells of rice leaves. Failing the efficient operation of an endogenous OAA transporter, it may be necessary to introduce a heterologous transporter such as the OAA transporter present in pea leaf mitochondria (Oliver and Walker 1984); importantly, no counter ion movement is required for OAA transport. Similarly, a plastidial PEP transporter has been identified in plants (Kubis and Rawsthorne 2000, Leegood 2000). The efficiency of C_4 photosynthesis in rice may be enhanced by the introduction of both of these transporters.

Conclusions

I still believe, as I did almost 20 years ago, that it is possible to increase the rate of photosynthesis (as judged by the rate of inorganic carbon fixation) in rice by manipulating the expression of foreign genes. The expression of carbonic anhydrase and PEP carboxylase in the cytosol and the expression of one decarboxylase and PPDK in the chloroplast may be the backbone of any C_4 acid cycle introduced into rice. Maximizing the operation of these enzymes may depend on several factors.

1. *Optimizing the amount of gene expression.*
   This might be achieved by transferring entire genes from C_4 plants into rice plants. An increasing number of reports indicate higher gene expression in maize and rice plants containing entire genes from C_4 plants, or at least the inclusion of an appropriate intron in or near the 5′-untranslated regions of the genes (Callis et al 1987, Fukayama et al 2001, Ku et al 1999, Ohta et al 2006, Tanaka et al 1990, Vasil et al 1989).
2. *Maximizing transport of photosynthetic intermediates between intracellular compartments.*
   The efficient transport of oxaloacetate (or malate) into chloroplasts and the movement of PEP (or pyruvate) out of the chloroplasts into the cytosol may be critical to maximizing the rate of any C_4 acid pathway in rice.
3. *Optimizing the regulation of introduced C_4 enzymes.*
   PEPC, PPDK, and PEPCK are all regulated by phosphorylation-dephosphorylation mechanisms that directly alter their activities or alter their sensitivities to photosynthetic intermediates. Since all three enzymes may be present in C_3 plants and function in nonphotosynthetic pathways, regulation of their activities by endogenous regulatory proteins may limit their coordinated activities. A clearer understanding of the regulation of exogenous PPDK in C_3 chloroplasts by endogenous PPDK regulatory protein is required as it is now recognized that PPDK may play an important role in the synthesis of aromatic amino acids (Chastain and Chollet 2003).

Postscript

In 1990, I wrote a document summarizing funding opportunities for research. Five research projects were included:

1. *Increased cold tolerance of C_4 plants by introduction of a gene coding for the cold-tolerant form of PPDK.*
2. *Molecular biology of the PPDK regulatory protein: its location, expression, and regulation.*
3. *Improvement of photosynthetic rates in C_3 plants by expression of genes coding for C_4-enzymes in specific intracellular compartments.*
4. *Development of C_4-specific herbicides.*
5. *A biochemical and molecular biological examination of the plasmodesmata of bundle sheath cells of C_4 plants (with a view to increasing resistance to viral infection).*

Of these projects, Projects 1 and 3 were funded by Japan Tobacco Inc. for six and three years, respectively, and the results are included above. In relation to Project 2, a long-standing collaboration with Dr. Chris Chastain has culminated in the identification of the gene sequence of pyruvate, Pi dikinase regulatory protein (Burnell and Chastain 2006). Regarding Project 4, the resistance to adopting plant gene technology in Australia stimulated a long-term collaboration with Dr. Lyndon Llewellyn at the Australian Institute of Marine Sciences, Townsville. Research jointly funded by James Cook University and Nufarm Pty Ltd. has resulted in the identification of a number of potential C_4-specific herbicides from marine organisms, which specifically target PPDK (Doyle et al 2005, Haines et al 2005). I have not been involved in any further research on plasmodesmata.

References

Badger MR, Price GD. 1994. The role of carbonic anhydrase in photosynthesis. Annu Rev. Plant Physiol. Plant Mol. Biol. 45:369-392.

Bowes G, Rao SK, Estavillo GM, Reiskind JB. 2002. C_4 mechanisms in aquatic angiosperms: comparisons with terrestrial C_4 systems. Functional Plant Biol. 29:379-392.

Burnell JN. 1990a. A comparative study of the cold-sensitivity of pyruvate, P_i dikinase in *Flaveria* species. Plant Cell Physiol. 31:295-297.

Burnell JN. 1990b. Immunological study of carbonic anhydrase in C_3 and C_4 plants using antibodies to maize cytosolic and spinach chloroplastic carbonic anhydrase. Plant Cell Physiol. 31:423-427.

Burnell JN. 2000. Carbonic anhydrases in higher plants: an overview. In: Chegwidden R, Carter ND, Edwards YH, editors. New horizons in carbonic anhydrase. Basel: Birkhauser Verlag. p 501-518.

Burnell JN, Chastain CJ. 2006. Cloning and expression of maize-leaf pyruvate, Pi dikinase regulatory protein gene. Biochem. Biophys. Res. Commun. 345:675-680.

Burnell JN, Gibbs MJ, Mason JG. 1990. Spinach chloroplastic carbonic anhydrase: nucleotide sequence analysis of cDNA. Plant Physiol. 92:37-42.

Burnell JN, Hatch MD. 1985. Light-dark modulation of leaf pyruvate, P_i dikinase. Trends Biochem. Sci. 111:288-291.

Burnell JN, Hatch MD. 1986. Activation and inactivation of an enzyme catalysed by a single bifunctional protein: a new example and why. Arch. Biochem. Biophys. 245:297-304.

Burnell JN, Hatch MD. 1988a. Photosynthesis in phospho*enol*pyruvate carboxykinase-type C_4 plants: photosynthetic activities of isolated bundle sheath cells from *Urochloa panicoides* Arch. Biochem. Biophys. 260:177-186.

Burnell JN, Hatch MD. 1988b. Photosynthesis in phospho*enol*pyruvate carboxykinase type-C_4 plants: pathways of C_4 acid decarboxylation in bundle sheath cells of *Urochloa panicoides*. Arch. Biochem. Biophys. 260:187-199.

Burnell JN, Hatch MD. 1988c. Low bundle sheath carbonic anhydrase is apparently essential for effective C_4 pathway operation. Plant Physiol. 86:1252-1256.

Burnell JN, Ludwig M. 1996. Plant carbonic anhydrases, In: Holmes RS, Lim HA, editors. Gene families: structure, function, genetics and evolution. Singapore: World Scientific. p 95-104.

Burnell JN, Ludwig M. 1997. Characterization of two cDNAs encoding carbonic anhydrase in maize leaves. Aust. J. Plant Physiol. 24:451-458.

Burnell JN, Suzuki I, Sugiyama T. 1990. Light induction and the effect of nitrogen status upon the activity of carbonic anhydrase in maize leaves. Plant Physiol. 94:384-387.

Callis J, Fromm M, Walbot V. 1987. Introns increase gene expression in cultured maize cells. Genes Dev. 1:1183-1200.

Cavallaro A, Ludwig M, Burnell JN. 1994. The nucleotide sequence of a complementary DNA encoding *Flaveria bidentis* carbonic anhydrase. FEBS Lett. 350:216-218.

Chastain CJ, Chollet R. 2003. Regulation of pyruvate, orthophosphate dikinase by ADP-/Pi-dependent reversible phosphorylation in C_3 and C_4 plants. Plant Physiol. Biochem. 41:523-532.

Chastain CJ, Fries JP, Vogel JA, Randklev CL, Vossen AP, Dittmer SK, Watkins EE, Fiedler LJ, Wacker SA, Meinhover KC, Sarath G, Chollet R. 2002. Pyruvate orthophosphate dikinase in leaves and chloroplasts of C_3 plants undergoes light/dark-induced reversible phosphorylation. Plant Physiol. 128:1368-1378.

Chollet R, Vidal J, O'Leary MH. 1996. Phospho*enol*pyruvate carboxylase: a ubiquitous, highly regulated enzyme in plants. Annu. Rev. Plant Physiol. Plant Mol. Biol. 47:273-298.

Doyle JR, Burnell JN, Haines DS, Llewellyn LE, Motti CA, Tapiolas DM. 2005. A rapid screening method to detect specific inhibitors of pyruvate orthophosphate dikinase as leads for C_4 plant-selective herbicides. J. Biomol. Screening 10:67-75.

Edwards G, Nakamoto H, Burnell JN, Hatch MD. 1985. Pyruvate, P_i dikinase and NADP-malate dehydrogenase in C_4 photosynthesis: kinetic properties and regulation. Annu. Rev. Plant Physiol. 36:255-286.

Fawcett TW, Browse JA, Volokita M, Barlett SG. 1990. Spinach carbonic anhydrase primary structure deduced from the sequence of a cDNA clone. J. Biol. Chem. 256:5414-5427.

Finnegan PM, Burnell JN. 1994. Isolation and sequence analysis of cDNAs encoding phospho*enol*pyruvate carboxykinase from the PCK-type C_4 grass *Urochloa panicoides*. Plant Mol. Biol. 27:365-376.

Finnegan PM, Suzuki S, Ludwig M, Burnell JN. 1999. Phospho*enol*pyruvate carboxykinase in the C_4 monocot *Urochloa panicoides* is encoded by four differentially expressed genes. Plant Physiol. 120:1033-1041.

Fukayama H, Tsuchida H, Agarie S, Nomura M, Onodera H, Ono K, Lee B-H, Hirose S, Toki S, Ku MSB, Makino A, Matsuoka M, Miyao M. 2001. Significant accumulation of C_4-specific pyruvate, orthophosphate dikinase in a C_3 plant, rice. Plant Physiol. 127:1136-1146.

Haines DS, Burnell JN, Doyle JR, Llewellyn LE, Motti CA, Tapiolas DM. 2005. Translation of in vitro inhibition by marine natural products of the C_4 acid cycle enzyme pyruvate, Pi dikinase to *in vivo* C_4 plant tissue death. J. Agric. Food Chem. 53:3856-3862.
Hatch MD, Burnell JN. 1990. Carbonic anhydrase activity in leaves and its role in the first step of C_4 photosynthesis. Plant Physiol. 93:825-828.
Ku MS, Agarie S, Nomura M, Fukayama H, Tsuchida H, Ono K, Hirose S, Toki S, Miyao M, Matsuoka M. 1999. High-level expression of maize phospho*enol*pyruvate carboxylase in transgenic rice plants. Nature Biotechnol. 17:76-80.
Kubis SE, Rawsthorne S. 2000. The role of plastidial transporters in developing embryos of seed rape (*Brassica napus* L.) for fatty acid synthesis. Biochem. Soc. Trans. 28:665-666.
Leegood RC. 2000. Transport during C_4 photosynthesis. In: Leegood RC, Sharkey TD, von Caemmerer S, editors. Photosynthesis: physiology and metabolism. Dordrecht (Netherlands): Kluwer. p 459-469.
Ludwig M, Burnell JN. 1995. Molecular comparison of carbonic anhydrase from *Flaveria* species demonstrating different photosynthetic pathways. Plant Mol. Biol. 29:353-365.
Majeau N, Coleman JR. 1991. Isolation and characterization of a cDNA encoding for pea chloroplastic carbonic anhydrase. Plant Physiol. 95:264-268.
Majeau N, Coleman JR. 1992. Nucleotide sequence of a complementary DNA encoding tobacco chloroplastic carbonic anhydrase. Plant Physiol. 100:1077-1078.
Ohta S, Ishida Y, Usami S. 2004 Expression of cold-tolerant pyruvate, orthophosphate dikinase cDNA, and heterotetramer formation in transgenic maize plants. Transgenic Res. 13:475-485.
Ohta S, Ishida Y, Usami S. 2006. High-level expression of cold-tolerant pyruvate, orthophosphate dikinase from a genomic clone with site-directed mutations in transgenic maize. Mol. Breed. 18(1):29-38.
Ohta S, Usami S, Ueki J, Kumashiro T, Komari T, Burnell JN. 1996. Identification of the amino acid residues responsible for cold tolerance in *Flaveria brownii* pyruvate, P_i dikinase. FEBS Lett. 396:152-156.
Okabe K, Yang S-Y, Tsuzuki M, Miyachi S. 1984. Carbonic anhydrase: its content in spinach leaves and its taxonomic diversity studied with anti-spinach leaf carbonic anhydrase antibody. Plant Sci. Lett. 33:145-153.
Oliver DJ, Walker GH. 1984. Characterisation of the transport of oxaloacetate by pea leaf mitochondria. Plant Physiol. 76:409-413.
Roeske CA, Ogren WL. 1990. Nucleotide sequence of pea cDNA encoding chloroplast carbonic anhydrase. Nucl. Acids Res. 18:3413.
Sugiharto B, Burnell JN, Sugiyama T. 1992. Cytokinin is required to induce the nitrogen-dependent accumulation of mRNAs for phospho*enol*pyruvate carboxylase and carbonic anhydrase in detached maize leaves. Plant Physiol. 100:153-156.
Sugiharto B, Suzuki I, Burnell JN, Sugiyama T. 1992. Glutamine induces the N-dependent accumulation of mRNAs encoding phospho*enol*pyruvate carboxylase and carbonic anhydrase in detached maize leaf tissue. Plant Physiol. 100:2066-2070.
Suzuki S, Burnell JN. 1995. Nucleotide sequence of a complementary DNA encoding rice chloroplastic carbonic anhydrase. Plant Physiol. 107:299-300.
Suzuki S, Burnell JN. 2003. The pck 1 promoter from *Urochloa panicoides* (a C_4 plant) directs expression differently in rice (a C_3 plant) and maize (a C_4 plant). Plant Sci. 165:603-611.

Suzuki S, Murai N, Burnell JN, Arai M. 2000. Alteration of photosynthetic flow in transgenic rice plants expressing C_4-type phospho*enol*pyruvate carboxykinase from *Urochloa panicoides*. Plant Physiol. 124:163-172.

Suzuki S, Murai N, Kasaoka K, Hiyoshi T, Imaseki I, Burnell JN, Arai M. 2006. Carbon metabolism in transgenic rice plants that express phospho*enol*pyruvate carboxylase and/or phospho*enol*pyruvate carboxykinase. Plant Sci. 170:1010-1019.

Tanaka A, Mita S, Ohata J, Kyozuka K, Shimamato K, Nakamura K. 1990. Enhancement of foreign gene expression by a dicot intron in rice but not in tobacco is correlated with an increased level of mRNA and an efficient splicing of the intron. Nucl. Acids Res. 18:6767-6770.

Taniguchi M, Izawa K, Ku MSB, Lin JH, Saito H, Ishida Y, Ohta S, Komari T, Matsuoka M, Sugiyama T. 2000. The promoter for the maize C_4 pyruvate, orthophosphate dikinase gene directs cell- and tissue-specific transcription in transgenic maize plants. Plant Cell Physiol. 41:42-48.

Usami S, Ohta S, Komari T, Burnell JN. 1995. Cold stability of pyruvate orthophosphate dikinase of *Flaveria brownii*. Plant Mol. Biol. 27:969-980.

Vasil V, Clancy M. Ferl RJ, Vasil KI, Hannah LC. 1989. Increased gene expression by the first intron of maize *shrunken-1* locus in grass species. Plant Physiol. 91:1575-1579.

von Caemmerer S, Quinn V, Hancock NC, Price GD, Furbank RT. Ludwig M. 2004. Carbonic anhydrase and C_4 photosynthesis: a transgenic analysis. Plant Cell Environ. 27:697-703.

Voznesenskaya EV, Franceschi VR, Kiirats O, Freitag H, Edwards GE. 2001. Kranz anatomy is not essential for terrestrial C_4 plant photosynthesis. Nature 414:543-546.

Notes

The editors would like to thank Jim Burnell for showing patent details for a method of enhancing photosynthetic activity.

Author's address: Department of Biochemistry and Molecular Biology, James Cook University, Townsville, Queensland 4811, Australia.

Acknowledgment: I wrote this chapter in response to a very generous invitation from Dr. John Sheehy (after initial contact with Dr. Peter Mitchell) to present some of the history behind the early endeavors to increase photosynthetic rates using molecular means.

Breaking the Kranz paradigm in terrestrial C_4 plants: does it hold promise for C_4 rice?

G.E. Edwards, E. Voznesenskaya, M. Smith, N. Koteyeva, Y.-I. Park, J.H. Park, O. Kiirats, T.W. Okita, and S.D.X. Chuong

A common feature of photosynthesis in practically all organisms is the assimilation of CO_2 into organic matter via a catalyst called ribulose 1,5-bisphosphate carboxylase oxygenase (Rubisco) in the carbon assimilation cycle. One of the constraints on the process in terrestrial plants is conditions where CO_2 becomes limiting because of high temperature, drought, or soil salinity. This can occur by restricting the entry of CO_2 into leaves, by decreased stomatal conductance, by decreased cytoplasmic solubility of CO_2, and by increased photorespiration (a process resulting from O_2 competing with CO_2 in Rubisco catalysis). In response to CO_2 limitations, some terrestrial plants evolved mechanisms to concentrate CO_2 around Rubisco through a C_4 cycle that requires spatial separation of fixation of atmospheric CO_2 into C_4 acids, and the donation of CO_2 from C_4 acids via decarboxylases to Rubisco (called C_4 plants). The paradigm for C_4 photosynthesis in terrestrial plants for more than 35 years was that a dual-cell system, called Kranz leaf anatomy, is required for spatial separation of these functions. Surprisingly, recent research on species in family Chenopodiaceae has shown that C_4 photosynthesis can occur within a single photosynthetic cell. Two very novel means of accomplishing this evolved in subfamily Suaedoideae. These systems function by spatial development of two cytoplasmic domains, which contain dimorphic chloroplasts. Emerging information on the biochemical and structural strategies for accomplishing C_4 has promise for improving the productivity of rice, which lacks a CO_2-concentrating mechanism, and for securing this important crop as a food supply under CO_2-limited conditions predicted with global warming.

Keywords: Chenopodiaceae, single-cell C_4, cytoskeleton, chloroplast differentiation

The distinguishing features of biochemistry and physiology of photosynthesis between C_3 and C_4 plants are well known (Edwards and Walker 1983, Sage and Monson 1999). C_3 plants directly assimilate atmospheric CO_2 into organic matter through the C_3 pathway (alternatively called the reductive pentose phosphate pathway or the Benson–Calvin cycle). In this process, fixation of CO_2 is catalyzed by the enzyme ribulose 1,5-bisphosphate carboxylase oxygenase (Rubisco), resulting in synthesis of two C_3 compounds (3-phosphoglyceric acids), which are used for synthesis of other products, such as starch and sucrose. Rubisco is a bifunctional enzyme, where CO_2 and O_2 are competitive substrates for reacting with ribulose bisphosphate (RuBP). Reaction of RuBP with CO_2 results in carbon assimilation; reaction with O_2 results in photorespiration, a counterproductive process involving the glycolate pathway and metabolism by chloroplasts, peroxisomes, and mitochondria. In terrestrial C_3 plants, in particular under drought, salinity, and/or high temperature, which are increasing on a global scale, CO_2 can become limiting for photosynthesis. Drought and salinity decrease the opening of stomata and increase diffusive resistance to CO_2. High temperature decreases the solubility of CO_2 in the cell liquid phase, and it alters the kinetic properties of Rubisco, favoring reaction of RuBP with O_2.

About 40 years ago, it was discovered that some plants have a mechanism for actively accumulating CO_2 from the atmosphere through a C_4 dicarboxylic acid cycle and donating it to the C_3 pathway. Species having this CO_2-concentrating mechanism are called C_4 plants, and, to date, they have been identified in 19 families among the approximately 500 families of plants (Sage 2004). C_4 photosynthesis is prevalent among the world's worst weeds, while it is notably lacking in rice and most other major crops. C_4 plants are well known to have an obvious advantage over C_3 plants under any conditions where CO_2 is limiting, resulting in higher rates of photosynthesis, and greater efficiency of use for nitrogen and water than in C_3 plants (Sage 2001). The largest number of C_4 species has been found in the monocot family Poaceace, whereas family Chenopodiaceae is currently the largest among dicots.

C_4 photosynthesis is accomplished by spatial separation of two phases of carbon assimilation. This involves the fixation of atmospheric CO_2 into C_4 acids in one compartment where pyruvate, Pi dikinase (PPDK) generates the substrate phospho*enol*pyruvate (PEP) for PEPC (PEP carboxylase), and transport of C_4 acids to another compartment where donation occurs of CO_2 by C_4 acid decarboxylases to the C_3 cycle (via Rubisco). There are three well-defined biochemical sub-types of C_4 plants based on differences in the C_4 acid decarboxylases used: NADP-malic enzyme (NADP-ME), NAD-malic enzyme (NAD-ME), and phospho*enol*pyruvate carboxykinase (PEPCK) (Kanai and Edwards 1999).

The dogma for more than 35 years was that the spatial separation of functions in terrestrial C_4 plants requires a dual-cell system called Kranz anatomy. This separation by cell type occurs with mesophyll cells fixing atmospheric CO_2 into C_4 acids (the carboxylation phase of the C_4 cycle, which is supported by the mesophyll chloroplasts), and donation of CO_2 from C_4 acids to the C_3 cycle in chloroplasts of bundle sheath cells where Rubisco is located. The minimum requirements for this CO_2-concentrating mechanism have been defined as follows: (1) cell-specific amplification of enzymes of C_4 photosynthesis (e.g., PEPC in mesophyll cells, C_4 acid decarboxylases and Rubisco in bundle sheath cells), with complementary adjustments of photosystem and electron transport activities; (2) novel, cell-specific organelle metabolite translocators; (3) symplastic connections between mesophyll and bundle sheath cells, the spatially separated sources and sinks of metabolite transfer in the C_4 cycle; and (4) barriers to CO_2 diffusion between the site of CO_2 fixation by PEPC in mesophyll cells and sites of CO_2 release and refixation by Rubisco in bundle sheath cells (Edwards et al 2001b).

In a striking contradiction to this paradigm for function of C_4 in terrestrial plants with Kranz anatomy, plants were discovered in family Chenopodiaceae that conduct C_4 photosynthesis within a single type of photosynthetic cell (Voznesenskaya et al 2001, 2002, Edwards et al 2004, Akhani et al 2005). In these cases, the requisite spatial partitioning occurs by development of two intracellular cytoplasmic compartments in the chlorenchyma cells. The discovery of C_4 plants approximately 40 years ago and their association with Kranz anatomy in terrestrial plants has led to numerous studies on occurrence, mechanism, structural and biochemical diversity, molecular control of development in two photosynthetic cells, and evolution (Edwards and Walker 1983, Hatch 1987, Sage and Monson 1999). With the paradigm that this occurs in land plants via development of Kranz anatomy, the finding that terrestrial species can conduct C_4 photosynthesis within individual chlorenchyma cells will change the way we think about requirements for the function of C_4 photosynthesis, the genetic control of development and chloroplast differentiation, evolution of C_4, and potential for engineering C_4 photosynthesis into C_3 crops.

Structural differences in terrestrial C_4 systems

Structural diversity in Kranz anatomy

C_4 photosynthesis evolved many times in terrestrial plants (see Sage and Sage, this volume) and diversity is considerable in the types of Kranz anatomy, especially in the families Poaceace and Chenopodiaceae. Poaceace has the most C_4 species; nine structural types of Kranz anatomy have been described based on differences in leaf anatomy, in structure and intracellular location of chloroplasts and mitochondria, and in biochemical sub-types. These include the three classical biochemical types accounting for many of the C_4 species in the family: NADP-ME, NAD-ME, and PEPCK; and six nonclassical types: Aristidoid (NADP-ME), Neurachneoid (NADP-ME or PEPCK), Arundinelloid (NADP-ME), Triodioid (NAD-ME), Eriachneoid (NADP-ME), and Stipagrostoid (NADP-ME) (Prendergast et al 1987, Dengler and Nelson 1999, Voznesenskaya et al 2005a,b).

Among dicot families, it is well established that family Chenopodiaceae has the largest number of C_4 species and also the greatest diversity in leaf anatomy, including C_3 and C_4 Kranz and C_4 single-cell types (Carolin et al 1975, Sage et al 1999, Edwards et al 2004). There are six major types of Kranz anatomy, based on differences in the position of primary and secondary vascular bundles in the leaf, and the position of two chlorenchyma layers in relation to water storage tissue and vascular bundles: Atriplicoid, Kochioid, Salsoloid, Kranz-Suaedoid, Conospermoid, and Kranz-Halosarcoid (Fig. 1), which occur in five tribes, Atripliceae, Camphorosmeae, Salsoleae, Suaedeae, and Salicornieae (Carolin et al 1975, Freitag and Stichler 2000, Schütze et al 2003, Kadereit et al 2003, Kapralov et al 2006). All anatomical types differ in the position and the level of development of water storage tissue: Atriplicoid type in laminate leaves and Conospermoid in semi-terete leaves do not have water storage

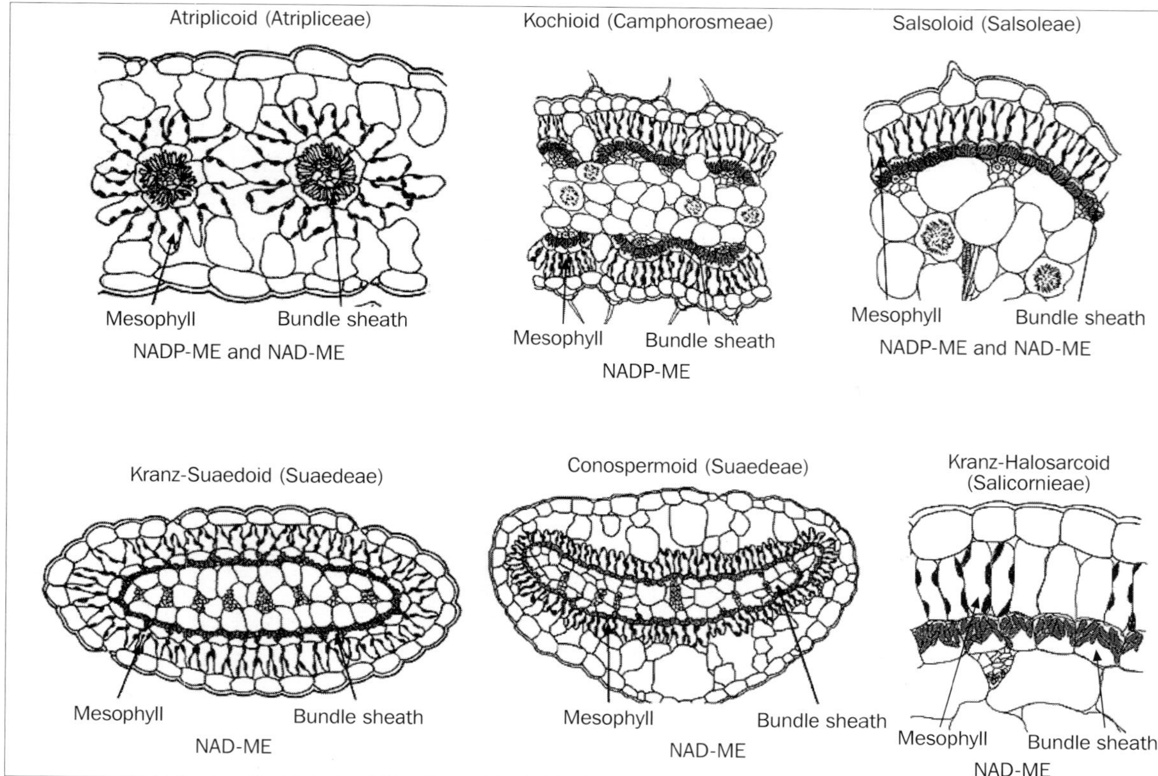

Fig. 1. Hand-drawn illustrations of types of Kranz anatomy in family Chenopodiaceae. Adapted from Gamaley and Voznesenskaya (1986), Freitag and Stichler (2002).

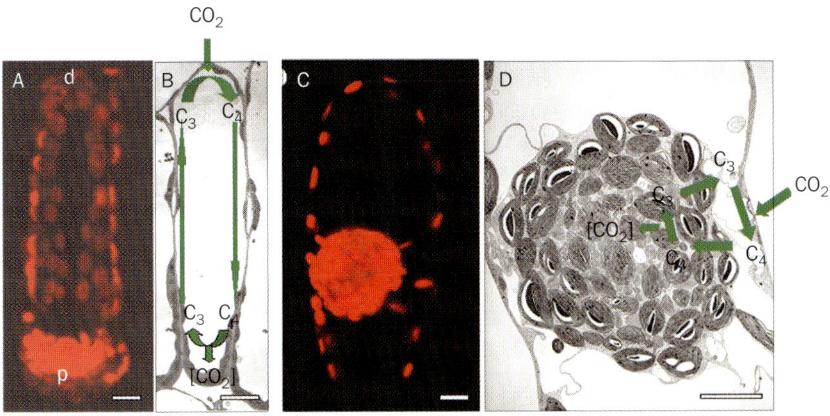

Fig. 2. Models for function of single-cell C_4 photosynthesis. (A, B) *Suaeda aralocaspica*. (A) Confocal microscopy showing a chlorenchyma cell with chloroplasts (fluorescence, red) concentrated at the distal end, and other chloroplasts around the periphery toward the distal end. (B) Atmospheric CO_2 enters the distal end of the cell, where it is fixed in the C_4 cycle, C_4 acids diffuse to the proximal end of the cell with decarboxylation, where they are captured by Rubisco. (C, D) *Bienertia* species. (C) Confocal microscopy showing chloroplasts in two domains, in a central and in a peripheral cytoplasmic compartment. (D) Atmospheric CO_2 is fixed into C_4 acids in the peripheral cytoplasm and C_4 acids diffuse through cytoplasmic channels to the central cytoplasmic compartment, where they are decarboxylated and the CO_2 fixed by Rubisco. d = distal, p = proximal end of cell, respectively. Adapted from Chuong et al (2006), Edwards et al (2004). Scale bars = 10 μm.

tissue as the hypodermis has this function. The Kochioid-type species, which have laminate or semi-terete leaves, and the Kranz-Suaedoid-type species, which have terete leaves, have only a few layers of water storage tissue, whereas terete cylindrical leaves of Salsoloid-type species have the largest volume of water storage tissue. C_4 species of the subfamilies Suaedoideae and Salsoloideae are predominant in biodiversity and biomass in deserts and semideserts of Central Asia, where they have adapted to grow under warm, dry, saline conditions.

Structural diversity in terrestrial C_4 species without Kranz anatomy

Single-cell C_4 species were found in subfamily Suaedoideae, which, as noted above, has two structural types of Kranz anatomy. There are two very different structural types of single-cell C_4 species (Fig. 2), one found in tribe Suaedeae (*Suaeda aralocaspica*, previously called *Borszczowia aralocaspica*) and the other in tribe Bienertieae (*Bienertia cycloptera* and *B. sinuspersici*). The spatial separation of functions into two domains in chlorenchyma cells in these species is proposed to serve an analogous function to mesophyll and bundle sheath cells in Kranz-type C_4 plants. This includes the exceptional development of dimorphic chloroplasts within the chlorenchyma cells. The models for C_4 function of each structural type are illustrated in Figure 2. *Suaeda aralocaspica* has elongated chlorenchyma cells, with the two chloroplast types partitioned toward the distal and proximal ends, separated by a thin layer of cytoplasm. The two *Bienertia* species have identical types of chlorenchyma cells, with chloroplast-containing central and peripheral cytoplasmic compartments connected by cytoplasmic channels. These models illustrate fixation of CO_2 into C_4 acids in one domain (functionally analogous to mesophyll cells in Kranz type) and donation of CO_2 from C_4 acids to Rubisco in the other domain (functionally analogous to bundle sheath cells in Kranz type). In these species, the mature chlorenchyma cells have two chloroplast types separated between two domains, along with control of spatial partitioning of mitochondria, peroxisomes, and the nucleus, in which the cytoskeleton is considered to have a central role.

Photosynthetic types, importance of control of chloroplast position, and role of the cytoskeleton

C_3 plants

Studies with various C_3 species, including macroalgae, ferns, mosses, and C_3 angiosperms, have shown that chloroplasts are dynamic, with movement in response to light. They exhibit an accumulation response along periclinal walls in low light to maximize photosynthesis, and an avoidance response along anticlinal walls in high light to minimize photodamage. In plants, this is controlled by blue-light photoreceptors (phototropin1 and phototropin2). Mutation in phototropin2, which prevents avoidance movement, causes photodamage in *Arabidopsis* (Kasahara et al 2002). Following the perception of light, actin filaments and Ca^{2+} fluxes are required for proper location of the chloroplasts (Kandasamy and Meagher 1999, Kasahara et al 2002, Wada et al 2003, Inada et al 2004, DeBlasio et al 2005, Matsuoka and Tokutomi 2005). The cytoskeleton has a key role in controlling the position of organelles. A role of actin filaments in the positioning of chloroplasts has been demonstrated in various studies, since anti-actin drugs inhibit chloroplast movement (Witztum and Parthasarathy 1985, Menzel and Schliwa 1986, Kadota and Wada 1992, Nagai 1993, Dong et al 1996, Kandasamy and Meagher 1999, Sato et al 2001, Oikawa et al 2003). Although most studies show that actin filaments are the predominant structure controlling chloroplast movement, some studies indicate that microtubules also have a role in organelle movement (Wada et al 2003, Wada and Suetsugu 2004).

C_4 Kranz plants

Although chloroplast mobility has commonly been observed in chlorenchyma cells in C_3 plants, in C_4 plants with Kranz anatomy, chloroplasts in bundle sheath cells generally show partitioning to a particular domain of the cell. For example, in family Poaceace, the classical NADP-ME species have chloroplasts located in the centrifugal position in bundle sheath cells, whereas the classical NAD-ME species have chloroplasts in the centripetal position, and the classical PEP-CK species have chloroplasts either in the centrifugal position or more evenly distributed around the cytoplasm. These positions are illustrated in Figure 3 (A-C), where the bundle sheath chloroplasts are labeled by immunolocalization with antibody to Rubisco. Under water stress when photosynthesis drops, the chloroplasts can change positions, which may be a form

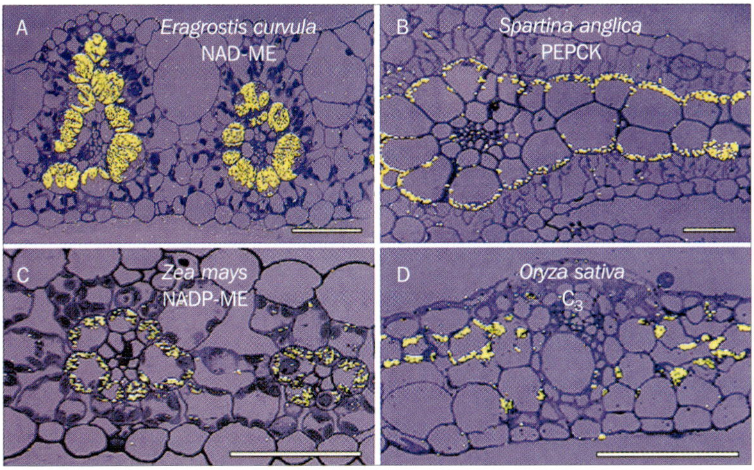

Fig. 3. Illustrations of different chloroplast positions in bundle sheath cells of C_4 plants as compared to rice, in family Poaceace. Shown are the three classical structural types, *Eragrostis curvula* = NAD-ME, *Spartina anglica* = PEPCK, *Zea mays* = NADP-ME, and *Oryza sativa* = C_3 leaf anatomy. Label indicating immunolocalization of Rubisco appears as yellow dots. Adapted from Edwards et al (2001a), Voznesenskaya et al (2006), and unpublished results (E. Voznesenskaya and G. Edwards). Scale bars = 50 μm.

of photoprotection (Lal et al 1996). By comparison, in rice, a C_3 plant (Fig. 3D), the chloroplasts are distributed around the periphery of the mesophyll cells. This positioning of chloroplasts in bundle sheath cells in different sub-types is important in considering the diffusive resistance to CO_2 from sites of C_4 acid decarboxylation to the intercellular air space (see the chapter by von Caemmerer et al, this volume, and von Caemmerer and Furbank 2003).

There is also control of the position of chloroplasts in bundle sheath cells in C_4 dicots. This can be illustrated in family Chenopodiaceae, subfamily Suaedoideae, between two Kranz-type species, of the Kranz-Suaedoid and Conospermoid types of anatomy (Figs. 1 and 4). In the Suaedoid type, the chloroplasts are in a centripetal position in bundle sheath cells, which is illustrated with *Suaeda arcuata* (Fig. 4A), whereas in the Conospermoid type the chloroplasts are in a centrifugal position, illustrated with *S. eltonica* and *S. cochlearifolia* (Fig. 4B,C). In the Conospermoid type, there is an interesting variation between *S. eltonica*, which has two layers of bundle sheath cells, and *S. cochlearifolia*, which has only one layer of bundle sheath cells in the center of the leaf, with chloroplasts that are polarized to opposite sides of the cell, adjacent to the mesophyll cells where C_4 acids will be imported. Thus, chloroplasts are located in two domains in the bundle sheath cell, but, in this case, they perform the same function (accept CO_2 from the mesophyll cells and assimilate it via Rubisco in the C_3 cycle). Again, the cytoskeleton must be involved in developing and maintaining these photosynthetic domains in the cell.

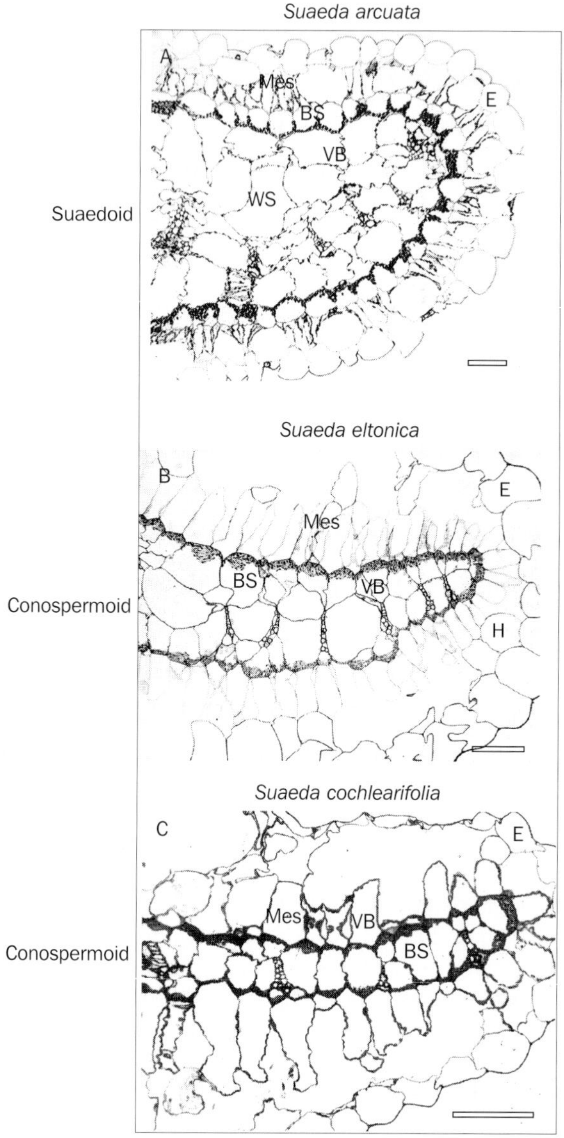

Fig. 4. Kranz-type C_4 species in family Chenopodiaceae, subfamily Suaedoideae, showing two types of Kranz anatomy. *Suaeda arcuata* has Kranz-Suaedoid anatomy with bundle sheath chloroplasts in the centripetal position. *S. eltonica* and *S. cochlearifolia* have Conospermoid anatomy with bundle sheath chloroplasts in the centrifugal position. Mes = mesophyll, BS = bundle sheath, WS = water storage, VB = vascular bundle, E = epidermis, H = hypodermis. Adapted from Kapralov et al (2006) and Voznesenskaya et al (n.d.). Scale bars represent 100 μm.

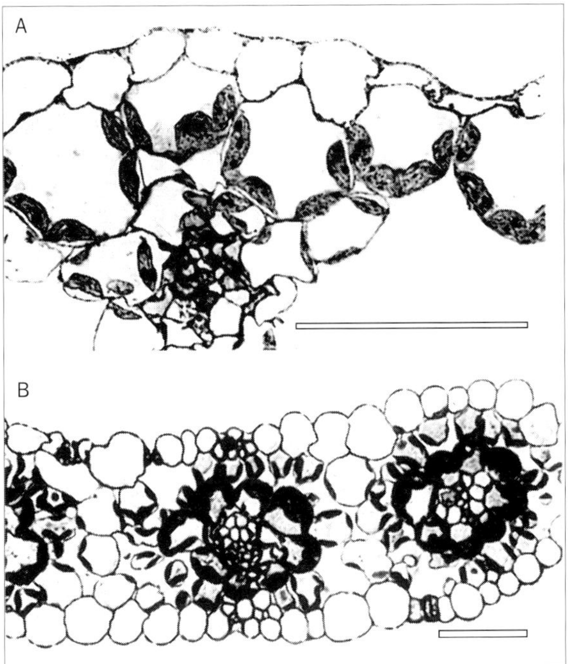

Fig. 5. Aquatic leaf anatomy (A) and terrestrial leaf anatomy (B) of *Orcuttia viscida*, family Poaceae. Adapted from Keeley (1998). Scale bars represent 50 μm.

Single-cell C_4 aquatic plants

Chloroplast position may also be important in single-cell-functioning C_4 photosynthesis, which occurs in some aquatic macrophytes. Some species of *Orcuttia,* family Poaceae, grow in seasonal pools formed by rain in California. They germinate and produce terete leaves when submerged, and then form laminate leaves when floating on the water and during continued growth as the pools dry up (Keeley 1998, 1999, Keeley and Rundel 2003). *Orcuttia viscida* functions as an NADP-ME C_4 plant in both aquatic and terrestrial leaves. The aquatic leaves have a single type of mesophyll-like chlorenchyma, with one type of chloroplast, which is located toward the centripetal position, whereas the terrestrial leaves have classical Poaceace NADP-ME Kranz anatomy (Fig. 5). In the aquatic form with one type of plastid, the proposed mechanism of photosynthesis is fixation of CO_2 in the cytosol by PEPC to form C_4 acids, then donation of CO_2 from malate via chloroplastic NADP-ME to Rubisco. The location of chloroplasts in the centripetal position may provide liquid-phase diffusive resistance to CO_2, enabling it to be concentrated and fixed by Rubisco. Interestingly, according to this model, chloroplasts that develop in mesophyll-like chlorenchyma cells in submerged leaves may be functionally more like the bundle sheath chloroplasts in Kranz-type leaves that develop under terrestrial conditions (both contain NADP-ME and Rubisco).

The most extensive studies on C_4 aquatic macrophytes have been made with the freshwater monocot *Hydrilla verticillata* (see Bowes et al, this volume). It grows under submerged conditions, and is induced under warm temperatures and high light in the summer to perform C_4 photosynthesis. It, like *O. viscida,* has a single-cell NADP-ME C_4 system, with monomorphic chloroplasts. It is not clear how CO_2 is concentrated around Rubisco in the chloroplast to prevent either leakage from the leaf or refixation by PEPC in a futile cycle (using the PEP regenerated from PPDK in the chloroplast). However, the position of chloroplasts in the cell might be important. There is evidence that the chloroplasts in leaves of *Hydrilla* change their position between dark or dim light and higher light. In the dark, the chloroplasts are dispersed, whereas in light (18 $W\ m^{-2}$) from a fluorescent source, which is approximately 5–10% of full sunlight) the chloroplasts form bands that extend across several cells, and are located on anticlinal and inner paradermal walls. Actin filaments are considered responsible for controlling this positioning of chloroplasts, since cytochalasin B, which disrupts actin, prevented this dark to light movement (Witztum and Parthasarathy 1985). Although this may be a photoprotection response in some species, movement of chloroplasts away from the cytoplasmic layer that faces the outer periclinal wall, to the cytoplasm on the anticlinal and inner paradermal walls in response to light, may also be favorable for C_4 photosynthesis in *Hydrilla.* Inorganic carbon entering the leaf could be captured by PEPC in the cytoplasm on the outer periclinal wall to form C_4 acids, which diffuse to the chloroplasts on the anticlinal and inner periclinal walls and donate CO_2 to Rubisco by chloroplastic NADP-ME. Bringing chloroplasts close together in one domain in the cytoplasm may provide their spatial separation from entry of CO_2 into the cell where C_4 acids are synthesized via PEPC. Also, having chloroplasts packed in one domain would help exclude the cytosolic PEPC in that domain and prevent futile recycling of CO_2 generated by NADP-ME. In both *O. viscida* and *H. verticillata*, it is possible that control of chloroplast position in the cytoplasm is important for the effective function of C_4 photosynthesis.

Single-cell C_4 terrestrial plants

In the two types of single-cell C_4 systems found in family Chenopodiaceae, *Suaeda aralocaspica* and two species of *Bienertia,* in very young leaves, the chlorenchyma cells have a centrally located nucleus surrounded by chloroplasts. The chloroplasts are one type, all containing Rubisco and having a similar ultrastructure. In mature leaves, the chlorenchyma cells have formed two cytoplasmic domains that are separated by a thin layer of cytoplasm in *S. aralocaspica,* and by cytoplasmic channels in *Bienertia.* The chloroplasts have become differentiated for specialized C_4 function. Maintaining the compartmentalization is considered essential for the function of C_4 photosynthesis.

We have analyzed the cytoskeleton in mature chlorenchyma cells of *S. aralocaspica* and *B. sinuspersici* by immunofluorescence techniques using anti-actin and anti-β-tubulin monoclonal antibodies with Oregon green-conjugated secondary an-

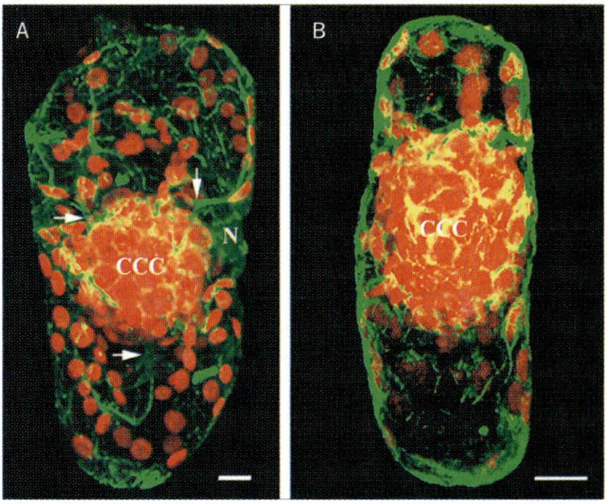

Fig. 6. Isolated chlorenchyma cells of *B. sinuspersici.* (A) Immunofluorescence using anti-actin antibody. Red = chlorophyll fluorescence, green = actin. Arrows: thick actin filament bundles connecting the central and peripheral cytoplasmic compartments. (B) Immunofluorescence using anti-tubulin antibody. Red = chlorophyll fluorescence, green = microtubules. CCC = central cytoplasmic compartment, N = nucleus. Adapted from Chuong et al (n.d.). Scale bars represent 10 μm.

tibodies. There is an extensive cytoskeleton network throughout the cytoplasm and surrounding chloroplasts in the two domains (illustrated for *B. sinuspersici* in Fig. 6). The results indicate that, once mature cells develop, it is the microtubules that are critical to maintaining the two cytosolic domains because oryzalin, which disrupts microtubules, disperses the chloroplasts, whereas cytochalasin D, which disrupts actin filaments, does not. Actin may function early in development to target organelles to specific areas. In various C_4 systems, Kranz or single-cell, it is expected that the cytoskeleton has an important role in controlling the spatial separation of CO_2 fixation by PEPC and donation of CO_2 from C_4 acids to Rubisco.

Differentiation to form dimorphic chloroplasts

Kranz type

More than 90% of chloroplast polypeptides are nucleus-encoded and thus must be imported into the chloroplast (Bedard and Jarvis 2005). Among enzymes in carbon assimilation, the Rubisco large subunit, encoded by the chloroplast genome, is an exception. In Kranz-type C_4, the cell-specific synthesis of certain photosynthetic enzymes in the two cell types requires differentially regulated expression of identical genomes (Walbot 1977) in nuclei and chloroplasts of adjacent cells (Sheen 1999). Regions of several nuclear-encoded C_4 genes responsible for controlling cell-specific expression and level of expression have been identified, with considerable evidence for transcriptional regulation (most research in maize, sorghum, *Flaveria,* and *Amaranthus*). For enzymes in the C_4 pathway, C_4 isoforms evolved for effective function of the CO_2 pump, with diversity in the control of expression by light and stage of development (Ku et al 1996, Berry et al 1997, Dengler and Nelson 1999, Sheen 1999, Blasing et al 2000, Dengler and Taylor 2000, Nomura et al 2000, Taylor 2000, Matsuoka et al 2001, Lai et al 2002, Gowik et al 2004). Expression of Rubisco genes for large (chloroplast genome) and small subunits (nuclear genome) is confined to bundle sheath cells, and repressed in mesophyll cells. There is also evidence for posttranscriptional regulation of *RbcS* (nuclear-encoded) and *RbcL* (chloroplast-encoded) in the selective synthesis of Rubisco in bundle sheath cells of C_4 plants (McCormac et al 2001, Patel et al 2004, 2006).

Single-cell C_4 terrestrial plants

We have identified three enzymes that are selectively expressed in chloroplasts of mature leaves of *Suaeda aralocaspica* and the *Bienertia* species: PPDK in one type of chloroplast (functioning in mesophyll cells of Kranz type) and Rubisco and ADPG pyrophosphorylase (ADPGase) in the other type of chloroplast (functioning in bundle sheath cells of Kranz type). Genes for the Rubisco small subunit, ADPGase large and small subunits, and PPDK are nuclear-encoded. Very young chlorenchyma cells in these plants have one type of chloroplast, all of which contain Rubisco, and they lack PPDK of the C_4 cycle. During development, differentiation occurs to form two types of chloroplasts that differ in structure and biochemistry.

This is analogous to the pattern of development in Kranz-type C_4 photosynthesis. In very young tissue, both mesophyll and bundle sheath cells have chloroplasts with low levels of Rubisco and similar structure. This is followed by repression of synthesis of Rubisco, and selective expression of the synthesis of PPDK in the mesophyll chloroplast. In both single-cell and Kranz-type C_4, the chloroplasts initially are in a C_3 default condition, and later they differentiate. By analogy, in the single-cell system, with partitioning of chloroplasts to two cytoplasmic compartments, repression of synthesis of Rubisco and ADPGase and expression of synthesis of PPDK occur in one domain, whereas the other domain develops a C_3-like chloroplast (with Rubisco, ADPGase, and starch biosynthesis). Although differential transcriptional control of expression of photosynthetic genes is considered a major means of developing dimorphic chloroplasts in Kranz-type C_4 plants, there must be other means for this differentiation in the single-cell C_4 chenopod species. In single-cell C_4 plants, research is needed to determine the mechanism of this differentiation of function, which may involve targeting of nuclear-encoded mRNA along the cytoskeleton to a specific cytoplasmic domain, uniform distribution of mRNA with control by posttranscriptional regulation, or selective polypeptide importers between the two chloroplast types.

mRNA targeting by cytoskeleton. Targeting of transcripts from the nucleus to specific domains is a possible means of controlling the intracellular development of two types of chloroplasts. RNA localization is recognized as a major cellular process in the asymmetric distribution of proteins, which lack their own sorting signals, within the cell. This is readily evident in structurally polarized somatic cells (Singer

1993, 1996, St Johnston 1995, Hesketh 1996, Bassell and Singer 1997) and during early embryo development (St Johnston 1995, Bogucka-Glotzer and Ephrussi 1996, Grunert and St Johnston 1996, King 1996), where it is responsible for embryonic polarity and cell fate determination (St Johnston 1995, Bogucka-Glotzer and Ephrussi 1996, King 1996). Localization of mRNA is also responsible for determining mating type in budding yeast, *Saccharomyces cerevisiae* (Long et al 1997, Takizawa et al 1997). In rice, RNA localization is required for targeting the storage protein RNAs to specific subdomains of the endoplasmic reticulum, which facilitates targeting of the coded proteins to specific compartments of the endomembrane system (Choi et al 2000, Crofts et al 2005, Washida et al 2006).

Although there are several mechanisms for localizing RNAs, the best characterized examples occur by active transport along the cytoskeleton, recently reviewed in St Johnston (2005). In metazoans, RNAs are almost invariably transported along microtubules (Bogucka-Glotzer and Ephrussi 1996, Jansen 1999, de Heredia and Jansen 2004). The few exceptions that use actin filaments include β-actin, which is transported to the growing edge of the lammellipodia in fibroblast cells, and *prospero* RNA, which is transported to the basal cortex in *Drosophila* neuroblasts (St Johnston 2005). In general, microtubules are used for long-distance transport, whereas actin filaments are used for short-distance transport (Kloc et al 2002).

In budding yeast (e.g., *Ash1* RNA) and the single documented case in higher plants (rice storage protein RNAs), RNAs are transported along actin filaments (Chartrand et al 2001, Hamada et al 2003), the cytoskeletal element used for transport of organelles in these organisms.

Posttranscriptional regulation. Another possible means for differentiation of chloroplasts in the single-cell C_4 system is posttranscriptional regulation. In the Kranz-type C_4 plants, it has been shown that untranslated regions of *RbcL* and *RbcS* mRNA control bundle sheath cell–specific expression (McCormac et al 2001, Patel et al 2004, 2006). In the single-cell C_4 system, domains could differ in factors interacting with untranslated regions on the mRNA, which could control translation or mRNA degradation.

Selective protein import into chloroplasts. Another potential level of control for differentiation of the two types of chloroplasts is the selective import of preproteins encoded by nuclear genes. There is evidence for multiple pathways for importing polypeptides at the outer membrane of chloroplasts (see Jarvis and Robinson 2004, Bedard and Jarvis 2005). Selective import may be controlled by differences in the N-terminal transit peptide (TP), by different isoforms of the TOC (translocons at the outer envelope surface) proteins, which recognize the TP, or possibly by cytosolic protein factors that interact with the TP. TPs are very variable in length, sequence, and amino acid composition. Thus, it is possible that the dimorphic chloroplasts in the single-cell C_4 species have differences in receptor components for protein import on the chloroplast envelope (e.g., for PPDK and the small subunit of Rubisco).

NAD-ME C_4 photosynthesis

In family Chenopodiaceae, all the C_4 plants examined in subfamily Suaedoideae (including two Kranz types and two single-cell C_4 types) are biochemically NAD-ME C_4 species, in which malate decarboxylation occurs in the mitochondria. Thus, the position of mitochondria is critical to the effective function of this C_4 system. In *S. aralocaspica* and the *Bienertia* species, the mitochondria are partitioned to the domain where Rubisco-containing chloroplasts are located (at the proximal end of the cell in *S. aralocaspica* and in the central cytoplasmic compartment of *Bienertia*), and immunolocalization studies show that NAD-ME is located in the mitochondria. The C_4 cycle is proposed to function by fixing atmospheric CO_2 by PEPC in one cytoplasmic domain, with diffusion of the C_4 acids to another domain, followed by donation of CO_2 to Rubisco (via NAD-ME) and return of the 3-C substrate for regeneration of PEP by PPDK. This model for the function of the C_4 cycle is illustrated for *Bienertia* in Figure 7.

Photorespiration

The compartmentation of enzymes of photorespiration and the glycolate pathway, which involves chloroplasts, mitochondria, and peroxisomes, is also important in the function of C_4 photosynthesis. In Kranz-type plants, any photorespiration that occurs as a consequence of O_2 reacting with RuBP is localized in bundle sheath cells, where it can contribute to the CO_2-concentrating mechanism (limiting the competitive reaction of O_2 with CO_2 by Rubisco) as well as allow for refixation of photorespired CO_2 by Rubisco. In the single-cell C_4 plants *B. sinuspersici* and *S. aralocaspica,* glycine decarboxylase is also localized in the mitochondria, which are associated with the Rubisco-containing chloroplasts.

Carbon isotope composition

Both types of single-cell C_4 species, *S. aralocaspica* and the *Bienertia* species, have C_4 isotope values (Voznesenskaya et al 2001, 2002, Freitag and Stichler 2002, Akhani et al 2005). In collections from natural habitats and from growth chamber-grown plants, the average $\delta^{13}C$ values were –13.2 and –12.4, respectively, for *S. aralocaspica*, and –13.9 and –14.2, respectively, for the two *Bienertia* species. In *Bienertia* species, more negative values (between –14 and –19) are observed in young leaves before full development of the C_4 system. By comparison, six different Kranz-type *Suaeda* species had average $\delta^{13}C$ values of –13.2 (Freitag and Stichler 2002).

Physiology

Table 1 shows an analysis of gas exchange of single-cell C_4 species compared with Kranz-type C_4 and C_3 relatives in subfamily Suaedoideae. The two single-cell C_4 species, *B. sinuspersici* and *S. aralocaspica*, have comparable rates with the average of three Kranz-type *Suaeda* species (*S. vermiculata, S. taxifolia,* and *S. eltonica*). The light-saturated rates of CO_2 fixation on a leaf area basis (the area exposed to incident

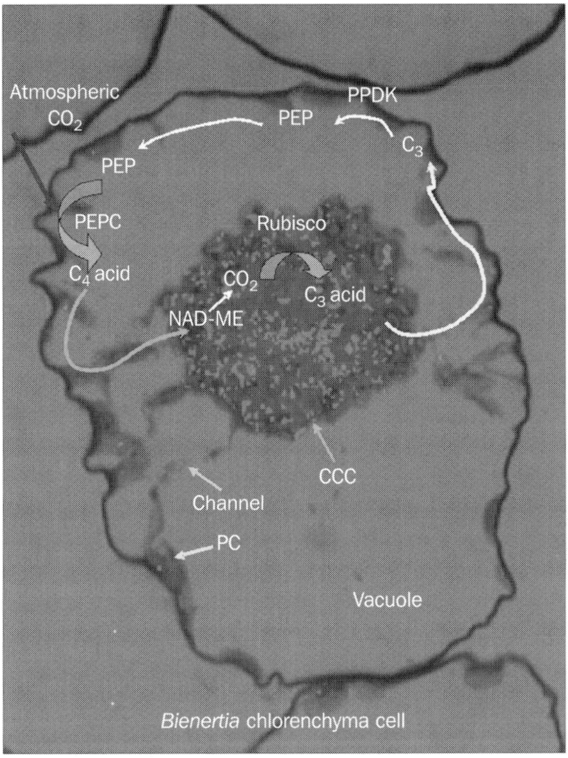

Fig. 7. Model for function of C_4 photosynthesis in the single-cell C_4 species of *Bienertia*. Pyruvate, Pi dikinase in chloroplasts in the peripheral cytoplasm (PC) converts pyruvate to PEP, PEP is used by PEPC to generate C_4 acids, which diffuse through the cytoplasmic channels to the central cytoplasmic compartment (CCC), where malate is decarboxylated by mitochondrial NAD-ME, the CO_2 donated to Rubisco, and pyruvate returned through the cytoplasmic channels to the peripheral chloroplasts. The width of the cell shown is about 30 μm.

light), under atmospheric CO_2 at 25 °C, were higher in *S. aralocaspica* than in *B. sinuspersici*, whereas the lowest rates were with the C_3 species *S. heterophylla*. The single-cell C_4 species and the Kranz-type *Suaeda* species have low CO_2 compensation points indicative of C_4 plants (compensation points measured by the method of Brooks and Farquhar 1985) from initial slopes of response to CO_2 at several light intensities. C_4 plants tend to have higher water-use efficiency than C_3 plants under a given condition. The water-use efficiency (μmol CO_2 $mmol^{-1}$ water) measured under saturating light and atmospheric CO_2 was about twofold higher in the single-cell C_4 and Kranz-type species than in the C_3 species *S. heterophylla*.

Table 1. Physiological comparisons of single-cell C_4 species and related Kranz and C_3 species in family Chenopodiaceae, subfamily Suaedoideae. The maximum rate of photosynthesis, A_{max}, is at saturating light, 25 °C, and normal atmospheric concentration of CO_2. The CO_2 compensation point, Γ^*, was measured by the method of Brooks and Farquhar (1985), taking the initial slopes in response to low concentrations of CO_2 at varying light intensities, in order to eliminate the effect of dark respiration. The stomatal conductance, g_s, is also given. Water-use efficiency, WUE, was measured under 25 °C and saturating light. The data for the Kranz C_4 is the average of three species (*Suaeda vermiculata, S. taxifolia,* and *S. eltonica*). For SE values (in parentheses), n = 3 (SE values for the three Kranz C_4 species were averaged). The surface area of the leaves exposed to light was calculated using a digital image program, after importing a digital image into Scion's Image J program (Scion Corporation, Frederick, Md., USA).

Species	*Bienertia sinuspersici*	*Suaeda aralocaspica*	*Suaeda species*	*Suaeda heterophylla*
Type	Bienertioid single-cell C_4	Borszczowoid single-cell C_4	Kranz C_4	C_3
A_{max} at sat. light (μmol m^{-2} s^{-1})	20.3 (0.9)	34.7 (0.6)	23.6 (2.0)	6.86 (1.25)
Γ^* (ppm)	2.5	7.5	4.9	68
g_s (mol m^{-2} s^{-1})	0.29 (0.03)	0.81 (0.04)	0.55 (0.11)	0.39 (0.22)
WUE (μmol mmol^{-1})	7.5 0.27)	6.3 (0.33)	7.6 (0.8)	3.2 (1.35)

Diffusive resistance analysis

Not only does the effective function of C_4 photosynthesis require proper spatial compartmentation of the biochemistry, it also requires efficient trapping of CO_2, generated by the C_4 pump, and refixation of photorespired CO_2 by Rubisco. Incomplete spatial separation of functions, or leakage of CO_2 from sites of decarboxylation, could result in futile cycles and lower efficiency of donation of CO_2 to Rubisco, greater discrimination against fixing $^{13}CO_2$, lower quantum yields, lower CO_2 fixation rates, and lower temperature optima. Thus, the diffusive resistance for CO_2 from sites of C_4 acid decarboxylation and donation of CO_2 to Rubisco in C_4 plants must be much higher than that between Rubisco and the intercellular air space in C_3 plants. Table 2 shows estimates of diffusive resistance of CO_2 between Rubisco and the intercellular air space in the leaf, using different methods. In *S. aralocaspica,* the very elongated chlorenchyma cells provide a long liquid-phase diffusion path from C_4 acids to Rubisco in the proximal ends of the chlorenchyma cells and back to the intercellular air space at the distal ends (mean distance approx. 50 μm). The calculated diffusive resistance to CO_2 in these cells through the liquid phase, on a leaf area basis, was 110 m^2 s^{-1} mol^{-1}, considering area of chlorenchyma exposed to intercellular space

Table 2. Estimates of diffusive resistance to loss of CO_2 in terrestrial single-cell C_4 plants versus Kranz-type C_4 species and C_3 species.

Species	Method	Value (leaf area basis) ($m^2 s^{-1} mol^{-1}$)
Suaeda aralocaspica[a]	Liquid to air phase pathway	112
Bienertia cycloptera[b]	PEPC inhibitor (DCDP)	63
Amaranthus edulis[c]	PEPC inhibitor (DCDP)	120
	Mutant, defective in C_4 pathway	113
Kranz-type C_4 species[d]	Analysis by microscopy	54–151
C_3 species, chloroplast to atmosphere[e, f]	CO_2 exchange—chlorophyll fluorescence, isotope method	1–3

[a]Voznesenskaya et al (2003). [b]Kiirats and Edwards, unpublished. [c]Kiirats et al (2002). [d]von Caemmerer and Furbank (2003). [e]Evans et al (1994). [f]Laisk and Loreto (1996).

at the distal ends (Kiirats et al 2002). This value is similar to that in the Kranz-type NAD-ME species *Amaranthus edulis,* determined by using plants in which the C_4 cycle is inactivated by mutation or chemically by the PEPC inhibitor 3,3-dichloro-2-(dihydroxyphosphinoyl-methyl)-propenoate (DCDP), and is in the range of values based on structural and liquid-phase differences in sites of decarboxylation in various Kranz-type C_4 species (von Caemmerer and Furbank 2003). The value determined for *B. cycloptera* of 63 $m^2 s^{-1} mol^{-1}$ by chemically inhibiting the C_4 cycle with DCDP is on the lower end of the range of values for various Kranz species, when calculated on the basis of analysis of physical barriers (von Caemmerer and Furbank 2003). In both *Bienertia* and *S. aralocaspica,* the chloroplasts containing Rubisco are often positioned "external" to the mitochondria, where CO_2 is generated from malate by NAD-ME and from glycine by GDC, which may facilitate further the capture of CO_2 by Rubisco. From these data, the diffusive resistance values for C_4-type photosynthesis are roughly about 50-fold higher than those for C_3 plants.

In summary, results on carbon isotope composition of leaves of the single-cell C_4 species (above), gas exchange analysis at 25 °C, and estimates of diffusive resistance to CO_2 loss from sites of decarboxylation to substomal air space are characteristic of C_4 plants. The results suggest that these single-cell C_4 plants have spatial separation of photosynthetic metabolism for efficient fixation of CO_2 by Rubisco.

Relevance to designing rice to perform C_4 or C_3-C_4 photosynthesis

The discovery of single-cell C_4 photosynthesis in family Chenopodiaceae shows that Kranz anatomy is not required for the function of C_4 in terrestrial plants, making it feasible to consider these systems in strategies to genetically introduce C_4 photosynthesis into rice. Initial studies on the physiology of the single-cell C_4 chenopods indicate that they have the same features of carbon assimilation as Kranz-type C_4 relatives that grow in Central Asian deserts and the Middle East, that is, C_4-type isotope composition, low CO_2 compensation points, higher water-use efficiency, and higher rates of CO_2 assimilation under low concentrations of CO_2, as compared to C_3 plants.

Analyses of both single-cell C_4 and Kranz C_4 systems indicate that control of positions of chloroplasts and mitochondria is important to the performance of C_4 photosynthesis. Mitochondria need to be positioned so that the maximum amount of CO_2, which is released in photorespiration, can be refixed by Rubisco. The cytoskeleton plays a critical role in spatial compartmentation in terrestrial single-cell C_4 photosynthesis; it may also be important for the function of C_4 in aquatic single-cell C_4, and in terrestrial Kranz-type C_4 species, by providing the necessary spatial separation between the C_4 carboxylation phase and C_4 donation of CO_2 to Rubisco. Thus, in considering strategies for engineering rice to perform C_4 photosynthesis via either a single-cell, or Kranz-type C_4 system, the control of organelle movement and positioning needs to be considered, along with biochemistry and anatomy.

In some C_3 plants, the position of chloroplasts can change from an accumulation response for maximum exposure under limiting light to an avoidance response by positioning along anticlinal walls under high light. Drought may cause chloroplasts and mitochondria to accumulate at the bottom of cells in the centripetal position, in part to increase refixation of photorespired CO_2 when availability of atmospheric CO_2 is limiting (a form of C_3-C_4 intermediate, see Fig. 8A). Under prolonged CO_2 limitations, perhaps this led to the evolution of single-cell C_4 photosynthesis in terrestrial plants, that is, *S. aralocaspica* and *B. sinuspersici*, with the cytoplasmic domain at the bottom of the cell developing as the site of donation of CO_2 from C_4 acids to Rubisco and chloroplasts in the anticlinal position evolving to support the capture of atmospheric CO_2 into C_4 acids.

Besides considering selection, or development, of rice functioning as a C_3-C_4 intermediate or with C_4 photosynthesis with Kranz anatomy, three forms of single-cell models can be envisioned, in which the cytoskeleton would function to separate intracellular function. In these models, concentrating CO_2 around chloroplasts in the centripetal position would depend in part on diffusive resistance in the aqueous phase and having minimal intercellular air space in that region.

These three models follow:

1. *A single-cell C_3-C_4 intermediate model* (Fig. 8A). In this case, most of the chloroplasts are in a centrifugal position performing C_3 photosynthesis, whereas mitochondria and some chloroplasts are in a centripetal position. Atmospheric CO_2 would be fixed by chloroplasts in the centrifugal position, and metabolism through the glycolate pathway would result in decarboxylation of CO_2 by mitochondria in the centripetal position in the cell, where it would be refixed by chloroplasts in that domain. An interesting example of polarization of C_3-type chloroplasts to opposite sides of the cells is seen in *Suaeda cochlearifolia* (Fig. 4). This arrangement would be analogous to the C_3-C_4 intermediates, which have spatial separation of fixation of atmospheric CO_2 in the mesophyll cells, and refixation of photorespired CO_2 in the bundle sheath cells.

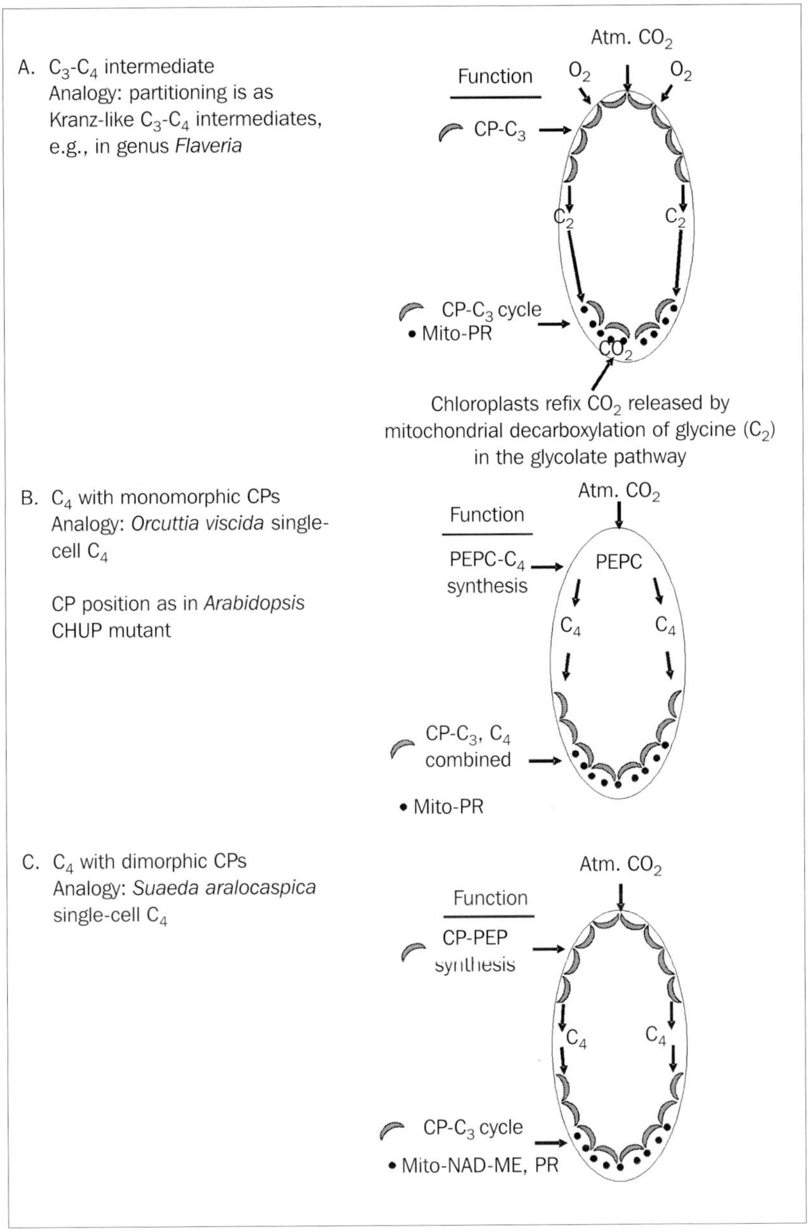

Fig. 8. Three single-cell C_4 models considering chloroplast position. (A) C_3-C_4 intermediate. (B) Single chloroplast type with C_4 decarboxylase to donate CO_2 to C_3 cycle. (C) Dimorphic chloroplasts analogous to Kranz NAD-ME-type C_4. Mito-PR = site of release of photorespired CO_2 in mitochondria. CP = chloroplast.

2. *The* Orcuttia viscida *monomorphic chloroplast model* (Fig. 8B). Having the chloroplasts and mitochondria in a centripetal position would provide separation of fixation of atmospheric CO_2 by PEPC and C_4 acid donation of CO_2 to Rubisco. Either NADP-ME/PPDK or PEPCK targeted to the chloroplast could function to donate CO_2 to Rubisco, and the mitochondria associated with chloroplasts would favor refixation of photorespired CO_2. Interestingly, Oikawa et al (2003) have isolated a mutant of *Arabidopsis* in which the chloroplasts are permanently located in a centrifugal position. Such a mutation in rice might locate chloroplasts in a position favorable for the function of this form of C_4 photosynthesis.
3. *The chenopod NAD-ME model* (Fig. 8C). This requires dimorphic chloroplasts with C_3-type chloroplasts and mitochondria in one domain and Kranz mesophyll-type chloroplasts in the other domain.

In conclusion, understanding the genetic control of development of the spatial compartmentation of organelles in chlorenchyma cells and research on single-cell C_4 systems, including the novel single-cell C_4 terrestrial plants, hold promise for C_4 rice.

References

Akhani H, Barroca J, Koteeva NK, Voznesenskaya E, Franceschi VR, Edwards G, Ziegler H. 2005. *Bienertia sinuspersici* (Chenopodiaceae): a new species from southwest Asia and discovery of a third terrestrial C_4 plant without Kranz anatomy. Systematic Bot. 30:290-301.

Bassell G, Singer RH. 1997. mRNA and cytoskeletal elements. Curr. Opin. Cell Biol. 9:109-115.

Bedard J, Jarvis PG. 2005. Recognition and envelope translocation of chloroplast preproteins. J. Exp. Bot. 56:2287-2320.

Berry J, McCormac D, Long J, Boinski JJ, Corey A. 1997. Photosynthetic gene expression in amaranth, an NAD-ME type C_4 dicot. Aust. J. Plant Physiol. 24:423-428.

Blasing O, Westhoff P, Svensson P. 2000. Evolution of C_4 phospho*enol*pyruvate carboxylase in *Flaveria*, a conserved serine residue in the carboxyl-terminal part of the enzyme is a major determinant for C_4-specific characteristics. J. Biol. Chem. 275:27917-27923.

Bogucka-Glotzer J, Ephrussi A. 1996. mRNA localization and the cytoskeleton. Semin. Cell Dev. Biol. 7:357-365.

Brooks A, Farquhar GD. 1985. Effect of temperature on the CO_2/O_2 specificity of ribulose-1,5-bisphosphate carboxylase/oxygenase and the rate of respiration in the light. Planta 165:397-406.

Carolin RC, Jacobs SWL, Vesk M. 1975. Leaf structure in Chenopodiaceae. Bot. Jahrb. Syst. Pflanzengesch. Pflanzengeogr. 95:226-255.

Chartrand P, Singer RH, Long RM. 2001. RNP localization and transport in yeast. Annu. Rev. Cell Dev. Biol. 17:297-310.

Choi SB, Wang C, Muench K, Ozawa K, Franceschi VR, Wu Y, Okita TW. 2000. Messenger RNA targeting of rice seed storage proteins to specific ER subdomains. Nature 407:765-767.

Chuong SDX, Franceschi VR, Edwards GE. 2006. The cytoskeleton maintains organelle partitioning required for single-cell C_4 photosynthesis in Chenopodiaceae species. Plant Cell 18:2207-2223.

Crofts AJ, Washida H, Okita TW, Ogawa M, Kumamaru T, Satoh H. 2005. The role of mRNA and protein sorting in seed storage protein synthesis, transport and deposition. Biochem. Cell Biol. 83:728-737.

de Heredia ML, Jansen R-P. 2004. mRNA localization and the cytoskeleton. Curr. Opin. Cell Biol. 16:80-85.

DeBlasio SL, Luesse DR, Hangarter RP. 2005. A plant-specific protein essential for blue-light induced chloroplast movement. Plant Physiol. 139:101-114.

Dengler NG, Nelson T. 1999. Leaf structure and development in C_4 plants. In: Sage RF, Monson RK, editors. C_4 plant biology. San Diego, Calif. (USA): Academic Press. p 133-172.

Dengler NG, Taylor WC. 2000. Developmental aspects of C_4 photosynthesis. In: Leegood RC, Sharkey TD, von Caemmerer S, editors. Photosynthesis: physiology and metabolism. Dordrecht (Netherlands): Kluwer Academic Publishers. p 471-495.

Dong X-J, Ryu J-H, Takagi S, Nagai R. 1996. Dynamic changes in the organization of microfilaments associated with the photocontrolled motility of chloroplasts in epidermal cells of *Vallisneria.* Protoplasma 195:18-24.

Edwards G, Franceschi VR, Ku MSB, Voznesenskaya EV, Pyankov VI, Andreo CS. 2001a. Compartmentation of photosynthesis in cells and tissues of C_4 plants. J. Exp. Bot. 52:577-590.

Edwards GE, Walker DA. 1983. C_3, C_4: mechanisms, and cellular and environmental regulation, of photosynthesis. Oxford (UK): Blackwell Scientific Publications. 542 p.

Edwards GE, Furbank RT, Hatch MD, Osmond CB. 2001b. What does it take to be C_4? Lessons from the evolution of C_4 photosynthesis. Plant Physiol. 125:46-49.

Edwards GE, Franceschi VR, Voznesenskaya EV. 2004. Single-cell C_4 photosynthesis versus the dual-cell (Kranz) paradigm. Annu. Rev. Plant Biol. 55:173-196.

Evans JR, von Caemmerer S, Setchell BA, Hudson GS. 1994. The relationship between CO_2 transfer conductance and leaf anatomy in transgenic tobacco with a reduced content of Rubisco. Aust. J. Plant Physiol. 21:475-495.

Freitag H, Stichler W. 2000. A remarkable new leaf type with unusual photosynthetic tissue in a central Asiatic genus of Chenopodiaceae. Plant Biol. 2:154-160.

Freitag H, Stichler W. 2002. *Bienertia cycloptera* Bunge ex Boiss., Chenopodiaceae, another C_4 plant without Kranz tissues. Plant Biol. 4:121-132.

Gamaley YV, Voznesenskaya EV. 1986. Structural-biochemical types of C_4 plants. Soviet Plant Physiol. 33:616-630.

Gowik U, Burscheidt J, Akyildiz M, Schlue U, Koczor M, Streubel M, Westhoff P. 2004. *cis*-regulatory elements for mesophyll-specific gene expression in the C_4 plant *Flaveria trinervia*, the promoter of the C_4 phospho*enol*pyruvate carboxylase gene. Plant Cell 16:1077-1090.

Grunert S, St Johnston D. 1996. RNA localization and the development of asymmetry during *Drosophila* oogenesis. Curr. Opin. Genet. Dev. 6:395-402.

Hamada S, Ishiyama K, Choi SB, Wang C, Singh S, Kawai N, Franceschi VR, Okita TW. 2003. The transport of prolamine RNAs to prolamine protein bodies in living rice endosperm cells. Plant Cell 15:2253-2264.

Hatch MD. 1987. C_4 photosynthesis: a unique blend of modified biochemistry, anatomy and ultrastructure. Biochim. Biophys. Acta 895:81-106.

Hesketh JE. 1996. Sorting of messenger RNAs in the cytoplasm: mRNA localization and the cytoskeleton. Exp. Cell Res. 225:219-236.

Inada S, Ohgishi M, Mayama T, Okada K, Sakai T. 2004. RPT2 is a signal transducer involved in phototropic response and stomatal opening by association with photropin1 in *Arabidopsis thaliana.* Plant Cell 16:887-896.

Jansen RP. 1999. RNA-cytoskeletal associations. FASEB J. 13:455-466.

Jarvis P, Robinson C. 2004. Mechanisms of protein import and routing in chloroplasts. Curr. Biol. 14:R1065-R1067.

Kadereit G, Borsch T, Weising K, Freitag H. 2003. Phylogeny of Amaranthaceae and Chenopodiaceae and the evolution of C_4 photosynthesis. Int. J. Plant Sci. 164:959-986.

Kadota A, Wada M. 1992. Photoinduction of formation of circular structures by microfilaments on chloroplasts during intracellular orientation in protonemal cells of the fern *Adiantum capillus-veneris.* Protoplasma 167:97-107.

Kanai R, Edwards G. 1999. The biochemistry of C_4 photosynthesis. In: Sage RF, Monson RK, editors. C_4 plant biology. San Diego, Calif. (USA): Academic Press. p 49-87.

Kandasamy MK, Meagher RB. 1999. Actin-organelle interaction: association with chloroplast in *Arabidopsis* leaf mesophyll cells. Cell Motility Cytoskeleton 44:110-118.

Kapralov MV, Akhani H, Voznesenskaya E, Edwards G, Franceschi VR, Roalson EH. 2006. Phylogenetic relationships in the Salicornioideae/Suaedoideae/Salsoloideae s.l. (Chenopodiaceae) clade and a clarification of the phylogenetic position of *Bienertia* and *Alexandra* using multiple DNA sequence datasets. Systematic Bot. 31:571-585.

Kasahara MK, Kagawa T, Oikawa K, Suetsugu N, Miyao M, Wada M. 2002. Chloroplast avoidance movement reduces photodamage in plants. Nature 420:829-832.

Keeley JE. 1998. C_4 photosynthesis modification in the evolutionary transition from land to water in aquatic grasses. Oecologia 116:85-97.

Keeley JE. 1999. Photosynthetic pathway diversity in a seasonal pool community. Funct. Ecol. 13:106-118.

Keeley JE, Rundel PW. 2003. Evolution of CAM and C_4 carbon-concentrating mechanisms. Int. J. Plant Sci. 164:55-77.

Kiirats O, Lea PJ, Franceschi VR, Edwards GE. 2002. Bundle sheath diffusive resistance to CO_2 and effectiveness of C_4 photosynthesis and refixation of photorespired CO_2 in a C_4 cycle mutant and wild-type *Amaranthus edulis.* Plant Physiol. 130:964-976.

King ML. 1996. Molecular basis for cytoplasmic localization. Dev. Genet. 19:183-189.

Kloc M, Dougherty MT, Bilinski S, Chan AP, Brey E, King ML, Patrick CW Jr, Etkin LD. 2002. Three-dimensional ultrastructural analysis of RNA distribution within germinal granules of *Xenopus*. Dev. Biol. 241:79-93.

Ku MSB, Kanomurakami Y, Matsuoka M. 1996. Evolution and expression of C_4 photosynthesis genes. Plant Physiol. 111:949-957.

Lai LB, Wang L, Nelson TM. 2002. Distinct but conserved functions for two chloroplastic NADP-malic enzyme isoforms in C_3 and C_4 *Flaveria* species. Plant Physiol. 128:125-139.

Laisk A, Loreto F. 1996. Determining photosynthetic parameters from leaf CO_2 exchange and chlorophyll fluorescence. Plant Physiol. 110:903-912.

Lal A, Ku MSB, Edwards GE. 1996. Analysis of inhibition of photosynthesis under water stress in the C_4 species *Amaranthus cruentus* and *Zea mays*: electron transport, CO_2 fixation and carboxylation capacity. Aust. J. Plant Physiol. 23:403-413.

Long RM, Singer RH, Meng X, Gonzalez I, Nasmyth K, Jansen R-P. 1997. Mating type switching in yeast controlled by asymmetric localization in *ASH1* mRNA. Science 277:383-387.
Matsuoka D, Tokutomi S. 2005. Blue-light regulated molecular switch of Ser/Thr kinase in phototropin. Proc. Natl. Acad. Sci. USA 102:13337-13342.
Matsuoka M, Furbank RT, Fukayama H, Miyao M. 2001. Molecular engineering of C_4 photosynthesis. Annu. Rev. Plant Physiol. Plant Mol. Biol. 52:297-314.
McCormac DJ, Litz H, Wang J, Gollnick PD, Berry JO. 2001. Light-associated and processing-dependent protein binding to 5′ regions of *rbcL* mRNA in the chloroplasts of a C_4 plant. J. Biol. Chem. 276:3476-3483.
Menzel D, Schliwa M. 1986. Motility in the siphonous green alga *Bryopsis*. II. Chloroplast movement requires organized arrays of both microtubules and actin filaments. Eur. J. Cell Biol. 40:286-289.
Nagai R. 1993. Regulation of intracellular movements in plant cells by environmental stimuli. Int. Rev. Cyt. 145:251-310.
Nomura M, Sentoku N, Nishimura A, Lin J-H, Honda C, Taniguchi M, Ishida Y, Ohta S, Komari T, Miyao-Tokutomi M, Kano-Murakami Y, Tajima S, Ku MSB, Matsuoka M. 2000. The evolution of C_4 plants: acquisition of *cis*-regulatory sequences in the promoter of C_4-type pyruvate, orthophosphate dikinase gene. Plant J. 22:211-221.
Oikawa K, Kasahara M, Kiyosue T, Kagawa T, Suetsugu N, Takahashi F, Kanegae T, Niwa Y, Kadota A, Wada M. 2003. CHLOROPLAST UNUSUAL POSITIONING1 is essential for proper chloroplast positioning. Plant Cell 15:2805-2815.
Patel M, Corey AC, Yin L-P, Ali S, Taylor WC, Berry JO. 2004. Untranslated regions from C_4 amaranth *AhRbcS1* mRNAs confer translational enhancement and preferential bundle sheath cell expression in transgenic C_4 *Flaveria bidentis*. Plant Physiol. 136:3550-3561.
Patel M, Siegel AJ, Berry JO. 2006. Untranslated regions of *FbRbcS1* mRNA mediate bundle sheath cell-specific gene expression in leaves of a C_4 plant. J. Biol. Chem. 281:25485-25491.
Prendergast HDV, Hattersley PW, Stone NE. 1987. New structural/biochemical associations in leaf blades of C_4 grasses (Poaceae). Aust. J. Plant Physiol. 14:403-420.
Sage RF, Li M, Monson RK. 1999. The taxonomic distribution of C_4 photosynthesis. In: Sage RF, Monson RK, editors. C_4 plant biology. San Diego, Calif. (USA): Academic Press. p 551-584.
Sage RF, Monson RK, editors. 1999. C_4 plant biology. San Diego, Calif. (USA): Academic Press. 596 p.
Sage RF. 2001. Environmental and evolutionary preconditions for the origin and diversification of the C_4 photosynthetic syndrome. Plant Biol. 3:202-213.
Sage RF. 2004. The evolution of C_4 photosynthesis. New Phytol. 161:341-370.
Sato Y, Wada M, Kadota A. 2001. Choice of tracks, microtubules and/or actin filaments, for chloroplast photo-movement is differentially controlled by photochrome and a blue light receptor. J. Cell Sci. 114:269-280.
Schütze P, Freitag H, Weising K. 2003. An integrated molecular and morphological study of the subfamily Suaedoideae Ulbr. (Chenopodiaceae). Plant Syst. Evol. 239:257-286.
Sheen J. 1999. C_4 gene expression. Annu. Rev. Plant Physiol. Plant Mol. Biol. 50:187-217.
Singer RH. 1993. RNA zipcodes for cytoplasmic addresses. Curr. Biol. 3:719-721.
Singer RH. 1996. RNA: traffic report. Trends Cell Biol. 6:486-489.
St Johnston D. 1995. The intracellular localization of messenger RNAs. Cell 81:161-170.

St Johnston D. 2005. Moving messages: the intracellular localization of mRNAs. Nat. Rev. Mol. Cell Biol. 6:363-375.

Takizawa PA, Sil A, Swedlow JR, Herskowitz I, Vale RD. 1997. Actin-dependent localization of an RNA encoding a cell-fate determinant in yeast. Nature 389:90-93.

Taylor WC. 2000. C_4 rice: What are the lessons from developmental and molecular studies? In: Sheehy JE, Mitchell PL, Hardy B, editors. Redesigning rice photosynthesis to increase yield. Makati City, Philippines: IRRI and Elsevier Science. p 87-98.

von Caemmerer S, Furbank RT. 2003. The C_4 pathway: an efficient CO_2 pump. Photosyn. Res. 77:191-207.

Voznesenskaya E, Franceschi VR, Chuong SDX, Edwards GE. 2006. Functional characterization of phospho*enol*pyruvate carboxykinase type C_4 leaf anatomy: immuno, cytochemical and ultrastructural analyses. Ann. Bot. 98:77-91.

Voznesenskaya EV, Franceschi VR, Kiirats O, Freitag H, Edwards GE. 2001. Kranz anatomy is not essential for terrestrial C_4 plant photosynthesis. Nature 414:543-546.

Voznesenskaya EV, Franceschi VR, Kiirats O, Artyusheva EG, Freitag H, Edwards GE. 2002. Proof of C_4 photosynthesis without Kranz anatomy in *Bienertia cycloptera* (Chenopodiaceae). Plant J. 31:649-662.

Voznesenskaya EV, Edwards GE, Kiirats O, Artyusheva EG, Franceschi VR. 2003. Development of biochemical specialization and organelle partitioning in the single celled C_4 system in leaves of *Borszczowia aralocaspica* (Chenopodiaceae). Am. J. Bot. 90:1669-1680.

Voznesenskaya EV, Chuong SDX, Kirrats O, Franceschi VR, Edwards GE. 2005a. Evidence that C_4 species in genus *Stipagrostis*, family Poaceae, is NADP-malic enzyme subtype with nonclassical type of Kranz anatomy (Stipagrostoid). Plant Sci. 168:731-739.

Voznesenskaya EV, Chuong SDX, Koteeva NK, Edwards GE, Franceschi VR. 2005b. Functional compartmentation of C_4 photosynthesis in the triple-layered chlorenchyma of *Aristida* (Poaceae). Funct. Plant Biol. 32:67-77.

Voznesenskaya EV, Chuong SDX, Koteyeva NK, Franceschi VR, Freitag H, Edwards GE. n.d. Structural, biochemical and physiological characterization of C_4 photosynthesis in species having two vastly different types of Kranz anatomy in genus *Suaeda* (Chenopodiaceae). Plant Biol. (In press.)

Wada M, Kagawa T, Sato Y. 2003. Chloroplast movement. Annu. Rev. Plant Biol. 54:455-468.

Wada M, Suetsugu N. 2004. Plant organelle positioning. Curr. Opin. Plant Biol. 7:626-631.

Walbot V. 1977. The dimorphic chloroplasts of the C_4 plant *Panicum maximum* contain identical genomes. Cell 11:729-737.

Witztum A, Parthasarathy MV. 1985. Role of actin in chloroplast clustering and banding in leaves of *Egeria, Elodia* and *Hydrilla.* Eur. J. Cell Biol. 39:21-26.

Notes

This work is dedicated to the late Dr. Vincent R. Franceschi (1953-2005), our friend and colleague, without whom this work would not have been possible.

Authors' addresses: G.E. Edwards, M. Smith, J.H. Park, O. Kiirats, and S.D.X. Chuong, School of Biological Sciences, Washington State University, Pullman, WA 99164-4236 USA; E. Voznesenskaya and N. Koteyeva, Laboratory of Anatomy and Morphology, V.L. Komarov Botanical Institute, Russian Academy of Sciences, 2 Prof. Popov St., 197376, St. Petersburg, Russia; Y.-I. Park, Division of Biosciences and Biotechnology, Chungnam National University, Daejeon 305-764, Korea; T.W. Okita, Institute of Biological Chemistry, Washington State University, Pullman, WA 99164-6340, USA.

Acknowledgments: This research was supported by National Science Foundation Grants IBN-0131098 and IBN-0236959, U.S. Civilian Research and Development Foundation RUB1-2829-ST-06, Russian Foundation of Basic Research 05-04-49622, and CFGC Grant CG2122. We also thank the Vincent Franceschi Microscopy and Imaging Center for use of facilities and staff help and C. Cody for plant growth management.

Hydrilla: retrofitting a C_3 leaf with a single-cell C_4 NADP-ME system

G. Bowes, S.K. Rao, J.B. Reiskind, G.M. Estavillo, and V.S. Rao

Ribulose bisphosphate carboxylase–oxygenase (Rubisco) is inhibited by O_2 and, as a consequence, atmospheric CO_2 does not saturate C_3 photosynthesis. The O_2 effect has two components: direct inhibition of carboxylation and an oxygenase reaction that initiates photorespiration. C_4 photosynthesis concentrates CO_2 for Rubisco, which minimizes both components, and increases photosynthesis up to 50%. Although atmospheric [CO_2] is projected to reach 550 μbar by 2050, it will not eliminate adverse O_2 effects. Rice yields will increase, but the benefit may be offset by projected higher temperatures and reductions in rice Rubisco protein. *Hydrilla verticillata* is a monocot that operates a facultative, single-cell C_4 system. Based on this single-cell premise, rice plants have been transformed with C_4-cycle enzymes to improve photosynthesis and yield, but the results have been disappointing. The *Hydrilla* system can provide clues to the essential elements needed for an effective CO_2-concentrating mechanism (CCM) because the C_4 and C_3 cycles operate in series in the same C_3 cell, without the bundle sheath anatomy of terrestrial C_4 plants. In *Hydrilla*, phospho*enol*-pyruvate carboxylase (PEPC) in the cytosol is segregated from Rubisco and the decarboxylase, NADP-dependent malic enzyme (NADP-ME), in the chloroplasts, where CO_2 is concentrated. Multiple isoforms of PEPC and NADP-ME exist in *Hydrilla*, but *hvpepc4* and *hvme1* are up-regulated in C_4 leaves and encode proteins with characteristics specific to C_4 photosynthesis. A β-carbonic anhydrase (CA) is also up-regulated, presumably in the cytosol to aid PEPC fixation, but we hypothesize that CA is down-regulated in C_4 chloroplasts. To maintain the $NADPH/NADP^+$ ratio in the granal chloroplasts of C_4 leaves, oxaloacetate and/or aspartate may be imported and reduced to malate for decarboxylation. A major unknown is how the *Hydrilla* chloroplasts, which in the C_3 state must maximize CO_2 conductance for Rubisco, minimize this permeability to reduce leakage from the CCM when the C_4 system is induced. The down-regulation of chloroplast envelope aquaporins may be involved, and the *Hydrilla* system provides a means to study this crucial component. If chloroplast leakage is not regulated in a single-cell C_4 rice plant, even a high-capacity C_4 pump will be ineffective, and low quantum yields will compromise productivity. *Hydrilla* studies indicate that transporter and permeability issues, and the nuances of enzyme regulation, should be incorporated in the design of a single-cell C_4 rice plant to produce an effective CCM.

The photosynthesis of most plants, including rice (*Oryza sativa*), operates at less than full capacity in the current atmosphere: even up to 50% less (Long et al 2006a). This is because O_2 competes with CO_2 for the active site of ribulose bisphosphate carboxylase–oxygenase (Rubisco), the initial carboxylating enzyme of species with the C_3 photosynthetic pathway (Ogren and Bowes 1971). The inhibitory effect of O_2 has two components: the major impact, around two-thirds of the inhibition at 25 °C, is due to O_2 in the active site impeding the carboxylation of RuBP (Laing et al 1974). The remaining one-third is due to Rubisco acting as a bifunctional enzyme and catalyzing an oxygenase reaction with RuBP to produce P-glycolate, the substrate for photorespiration (Bowes et al 1971). It is the subsequent release of previously fixed CO_2 from the photorespiratory pathway that further reduces net photosynthesis. Distinguishing the two components is important, as systems that recycle only photorespiratory CO_2 cannot eliminate all the inhibitory effect of O_2 on photosynthesis.

O_2 inhibition of photosynthesis is an underlying rationale for C_4 species

The bifunctionality of Rubisco has led to determinations of its CO_2/O_2 specificity ($S_{c/o}$), which measures the degree to which Rubisco discriminates between the competing substrates of O_2 and CO_2, with higher values indicating greater specificity for CO_2. Species differ in $S_{c/o}$ values, but in no cases is the oxygenase activity eliminated, and, due to the chemistry of catalysis, Rubiscos without any oxygenase activity probably do not exist (Tcherkez et al 2006). This has implications for improving rice photosynthesis by introducing a high-specificity Rubisco, as this suggests that elimination of the O_2 effects by this method is unlikely, though improvements may be possible. In this context, Rubisco from the red alga *Griffithsia monilis* has a high $S_{c/o}$ value of 167, which is twice that of C_3 plants, and it also has a respectable carboxylation rate (k^c_{cat} = 2.6 s^{-1}). If it could be successfully introduced into rice, a *Griffthsia*-like Rubisco could enhance photosynthesis, but, as temperatures rise above 25 °C, the gains may be eroded (Raines 2006, Tcherkez et al 2006).

Under current atmospheric conditions, species that perform C_4 photosynthesis often have a distinct photosynthetic advantage over their C_3 counterparts, and the advantages translate into higher growth rates and yields. This is because C_4 species use phospho*enol*pyruvate carboxylase (PEPC) as the initial fixation enzyme and the crucial difference between PEPC and Rubisco is that PEPC activity is not inhibited by O_2 (Bowes and Ogren 1972). In C_4 species, PEPC initiates a C_4 acid cycle that functions as a CO_2-concentrating mechanism (CCM) to concentrate CO_2 around Rubisco and essentially eliminate the inhibitory effects of O_2. It is sometimes erroneously suggested that the C_4 advantage is because the carboxylases differ in Michaelis constants, with PEPC having a greater affinity for CO_2 than Rubisco (Osborne and Beerling 2006). In fact, the reverse is true. For example, Rubisco extracted from *Hydrilla verticillata* has a $K_m(CO_2)$ or K_c value of 26–28 μM, but the PEPC $K_m(HCO_3^-)$ is approximately tenfold higher at 220–330 μM (Ascencio and Bowes 1983, Bowes and Salvucci 1984). Rather than the K_m per se, the major advantages of PEPC are its lack of O_2 inhibition and its high k_{cat}, which together enable it to operate at a higher rate than Rubisco.

Rubisco K_c values vary among species, and this is true for terrestrial C_3 and C_4 plants, as the former tend to have the lower values (Yeoh et al 1980, Yeoh and Hattersley 1985). For example, C_3 rice and C_4 maize (*Zea mays*) have reported K_c values of 17 and 56 μM, respectively. There is a trade-off among K_c, $S_{c/o}$, and k^c_{cat} in that high affinity and specificity for CO_2 tend to be associated with lower k^c_{cat} values. When the [CO_2] at the Rubisco active site is subsaturating, as in C_3 plants, high affinity and specificity are essential, but, in C_4 species with CO_2 concentrated around Rubisco, the ability to operate at a high maximum rate takes precedence, and they tend to have lower CO_2 specificities (Seemann et al 1984, Tcherkez et al 2006).

In the majority of terrestrial C_4 species, Rubisco is localized in chloroplast-containing bundle sheath cells (BSC), where CO_2 concentrations estimated to be in the 70 μM range occur (Leegood and Edwards 1996). PEPC in the mesophyll cell (MC) cytosol is the source of C_4 acids that are transported to the BSC for subsequent decarboxylation. It is this decarboxylation event that provides the high [CO_2] for Rubisco. A major unresolved issue is how the CCM of C_4 species maintains a high [CO_2] at the Rubisco fixation site; in essence, how leakage losses are minimized and concomitantly energetic efficiency is maximized. The conductance estimates for C_4 leaves range from 0.0016 to 0.0056 cm s^{-1}, which are lower than those of C_3 plants by up to a 100-fold (Jenkins et al 1989). Likewise, isolated BSC are one to two orders of magnitude less permeable to CO_2 than C_3 cells (Furbank et al 1989). The BSC wall thickness and the presence of suberin have been implicated in retarding CO_2 diffusion into the surrounding MC, but these two features are not universal in C_4 species, and cytosol resistances may be as important (von Caemmerer and Furbank 2003). There are also uncertainties about the diffusion properties of BSC membranes, and in this context it is interesting that, although C_4 BSC exhibit relatively low CO_2 permeability, microalgal and cyanobacterial values may be far lower, in the range of 10^{-4} to 10^{-7} cm s^{-1} (Badger 1987).

The potential importance of cytosolic features in regulating CO_2 conductance has been borne out by recent discoveries of "single-cell" terrestrial chenopods that lack BSC but yet show the biochemical and physiological features of C_4 plants (Edwards et al 2004). The chloroplasts are dimorphic and spatially segregated in single elongated cells. The Rubisco-containing chloroplasts are associated with mitochondria that contain the decarboxylase, NAD-dependent malic enzyme (NAD-ME), but are segregated from other chloroplasts that lack Rubisco and are located near cytosolic PEPC. The long liquid-phase diffusion path and the position of the Rubisco-containing chloroplasts relative to the decarboxylase minimize futile cycling and apparently provide low conductance (von Caemmerer and Furbank 2003, Edwards et al 2004).

The first single-cell C_4 system to be described was that of the aquatic monocot *Hydrilla* (Holaday and Bowes 1980, Salvucci and Bowes 1981, 1983a). *Hydrilla* is in the order Alismatales, which has an ancient lineage. When first described, the *Hydrilla* C_4 system was unique and not readily accepted because it lacks BSC compartmentation. Despite this, it maintains a high [CO_2] in the chloroplasts, estimated to be around

400 μM. The particulars of the *Hydrilla* system and its applicability to engineering a C_4 rice plant will be described later. There are indications that two other species in the Alismatales, *Egeria densa* and *Sagittaria subulata,* also perform single-cell C_4 photosynthesis, and in the case of *Egeria* the decarboxylase is NADP-ME, the same as in *Hydrilla* (Casati et al 2000, Bowes et al 2002).

Other single-cell aquatic C_4 systems from older lineages than the angiosperms have been described. They include the marine macroalga *Udotea flabellum* in the Chorophyta, which uses phospho*enol*pyruvate carboxykinase (PEPCK) in the carboxylating direction, instead of PEPC (Reiskind et al 1988, Reiskind and Bowes 1991). In contrast, a related green macroalga, *Codium fragile*, whose marine range overlaps that of *Udotea*, has C_3 photosynthesis. Most recently, it has been suggested that the smallest eukaryotic unicellular organism known, *Ostreococcus tauri*, a green picoalga, may operate a C_4 system (Derelle et al 2006). Genes putatively encoding all of the C_4 photosynthetic enzymes are in the *Ostreococcus* genome, but no genes comparable to those of the CCM in the green alga *Chlamydomonas reinhardtii* are present. In addition to genes for PEPC, pyruvate orthophosphate dikinase (PPDK), and NADP-dependent malate dehydrogenase (NADP-MDH), *Ostreococcus* has two NADP-ME orthologs similar to those of *Hydrilla*, with at least one apparently targeted to the chloroplast. A genome analysis is not conclusive evidence for a C_4-based CCM; confirmation requires biochemical and physiological data. There is physiological and biochemical evidence for C_4-type photosynthesis in the marine diatom *Thalassiosira weissflogii* (Reinfelder et al 2004), which makes the report for *Ostreococcus* even more intriguing, and brings to the forefront the issue of how high [CO_2] might be maintained in small single-cell organisms.

Rising atmospheric CO_2 and rice

It is predicted that by 2050 atmospheric [CO_2] will have risen from the 2006 annual mean of 382 to about 550 μbar (Prentice et al 2001). This raises a legitimate issue as to whether rice needs a C_4-based CCM given the inevitability of the rise and its potential to reduce the inhibitory effects of O_2 on photosynthesis. A number of studies have examined how rice responds to increased [CO_2] and temperatures. In experiments where rice (cv. IR30) was grown season-long at [CO_2] from 160 to 900 μbar, leaf and canopy photosynthesis rates increased with [CO_2], but only up to around 500 μbar, which is close to the predicted value for 2050. Concomitantly, shoot and root biomass and grain yield increased, but leveled off at 500–600 μbar CO_2, even though fertilizer was applied regularly to minimize acclimation (Baker et al 1990b). In contrast, water-use efficiency (WUE) continued to increase up to 900 μbar (Baker et al 1990a). These results for rice differ from those of soybean (*Glycine max*), whose photosynthesis continued to respond positively up to 990 μbar CO_2 (Gesch et al 2001). Horie et al (2000) concluded from several data sets that an overall 30% improvement in rice grain yield could occur with a doubling in atmospheric CO_2. But even a 30% gain does not keep pace with projected population increases for 2025. Furthermore, crop studies at FACE (Free Air CO_2 Enrichment) sites to emulate agricultural conditions have produced a less optimistic outlook, with projections some 50% lower than those from enclosure studies, particularly if temperature is factored in (Long et al 2005, 2006b).

Rice Rubisco has a K_c of 17 μM and is not saturated by 550 μbar CO_2 in air, especially given the presence of more than 250 μM O_2 in the chloroplast when PSII is operating. In addition, Rubisco activity and protein are down-regulated in a linear fashion with increasing growth [CO_2], such that there was one-third less Rubisco protein in leaves of rice (cv. IR30) grown at 900 versus 160 μbar CO_2 (Rowland-Bamford et al 1991). This is the major reason why rice photosynthesis did not increase above 500 μbar CO_2 as gains from more substrate CO_2 were negated by the loss of Rubisco capacity. Down-regulation of Rubisco is probably related to rapid changes in gene expression (Gesch et al 1998, 2001, 2003). When leaves of rice at the late vegetative stage were switched from 350 to 700 μbar CO_2, they showed, within 24 hours, up to a 23% decrease in rbcS mRNA, while the opposite was true for high-CO_2 leaves switched to ambient concentrations.

Based on the preceding, it would seem that Rubisco in an optimized transgenic C_4 rice with a fully functional CCM should have an up-regulated k^c_{cat}, and somewhat less protein to improve nitrogen-use efficiency. Protein down-regulation, however, should not be to the extent seen in IR30 rice plants (Rowland-Bamford et al 1991), and this is a consideration when selecting cultivars for genetic engineering. There is reason to be optimistic that not all rice cultivars exhibit the same degree of Rubisco down-regulation. Ziska et al (1996) in work performed at IRRI found that rice cultivars differ in the degree to which elevated CO_2 and temperature influence growth and yield. Similarly, Gesch et al (2001) have shown that IR72 (indica) and M-103 (japonica) cultivars differ in rcbS expression response, with the indica cultivar having less down-regulation at high CO_2.

Transgenic C_4 photosynthesis

In a single-cell approach to making a transgenic C_4 plant, a number of laboratories have overexpressed one or several C_4 genes in heterologous C_3 systems, but with very mixed results (Takeuchi et al 2000, Fukayama et al 2001, Matsuoka et al 2001, Miyao 2003, Miyao and Fukyama 2003). This hit-and-miss approach has produced useful information, but in the main it results in an unregulated system with the metabolome disrupted to a lesser or greater degree. For example, overproduction of maize PEPC in rice MC is reported to increase photosynthesis (Jiao et al 2002), but regulatory phosphorylation was abnormal, and PEPC may just be recycling CO_2 or enhancing anaplerosis (Fukayama et al 2003, Miyao 2003). The overexpression of PPDK in rice failed to enhance photosynthesis (Fukayama et al 2001). Overproduction of maize NADP-ME in rice caused substantial photoinhibition and agranal chloroplasts, probably due to an increase in the NADPH/$NADP^+$ ratio (Takeuchi et al 2000). A dual PEPC and PPDK transformant of rice apparently had enhanced photosynthesis and grain yield (Ku et al 2001), but again it may just be recycling photorespiratory CO_2. Photorespiration appeared to be less in dual PEPC/NADP-ME transformants of

Solanum tuberosum (potato), but similar *Nicotiana tabacum* (tobacco) transformants showed no real improvement in photosynthesis, indicating that the responses may vary with species (Häusler et al 2001, 2002).

Various strategies to improve transformant performance have been suggested (Miyao 2003). They include incorporation of the entire gene with its promoter and terminator sequences, phylogenetic relatedness, addition of a strong target-cell promoter, the use of specific targeting signals, and more consideration given to the regulatory elements of the encoded proteins. However, this assumes that we understand the components and regulation of a "natural" C_4 system, which we do not. In *Hydrilla*, C_4 gene expression is up-regulated by low $[CO_2]$, but without BSC differentiation, and, since this single-cell C_4 cycle occurs only in MC, a better understanding of its operation should provide insight into the requirements for successfully moving C_4 traits into C_3 plants (Edwards 1999, Takeuchi et al 2000, Matsuoka et al 2001).

The *Hydrilla* single-cell system

The objective of the *Hydrilla* project is to identify and characterize the components for a single-cell C_4 system to effectively concentrate CO_2. As a model system, it has several advantages. It is easily manipulated and studied in the laboratory. It is facultative, changing from a C_3 to a C_4 mode, and thus the two states and the intervening steps can be investigated in the same species. This avoids confounding species effects that can occur when comparing even closely related C_3 and C_4 species. Unlike modern "well-bred" C_4 crops, *Hydrilla* is a natural and relatively ancient form of angiosperm C_4 photosynthesis, possibly predating terrestrial C_4 systems. Thus, it may provide clues to factors involved in the origination of C_4 photosynthesis among flowering plant groups (Bowes et al 2002).

In many respects, *Hydrilla*'s facultative system is unique. As a single-cell C_4 NADP-ME species, it differs from BSC-type NADP-ME C_4 plants in ways that go beyond anatomy. It typically has C_3 gas-exchange and biochemical characteristics, but exposure to low $[CO_2]$ induces a C_4-based CCM within 10 to 12 days (Salvucci and Bowes 1981, Holaday et al 1983, Magnin et al 1997). Induction occurs naturally in the lake under adverse photosynthetic conditions of low $[CO_2]$ and high $[O_2]$, pH, and temperature. Thus, in conditions that favor photorespiration, C_4 genes are up-regulated and expressed in a C_3 MC background, while other genes are down-regulated to achieve coordination between the C_4 and C_3 cycles (Rao et al 2006a).

Table 1 is an updated summary of some of the differences that have been identified between C_3 and C_4 leaves. The maximum photosynthesis rate increases when the C_4 state is induced, with reported values as high as 124 $\mu mol\ g^{-1}$ fresh wt h^{-1} (Spencer et al 1994), and rates at subsaturating $[CO_2]$ are also higher. The C_4 leaves exhibit low CO_2 compensation points (the $[CO_2]$ at which photosynthetic CO_2 uptake and photorespiratory CO_2 release are equal), similar to those of terrestrial C_4 plants, and the O_2 inhibition of photosynthesis is virtually eliminated. These latter two physiological characteristics are indicators of a functional CCM that is able to substantially reduce photorespiration. The presence of a CCM is borne out by measurements of leaf internal

Table 1. The facultative C_3 and C_4 photosynthetic systems of Hydrilla.

Characteristic	C_3 leaf	C_4 leaf
Induction by low $[CO_2]$	No	Yes
Bundle sheath anatomy	None	None
CO_2 compensation point (μbar)	>40	<10
O_2 inhibition of photosynthesis (%)	>28	<2
Net photosynthesis, limiting $[CO_2]$ (μmol g^{-1} FW h^{-1})	2	44
Net photosynthesis, saturating $[CO_2]$	–	124
Ratio internal/external inorganic carbon	0.8	4.2
Estimated chloroplast $[CO_2]$ (μM)	7	400
Rubisco activity (μmol g^{-1} FW h^{-1})	45	40
Rubisco location in leaf cells	Chloroplast	Chloroplast
PEPC activity (μmol g^{-1} FW h^{-1})	<10	>150
Major PEPC isoform	HVPEPC3	HVPEPC4
PEPC isoform location in leaf cells	Cytosol	Cytosol
PEPC light-activated and phosphorylated	No	Yes
Ratio PEPC/Rubisco activity	0.2	>4.2
NADP-ME activity (μmol g^{-1} FW h^{-1})	16	44
Major NADP-ME isoform	HVME3	HVME1
NADP-ME isoform location in leaf cells	Cytosol	Chloroplast
PPDK activity (μmol g^{-1} FW h^{-1})	3	35
PPDK isoform	Unknown	HVPPDK1
PPDK location in leaf cells	Unknown	Chloroplast
50% ^{14}C-malate + ^{14}C-aspartate turnover (s)	N/A	<180
CA activity (EU g^{-1} FW h^{-1})[a]	246	1,365
External leaf CA activity	No	No
CA location in leaf cells	Unknown	Unknown
CA isoform	Unknown	HVCA1
Leaf abaxial surface pH	3.6	4.7

[a]EU = enzyme unit, a measure of the activity of CA using the catalyzed and uncatalyzed rates of reaction.

$[CO_2]$. C_3 leaves have $[CO_2]$ lower than the surrounding medium, as expected from simple diffusion. In contrast, C_4 leaves have a high internal $[CO_2]$ in the chloroplasts, with an estimated 400 μM that surpasses terrestrial C_4 leaf values (Leegood and Edwards 1996, Reiskind et al 1997). C_4 and C_3 leaves exhibit pH polarization, such that the abaxial boundary layer is acidified to around pH 4 in the light, while the adaxial side is alkaline (van Ginkel et al 2001). As both leaf types show pH polarization, it is evident that the C_4 leaf CCM cannot operate at the plasma membrane, but internally, which is consistent with a chloroplastic CCM. Consequently, the predominant inorganic carbon form entering both C_3 and C_4 leaf cells is dissolved CO_2, not HCO_3^-, and it enters by diffusion. Likewise, in terrestrial leaf cells, CO_2 dissolves in the acidic solution of the MC walls and diffuses in solution to the chloroplasts.

Two physiological objections to using the *Hydrilla* model to engineer a single-cell terrestrial C_4 system involve apparently low photosynthesis rates and photon-use efficiency. Because of a 10^4-fold difference in the CO_2 diffusion resistance of aqueous

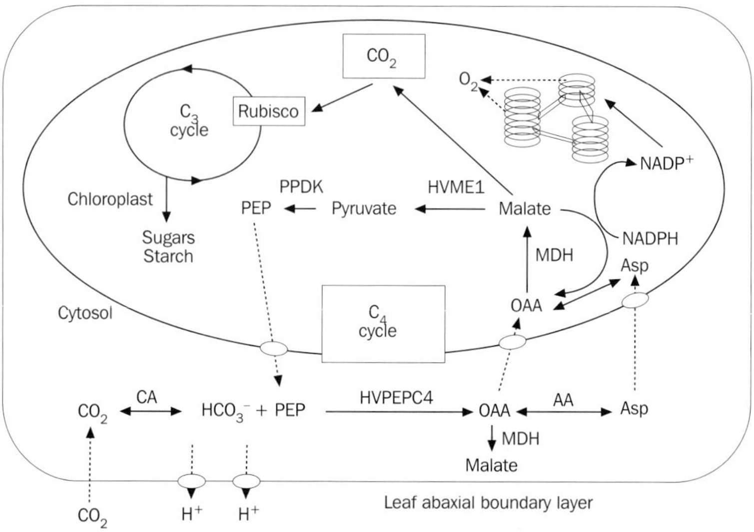

Fig. 1. Schematic representation of how the C_4 cycle potentially operates between the cytosol and chloroplast to concentrate CO_2 in the chloroplast of a *Hydrilla* C_4 leaf cell. The dotted lines indicate components that are transported across the plasma or chloroplast membranes. AA: aminotransferase; Asp: aspartate; CA: carbonic anhydrase; HVME1: NADP-dependent malic enzyme, photosynthetic isoform; HVPEPC4: phospho*enol*pyruvate carboxylase, photosynthetic isoform; MDH: NADP-dependent malate dehydrogenase; OAA: oxaloacetate; PEP: phospho*enol*pyruvate; PPDK: pyruvate orthophosphate dikinase; Rubisco: ribulose bisphosphate carboxylase–oxygenase.

and gaseous media, photosynthesis rate comparisons between terrestrial and aquatic plants are problematic, but on a cell-area basis the *Hydrilla* rates may exceed terrestrial values. The reported low efficiency is based on two apparent quantum yield (QY) values for *Hydrilla* plants from a lake, with the C_4 value reported to be 50% lower than the C_3 value (Spencer et al 1994). If accurate, this suggests that the C_4 leaf CCM is very leaky. But, the C_3 and C_4 plants may be from different irradiance regimes, which would influence QY irrespective of the pathway, and the QY was not based on absorbed photons. In contrast, the high chloroplastic [CO_2] reported for *Hydrilla* (Reiskind et al 1997), higher than in terrestrial C_4 species, is not commensurate with a leaky system. These issues clearly require more study.

Figure 1 is a schematic of how the C_4 system may operate in a *Hydrilla* cell. When the C_4 system is induced, PEPC activity increases up to 15-fold, along with other C_4 enzymes (Salvucci and Bowes 1981, 1983a,b, Ascencio and Bowes 1983, Magnin et al 1997). Malate and aspartate are initial products with a rapid turnover into the C_3 cycle. Incubation of C_4 leaves with diethyloxalacetate (DOA), a PEPC inhibitor, reduces the photosynthesis rate and substantially increases O_2 inhibition (Magnin et al 1997). In contrast, DOA has no effect on C_3 leaf photosynthesis. These results are similar to those for terrestrial BSC-type C_4 species with the PEPC inhibi-

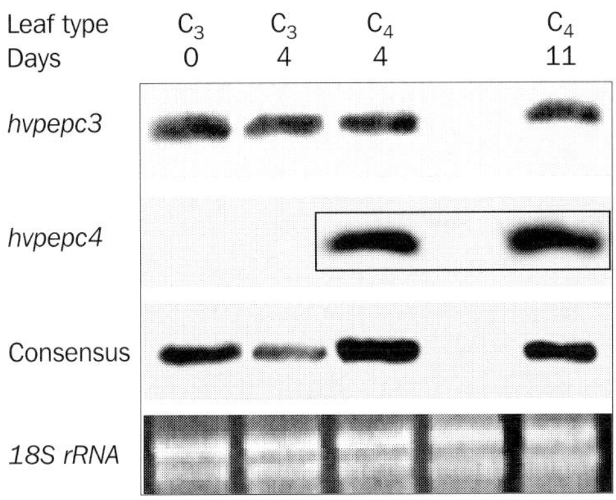

Fig. 2. Northern analysis showing the transcript expression of two leaf phospho*enol*pyruvate carboxylase (PEPC) isoforms as a function of time in days for *Hydrilla* plants maintained in C_3 or under C_4 induction conditions. The increase in signal of the photosynthetic isoform under C_4 conditions is outlined by the rectangle.

tor 3,3-dichloro-2-(dihydroxyphosphinoyl-methyl)-propenoate (Jenkins et al 1989). Fluorescence and immunogold labeling show that PEPC and Rubisco are in the same cell, but compartmentalized in the cytosol and chloroplast (Reiskind et al 1989). Localization studies with isolated chloroplasts have also shown that the decarboxylase, NADP-ME, is in C_4 leaf chloroplasts along with PPDK and NADP-MDH (Magnin et al 1997). Four key C_4 enzymes have been a focus of study as *Hydrilla* leaves change their photosynthetic mode.

Phospho*enol*pyruvate carboxylase

We have identified and sequenced three *Hydrilla* PEPC isoforms, of which two play major roles in leaves. When *Hydrilla* was maintained for 12 days under low [CO_2] conditions, the PEPC-specific activity in leaf extracts increased in a linear manner, unlike the activity in plants that were incubated under C_3-type conditions (Rao et al 2002). Northern analyses at days 4 and 11 after the start of induction showed that mRNA expression for a particular PEPC isoform, *hvpepc4*, was up-regulated (Fig. 2). In contrast, *hvpepc3* was constitutive in both leaf types, and was not up-regulated. When expression was determined as a function of the photoperiod, up-regulation of *hvpepc4* in C_4 leaves was far more pronounced during the light period, whereas light had very little effect on *hvpepc3*.

In terms of protein, western analyses showed that the HVPEPC4 monomer comprised approximately half of the total PEPC signal in C_3 leaves, but this increased to 88% in C_4 leaves in the light, which corresponds with the light-dependent increase in mRNA expression (Rao et al 2006b). Western analyses using an anti-P serine/threonine

Akt substrate and an anti-P site-specific IgG antibody demonstrated that HVPEPC4, but not HVPEPC3, underwent posttranslational phosphorylation in the light (Rao et al 2006b). Furthermore, incubation with an alkaline phosphatase caused de-phosphorylation of the HVPEPC4 protein.

Kinetic analyses of PEPC extracted from C_3 and C_4 leaves showed that they differed biochemically (Rao et al 2002, 2006b). Thus, when assayed at a cytosol-like pH of 7.3, PEPC from C_4 leaves exhibited high activity, light-activation, and a sigmoidal response to [PEP]; whereas, from C_3 leaves, the activity was low, showed no light effect, and had Michaelis–Menten kinetics. The two also differed in response to the negative effector malate. The C_4 leaf PEPC had I_{50} (malate) values of 0.4 and 0.2 mM in the light and dark, respectively, which are in the range reported for other C_4 species (Echevarria et al 1994, Gupta et al 1994, Casati et al 2000), including *Sorghum bicolor* (1.35 and 0.5 mM) and *Egeria* (1.0 and 0.4 mM). The C_3 leaf PEPC had an unexpected response to malate. Despite its lack of phosphorylation and light stimulation, it was virtually insensitive to malate, with an I_{50} value of around 5 mM, thus differing from other anaplerotic PEPC enzymes (Leport et al 1996, Bläsing et al 2002). As reported for photosynthetic PEPC isoforms, glucose-6-P was a positive effector that overcame malate inhibition of the C_4 leaf PEPC and reduced by sixfold the $K_{0.5}$ (PEP). In contrast, glucose-6-P increased the C_3 leaf $K_{0.5}$ (PEP) by more than twofold (Rao et al 2006b).

These data indicate that a particular isoform, HVPEPC4, functions as the initial photosynthetic carboxylase in C_4 leaves, and it does so because of specific regulatory and kinetic characteristics. HVPEPC3 likely has an anaplerotic function, and may refix respiratory CO_2 at night to conserve inorganic carbon when daytime [CO_2] is limiting. Dark fixation does occur in *Hydrilla* leaves, with subsequent malate accumulation (Holaday et al 1983, Rao et al 2006b).

Elegant work with recombinant PEPC from *Flaveria* has shown the kinetic importance for C_4 isoforms of a serine residue at *Flaveria* 774, which is replaced by alanine in C_3-type isoforms (Bläsing et al 2002). This serine appears to be a "C_4 signature" residue, as catalysis is influenced by its proximity to the PEP binding site. However, the photosynthetic isoform of *Hydrilla*, despite its C_4-type kinetics, does not contain serine at the equivalent site, but instead all three *Hydrilla* sequences retain the "C_3" alanine (Rao et al 2002). It is possible that protein phosphorylation, rather than this serine residue, is the important determinant of the C_4 isoform characteristics in *Hydrilla*. Another C_4 signature residue is lysine, as opposed to arginine, at *Flaveria* 347 (Westhoff and Gowik 2004). With *Hydrilla*, a lysine residue occurs at the comparable site in HVPEPC4 but not in HVPEPC3; however, the kinetic importance of this difference is unknown (Rao et al 2002). Recombinant proteins for HVPEPC4 and HVPEPC3 have been generated and purified from *Escherichia coli*, and their respective kinetic characteristics are being characterized. Using site-directed mutagenesis, we plan to change the native alanine to serine to determine whether this affects the kinetics. These experiments should enhance our understanding of the role of phosphorylation and the alanine/serine substitution in the regulation of PEPC kinetics and

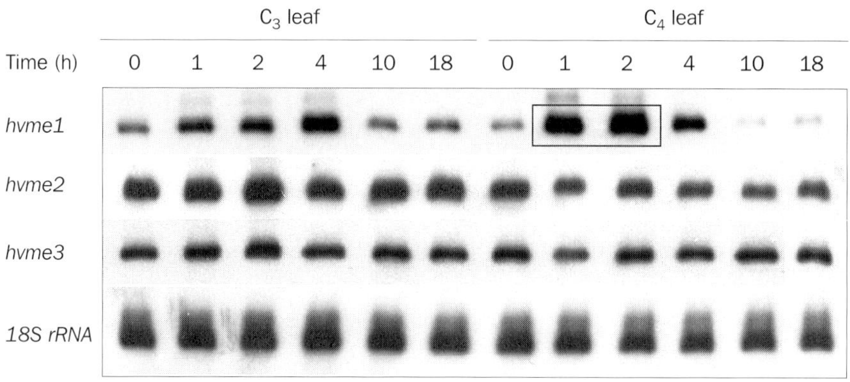

Fig. 3. Northern analysis showing the transcript expression of two leaf NADP-dependent malic enzyme (NADP-ME) isoforms during the light and dark periods for *Hydrilla* plants maintained in C_3 or under C_4 induction conditions. The increase in signal of the photosynthetic decarboxylase under C_4 conditions at the start of the photoperiod is outlined by the rectangle.

provide insight into what features might enhance PEPC performance in a transgenic C_4 species.

NADP-dependent malic enzyme

The activity of this decarboxylating enzyme increases several-fold during the induction of C_4 photosynthesis (Magnin et al 1997). We have identified a small NADP-ME gene family in *Hydrilla* composed of three isoforms, and possibly a fourth (Estavillo 2006). Three full-length cDNA sequences, *hvme1* (GenBank Locus AY594687), *hvme2* (AY594688), and *hvme3* (AY594689), were obtained that ranged from 2.17 to 2.59 kb, with the number of deduced amino acid residues being 654, 614, and 575, respectively. One sequence, HVME1, had a predicted transit peptide of 75 amino acids and it appears to be the photosynthetic decarboxylase in the chloroplast. HVME2 and HVME3 shared 76% and 84% identity, respectively, with HVME1, and they appear to be cytosolic isoforms that may supply NADPH for cytosolic metabolism or function in pH regulation and defense. All the *Hydrilla* isoforms showed the characteristic domains and conserved catalytic residues common to NADP-ME proteins.

Transcript abundance, analyzed by northern blots and semiquantitative RT-PCR, indicated that *hvme1* was present in both C_3 and C_4 leaves, but its expression was up-regulated in the C_4 leaves. In contrast, *hvme3* was constitutively expressed in C_3 and C_4 leaves, while *hvme2* was in low abundance but had its greatest signal in C_3 leaves. Of the three, only *hvme1* showed appreciable light regulation, with the highest expression occurring in C_4 leaves during the first 2 hours of the light period (Fig. 3).

Although *hvme1* transcripts were up-regulated in the light, a western analysis did not show light regulation of the HVME1 protein (Estavillo 2006). An antibody against a rice recombinant chloroplastic NADP-ME (rice anti-ME) recognized a 64

Table 2. Molecular mass and kinetics of recombinant isoforms from *Hydrilla* and rice.

NADP-ME isoform	Molecular mass (kDa)	pH optimum	V_{max} (µmol min^{-1} mg^{-1} protein)	K_m Malate (mM)	K_m NADP$^+$ (mM)	k_{cat} (s^{-1})
Hydrilla						
HVME1[a]	64	7.8	34	0.62	0.01	48
HVME3	63	7.5	13	0.85	0.02	18
Rice[b]						
OsCyME2	65	7.3	–	2.60	0.08	92
OsCyME3	62	7.7	–	3.10	0.09	97

[a]Photosynthetic NADP-ME isoform of *Hydrilla*. [b]Cytosolic isoforms of rice NADP-ME. Data from Cheng et al (2006).

kDa band in the C_3 and C_4 samples, which is the predicted mass of HVME1. In a maize extract, it also recognized two bands of 62 and 66 kD corresponding to the maize photosynthetic and nonphotosynthetic isoforms (Saigo et al 2004). To determine the subcellular locations of the two most abundant NADP-ME isoforms, intact chloroplasts from C_4 leaves were isolated on a Percoll gradient and subjected to western analyses. The rice anti-ME detected the 64-kDa band in the chloroplast and total extracts, but not in the cytosol (Estavillo 2006).

Blue native polyacrylamide gel electrophoresis (BN-PAGE) showed that the *Hydrilla* NADP-ME isoforms have different oligomeric states. The chloroplastic isoform appeared to be a tetramer, which is the common oligomeric association reported for other photosynthetic isoforms. The pI of the chloroplastic NADP-ME band coincided with the predicted value of the *hvme1* encoded polypeptide (5.9).

Recombinant proteins for the two major isoforms were generated in *E. coli* in order to compare their kinetics (Table 2). The photosynthetic isoform exhibited properties that corresponded to the chloroplastic isoforms of other species. They included a higher pH optimum than HVME3, and affinities for malate and NADP$^+$ that were somewhat higher than those of HVME3. The k_{cat} of the photosynthetic isoform was more than twice that of the cytosolic form. Chi et al (2004a) have demonstrated that rice also has a small NADP-ME gene family that contains four members, three of which are cytosolic and the fourth (*OsChlME*) a nonphotosynthetic plastidic form. Table 2 also summarizes the kinetic data of the recombinant rice cytosolic NADP-ME isoforms, *OsCyME2* and *OsCyME3*, from a study by Cheng et al (2006). Unfortunately, the chloroplastic form was not included in the study, and thus cannot be compared. The malate and NADP$^+$ affinities of the rice cytosolic isoforms were several-fold lower than the *Hydrilla* values, but the rice k_{cat} values were higher than those of *Hydrilla*, including that of HVME1.

A phylogenetic analysis examined the relationships of the *Hydrilla* NADP-ME sequences to 43 other plant sequences, including monocots, eudicots, a basal angio-

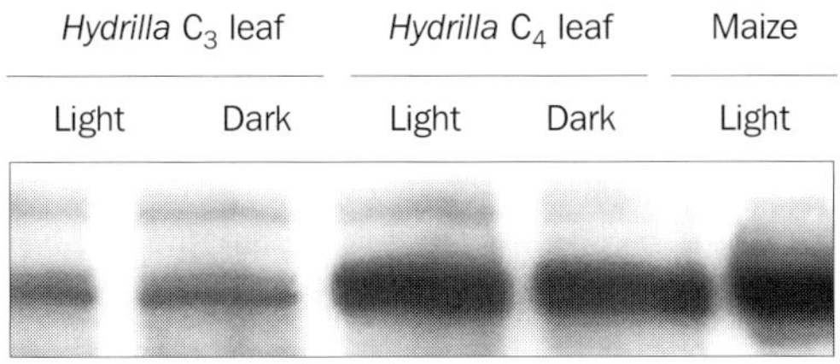

Fig. 4. Western analysis showing the protein expression of leaf pyruvate orthophosphate dikinase (PPDK) enzyme in the light and dark period for *Hydrilla* plants maintained in C_3 or under C_4 induction conditions. Maize in the light is included for reference.

sperm, and a gymnosperm (*Pinus*) that was used to root the tree (Estavillo 2006). The chloroplastic isoforms of the eudicots and the grasses each formed a monophyletic group, with each clade derived independently from cytosolic ancestors. The *Hydrilla* isoforms comprised a separate clade that was sister to all the other monocot sequences, with the cytosolic *hvme3* being sister to *hvme2* and the chloroplastic *hvme1*. The placement of the *Hydrilla* sequences in a single clade suggests that there was an additional independent origin for the chloroplastic isoform in the monocots, apart from the grasses. Likewise, a phylogenetic analysis of the PEPC isoforms in *Hydrilla* indicates that they arose independently (Rao et al 2002). Origins that are independent from the grasses might be expected based on the ancient lineage of the Alismatales, which is the order that contains *Hydrilla*.

Pyruvate orthophosphate dikinase

PPDK catalyzes the regeneration of PEP from pyruvate for PEPC in the photosynthetic C_4 cycle. PPDK was originally thought to be restricted to C_4 species, because its activity in C_3 species was difficult to detect, but it is now known to be widely distributed. In rice, two PPDK isoforms have been identified (Glackin and Grula 1990, Imaizumi et al 1997). In terrestrial C_4 plants, such as maize, the predominant isoform is in the MC chloroplasts, while a second form is cytosolic.

The activity of this enzyme is low in the C_3 leaves of *Hydrilla* but increases upon induction of C_4 photosynthesis (Table 1), and 82% of the total activity is associated with the chloroplasts (Magnin et al 1997). A differential display technique has shown that *Hydrilla* contains two PPDK isoforms that share 84% identity (Rao et al 2006a). Northern analysis indicated transcript sizes of 3.0 and 3.1 kb that potentially encode 95-kDa proteins. The transcript expression of both was up-regulated in C_4 leaves. A western analysis indicated that PPDK protein was overexpressed in C_4 leaves relative to C_3 leaves, and the amount was not modulated by light (Fig. 4). However, the activity was increased by light (Magnin et al 1997). The identity of the photosynthetic isoform remains to be determined.

Carbonic anhydrase

In C_3 species, most of the CA activity is localized in the stroma of the MC chloroplasts, whereas, in C_4 plants, it is mainly in the MC cytosol, though there does appear to be some activity in the BSC (Ludwig et al 1998). For C_3 species, CA facilitates the entry of CO_2 into the chloroplast; in C_4 plants, it catalyzes the hydration of CO_2 to HCO_3^- and thus provides the specific inorganic carbon substrate for PEPC.

This enzyme has an important role in *Hydrilla*, especially in the C_4 leaves (Salvucci and Bowes 1983b). Its activity increases as much as fivefold when C_4 photosynthesis is induced (Table 1). Exposure of *Hydrilla* to ethoxyzolamide (EZ), a potent inhibitor of CA that can penetrate the cells, reduced the photosynthesis rate of C_4 leaves by 40%, but did not inhibit that of C_3 leaves. EZ also increased the CO_2 compensation point and O_2 inhibition in C_4 leaves, but high [CO_2] partially overcame these adverse effects (Salvucci and Bowes 1983b). These observations are consistent with CA activity in the cytosol facilitating the conversion of the entering CO_2 to HCO_3^- for use by PEPC. We have identified a β-CA, *hvca1*, whose transcript expression occurs only under C_4 induction conditions in the light (Rao et al 2005). Alignments with β-type sequences from various species showed that it shares 85% identity with a rice CA. From antibody studies, it appears that there are at least two CA types in *Hydrilla*. CA activity cannot be detected on the outside of the leaf, but its intracellular locations have not yet been determined.

The location of CA in *Hydrilla* is of interest because in the C_3 leaf it should be in the chloroplast stroma, as in C_3 species, but in the C_4 mode it should predominate in the cytosol where PEPC is located. Our working hypothesis is that during C_4 induction *hvca1* is up-regulated in the cytosol, while stromal CA is down-regulated. If substantial activity remains in the alkaline chloroplast stroma of the C_4 leaves, it might compete with Rubisco by catalyzing the conversion of CO_2 to HCO_3^-. Ludwig et al (1998) have shown that overexpression of CA in the BSC cytosol of the C_4 eudicot *Flaveria bidentis* produces a leaky system. It is not clear whether this applies to CA activity in the chloroplast stroma of *Hydrilla*. An understanding of the location and roles of CA in *Hydrilla* has implications for a single-cell C_4 transgenic rice plant, including whether this enzyme needs to be part of the design.

The *Hydrilla* conductance conundrum

The facultative nature of the *Hydrilla* system poses a fascinating conundrum. In the C_3 mode, CO_2 conductance from the medium into the chloroplasts must be maximized to support the fixation rate by Rubisco. In contrast, during induction of the C_4 system, CO_2 conductance at the chloroplast membrane, but not at the plasma membrane, must decrease substantially to minimize leakage from the high CO_2 pool accumulating in the chloroplasts. The mechanism enabling the C_4 leaf chloroplasts to retain a [CO_2] sufficient to eliminate O_2 inhibition of photosynthesis and photorespiration is unknown. The problem is compounded by many small chloroplasts that circulate in *Hydrilla* cells, not one large entity as in some algae; also, pyrenoid-like or carboxysome-like structures have not been observed in *Hydrilla* chloroplasts. Leakage is restricted inside the cell, not at the leaf surface, because in both C_3 and C_4 leaves the abaxial surface is acidified in the light to facilitate the diffusive uptake of CO_2 (van Ginkel et al 2001). Thus, as in terrestrial MC with an acidic cell wall solution, dissolved CO_2 diffuses into the cytosol, which would not produce a net influx if it were against a CO_2 gradient.

Aquaporins are possible candidates to explain the putative change in chloroplast CO_2 permeability. Aquaporins are membrane MIPs (major intrinsic proteins) of 26 to 29 kDa mass that form water-permeable complexes for processes that require regulated water flow (Weig et al 1997). Evidence has been accumulating that they also facilitate CO_2 transport and play a role in photosynthesis. In tobacco, the plasma membrane aquaporin NtAQP1 reportedly facilitates CO_2 transport, possibly in a unidirectional manner (Uehlein et al 2003). Suppression of NtAQP1 expression reduces CO_2 permeability and the photosynthesis rate. Prasad et al (1998) reported that the presence of aquaporins increased the CO_2 permeability of the plasma membrane and the chloroplast envelope. Aquaporins are associated with the outer membrane of the chloroplast, which has a greater CO_2 permeability than the inner membrane (Raven et al 2002, Tetlow et al 2005). If aquaporins are a factor in the CO_2 conductance of *Hydrilla* leaves, we postulate that their abundance and expression should change during the shift from C_3 to C_4 photosynthesis. Such a change ought to be evident in a down-regulation at the C_4 leaf chloroplast envelope, but not at the plasma membrane.

C_4 acid transporters

On the other side of the CCM problem, it is important to identify the dicarboxylic acids transported into the chloroplasts to generate the CO_2 pool, and the compound returning to the cytosol to complete the C_4 cycle. Chloroplasts in the C_4 *Hydrilla* leaf are granal, which differs from the agranal BSC chloroplasts of terrestrial C_4 NADP-ME species. Despite this, the C_4 *Hydrilla* leaf was not particularly susceptible to photoinhibition (White et al 1996), unlike the situation in rice chloroplasts when NADP-ME was overexpressed (Takeuchi et al 2000, Chi et al 2004b). The *Hydrilla* results suggest that oxaloacetate and/or aspartate, not malate, is the major imported acid. Such a scenario would enable NADP-MDH to recycle NADPH produced by NADP-ME, and avoid high NADPH/$NADP^+$ ratios and photoinhibition. As a consequence, we hypothesize that up-regulation of specific dicarboxylic acid transporters, potentially for oxaloacetate and/or aspartate, should occur during induction of the C_4 system. In terrestrial plants, such transporters are associated with the inner membrane of the chloroplast envelope. Recently, orthologs that encode general dicarboxylate transporters with broad substrate specificities (including oxaloacetate), but differing expression patterns, have been reported (Renne et al 2003, Taniguchi et al 2004, Weber et al 2005). It will be important to determine whether the expression and distribution of dicarboxylate translocators differ in chloroplasts of C_3 and C_4 *Hydrilla* leaves. There is evidence from a C_3 and C_4 leaf differential display that they do, as the transcript for an ABC transporter was up-regulated during the induction of C_4 photosynthesis (Rao et al 2006a). In regard to the compound returning to the cytosol, the high activity of PPDK in C_4 *Hydrilla* leaves points to PEP as the entity exported from the chloroplasts.

Implications of the *Hydrilla* system for a transgenic C_4 rice plant

The attempts in recent years to produce transgenic C_4 rice plants by inserting one or several components of the C_4 cycle in a C_3 MC are all based on the *Hydrilla* premise: that a minimalist C_4 system can operate effectively in a single C_3 cell without BSC anatomy. The issue that is unresolved is to what extent a system like that in *Hydrilla* can function effectively in air. In the few transgenic cases where improvements in photosynthesis have been reported, it appears unlikely that it is the result of a C_4 CCM. More likely it is due to an enhanced capacity to recapture photorespiratory CO_2 from the overexpression of PEPC. Even if all the key enzymes shown in Figure 1 (e.g., CA, PEPC, NADP-ME, and PPDK) were expressed in the correct locations and a C_4 cycle initiated, it would probably be an ineffective chloroplastic CCM. Two factors that have been largely overlooked in the race to develop a transgenic C_4 plant are the nature of the organic acids transported into and out of the chloroplast, and the permeability of the MC chloroplast envelope to CO_2.

Transgenic attempts to introduce NADP-ME into grana-containing rice chloroplasts have failed, probably because of a severe redox disruption. However, the *Hydrilla* system demonstrates that it is possible to have an effective C_4 NADP-ME CCM that retains granal chloroplasts, possibly by importing oxaloacetate or aspartate or both, instead of malate, thereby allowing the chloroplastic NADP-MDH to recycle NADPH and provide sufficient $NADP^+$ for linear electron transport (Fig. 1). This may require expression or up-regulation of specific transporters in the rice chloroplast envelope. An alternative method is to overexpress PEPCK as the chloroplastic decarboxylase (Suzuki et al 2000), which is less likely to disrupt the chloroplast redox state. But this requires oxaloacetate generation in the chloroplasts, if malate is imported, and the presence of a transporter with the capacity to export PEP to the cytosol.

Even with the appropriate transporters and decarboxylase in place, the leakage problem must still be solved. Without a means to regulate leakage, even with a large-capacity C_4 pump, the QY will rise due to overcycling, and energy limitations on photosynthesis will increase at low irradiance. As a C_3 plant, rice MC chloroplasts are optimized to allow rapid permeation of CO_2 through the envelope and to maximize its availability for Rubisco. This is in direct contrast to what is required if the chloroplast becomes a CO_2-concentrating site. The fact that the *Hydrilla* C_4 system functions as an effective CCM provides some evidence that substantial changes in membrane permeability may occur. Likewise, the facultative *Hydrilla* system seems to be the best at present to investigate whether aquaporins are a key to the CO_2 permeability issue and whether they can be differentially regulated to reduce conductance in a chloroplast membrane. In most terrestrial C_4 plants, the BSC appear to solve the leakage issue, but even this is not a simple solution as a number of factors are involved (von Caemmerer and Furbank 2003), so that the presence of BSC in a transgenic rice plant does not guarantee a low-leakage system.

A further concept that the *Hydrilla* studies confirm is that isoforms with particular regulatory and kinetic characteristics are needed to optimize a C_4 transgenic system. The induction of the specific isoforms, HVPEPC4 and HVME1, to function as the initial carboxylase and decarboxylase, respectively, implies that the choice of isoforms used in the design may be critical to ensure that the system is appropriately modulated and that all the components operate in a concerted and homeostatic manner.

Likewise, the choice of a rice cultivar is important not only for transformability but also in regard to the down-regulation of Rubisco. It is noteworthy that Rubisco protein and activity are somewhat down-regulated as the *Hydrilla* C_4 system is induced, and this should improve nitrogen-use efficiency. However, if the loss of Rubisco reported in some CO_2-enriched rice cultivars were to occur with a transgenic C_4 rice cultivar, due to the C_4 system raising the leaf [CO_2], then this might reduce gains in photosynthesis and productivity.

In summary, although we have gained a substantial amount of information about C_4 photosynthesis since its discovery in the 1960s, there are still important components whose modulation remains elusive. These include transporter and permeability issues, and the nuances of enzyme regulation. As such, it appears that the technology to produce a transgenic C_4 rice plant may be ahead of our understanding of the basic biology. The biological and technical challenges to designing a C_4 rice plant are large, but the enormous potential benefits, both intellectual and practical in terms of increased food production, make the effort worthwhile.

References

Ascencio J, Bowes G. 1983. Phospho*enol*pyruvate carboxylase in *Hydrilla* plants with varying CO_2 compensation points. Photosynth. Res. 4:151-170.

Badger MR. 1987. The CO_2 concentrating mechanism in aquatic phototrophs. In: Hatch MD, Boardman NK, editors. The biochemistry of plants. Vol. 10. New York, N.Y. (USA): Academic Press. p 219-274.

Baker JT, Allen LH Jr, Boote KJ, Jones P, Jones JW. 1990a. Rice photosynthesis and evapotranspiration in subambient, ambient and superambient carbon dioxide concentrations. Agron. J. 82:834-840.

Baker JT, Allen LH Jr, Boote KJ. 1990b. Growth and yield responses of rice to carbon dioxide concentration. J. Agric. Sci. 115:313-320.

Bläsing OE, Ernst K, Streubel M, Westhoff P, Svensson P. 2002. The nonphotosynthetic phosphoe*nol*pyruvate carboxylases of the C_4 dicot *Flaveria trinervia*: implications for the evolution of C_4 photosynthesis. Planta 215:448-456.

Bowes G, Ogren WL, Hageman RH. 1971. Phosphoglycolate production catalyzed by ribulose diphosphate carboxylase. Biochem. Biophys. Res. Commun. 45:716-722.

Bowes G, Ogren WL. 1972. Oxygen inhibition and other properties of soybean ribulose 1,5-diphosphate carboxylase. J. Biol. Chem. 247:2171-2176.

Bowes G, Salvucci ME. 1984. *Hydrilla*: inducible C_4-type photosynthesis without Kranz anatomy. In: Sybesma C, editor. Advances in photosynthesis research, Vol. III. The Hague (The Netherlands): Martinus Nijhoff/Dr. W. Junk. p 829-832.

Bowes G, Rao SK, Estavillo GM, Reiskind JB. 2002. C_4 mechanisms in aquatic angiosperms: comparisons with terrestrial C_4 systems. Funct. Plant Biol. 29:379-392.

Casati P, Lara MV, Andreo CS. 2000. Induction of a C_4-like mechanism of CO_2 fixation in *Egeria densa*, a submersed aquatic species. Plant Physiol. 123:1611-1621.

Cheng YX, Takano T, Zhang XX, Yu S, Liu DL, Liu SK. 2006. Expression, purification, and characterization of two NADP-malic enzymes of rice (*Oryza sativa* L.) in *Escherichia coli*. Protein Express. Purific. 45:200-205.

Chi W, Yang J, Wu N, Zhang F. 2004a. Four rice genes encoding NADP malic enzyme exhibit distinct expression profiles. Biosci. Biotechnol. Biochem. 68:1865-1874.

Chi W, Zhou J-S, Zhang F, Wu N-H. 2004b. Photosynthetic features of transgenic rice expressing sorghum C_4 type NADP-ME. Acta Bot. Sin. 46:873-882.

Derelle E, Ferraz C, Rombauts S, Rouzé P, Worden AZ, Robbens S, Partensky F, Degroeve S, Echeynié S, Cooke R, Saeys Y, Wuyts J, Jabbari K, Bowler C, Panaud O, Piégu B, Ball SG, Ral J-P, Bouget F-Y, Piganeau G, De Baets B, Picard A, Delseny M, Demaille J, Van de Peer Y, Moreau H. 2006. Genome analysis of the smallest free-living eukaryote *Ostreococcus tauri* unveils many unique features. Proc. Nat. Acad. Sci. USA 103:11647-11652.

Echevarria C, Pacquit V, Bakrim N, Osuna L, Delgado M, Arrio-Dupont M, Vidal J. 1994. The effect of pH on the covalent and metabolic control of C_4 phospho*enol*pyruvate carboxylase from *Sorghum* leaf. Arch. Biochem. Biophys. 315:425-430.

Edwards GE. 1999. Tuning up crop photosynthesis. Nat. Biotechnol. 17:22-23.

Edwards GE, Franceschi VR, Voznesenskaya EV. 2004. Single-cell C_4 photosynthesis versus the dual-cell (Kranz) paradigm. Annu. Rev. Plant Biol. 55:173-196.

Estavillo GM. 2006. NADP-dependent malic enzyme gene family in a facultative single-cell C_4 photosynthesis system. PhD dissertation. University of Florida, Gainesville, Florida, USA. 127 p.

Fukayama H, Tsuchida H, Agarie S, Nomura M, Onodera H, Ono K, Lee BH, Hirose S, Toki S, Ku MSB, Makino A, Matsuoka M, Miyao M. 2001. Significant accumulation of C_4-specific pyruvate, orthophosphate dikinase in a C_3 plant, rice. Plant Physiol. 127:1136-1146.

Fukayama H, Hatch MD, Tamai T, Tsuchida H, Sudoh S, Furbank RT, Miyao M. 2003. Activity regulation and physiological impacts of maize C_4-specific phospho*enol*pyruvate carboxylase overproduced in transgenic rice plants. Photosynth. Res. 77:227-239.

Furbank RT, Jenkins CLD, Hatch MD. 1989. CO_2 concentrating mechanism of C_4 photosynthesis: permeability of isolated bundle sheath cells to inorganic carbon. Plant Physiol. 91:1364-1371.

Gesch RW, Boote KJ, Vu JCV, Allen LH, Bowes G. 1998. Changes in growth CO_2 result in rapid adjustments of ribulose-1,5-bisphosphate carboxylase/oxygenase small subunit gene expression in expanding and mature leaves of rice. Plant Physiol. 118:521-529.

Gesch RW, Vu JCV, Allen LH Jr, Boote KJ, 2001. Photosynthetic responses of rice and soybean to elevated CO_2 and temperature. Recent Res. Dev. Plant Physiol. 2:125-137.

Gesch RW, Kang I-H, Gallo-Meagher M, Vu JCV, Boote KJ, Allen LH Jr, Bowes G. 2003. Rubisco expression in rice leaves is related to genotypic variation of photosynthesis under elevated growth CO_2 and temperature. Plant Cell Environ. 26:1941-1950.

Glackin CA, Grula JW. 1990. Organ-specific transcripts of different size and abundance derive from the same pyruvate, orthophosphate dikinase gene in maize. Proc. Natl. Acad. Sci. USA 87:3004-3008.

Gupta SK, Ku M.B, Lin J-H, Zhang D, Edwards GE. 1994. Light/dark modulation of phospho*enol*pyruvate carboxylase in C_3 and C_4 species. Photosynth. Res. 42:133-143.

Häusler RE, Rademacher T, Li J, Lipka V, Fischer KL, Schubert S, Kreuzaler F, Hirsch HJ. 2001. Single and double overexpression of C_4 cycle genes had differential effects on the pattern of endogenous enzymes, attenuation of photorespiration and on contents of UV protectants in transgenic potato and tobacco plants. J. Exp. Bot. 52:1785-1803.

Häusler RE, Hirsch HJ, Kreuzaler F, Peterhänsel C. 2002. Overexpression of C_4 cycle enzymes in transgenic C_3 plants: a biotechnological approach to improve C_3 photosynthesis. J. Exp. Bot. 53:591-607.

Holaday AS, Bowes G. 1980. C_4 acid metabolism and dark CO_2 fixation in a submersed aquatic macrophyte *(Hydrilla verticillata).* Plant Physiol. 65:331-335.

Holaday AS, Salvucci ME, Bowes G. 1983. Variable photosynthesis/photorespiration ratios in *Hydrilla* and other submersed aquatic macrophyte species. Can. J. Bot. 61:229-236.

Horie T, Baker JT, Nakagawa H, Matsui T. 2000. Crop ecosystem responses to climatic change: rice. In: Reddy KR, Hodges HF, editors. Climatic change, plant productivity and global implications. New York, N.Y. (USA): CABI Publishing. p 81-106.

Imaizumi N, Ku MSB, Ishihara K, Samejima M, Kaneko S, Matsuoka M. 1997. Characterization of the gene for pyruvate, orthophosphate dikinase from rice, a C_3 plant, and a comparison of structure and expression between C_3 and C_4 genes for this protein. Plant Mol. Biol. 34:701-716.

Jenkins CLD, Furbank RT, Hatch MD. 1989. Inorganic carbon diffusion between C_4 mesophyll and bundle sheath cells. Plant Physiol. 91:1356-1363.

Jiao D, Huang X, Li X, Chi W, Kuang T, Zhang Q, Ku MSB, Cho D. 2002. Photosynthetic characteristics and tolerance to photo-oxidation of transgenic rice expressing C_4 photosynthesis enzymes. Photosynth. Res. 72:85-93.

Ku MSB, Cho DH, Li X, Jiao DM, Pinto M, Miyao M, Matsuoka M. 2001. Introduction of genes encoding C_4 photosynthesis enzymes into rice plants: physiological consequences. In: Rice biotechnology: improving yield, stress tolerance and grain quality. Novartis Foundation Symposium 236:100-116.

Laing WA, Ogren WL, Hageman RH. 1974. Regulation of soybean net photosynthetic CO_2 fixation by the interaction of CO_2, O_2 and ribulose1,5-diphosphate carboxylase. Plant Physiol. 54:678-685.

Leegood RC, Edwards GE. 1996. Carbon metabolism and photorespiration: temperature dependence in relation to other environmental factors. In: Baker NR, editor. Photosynthesis and the environment. Dordrecht (The Netherlands): Kluwer Academic Publishers. p 191-221.

Leport L, Kandlbinder A, Baur B, Kaiser WM. 1996. Diurnal modulation of phospho*enol*pyruvate carboxylation in pea leaves and roots as related to tissue malate concentrations and to the nitrogen source. Planta 198:495-501.

Long SP, Ainsworth EA, Leakey ADB, Morgan PB. 2005. Global food insecurity: treatment of major food crops with elevated carbon dioxide or ozone under large-scale fully open-air conditions suggests recent models may have overestimated future yields. Phil. Trans. R. Soc. Biol. Sci. 360:2011-2020.

Long SP, Zhu X-G, Naidu SL, Ort DR. 2006a. Can improvement in photosynthesis increase crop yields? Plant Cell Environ. 29:315-330.

Long SP, Ainsworth EA, Leakey ADB, Nosberger J, Ort DR. 2006b. Food for thought: lower-than-expected crop yield stimulation with rising CO_2 concentrations. Science 312:1918-1921.

Ludwig M, von Caemmerer S, Price GD, Badger MR, Furbank RT. 1998. Expression of tobacco carbonic anhydrase in the C_4 dicot *Flaveria bidentis* leads to increased leakiness of the bundle sheath and a defective CO_2-concentrating mechanism. Plant Physiol. 117:1071-1081.

Magnin NC, Cooley BA, Reiskind JB, Bowes G. 1997. Regulation and localization of key enzymes during induction of Kranz-less C_4-type photosynthesis in *Hydrilla verticillata*. Plant Physiol. 115:1681-1689.

Matsuoka M, Furbank RT, Fukayama H, Miyao M. 2001. Molecular engineering of C_4 photosynthesis. Annu. Rev. Plant Physiol. Mol. Biol. 52:297-314.

Miyao M. 2003. Molecular evolution and genetic engineering of C_4 photosynthetic enzymes. J. Exp. Bot. 54:179-189.

Miyao M, Fukayama H. 2003. Metabolic consequences of overproduction of phospho*enol*pyruvate carboxylase in C_3 plants. Arch. Biochem. Biophys. 414:197-203.

Ogren WL, Bowes G. 1971. Ribulose diphosphate carboxylase regulates soybean photorespiration. Nature New Biol. 230:159-160.

Osborne CP, Beerling DJ. 2006. Nature's green revolution: the remarkable evolutionary rise of C_4 plants. Phil. Trans. R. Soc. B 361:173-194.

Prasad GVR, Coury LA, Finn F, Zeidel ML. 1998. Reconstituted aquaporin 1 water channels transport CO_2 across membranes. J. Biol. Chem. 273:33123-33126.

Prentice IC, Farquhar GD, Fasham MJR, Goulden M, Heinmann M, Jaramillo VJ, Kheshgi HS, Le Quéré C, Scholes RJ, Wallace DWR. 2001. The carbon cycle and atmospheric carbon dioxide. In: Houghton JT, Ding Y, Griggs DJ, Noguer M, van der Linden PJ, editors. Climate Change 2001: The Scientific Basis. Contributions of the Working Group I to the Third Assessment Report of the Intergovernmental Panel on Climate Change. Cambridge (UK): Cambridge University Press. p 183-238.

Raines CA. 2006. Transgenic approaches to manipulate the environmental responses of the C_3 carbon fixation cycle. Plant Cell Environ. 29:331-339.

Rao SK, Magnin NC, Reiskind JB, Bowes G. 2002. Photosynthetic and other phospho*enol*pyruvate carboxylase isoforms in the single-cell, facultative C_4 system of *Hydrilla verticillata*. Plant Physiol. 130:876-886.

Rao VS, Rao SK, Reiskind JB, Bowes G. 2005. Carbonic anhydrase isoforms in the C_4 CCM of *Hydrilla*. In: Estvander A, Bruce D, editors. Photosynthesis: fundamental aspects to global perspectives, Kansas (USA): Allen Press, Inc. p 954-955.

Rao SK, Fukayama H, Reiskind JB, Miyao M, Bowes G. 2006a. Identification of C_4 responsive genes in the facultative C_4 plant *Hydrilla verticillata*. Photosynth. Res. 88:173-183.

Rao SK, Reiskind JB, Bowes G. 2006b. Light regulation of the photosynthetic phospho*enol*-pyruvate carboxylase (PEPC) in *Hydrilla verticillata*. Plant Cell Physiol. 47: in press.

Raven JA, Johnson AM, Kübler JE, Korb R, McInory SG, Handley LL, Scrimgouer CM, Walker DJ, Beardall J, Vanderklift M, Fredriksen S, Dunton KH. 2002. Mechanistic determination of carbon isotope discrimination by marine macroalgae and seagrasses. Funct. Plant Biol. 29:355-378.

Reinfelder JR, Milligan AJ, Morel FMM. 2004. The role of the C_4 pathway in carbon accumulation and fixation in a marine diatom. Plant Physiol. 135:2106-2111.

Reiskind JB, Seamon PT, Bowes G. 1988. Alternative methods of photosynthetic carbon assimilation in marine macroalgae. Plant Physiol. 87:686-692.

Reiskind JB, Berg RH, Salvucci ME, Bowes G. 1989. Immunogold localization of primary carboxylases in leaves of aquatic and a C_3-C_4 intermediate species. Plant Sci. 61:43-52.

Reiskind JB, Bowes G. 1991 The role of phospho*enol*pyruvate carboxykinase in a marine macroalga with C_4-like photosynthetic characteristics. Proc. Natl. Acad. Sci. USA 88:2883-2887.

Reiskind JB, Madsen TV, Van Ginkel LC, Bowes G. 1997. Evidence that inducible C_4-type photosynthesis is a chloroplastic CO_2-concentrating mechanism in *Hydrilla*, a submersed monocot. Plant Cell Environ. 20:211-220.

Renne P, Dressen U, Hebbeker U, Hille D, Flügge UI, Westhoff P, Weber AP. 2003. The *Arabidopsis* mutant *dct* is deficient in the plastidic glutamate/malate translocator DiT2. Plant J. 35:316-319.

Rowland-Bamford AJ, Baker JT, Allen LH, Bowes G. 1991. The acclimation of rice to changing atmospheric carbon dioxide concentration. Plant Cell Environ. 14:577-584.

Saigo M, Bologna FP, Maurino VG, Detarsio E, Andreo CS, Drincovich MF. 2004. Maize recombinant non-C_4 NADP-malic enzyme: a novel dimeric malic enzyme with high specific activity. Plant Mol. Biol. 55:97-107.

Salvucci ME, Bowes G. 1981. Induction of reduced photorespiratory activity in submersed and amphibious aquatic macrophytes. Plant Physiol. 67:335-340.

Salvucci ME, Bowes G. 1983a. Two photosynthetic mechanisms mediating the low photorespiratory state in submersed aquatic angiosperms. Plant Physiol. 73:488-496.

Salvucci ME, Bowes G. 1983b. Ethoxyzolamide repression of the low photorespiration state in two submersed angiosperms. Planta 158:27-34.

Seemann JR, Badger MR, Berry JA. 1984 Variations in the specific activity of ribulose-1,5-bisphosphate carboxylase between species utilizing differing photosynthetic pathways. Plant Physiol. 74:791-794.

Spencer WE, Teeri J, Wetzel RG. 1994. Acclimation of photosynthetic phenotype to environmental heterogeneity. Ecology 75:301-314.

Suzuki S, Murai N, Burnell JN, Arai M. 2000. Changes in photosynthetic carbon flow in transgenic rice plants that express C_4-type phospho*enol*pyruvate carboxykinase from *Urochloa panicoides*. Plant Physiol. 124:163-172.

Takeuchi K, Akagi H, Kamasawa N, Osumi M, Honda H. 2000. Aberrant chloroplasts in transgenic rice plants expressing a high level of maize NADP-dependent malic enzyme. Planta 211:265-274.

Taniguchi Y, Nagasaki J, Kawasaki M, Miyake H, Sugiyama T, Taniguchi M. 2004. Differentiation of dicarboxylate transporters in mesophyll and bundle sheath chloroplasts of maize. Plant Cell Physiol. 45:187-200.

Tcherkez GGB, Farquhar GD, Andrews TJ. 2006. Despite slow catalysis and confused substrate specificity, all ribulose bisphosphate carboxylases may be nearly perfectly optimized. Proc. Natl. Acad. Sci. USA 103:7246-7251.

Tetlow IJ, Rawsthorne S, Raines C, Emes MJ. 2005. Plastid metabolic pathways. In: Annu. Plant Rev. 13. Oxford (UK): CRC Press, Blackwell Publishing. p 60-125.

Uehlein N, Lovisolo C, Siefritz E, Kaldenhoff R. 2003. The tobacco aquaporin NtAQP1 is a membrane CO_2 pore with physiological functions. Nature 425:734-737.

van Ginkel LC, Bowes G, Reiskind JB, Prins HBA. 2001. A CO_2-flux mechanism operating via pH-polarity in *Hydrilla verticillata* leaves with C_3 and C_4 photosynthesis. Photosynth. Res. 68:81-88.

von Caemmerer S, Furbank RT. 2003. The C_4 pathway: an efficient CO_2 pump. Photosynth. Res. 77:191-207.

Weber APM, Schwacke R, Flügge UI. 2005. Solute transporters of the plastid envelope membrane. Annu. Rev. Plant Biol. 56:133-164.

Weig A, Deswarte C, Chrispeels MJ. 1997. The major intrinsic protein family of *Arabidopsis* has 23 members that form three distinct groups with functional aquaporins in each group. Plant Physiol. 114:1347-1357.

Westhoff P, Gowik U. 2004. Evolution of C_4 phospho*enol*pyruvate carboxylase. Genes and proteins: a case study with the genus *Flaveria.* Ann. Bot. 93:13-23.

White A, Reiskind JB, Bowes G. 1996. Dissolved inorganic carbon influences the photosynthetic responses of *Hydrilla* to photoinhibitory conditions. Aquat. Bot. 53:3-13.

Yeoh H-H, Badger MR, Watson L. 1980. Variations in $K_m(CO_2)$ of ribulose-1,5-bisphosphate carboxylase among grasses. Plant Physiol. 66:1110-1112.

Yeoh H-H, Hattersley P. 1985. K_m (CO_2) values of ribulose-1,5-bisphosphate carboxylase in grasses of different C_4 type. Phytochemistry 24:2277-2279.

Ziska LH, Manalo PA, Ordonez RA. 1996. Intraspecific variation in the response of rice (*Oryza sativa* L.) to increased CO_2 and temperature: growth and yield response of 17 cultivars. J. Exp. Bot. 47:1353-1359.

Notes

Authors' address: Department of Botany, University of Florida, Gainesville, FL 32611, USA.

Acknowledgments: We thank Nicole Boyle for work on the PPDK protein. The *Hydrilla* project was supported by the U.S. Department of Agriculture National Research Initiatives Competitive Grants Program, grant nos. 98-35306-6449 and 2002-35318-12540.

The ecology and evolution of single-cell C_4-like photosynthesis in diatoms: relevance to C_4 rice

J.A. Raven, K. Roberts, E. Granum, and R.C. Leegood

The 10,000 or more species of diatoms are microscopic photosynthetic organisms of the class Bacillariophyceae in the phylum Heterokontophyta. They are dominant primary producers in marine and inland water habitats, and may account for up to 20% of global primary productivity. The core carboxylation enzyme in their photosynthesis is Form ID Rubisco (ribulose bisphosphate carboxylase–oxygenase), which, if it replaced rice Form IB Rubisco on a molecule-for-molecule basis, would give slightly lower rates of photosynthesis at extant CO_2 concentrations. These kinetic characteristics, along with the low conductance for CO_2 of aqueous boundary layers, rationalize the occurrence of CCMs (inorganic carbon-concentrating mechanisms) in all diatoms investigated. It was assumed that these mechanisms, which increase the CO_2 concentration around Rubisco, were all based on active transport of CO_2, HCO_3^-, or H^+ across membranes. It now appears, from recent extensions of earlier work, that there is a C_4-like photosynthetic carbon metabolism in certain diatoms. However, more work is needed to determine the extent to which diatoms have photosynthesis analogous to that of single-cell C_4 higher plants. The relevance of this work to producing C_4 rice probably comes more from concepts than from the direct introduction of diatom genes in rice. One such concept is the possibility that C_4-like photosynthesis in diatoms involves no carbonic anhydrases (CAs), and so needs less Zn. However, this requires HCO_3^- entry, so decreased Zn costs of growth may be less readily achieved in rice unless phospho*enol*pyruvate carboxykinase (using CO_2) replaces phospho*enol*pyruvate carboxylase (using HCO_3^-) as the C_4 carboxylase.

Keywords: diatom, carbon-concentrating mechanism, carbonic anhydrase, Rubisco ID, *Thalassiosira*, zinc

The concept of introducing C_4 photosynthetic metabolism into *Oryza sativa* L. poses several problems, including the decision as to which extant, "natural" C_4 organisms should be examined for guidance, and probably as a source of specifically C_4 versions of genes. The "obvious" model would be a C_4 grass, all of which adhere, like the great majority of C_4 flowering plants, to the "mesophyll/bundle sheath" mechanism involving

Kranz anatomy (Sage and Monson 1999). Here, the $C_3 + C_1$ carboxylation catalyzed by phospho*enol*pyruvate carboxylase (PEPC) is in the mesophyll cells, which have ready diffusive access to CO_2 in the intercellular gas spaces of the leaves. These cells have chloroplasts that are involved, at least, in supplying the ATP used by the pyruvate, phosphate dikinase (PPDK), which converts pyruvate to phospho*enol*pyruvate; they can also generate NADPH. The subsequent $C_4 - C_1$ decarboxylation takes place in the bundle sheath cells, where Rubisco (ribulose bisphosphate carboxylase-oxygenase) assimilates the released CO_2, with further metabolism to the carbohydrate level of reduction using the PCRC (photosynthetic carbon reduction cycle). Depending on the organism, the $C_4 - C_1$ decarboxylation involves NADP-ME (NADP malic enzyme), NAD-ME (NAD malic enzyme), or PEPCK (phospho*enol*ypyruvate carboxykinase, invariably with some functioning of NAD-ME or NADP-ME) (Sage and Monson 1999, Granum et al 2005). The bundle sheath cells also have chloroplasts that may produce mainly ATP (some NADP-ME plants) or, more usually, ATP and NADPH. The bundle sheath cells have limited access to gas spaces, and may also have other mechanisms that limit CO_2 leakage.

However, some C_4 photosynthetic systems involve only a single-cell type, and these mechanisms might be simpler to engineer into rice. In terrestrial flowering plants, the single-cell type is exemplified by certain members of the Chenopodiaceae (Edwards et al 2004). Here, in a cylindrical photosynthetic structure, the photosynthetic cells are radially elongate with, in one case, the centrifugal end (with its plastids) functioning as mesophyll cells and the centripetal end (with its plastids) functioning as bundle sheath cells. In all cases, there are two functional types of plastids, as in C_4 plants with mesophyll and bundle sheath cells.

Among aquatic plants, Bowes and co-workers showed that the freshwater monocotyledon *Hydrilla verticillata* shows increasing induction of a single-cell C_4 photosynthetic metabolism when inorganic carbon becomes particularly limiting for growth (Edwards et al 2004); here, and in other cases of aquatic C_4, there seems to be only one functional plastid type. Marine algae also provide evidence of single-cell C_4 photosynthesis. The acellular green macroalga *Udotea flabellum* has a C_4-like mechanism that involves $C_3 + C_1$ carboxylation in the cytosol and $C_4 - C_1$ decarboxylation, and Rubisco activity, in the plastids. Interestingly, it appears that $C_3 + C_1$ carboxylation and probably $C_4 - C_1$ decarboxylation both involve PEPCK (Reiskind et al 1988, Reiskind and Bowes 1991). Chen et al (2002) showed PEPCK to have a CO_2 affinity similar to that of some of the higher-affinity Rubiscos when assayed with concentrations of divalent cations more closely reflecting the *in vivo* values. Such a mechanism would clearly require differential concentrations of substrates and regulators of PEPCK in the cytosol and the plastid, and perhaps different molecular forms, so that it can operate as a carboxylase in the cytosol and as a decarboxylase in the plastid. However, the use of PEPCK as the sole decarboxylase in C_4 photosynthesis has significant energetic implications (Granum et al 2005). Thus, there is only an input of one ATP (PEPC as carboxylase) or no ATP (PEPCK as carboxylase) per cycle if PEP is the C_3 compound transferred from the plastid to the cytosol (Table 1). Even

Table 1. Net ATP consumption in the C_3-C_4 cycle of C_4 photosynthesis with different carboxylase and decarboxylase enzymes, and either PEP or pyruvate as the C_3 compound moving from the $C_4 - C_1$ decarboxylation site to the $C_3 + C_1$ carboxylation site.

Carboxylase	C_3 compound transferred	Decarboxylase	Net ATP used in C_3-C_4 cycle
PEPC	PEP	PEPCK	1
PEPCK	PEP	PEPCK	0
PEPC	Pyruvate	PEPCK	2
PEPCK	Pyruvate	PEPCK	1
PEPC	Pyruvate	NAD(P)-ME	2
PEPCK	Pyruvate	NAD(P)-ME	1

Sources: Sage and Monson (1999), Granum et al (2005).

with pyruvate as the C_3 compound transferred, there is only an input of one ATP per cycle when PEPCK is the carboxylase.

The other marine example of single-cell C_4-like metabolism in marine algae is the diatom *Thalassiosira weissflogii*, which has PEPC and PEPCK (Reinfelder et al 2000, 2004, Morel et al 2002, cf. Johnston et al 2001, Granum et al 2005). The rest of this paper considers the evidence for C_4 photosynthesis in diatoms, and its possible relevance to producing C_4 rice. We begin by giving a background to diatoms and their role in biogeochemistry.

The evidence for CCMs and C_4 photosynthesis in diatoms

The diatoms

The diatoms comprise the class Bacillariophyceae in the phylum (division) Heterokontophyta, with classes such as their closest relatives, the recently discovered very small-celled Bolidophyceae, and the Fucophyceae (brown algae), Chrysophyceae, and Synurophyceae (golden algae) (van den Hoek et al 1995, Falkowski and Raven 2006, Sims et al 2006). They are characterized by an absolute requirement for silicon for growth, with most of the cellular silicon occurring in the silicified bipartite cell wall. This mineralization accounts for the excellent fossil record of diatoms stretching back 174 million years to the early Jurassic (i.e., rather before the earliest known fossils of flowering plants) (Raven and Waite 2004, Sims et al 2006). The molecular phylogeny of diatoms suggests that they originated (in a nonpreserved form) up to 240 million years ago, and the secondary endosymbiosis (or endosymbioses) with a red alga that gave rise to the plastids of the heterokonts (and some other algal phyla) occurred even earlier, in the Proterozoic (Falkowski et al 2004, Yoon et al 2004).

The diatoms today have a size range (equivalent spherical radius) for individual cells of 1 μm to 1 mm. They are found in the ocean, inland waters, and the soil, and are the dominant primary producers in the plankton. It is estimated that diatoms worldwide assimilate at least 15 Pg and probably 20 Pg inorganic C into organic C

Table 2. A comparison of some photosynthetic characteristics of diatoms with those of a C_3 vascular plant such as rice.

Characteristic	Diatoms	C_3 plants, e.g., *Oryza*
Rubisco form	ID	IB
Physiology of gas exchange	C_4-like	C_3
Biochemistry of inorganic C assimilation	C_3- or C_4-like	C_3
Occurrence of CCM	Present, with entry of CO_2 and HCO_3^-	Absent; CO_2 entry by diffusion
$\delta^{13}C$	C_3-like to C_4-like	C_3

Sources: Fielding et al (1998), Laws et al (1998), Reinfelder et al (2000, 2004), Burkhardt et al (2001a,b), Morel et al (2002), Giordano et al (2005), Granum et al (2005), and Roberts, Granum, Raven, and Leegood (unpublished).

each year out of about 50 Pg C in the oceans as a whole, and about 110 Pg C for the whole Earth (Field et al 1998, cf. Granum et al 2005).

Evidence of CCMs in diatoms

The Form ID Rubisco of diatoms (as well as other chromists, and red algae) has a range of affinities for CO_2 and O_2, CO_2-O_2 selectivities, and maximum specific reaction rates (Badger et al 1998, Shiraiwa et al 2004, Tcherkez et al 2006). Whitney et al (2001) showed that, although a red algal Rubisco replacing the native Form IB enzyme in a terrestrial C_3 vascular plant on a molecule-for-molecule basis would give higher net CO_2 fixation rates at all relevant CO_2 concentrations, the corresponding operation with a diatom Form ID Rubisco would give lower net CO_2 fixation rates, and even lower rates for diatom Rubisco in a diatom cell in natural waters relying on diffusive CO_2 supply. Although the concentration of CO_2 in solution is similar (within a factor of two or so) to that in an equilibrium atmosphere, with variations as a function of temperature and salinity, the diffusion coefficient for CO_2 is 10,000-fold lower, offset to only a limited extent (10–100-fold) by the smaller diffusion boundary layer thicknesses in aquatic environments (Raven 1970, 1984, Falkowski and Raven 2006). An additional complication is that the CO_2 in surface waters is rarely at air equilibrium, especially in inland waters (Maberly 1996).

The observed physiological properties of photosynthesis and growth by diatoms (Tables 2 and 3) are consistent with the occurrence of CCMs. The affinity for inorganic C *in vivo*, expressed in terms of CO_2, is much higher than that of isolated Rubisco, and this is very unlikely to be a consequence of a large excess of Rubisco activity in diatom cells. This conclusion is supported by the small O_2 sensitivity of photosynthesis by diatoms, the lower rates of glycolate synthesis and metabolism than expected for diffusive CO_2 entry except under extreme conditions, and the low CO_2 compensation concentrations (Birmingham and Colman 1979, Birmingham et al 1982, Badger et al 1998, Parker et al 2004, Parker and Armbrust 2005, Allen et al 2006).

Table 3. Characteristics of photosynthesis and growth in relation to inorganic carbon in two species of the diatom *Thalassiosira*. References as for Table 2, plus Clark and Flynn (2000). Errors quoted are 95% confidence limits.

Characteristic	*Thalassiosira pseudonana*	*T. weissflogii*
Biochemistry of inorganic C assimilation	C_3	C_4-like
$K_{0.5}$ for inorganic C for growth (mM)	273 ± 7 (NO_3^--grown) 333 ± 6 (NH_4^+-grown)	258 ± 4 (NO_3^--grown) 135 ± 2 (NH_4^+-grown)
Specific growth rate at inorganic C saturation (day^{-1})	1.33 ± 0.11 (NO_3^--grown) 1.75 ± 0.14 (NH_4^+-grown)	1.55 ± 0.08 (NO_3^--grown) 1.52 ± 0.05 (NH_4^+-grown)
$\delta^{13}C$ of diatom relative to $\delta^{13}C$ of dissolved CO_2	–12.5‰ (low inorganic C for growth) to –20.5‰ (high inorganic C for growth)	–4.0‰ (low inorganic C for growth) to –14.3‰ (high inorganic C for growth)

Since the occurrence of a CCM requires a higher concentration of CO_2 around Rubisco than in air-equilibrium solution, a conclusive marker for CCMs is the measurement of this concentration difference. Such differences have been measured at the level of total inorganic carbon, but the accumulation ratio is relatively small (twofold or so) when measurements are made for marine diatoms in a normal seawater medium (Burns and Beardall 1987, Colman and Rotatore 1988, Rotatore and Colman 1992, Rotatore et al 1995, Johnston and Raven 1996, Mitchell and Beardall 1996, Reinfelder et al 2004). Translating this into the CO_2 concentration available to Rubisco requires knowledge of the location of the accumulated inorganic carbon (whole cell, chloroplast, or a subplastid compartment such as the pyrenoid: Schmid 2001) and the pH in the compartment in which the inorganic carbon is accumulated. Compartmentation within the cell and a pH at the site of accumulation, which is lower than that of seawater, would give the required CO_2 concentration.

Mechanisms of CCMs in diatoms

The inorganic C species entering the cells. The form(s) of inorganic carbon entering cells can be assessed using membrane inlet mass spectrometry and isotope disequilibrium, provided there is no extracellular carbonic anhydrase (CA) activity. Martin and Tortell (2006) point out that the expression of extracellular CA is low in diatoms (and other marine phytoplankton) in the ocean. Determination of the form(s) in which inorganic carbon enters cells in laboratory cultures, with greater expression of extracellular CA, requires inhibition of this carbonic anhydrase; Martin and Tortell (2006) use their own and published data to show that inhibition of this CA may also inhibit HCO_3^- transport. Although this does not quantify HCO_3^- use in relation to total inorganic carbon assimilation, pH drift experiments show that the diatoms tested can use HCO_3^- (e.g., Allen and Spence 1981).

These problems notwithstanding, the data show that both CO_2 and HCO_3^- are taken up by all of the diatoms tested, and that generally more CO_2 than HCO_3^- is taken up (Colman and Rotatore 1988, Rotatore and Colman 1992, Rotatore et al 1995, Korb et al 1997, Burkhardt et al 2001b, Martin and Tortell 2006, Rost et al 2003, Tortell et al 2006, Chen and Gao 2004, cf. Cassar et al 2002). It is not clear whether the trans-plasmalemma transport of either of these species is energized and thus potentially an inorganic carbon-accumulating step in the CCM. It should be borne in mind that the energy requirements for active transport of HCO_3^- as a single anion involve the inside-negative electrical potential (Boyd and Gradmann 1999) component of the electrochemical potential difference of the anion as well as the concentration difference (Raven 1984).

The affinity for inorganic carbon for the combined uptake systems in marine diatoms means that inorganic carbon transport is saturated by normal seawater (see also Fielding et al 1998); the same is true of diatom growth (Clark and Flynn 2000) (Tables 2, 3). Inorganic carbon affinity of photosynthesis increases with decreasing inorganic carbon availability for growth (Fielding et al 1998, Burkhardt et al 2001b, Rost et al 2003). Furthermore, the ratio of CO_2 uptake to HCO_3^- uptake increases with decreasing inorganic supply for growth (Burkhardt et al 2001b, Rost et al 2003).

The biochemistry of inorganic carbon assimilation. If the initial carboxylation reaction is catalyzed by Rubisco, that is, C_3 biochemistry, then the CCM must involve active transport of HCO_3^-, CO_2, and/or H^+ at one or more membranes (Giordano et al 2005). However, it is possible that the initial carboxylation reaction is catalyzed by a $C_3 + C_1$ carboxylase to produce a C_4 dicarboxylic acid in a compartment other than the one in which Rubisco occurs, and the dicarboxylic acid is subsequently decarboxylated by a $C_4 - C_1$ decarboxylase at the site of Rubisco activity, constituting a C_4 mechanism of photosynthesis. C_4 photosynthesis can constitute a CCM on its own, or can supplement CCMs based on active transport across membranes.

Earlier short-term ^{14}C labeling experiments suggested that diatoms assimilated inorganic carbon mainly through the C_3 pathway, or less frequently had C_4 dicarboxylic acids as the main short-term products, consistent with a C_4 pathway (Coombs and Volcani 1968, Raven 1974, Beardall et al 1976, Holdsworth and Colbeck 1976, Mortain-Bertrand et al 1987). However, some of these investigations used rather long incubation times (10 seconds or longer); 2 seconds would be preferable (e.g., Beardall et al 1976). After a period in which it was implicitly or explicitly assumed that C_3 biochemistry predominates in diatom photosynthesis, Reinfelder et al (2000, 2004) and Morel et al (2002) found significant evidence for C_4 photosynthesis in the marine diatom *Thalassiosira weissflogii* (see critiques by Johnston et al 2001 and Granum et al 2005) (Table 3).

These findings have been followed up by Roberts, Granum, Leegood, and Raven (unpublished), who, among other things, compared the short-term (shortest 2 seconds) inorganic ^{14}C labeling products of *T. weissflogii* with those of *T. pseudonana* for which the complete genome sequence is known (Armbrust et al 2004) but which is relatively remotely related to *T. weissflogii* in the Thalassiosirales (Kaczmarska et al 2006). The investigation of *T. weissflogii* broadly confirmed the results of Reinfelder et al (2000, 2004) and Morel et al (2002). However, the ratio of C_4 dicarboxylic acid (malate) to organic phosphates (mainly 3-phosphoglycerate) at the shortest times was rather lower than that found by Reinfelder and collaborators. There was no change in the labeling pattern when cells were grown on low inorganic carbon (significantly below the present seawater concentrations) rather than on high inorganic carbon (about double present seawater), that is, the C_4 component does not increase as inorganic C for growth decreases. In contrast, *T. pseudonana* shows a strictly C_3 labeling pattern, again regardless of whether growth involved a higher or lower inorganic carbon concentration than is found in seawater. Armbrust et al (2004) speculated that *T. pseudonana*, like *T. weissflogii*, has C_4-like photosynthesis on the basis of the occurrence of genes for PEPC, PEPCK, and PPDK in its genome; the results of the labeling experiments show that it has strictly C_3 biochemistry.

These data suggest that, although *T. pseudonana* must have a CCM based entirely on active transport across membranes, *T. weissflogii* probably has a C_4-like fixation pathway contributing to, or perhaps even supplanting, a CCM based on active transport across membranes. If the CCM is indeed entirely based on a C_4 pathway with HCO_3^- entering the cells, PEPC as the ($C_3 + C_1$) carboxylase and any ($C_4 - C_1$) decarboxylase, then there is no need for carbonic anhydrase, with a consequent economy in the Zn need for growth (Reinfelder et al 2000, 2004, Morel et al 2002). This will be returned to later in the context of "C_4 rice." However, several additional experiments are needed to establish whether C_4 photosynthesis is really happening (Granum et al 2005).

Whatever the outcome of additional experiments, the data on resource-saturated growth rates and inorganic carbon affinity for growth (Clark and Flynn 2000; Table 3) show no significant differences between the two species of *Thalassiosira*. Although the $\delta^{13}C$ of the cells of both species shows large variations as a function of the inorganic carbon availability for growth, the C_4-like *T. weissflogii* has smaller discrimination values than *T. pseudonana*. Before these data are interpreted mechanistically (see Raven et al 2002a,b), it should be noted that the data were obtained in different laboratories with different growth techniques, and further work with directly comparable cultures is needed.

Spatial organization of CCM components. The C_4-like photosynthesis of *T. weissflogii* is thought to involve PEPC (and PPDK?) in the cytosol, and PEPCK (and pyruvate kinase?) in the plastids, although more data are needed to clarify this (Granum et al 2005), particularly to determine the intracellular location of the component enzymes.

CCMs based entirely on active transport across membranes of CO_2 or HCO_3^- could involve either the plasmalemma or the plastid envelopes, or both. This has not been clarified, nor has the species of inorganic carbon crossing the plastid envelope. Analogy with the much-better-known cyanobacterial system requires that HCO_3^- be the species accumulated in the plastid stroma (cytosol of cyanobacteria), and that the HCO_3^- be converted to CO_2 by CA, with assimilation by Rubisco, in the pyrenoids (carboxysomes in cyanobacteria) (Badger et al 1998, Schmid 2001, Giordano et al 2005). However, the best-characterized CA of diatoms, a β-CA of *Phaeodactylum tricornutum*, is in aggregates arranged peripherally in the plastid stroma (Tanaka et al 2005). Perhaps these aggregates are analogues of pyrenoids that are lacking in many strains of *P. tricornutum* (Tanaka et al 2005). There is also evidence of a cytosolic form of CA in diatoms (Morel et al 2002). Clearly, a lot more experimentation is required.

An alternative means of concentrating CO_2 around Rubisco is to have HCO_3^- entry to an adjacent acidic compartment, followed by conversion to the high-equilibrium concentration of CO_2, followed by diffusion of CO_2 to Rubisco. This model, which does not require active transport of inorganic carbon, was applied to the thylakoid lumen by Pronina and Semenenko (1992), and extended and quantified by Raven (1997) in the context of the green microalga *Chlamydomonas reinhardti.* There seems to be no information on key requirements of this mechanism for diatoms, for example, the occurrence of a CA in the thylakoid lumen, although it is possible that the lumen-located photosystem II *PsbO* gene product, universal in oxygen-evolvers, acts as a CA (Enami et al 2005, Lu et al 2005, cf. Hillier et al 2006). However, in *C. reinhardti*, it is another luminal carbonic anhydrase, the α-CA *Cah3*, that is involved in the CCM (Hanson et al 2003). Lee and Kugrens (2000) suggested that the chloroplast endoplasmic reticulum of diatoms (and other organisms whose plastids arose from secondary endosymbiosis) is involved in this sort of CCM; again, there is no direct evidence bearing on this suggestion. Wittpoth et al (1998) found that inorganic carbon-dependent oxygen evolution in isolated plastids of diatoms occurs even when the chloroplast endoplasmic reticulum has been lost in experiments carried out at inorganic carbon saturation for photosynthesis; such an experimental system would not be relevant to testing the hypothesis of Lee and Kugrens (2000).

Location and nature of the barrier to CO_2 leakage. A problem common to the effective operation of all CCMs is the restriction of CO_2 leakage from the pool drawn on by Rubisco (von Caemmerer 2003). That such a restriction occurs in diatoms is seen from the high photosynthetic rates, and corresponding high specific growth rates, of diatoms under resource-saturating conditions. Another indicator of restricted leakage is the high photon yield (absorbed photon basis) of diatoms presumably expressing CCMs (e.g., Geider et al 1985, 1986). Such a barrier must be located between the site of accumulation of CO_2 and the medium. Data on the quantity of CO_2 in photosynthesizing diatom cells suggest that, on a whole-cell volume basis, the CO_2 concentration is rather low relative to what is needed to account for the observed gas exchange, granted the observed kinetics of diatom Form ID Rubisco and the apparent absence of a large excess of Rubisco over what is needed to account for the resource-saturated rate of photosynthesis. This suggests that the elevated CO_2 pool is restricted to a subcellular compartment, such as the chloroplast stroma or pyrenoid.

If the CO_2 pool is in the pyrenoids, then the previously mentioned analogy with the cyanobacterial carboxysome is of relevance. Schmid (2001) showed that most diatom pyrenoids have a bounding "membrane" of unknown composition. If this corresponds to the protein shell of the carboxysome, it is possible that it has transport properties similar to those of the anion channels of the carboxysome shell (Kerfeld et al 2005). These anion channels would allow anionic substrates (ribulose bisphosphate, HCO_3^-) to enter and anionic products (3-phosphoglycerate) to leave, as well as any OH^- fluxes needed to account for charge and acid-base regulation. It has been suggested (Kerfeld et al 2005) that these anion channels cannot transport CO_2, thus restricting CO_2 leakage. Such a barrier could also limit the rate of O_2 entry, thus maintaining a very low steady-state O_2 concentration in the carboxysome, with slow entry balanced by a low rate of consumption by Rubisco oxygenase; there is no O_2 source in the carboxysomes. Whether the carboxysome shell (or any analogue in the pyrenoid membrane of diatom plastids) is quantitatively adequate to account for the restriction to leakage remained to be determined.

A case can be made for regulated leakage: at very high fluxes of photosynthetically active radiation, diatoms show a net efflux of CO_2, whereas, at lower radiation fluxes, there is a net influx of CO_2, as well as of the HCO_3^- that supports not only photosynthesis but also the net efflux of CO_2 at higher radiation fluxes (Tchernov et al 2003). Such inorganic C cycling at high fluxes of photosynthetically active radiation could help to alleviate photoinhibition. Regulated leakage could relate to a barrier in the plastid envelope (von Caemmerer 2003), especially since the diatoms have four envelope membranes.

Relevance to C_4 rice

Zinc-use efficiency

Zinc has been identified as an essential element that can limit rice growth under some culture conditions, and that can be deficient in humans with a high-rice diet (MacLean et al 2002, Fageria and Baligar 2005, Gao et al 2006). The relevance of diatoms here is that part of the motivation for the work of Reinfelder et al (2000, 2004) and Morel et al (2002) was apparently the hope of discovering whether diatoms could have a low Zn requirement for growth by decreasing the need for Zn-containing CA (Lane and Morel 2000), since Zn is limiting (or co-limiting) for growth of phytoplankton in parts of the ocean (see Crawford et al 2003, Franck et al 2003).

The C_4 pathway that Reinfelder et al (2000, 2004) and Morel et al (2002) propose, if HCO_3^- is the form in which inorganic carbon enters the cells, requires no CA in the mechanism of inorganic carbon movement from the medium to Rubisco. Transport of HCO_3^- from the medium to the cytosol, assimilation of HCO_3^- by PEPC in the cytosol, and release of CO_2 in the plastid stroma by PEPCK (or any other $C_4 - C_1$ decarboxylase) with subsequent assimilation by Rubisco do not need CA activity and thus economize in Zn (or one CA in *T. weissflogii*, using cadmium: Lane et al 2005). This pathway could occur in *T. weissflogii* (Table 3). Such a pathway could help to decrease the photosynthetic requirement for Zn in rice, thus helping to make the limitations on Zn availability in the rooting medium compatible with the requirement to increase the content in the grain of Zn available for human nutrition.

However, the *T. weissflogii* mechanism not involving CA could not work in rice, since there is no available source of HCO_3^- in the leaf medium. Rice photosynthesis depends on atmospheric CO_2, and either cell wall CA is needed to convert CO_2 into HCO_3^- followed by HCO_3^- entry to the cytosol with subsequent assimilation by PEPC, or, as in extant terrestrial C_4 plants (Sage and Monson 1999), CO_2 enters and is converted to HCO_3^- using CA in the cytosol, followed by assimilation using

PEPC. Furthermore, the first alternative requires the occurrence of a cell wall CA and HCO_3^- transport from the cell wall to the cytosol: these are not known attributes of the leaves of terrestrial flowering plants, although they are both found in secondarily aquatic flowering plants (Newman and Raven 1993, 1999, Beer et al 2002, Bowes et al 2002, Maberly and Madsen 2002, Uka et al 2005).

The use, as the sole mechanism of a CCM, of PEPCK as the $(C_3 + C_1)$ carboxylase and PEPCK, NAD-ME, or NADP-ME as the $(C_4 - C_1)$ decarboxylase would avoid the need for Zn in carbonic anhydrase since CO_2 is supplied to PEPCK from the atmosphere, and the $(C_4 - C_1)$ decarboxylase provides CO_2, the inorganic carbon substrate for Rubisco. However, there could be energetic problems with running a CCM of this type (Table 1; Granum et al 2005). Even this "zero Zn" pathway would have a smaller effect on the overall Zn budget of rice than of a diatom, since rice has superoxide dismutases containing copper plus zinc as well as superoxide dismutases with iron or manganese, while diatoms have only the Fe and Mn versions (Raven et al 1999, Wolf-Simon et al 2005). However, there could still be a significant savings in Zn from having a C_4 CCM with PEPCK as the carboxylase, replacing the C_3 mechanism with the use of CA, as a way of facilitating inorganic C fluxes through the chloroplast stroma (Raven et al 1999).

General points

A major aspect of the work on diatoms in the context of C_4 rice is that it provides additional evidence of high-capacity, low-leakage (and hence energetically efficient) CCMs in single cells. This suggests that a "single-cell" C_4 mechanism is a viable aim in engineering C_4 rice, even if the individual components of a diatom CCM are not appropriate for transfer to rice. In the case of an engineered CCM in rice involving active transport across membranes rather than C_4 metabolism, it could be argued that such a mechanism is much less common in terrestrial photosynthetic organisms than is C_4 photosynthesis. This is true, but it should be noted that there are terrestrial embryophytes with CCMs that are thought to be based on active transport across membranes, that is, those hornworts (anthoceropsids) with pyrenoids. In these organisms, the system is seen in the poikilohydric gametophytes lacking the homoiohydric cuticle, stomata, and intercellular gas spaces, and sporophytes with homoiohydric mechanisms and homologous with the homoiohydric sporophytic shoots of vascular plants, although lacking an endohydric (dead cells inside the plants) water-conducting tissue (Smith and Griffiths 1996a,b, 2000, Hanson et al 2002).

There are also a number of algae and cyanobacteria that have CCMs that are apparently based on active transport across membranes rather than C_4 mechanisms (although direct evidence of this is not available for all of the organisms considered: Raven 1970, 1984, Johnston 1991) and that can photosynthesize in air. One group of examples is the terrestrial free-living and lichenized cyanobacteria, and many free-living and lichenized green algae (Badger et al 1993, Palmqvist 1993, 1995, Palmqvist et al 1994a,b, 1995, 1997, Palmqvist and Badger 1996, Smith et al 1998, Lange et al 1999). The other group of examples is most intertidal macroalgae and lichens when still hydrated and metabolically active when exposed to the air at low tide (Johnston and Raven 1986, Maberly and Madsen 1990, Madsen and Maberly 1990, Raven et al 1990, Surif and Raven 1990). In both groups, any uptake of atmospheric CO_2 into the cells as HCO_3^- requires a cell wall carbonic anhydrase (Johnston and Raven 1986, Raven et al 1990, Surif and Raven 1990), whose activity is found in many of the organisms involved.

Conclusions

Studies on diatoms confirm other work indicating high-capacity, energetically efficient CCMs in single-cell systems, and suggest that single-cell approaches could be useful in producing rice with C_4 photosynthesis, or CCMs based on active transport across membranes.

References

Allen AE, Vardi A, Bowler C. 2006. An ecological and evolutionary context for integrating nitrogen metabolism and related signalling pathways in marine diatoms. Curr. Opin. Plant Biol. 9:264-273.

Allen ED, Spence DHN. 1981. The differential ability of aquatic plants to utilize the inorganic carbon supply in fresh waters. New Phytol. 87:269-283.

Armbrust EV, Berges JA, Bowler C, Green NR, Martinez D, Putnam NH, Zgou S, Allen AE, Apte KE, Bechner M, Brzezinski MA, Cheel BK, Ciovitti A, Davis AK, Desmarest MS, Detter JC, Glavina T, Goodstein D, Hadi MZ, Hellsten U, Hildebrand U, Hildebrand M, Jenkins BD, Jurka J, Kapitonov VV, Kröger N, Lau WW, Lane TW, Larimer FW, Lippmeier JC, Lucas S, Medina M, Montsant A, Obornik M, Parker MS, Palenik B, Pazour GJ, Richardson PM, Rynearson TA, Saito MA, Schwartz DC, Thamatrakoln K, Valentin K, Vardi A, Wilkerson F, Rokhsar DS. 2004. The genome of the diatom *Thalassiosira pseudonana*: ecology, evolution and metabolism. Science 306:79-86.

Badger MR, Pfanz H, Budel U, Lange OL. 1993. Evidence for the functioning of photosynthetic CO_2-concentrating mechanisms in lichens containing green algal and cyanobacterial photobionts. Planta 191:57-70.

Badger MR, Andrews TJ, Whitney SM, Ludwig M, Yellowlees C, Leggat W, Price GD. 1998. The diversity and coevolution of Rubisco, plastids, pyrenoids and chloroplast-based CO_2-concentrating mechanisms in algae. Can. J. Bot. 76:1052-1071.

Beardall J, Mukerji D, Glover HE, Morris I. 1976. The path of carbon in photosynthesis by marine phytoplankton. J. Phycol. 12:409-417.

Beer S, Bjork M, Hellblom F, Axelsson L. 2002. Inorganic carbon utilization in marine angiosperms (seagrasses). Funct. Plant Biol. 29:349-354.

Birmingham BC, Colman B. 1979. Measurement of carbon dioxide compensation points of freshwater algae. Plant Physiol. 64:892-895.

Birmingham BC, Coleman JR, Colman B. 1982. Measurement of photorespiration in algae. Plant Physiol. 69:259-262.

Bowes G, Rao SK, Estavillo GM, Reiskind JB. 2002. C-4 mechanisms in aquatic angiosperms: comparisons with terrestrial C-4 systems. Funct. Plant Biol. 29:379-392.

Boyd CM, Gradmann D. 1999. Electrophysiology of the marine diatom *Coscinodiscus wailesii*. I. Endogenous changes of membrane voltage and resistance. J. Exp. Bot. 50:445-452.

Burkhardt S, Riebesell U, Zondervan I. 2001a. Stable carbon isotope fractionation by marine phytoplankton in response to daylight, growth rate and CO_2 availability. Mar. Ecol. Progr. Ser. 46:31-41.

Burkhardt S, Amoroso G, Riebesell U, Sültemeyer D. 2001b. CO_2 and HCO_3^- uptake in marine diatoms acclimated to different CO_2 concentrations. Limnol. Oceanogr. 46:1378-1391.

Burns BD, Beardall J. 1987. Utilization of inorganic carbon by marine microalgae. J. Exp. Mar. Biol. Ecol. 107:75-86.

Cassar N, Laws EA, Popp BN, Bidigare RR. 2002. Sources of inorganic carbon for photosynthesis in a strain of *Phaeodactylum tricornutum*. Limnol. Oceanogr. 47:1192-1197.

Chen KW, Gao KS. 2004. Photosynthetic utilization of inorganic carbon and its regulation in the marine diatom *Skeletonema costatum*. Funct. Plant Biol. 31:1027-1033.

Chen ZH, Walker RP, Acheson RM, Leegood RC. 2002. Phospho*enol*pyruvate carboxykinase assayed at physiological concentrations of metal ions has a high affinity for CO_2. Plant Physiol. 128:160-164.

Clark DR, Flynn KJ. 2000. The relationship between the dissolved inorganic carbon concentration and growth rate in phytoplankton. Proc. Roy. Soc. London B 267:953-959.

Colman B, Rotatore C. 1988. Uptake and accumulation of inorganic carbon by a freshwater diatom. J. Exp. Bot. 39:1025-1032.

Coombs J, Volcani BE. 1968. Studies on the biochemistry and fine structure of silica shell formation in diatoms: silicon-induced metabolic transients in *Navicula pelliculosa* (Bréb.) Hilse. Planta 80:264-279.

Crawford DW, Lipsen MS, Purdie DA, Lohan MC, Statham PJ, Whitney, FA, Putland JN, Johnson WK, Sutherland N, Peterson TD, Harrison PJ, Wong CS. 2003. Influence of zinc and iron enrichments on phytoplankton growth in the northeastern subarctic Pacific. Limnol. Oceanogr. 48:1583-1600.

Edwards GE, Franceschi VR, Vosnesenskaya EV. 2004. Single-cell C_4 photosynthesis versus the dual-cell (Kranz) paradigm. Annu. Rev. Plant Biol. 55:173-196.

Enami I, Suzuki T, Tada O, Nakada Y, Nakamura K, Tohri A, Ohta H, Inoue I, Shen J-R. 2005. Distribution of the extrinsic proteins as a potential marker for the evolution of photosynthetic oxygen-evolving photosystem II. FEBS J. 272:5020-5030.

Fageria NM, Baligar VC. 2005. Growth components and zinc recovery efficiency of upland rice genotypes. Pesqu. Agropecu. Bras. 40:1211-1215.

Falkowski PG, Raven JA. 2006. Aquatic photosynthesis. 2nd Edition. Princeton, N.J. (USA): Princeton University Press. (In press.)

Falkowski PG, Katz ME, Knoll AH, Quigg A, Raven JA, Schofield O, Taylor FJR. 2004. The evolution of modern eukaryotic phytoplankton. Science 305:354-360.

Field CB, Behrenfeld MJ, Randerson JT, Falkowski P. 1998. Primary production in the ecosphere: integrating terrestrial and oceanic components. Science 2812:337-380.

Fielding AS, Turpin DH, Guy RD, Calvert SE, Crawford DW, Harrison PJ. 1998. Influence of the carbon concentrating mechanism on carbon stable isotope discrimination by the marine diatom *Thalassiosira pseudonana*. Can. J. Bot. 76:1098-1103.

Franck VM, Bruland KW, Hutchins DA, Brzezinski MA. 2003. Iron and zinc effect on silicic acid and nitrate uptake kinetics in three high-nutrient, low-chlorophyll (HNLC) regions. Mar. Ecol. Progr. Ser. 252:15-33.

Gao XP, Zou CQ, Fan XY, Zhang FS, Hoffland E. 2006. From flooded to aerobic conditions in rice cultivation: consequences for zinc uptake. Plant Soil 280:41-77.

Geider RJ, Osborne BA, Raven JA. 1985. Light dependence of growth and photosynthesis in the diatom *Phaeodactylum tricornutum*. J. Phycol. 21:609-619.

Geider RJ, Osborne BA, Raven JA. 1986. Growth, photosynthesis and maintenance costs in the diatom *Phaeodactylum tricornutum* at very low light levels. J. Phycol. 22:24-30.

Giordano M, Beardall J, Raven JA. 2005. CO_2 concentrating mechanisms in algae: mechanisms, environmental modulation, evolution. Annu. Rev. Plant Biol. 56:99-131.

Granum E, Raven JA, Leegood RC. 2005. How do marine diatoms fix 10 billion tonnes of inorganic carbon per year? Can. J. Bot. 83:898-908.

Hanson D, Andrews TJ, Badger MR. 2002. Variability of the pyrenoid-based CO_2 concentrating mechanism in hornworts (Anthocerotophyta). Funct. Plant Biol. 29:407-416.

Hanson DT, Franklin LA, Samuelssson G, Badger MR. 2003. The *Chlamydomonas reinhardtii cia3* mutant lacking a thylakoid lumen-localized carbonic anhydrase is limited by CO_2 supply to Rubisco and not photosystem II function in vivo. Plant Physiol. 132:2267-2275.

Hillier W, McConnell I, Badger MR, Boussac A, Klimov VV, Dismukes GC, Wydrynski T. 2006. Quantitative assessment of intrinsic carbonic anhydrase activity and capacity for bicarbonate oxidation in photosystem II. Biochemistry 45:2094-2102.

Holdsworth ES, Colbeck J. 1976. The pattern of carbon fixation in the unicellular alga *Phaeodactylum tricornutum*. Mar. Biol. 38:189-199.

Johnston AM. 1991. The acquisition of inorganic carbon by marine macroalgae. Can. J. Bot. 69:1123-1132.

Johnston AM, Raven JA. 1986. The analysis of photosynthesis in air and water by *Ascophyllum nodosum* (L.) Le Jol. Oecologia 69:288-295.

Johnston AM, Raven JA. 1996. Inorganic carbon accumulation by the marine diatom *Phaeodactylum tricornutum*. Eur. J. Phycol. 31:988-990.

Johnston AM, Raven JA, Beardall J, Leegood RC. 2001. Photosynthesis in a marine diatom. Nature 412:40-41.

Kcazmarska I, Beaton M, Benoit AC, Medlin LK. 2006. Molecular phylogeny of selected members of the order Thalassiosirales (Bacillariophyta) and evolution of the fultoportula. J. Phycol. 42:121-138.

Kerfeld CA, Saway MR, Tanaka S, Nguyen CV, Phillips M, Beeby M, Yeates TO. 2005. Protein structures forming the shell of primitive bacterial organelles. Science 309:936-938.

Korb RE, Saville PJ, Johnston AM, Raven JA. 1997. Sources of inorganic carbon for photosynthesis by three species of marine diatom. J. Phycol. 3:433-440.

Lane TW, Morel FMM. 2000. Regulation of carbonic anhydrase expression by zinc, cobalt and carbon dioxide in the marine diatom *Thalassiosira weissflogii*. Plant Physiol. 123:345-352.

Lane TW, Saito MA, George GN, Pickering IJ, Prince RJ, Morel FMM. 2005. A cadmium enzyme from a marine diatom. Nature 435:42.

Lange OL, Green TGA, Reichelberger H. 1999. The response of lichen photosynthesis to external CO_2 concentration and its interaction with thallus water-status. J. Plant Physiol. 154:157-166.

Laws EA, Thompson PA, Popp BN, Bidigare RR. 1998. Sources of inorganic carbon for marine microalgal photosynthesis: a reassessment of $\delta^{13}C$ data from batch culture studies of *Thalassiosira pseudonana* and *Emiliania huxleyi*. Limnol. Oceanogr. 43:136-142.

Lee RE, Kugrens P. 2000. Ancient atmospheric CO_2 and the timing of the evolution of secondary endosymbiosis. Phycologia 39:167-172.

Lu YK, Theg SM, Stemler AJ. 2005. Carbonic anhydrase activity of the photosystem II OEC33 protein from pea. Plant Cell Physiol. 46:1944-1953.

Maberly SC. 1996. Diel, episodic and seasonal changes in pH and concentrations of inorganic carbon in a productive lake. Freshwater Biol. 35:579-598.

Maberly SC, Madsen TV. 1990. Contribution of air and water to the carbon balance of *Fucus spiralis*. Mar. Ecol. Progr. Ser. 62:175-183.

Maberly SC, Madsen TV. 2002. Freshwater angiosperm carbon concentrating mechanisms: processes and patterns. Funct. Plant Biol. 29:393-405.

Maclean JL, Dawe DC, Hardy B, Hettel GP, editors. 2002. The rice almanac. Source book for the most important economic activity on Earth. Manila (Philippines): International Rice Research Institute. 253 p.

Madsen TV, Maberly SC. 1990. A comparison of air and water as environments for photosynthesis by the intertidal alga *Fucus spiralis*. J. Phycol. 26:24-30.

Martin CL, Tortell PD. 2006. Bicarbonate transport and extracellular carbonic anhydrase activity in Bering Sea phytoplankton assemblages: results from isotope disequilibrium experiments. Limnol. Oceanogr. 51. (In press.)

Mitchell C, Beardall J. 1996. Inorganic C uptake by an Antarctic sea-ice diatom *Nitzschia frigida*. Polar Biol. 16:95-99.

Morel FMM, Cox EH, Kraepiel AML, Lane TW, Milligan AJ, Schaperdoth I Reinfelder JR, Tortell PD. 2002. Acquisition of inorganic carbon by the marine diatom *Thalassiosira weissflogii*. Funct. Plant Biol. 29:301-308.

Mortain-Bertrand A, Descolas-Gros C, Jupin H. 1987. Short-term ^{14}C incorporation in *Skeletonema costatum* (Greville) Cleve (Bacillariophyceae) as a function of light regime. Phycologia 26:262-269.

Newman JR, Raven JA. 1993. Carbonic anhydrase in *Ranunculus penicillatus* ssp. *pseudofluitans*: activity, location, and implications for carbon assimilation. Plant Cell Environ. 16:491-500.

Newman JR, Raven JA. 1999. CO_2 is the main inorganic C species entering photosynthetically active leaf protoplasts of the freshwater macrophyte *Ranunculus penicillatus* ssp. *pseudofluitans*. Plant Cell Environ. 22:1019-1026.

Palmqvist K. 1993. Photosynthetic CO_2-use efficiency in lichens and their isolated photobionts: the possible role of a CO_2-concentrating mechanism. Planta 191:48-56.

Palmqvist K. 1995. Uptake and fixation of CO_2 in lichen photobionts. Symbiosis 18:95-109.

Palmqvist K, Ogren E, Lernmark U. 1994a. The CO_2-concentrating mechanism is absent in the green alga *Coccomyxa*: a comparative study of photosynthetic CO_2 and light responses of *Coccomyxa, Chlamydomonas reinhardtii* and barley protoplasts. Plant Cell Environ. 17:65-72.

Palmqvist K, Samuelsson G, Badger MR. 1994b. Photobiont-related differences among green-algal lichens. Planta 195:70-79.

Palmqvist K, Sültemeyer D, Baldet P, Andrews TJ, Badger MR. 1995. Characterization of inorganic carbon fluxes, carbonic anhydrase(s) and ribulose-1,5-bisphosphate carboxylase-oxygenase in the green unicellular *Coccomyxa*: comparisons with low-CO_2 cells of *Chlamydomonas reinhardtii*. Planta 197:352-361.

Palmqvist K, Badger MR. 1996. Carbonic anhydrase(s) associated with lichens: in vivo activities, possible locations and putative roles. New Phytol. 132:627-639.

Palmqvist K, de los Rios A, Ascaso C, Samuelsson G. 1997. Photosynthetic carbon acquisition in the lichen photobionts *Coccomyxa* and *Trebouxia*. Physiol. Plant 101:67-76.

Parker MS, Armbrust EV. 2005. Synergistic effect of light, temperature, and nitrogen source on transcription of genes for carbon and nitrogen metabolism in the centric diatom *Thalassiosira pseudonana* (Bacillariophyceae). J. Phycol. 41:1142-1149.

Parker MS, Armbrust EV, Piova-Scott J, Keil RG. 2004. Induction of photorespiration by light in the centric diatom *Thalassiosira weissflogii* (Bacillariophyceae): molecular characterization and physiological consequences. J. Phycol. 40:557-567.
Pronina NA, Semenenko VE. 1992. Role of the pyrenoid in concentration, generation and fixation of CO_2 in the chloroplast of microalgae. Soviet Plant Physiol. 39:470-476.
Raven JA. 1970. Exogenous inorganic carbon sources in plant photosynthesis. Biol. Rev. 45:167-221.
Raven JA. 1974. Carbon dioxide fixation. In: Stewart WDP, editor. Algal physiology and biochemistry. Oxford (UK): Blackwell. p 434-455.
Raven JA. 1984. Energetics and transport in aquatic plants. New York, N.Y. (USA): AR Liss. 587 p.
Raven JA. 1997. CO_2 concentrating mechanisms: a direct role for thylakoid lumen acidification? Plant Cell Environ. 20:147-154.
Raven JA, Waite A. 2004. The evolution of silicification in diatoms: inescapable sinking and sinking as escape? New Phytol. 162:45-61.
Raven JA, Johnston AM, Handley LL, McInroy SG. 1990. Transport and assimilation of inorganic carbon by *Lichina pygmaea* under emersed and submersed conditions. New Phytol. 114:407-417.
Raven JA, Evans MCW, Korb RE. 1999. The role of trace metals in photosynthetic electron transport in O_2-evolving organisms. Photosynth. Res. 60:111-149.
Raven JA, Johnston AM, Kübler JE, Korb RE, McInroy SG, Handley LL, Scrimgeour CM, Walker DI, Beardall J, Vanderklift M, Fredriksen J, Dunton KH. 2002a. Mechanistic interpretation of carbon isotope discrimination by marine macroalgae and seagrasses. Funct. Plant Biol. 29:355-378.
Raven JA, Johnston AM, Kübler JE, Korb RE, McInroy SG, Handley LL, Scrimgeour CM, Walker DI, Beardall J, Clayton MN, Vanderklift M, Fredriksen S, Dunton KH. .2002b. Seaweeds in cold seas: evolution and carbon acquisition. Ann. Bot. 90:525-536.
Reinfelder JR, Kraepiel AML, Morel FMM. 2000. Unicellular C_4 photosynthesis in a marine diatom. Nature 407:996-999.
Reinfelder JR, Milligan AJ, Morel FMM. 2004. The role of the C_4 pathway in carbon accumulation and fixation in a marine diatom. Plant Physiol. 135:2106-2111.
Reiskind JB, Bowes G. 1991. The role of phospho*enol*pyruvate carboxykinase in a marine macroalga with C_4-like photosynthetic characteristics. Proc. Natl. Acad. Sci. USA 88:2882-2887.
Reiskind JB, Seamon PT, Bowes G. 1988. Alternative modes of photosynthetic carbon assimilation in marine macroalgae. Plant Physiol. 87:686-692.
Rost B, Riebesell U, Burkhardt S, Sültemeyer D. 2003. Carbon acquisition by bloom-forming marine phytoplankton. Limnol. Oceanogr. 48:55-67.
Rotatore C, Colman B. 1992. Active uptake of CO_2 by the diatom *Navicula pelliculosa*. J. Exp. Bot. 43:571-576.
Rotatore C, Colman B, Kuzuma M. 1995. The active uptake of carbon dioxide by the marine diatoms *Phaeodactylum tricornutum* and *Cyclotella* sp. Plant Cell Environ. 18:913-918.
Sage RF, Monson RK, editors. 1999. C_4 plant biology. San Diego, Calif. (USA): Academic Press. 596 p.
Schmid AM. 2001. Value of pyrenoids in the systematics of the diatoms: their morphology and ultrastructure. In: Economou-Amilli A, editor. Proceedings of the 16th International Diatom Symposium. Athens (Greece): University of Athens Press. p 1-32.

Shiraiwa Y, Danbara A, Yoke K. 2004. Characterization of highly oxygen-sensitive photosynthesis in coccolithophorids. Jpn. J. Phycol. 52(supplement):87-94.

Sims PA, Mann DG, Medlin LK. 2006. Evolution of the diatoms: insights from fossil, biological and molecular data. Phycologia 45:361-402.

Smith EC, Griffiths H. 1996a. The occurrence of the chloroplast pyrenoid is correlated with the activity of a CO_2-concentrating mechanism and carbon isotope discrimination in lichens and bryophytes. Planta 198:6-16.

Smith EC, Griffiths H. 1996b. A pyrenoid-based carbon-concentrating mechanism is present in terrestrial bryophytes of the class Anthocerotae. Planta 200:203-212.

Smith EC, Griffiths H, Wood L, Gillon J. 1998. Intra-specific variation in the photosynthetic responses of cyanobiont lichens from contrasting habitats. New Phytol. 138:213-224.

Smith EC, Griffiths H. 2000. The role of carbonic anhydrase in photosynthesis and the activity of the carbon-concentrating-mechanism in bryophytes of the class Anthocerotae. New Phytol. 145:29-37.

Surif MB, Raven JA. 1990. Photosynthetic gas exchange under emersed conditions in eulittoral and normally submersed members of the Fucales and Laminariales: interpretation in relation to C isotope ratio and N and water use efficiency. Oecologia 82:68-80.

Tanaka Y, Nakatsuma D, Harada H, Ishida M, Matsuda Y. 2005. Localization of soluble beta-carbonic anhydrase in the marine diatom *Phaeodactylum tricornutum*: sorting to the chloroplast and cluster formation on the girdle lamellae. Plant Physiol. 138:207-217.

Tcherkez GGB, Farquhar GD, Andrews TJ. 2006. Despite slow catalysis and confused substrate specificity, all ribulose bisphosphate carboxylases may be nearly perfectly optimized. Proc. Natl. Acad. Sci. USA 103:7246-7251.

Tchernov D, Silverman J, Luz B, Reinhold L, Kaplan A. 2003. Massive light-dependent cycling between oxygenic photosynthetic microorganisms and their surroundings. Photosynth. Res. 77:95-103.

Tortell PD, Martin CL, Corkum ME. 2006. Inorganic carbon uptake and intracellular assimilation by subarctic Pacific phytoplankton assemblages. Limnol. Oceanogr. 51. (In press.)

Uku U, Beer S, Bjork M. 2005. Buffer sensitivity of photosynthetic carbon utilization in eight tropical seagrasses. Mar. Biol. 147:1085-1090.

van den Hoek C, Mann DG, Jahns HM. 1995. Algae: an introduction to phycology. Cambridge (UK): Cambridge University Press. 623 p.

von Caemmerer S. 2003. C_4 photosynthesis in a single C_3 cell is theoretically inefficient but may ameliorate internal CO_2 diffusion limitations of C_3 leaves. Plant Cell Environ. 26:1191-1197.

Whitney SM, Baldett P, Hudson GS, Andrews TJ. 2001. Form I Rubiscos from non-green algae are expressed abundantly but not assembled in tobacco chloroplasts. Plant J. 26:535-547.

Wittpoth C, Kroth PG, Weyrauch KV, Strotmann H. 1998. Functional characterization of isolated plastids from two diatoms. Planta 206:79-85.

Wolf-Simon F, Grzebyk D, Schofield O, Falkowski PG. 2005. The role and evolution of superoxide dismutase in algae. J. Phycol. 41:453-465.

Yoon HS, Hackett YD, Ciniglia C, Pinto G, Bhattacharya D. 2004. A molecular timeline for the origin of photosynthetic eukaryotes. Mol. Biol. Evol. 21:809-818.

Notes

Authors' addresses: J.A. Raven and K. Roberts, Plant Research Unit, University of Dundee at SCRI, Scottish Crop Research Institute, Invergowrie, Dundee DD2 5DA, UK, email: j.a.raven@dundee.ac.uk, karenbrown600@hotmail.com; E. Granum and R.C. Leegood, Department of Animal and Plant Sciences, University of Sheffield, Sheffield S10 2TN, UK, email: e.granum@sheffield.ac.uk, r.leegood@sheffield.ac.uk.

Acknowledgments: Work in the authors' laboratories on diatom inorganic carbon acquisition was supported by the Natural Environment Research Council, UK (NER/A/S/2001/01130). We thank Prof. D.G. Mann for pointing out the work of Schmid (2001).

Section 4:

The background and how C_4 rice can be delivered

The promise of systems biology for deciphering the control of C_4 leaf development: transcriptome profiling of leaf cell types

T. Nelson, S.L. Tausta, N. Gandotra, T. Liu, T. Ceserani, M. Chen, Y. Jiao, L. Ma, X.-W. Deng, N. Sun, M. Holfold, N. Li, and H. Zhao

C_4 photosynthesis is a system that uses resources present in C_3 plants. C_4 photosynthesis has evolved numerous times, in widely separated phylogenetic groups. All existing species that are able to fix carbon dioxide through one of the variety of C_4 schemes appear to rely on enzymatic activities and other factors present in most or all plant species, but regulated to exhibit an extreme intercellular or intracellular compartmentalization that supports the delivery of CO_2-derived carbon to Rubisco in an environment that disfavors competition from oxygen. C_4 species are particularly numerous in certain subfamilies of grasses, suggesting that the resources required for C_4 physiology are present and predisposed to this re-regulation. Rice lacks the dense leaf venation, bundle sheath (BS) differentiation, high BS plasmodesmatal density, and compartmentalization of photosynthetic activities that characterize nearly all C_4 grass species. To what extent are these C_4 resources already networked together in C_3 grasses such as rice and how might we find the targets and means for engineering the re-regulation of this network? A systems biology approach that compares the development of cell types in rice leaves to those in C_4 grasses could provide these targets and means. Emerging techniques such as laser microdissection of cell types and microarray profiling can provide the comprehensive data needed for a systems approach.

Keywords: bundle sheath, distinctive cell, interveinal distance, mesophyll, regulatory network

Independently evolved C_4 grasses demonstrate a collection of common traits. C_4 species have appeared multiple times in grass genera. Among grass C_4 species, the scheme is a two-cell metabolic cooperation between specialized mesophyll (M) and bundle sheath (BS) cells. It is most often accompanied by an increase in density of leaf secondary venation (reduced interveinal distance) (Ueno et al 2006) and an increase in the density of plasmodesmata joining mesophyll and BS cells (Dengler and Nelson 1999). Most C_4 grass species exhibit variations of Kranz anatomy, in which the two cooperating cell types form successive layers around a dense pattern of minor, intermediate, and major veins. However, some grass species (e.g., *Arundinella hirta*) achieve a similar adjacency of two complementary cellular compartments by the regular interruption of mesophyll cell regions with files of distinctive cells (DC), effectively files of BS cells not surrounding secondary veins (Dengler and Dengler 1990, Dengler et al 1997). The stepwise evolution of C_4 species from C_3 progenitors that is suggested by the characteristics of existing C_3-C_4 intermediate species also gives support to the view that the steps have been regulative and large rather than an accumulation of numerous discrete stepwise changes in the genes for specific enzymes, transcription factors, cellular properties, physiological regulatory systems, etc. (Sage 2004). These regulative points and the networks they regulate should be a major target for understanding and eventual engineering.

Developmental features limiting rice from C_4 physiology

It has been pointed out by authors of earlier reviews and perspective essays (Taylor 2000, Edwards et al 2001, 2004, Matsuoka et al 2001, Leegood 2002) that at least four of the features that distinguish most C_4 grasses from C_3 grasses are likely to require direct or indirect modification to permit C_4 physiology in rice.

1. High density of minor venation in the leaf blade; interveinal spacing of few photosynthetic cells.
2. Differentiation of a photosynthetic bundle sheath (or DC files) with a gas-impermeable boundary.
3. High density of plasmodesmata at the BS–mesophyll interface (and BS–vascular).
4. Compartmentalization of carbon-fixation and other photosynthetic activities between BS and mesophyll cells to produce C_4 biochemistry and physiology.

All four of these are complex traits that are likely to rely on the regulated interaction of the products of many genes, including many not yet identified. C_4 physiology is a syndrome of interrelated developmental, anatomical, cellular, and biochemical traits that almost unavoidably must rely on regulatory networks at levels higher than have been characterized to date. Efforts thus far to modify or to introduce one or more elements of the C_4 pathway into rice or other C_3 species, summarized in several excellent reviews (Matsuoka et al 2001, Häusler et al 2002, Miyao 2003), have revealed valuable information about the effects of perturbing the endogenous biochemistry and physiology, but have also served to reveal the regulative nature of the system and particularly the effect of host leaf anatomy. This paper will attempt to address the regulation of each of the four features listed above, briefly reviewing what is known and what needs to be known further to bring C_4 features to rice. More attention will focus on background discussion of venation (item 1), since items 2 and 3 are less well understood and 4 is well covered by other authors in this volume. The underlying constancy of the C_4 syndrome in so many independently evolved C_4 grass species suggests that a "systems biology" approach should be fruitful in revealing the networks and control points that link together the numerous genes and regulatory patterns that lie behind these traits.

Regulation of leaf venation

Polar auxin transport has a key role in the patterning of dicot leaf venation. By far, the majority of our mechanistic and molecular understanding of vein patterning and vascular cell development comes from studies of *Arabidopsis* and other dicots. Recent genetic and molecular studies in *Arabidopsis* have generally confirmed the empirically derived "canalization of auxin flow" hypothesis proposed in the 1970s by Tsvi Sachs. Sachs proposed that discrete veins form from a field of equivalent cells in a self-reinforcing fashion to drain auxin from the tissue in the direction of a sink (Sachs 1981, 1991). The stochastically better auxin flux in certain cells stimulates their differentiation into cells with stably improved flux capacity and prevents neighbors from experiencing the same conditions. In the last decade, observations on the effects of inhibitors of polar auxin transport (PAT), the cellular localization of PIN (PIN-FORMED) auxin efflux carriers during leaf development, and the phenotypes of mutants with defects in venation patterning and/or auxin signaling or transport have all tended to confirm the essential role that Sachs proposed for PAT in the formation of the venation pattern, and for auxin and other hormone signaling in the differentiation of specific vascular cells (Berleth and Sachs 2001). A key recent observation in *Arabidopsis* has been that auxin produced elsewhere in the plant enters leaf primordia through the epidermis (protoderm), accumulates at the tip and exits through the center, establishing the site of the midvein, all accompanied by polar localization of the PIN auxin efflux carriers in provascular cells that form along these routes (Benckova et al 2003, Scarpella et al 2006). Subsequent marginal loops and bridges in the vein pattern are associated with reorientations of auxin flux and PIN proteins at points ("convergences") in the leaf margins and existing venation. Although PAT and the accompanying localization of PIN proteins are well correlated with the formation of venation, it is currently unknown whether their roles are permissive or instructive to the process.

Mathematical models suggest that dicot venation patterns can be formed progressively and spontaneously. As molecular properties of auxin responses and transport have been defined by experimentation, mathematical models that seek to recreate a PAT-based ontogeny of vein patterns have been able to incorporate such biologically realistic parameters as leaf growth patterns and auxin source–sink gradients. Current models postulate that veins form at a particular condition of auxin concentration, gradient, or flux. Some such models can now recreate developmental features such as the acropetal development of the midvein, the bidirectional formation of lateral loops, the progressive bridging of aureoles, and the terminal formation of freely ending veinlets, all as observed normally in *Arabidopsis* leaf development (Kramer 2004, Rolland-Lagan and Prusinkiewicz 2005, Runions et al 2005, Dimitrov and Zucker 2006). The success of mathematical models in creating vein patterns based only on auxin fluxes combined with the pattern of leaf growth suggests that these changing features may be instructive—that venation patterns are produced spontaneously by the geometry of the growing leaf and the molecular properties of the PAT system. Some of these models can produce patterns resembling monocot leaf shapes and venation, although it is currently unknown whether the input parameters that they require for the localization of auxin and PAT correspond to the actual biology (e.g., spatial distributions of auxin).

Vein pattern formation differs by extent in the C_3 and C_4 grasses. How does the emerging picture of the regulation of dicot vein patterning apply to monocots? The formation of monocot leaf venation is largely understood at a descriptive level only. Most grasses share a simple ontogeny of leaf vein pattern (Nelson and Dengler 1997). The midvein and lateral veins are established early and in parallel. Most or all exit the base of the leaf and anastomose at their other ends toward the distal leaf margins and tip. Intermediate and minor veins form between and parallel to existing major (lateral) veins as the blade expands and interveinal distance increases. The minor veins are formed initially without attachment at either end to the existing network of major and minor veins. Commissural or transverse veins subsequently join new minor veins to each other and to the network. Information is currently lacking about the correlation of PIN and auxin localization with these developmental events. Generally, minor veins terminate within the blade or sheath of the grass leaf or anastomose with higher order veins. If a target feature for C_4 enhancement of rice is to increase the density of minor veins and to decrease concomitantly the interveinal distance, the regulation of the succession of lateral (major) and minor veins must be understood.

Why does the initiation of minor veins cease early in the development of C_3 grasses, producing a large interveinal distance, but continue in C_4 leaf blades? The highly vascularized leaves of C_4 grasses such as maize exhibit variation in the number of intercalary minor veins between lateral (major) veins. In the blade region, which is most active in C_4 metabolism, secondary venation is dense, while in the sheath region it is sparse or absent. In most grasses, C_3 and C_4, there is variation in the regulation of minor vein density between blade and sheath. In maize, the interveinal cell number varies between 4 (blade) and 10–20 (sheath) photosynthetic cells. In rice, the density in both the blade and sheath averages 10–12 cells, depending on leaf number and genetic background. If the factors that regulate this developmental variation can be identified, possibly by comparison between blade and sheath and between rice and maize, it should be possible to engineer an increase in minor vein density in rice.

Genetic variation has not yet provided tools for the analysis of grass venation. Although an abundance of natural and experimental variation in venation pattern can be recovered within and among dicot species (Turner and Sieburth 2002), the venation of monocots (and grasses in particular) has not yet yielded to a mutational analysis, nor is there yet a substantial literature on the effects in leaf primordia of hormone application or inhibition (see below). In part, this is due to the relative inaccessibility of grass leaf primordia at critical stages. In contrast to the accessible and flat leaf primordia of *Arabidopsis* and many other dicots, the primordia that wrap around the shoot apices of grasses such as rice and maize require manual dissection and flattening for inspection. The cell files of grass leaves remain strongly aligned with the apical–basal axis throughout development, leaving little opportunity for vascular cells to be misaligned, interrupted with other cell types, or apolar in a way that would produce easily recognizable defects in pattern. Nonetheless, it is surprising that few vein-defective monocot mutants have become available, beyond some with variations in frequency of transverse veins. Even the *tangled* mutant of maize, in which the strictly transverse cell divisions of wild-type leaf cell files are frequently oblique, exhibits little or no defect in venation pattern (Smith et al 1996). Possibly, a brute-force microscopic screen of the dissected plumules of embryo-lethal or seedling-lethal maize or rice mutants would yield more vein-defective mutants.

PAT may have a role in the patterning of grass leaf venation. It is currently unknown how the initiation of venation in grass leaves corresponds (if at all) to local concentrations and fluxes of auxin. Are the parallel venation of the grasses and the blade–sheath differences in vein density reflected in patterns of auxin synthesis, availability, transport, or responsiveness, as might be predicted from the emerging picture in dicots? When an inhibitor of PAT was applied to developing maize seedlings, effects on leaf development were pleiotropic, including twisting, ligule displacement, and vascular hypertrophy (Tsiantis et al 1999). No effects were apparent on the number or location of veins, however. The phenotype is similar to that observed in mutants of ROUGHSHEATH2, a negative regulator of *knox* homeobox genes that has a provascular expression pattern in leaves. In maize and rice, it may yet be possible to associate the persistence or cessation of minor vein initiation with a patterning signal of auxin (perhaps from leaf margins). If technical hurdles could be overcome, it would be of great interest to detail the cellular localization patterns of the PAT-associated PIN proteins throughout the ontogeny of the rice and maize venation, as has been done so profitably for *Arabidopsis*. If grass leaf venation patterning does indeed appear to be tightly associated with PAT, the engineering of a more persistent auxin flux during rice leaf development might encourage the continuous initiation of minor veins for a higher vein density.

Leaves of some dicot mutants exhibit monocot-like venation patterns. Clues that suggest that the pattern of auxin flux is crucial for vein parallelism and density in monocots come from treatments and mutants in *Arabidopsis* that mimic the monocot parallel venation pattern. The *parallel vein* (*par*) mutants of *Arabidopsis* produce secondary and tertiary vein orders that are nearly parallel to the midvein, instead of joining in the form of bridges and loops. All vein orders run parallel through the petiole rather than anastomosing with the midvein. The *par1* and *par2* mutants have been characterized in detail (J. Petricka and T. Nelson, unpublished results), and both exhibit alterations in the distribution of auxin across developing leaves, with significant displacement of the maximum normally at the tip, as revealed indirectly by DR5-GUS expression. The shape of *par1* and *par2* leaves is lanceolate and elongated. Remarkably, in the presence of the PAT inhibitor NPA (N-(1-naphthyl) phthalamic acid), the leaves become blunt, strap-shaped, and filled with parallel venation. A similar phenomenon has been reported following NPA treatment of the ARF-GAP mutant *scarface* (*sfc*/*van3*) (Deyholos et al 2000). Since NPA treatment is likely to cause the accumulation of auxin at the tip and margins of *Arabidopsis* leaves, this suggests that the density and parallelism of leaf venation are highly responsive to the site and timing of auxin accumulations or fluxes.

Many *Arabidopsis* mutants exhibit a reduction in leaf venation, in some cases lacking entire vein orders from the venation pattern. In addition to the *par1* and *par2* mutants described above, the leaf venation patterns of *monopteros, gnom, hemivenata, scarface/van3, axr1*, and many other mutants exhibit a reduction in the last veins normally initiated in wild-type leaves (Turner and Sieburth 2002). It is striking that nearly all of these mutants have been associated with defects in auxin transport or responses. It is tempting to speculate that the failure of late vein orders in these mutants is a direct or indirect consequence of a premature cessation in auxin signals or an inability of the available auxin flux to keep up with the increase in blade area.

More fundamental information is needed about grass vein patterning. A large number of genes have been associated with defects in the venation pattern of *Arabidopsis*. Many of these have been directly or indirectly associated with auxin signaling or PAT, and a surprising number with the endomembrane cycling system. Endomembrane traffic regulates the residency and localization of PIN proteins and certainly many other receptors and channels in polarized provascular and developing vascular cells. Many other venation-related genes have been identified by their mutant phenotypes or provascular expression patterns and are not yet associated with specific roles in vein patterning and differentiation. A comparative approach, in which the patterns of gene expression and protein localization for the rice orthologs of genes with known roles in *Arabidopsis* vein patterning, and the phenotypes of rice (and maize) gene knockouts are characterized, might prove extremely useful for accumulating tools and knowledge rapidly for rice vein patterning. It would be particularly interesting to follow the expression patterns of provascular gene orthologs in *Arundinella*, in which minor vein patterning seems to have "degenerated" into a patterning mechanism for DC files.

Differentiation of a gas-impermeable and photosynthetic BS layer

The development of a parenchymal BS cell layer is common to grasses, although its clonal derivation may vary among species (Dengler et al 1985, Dengler and Nelson 1999). The BS of C_3 grasses is generally populated with chloroplasts that are far more limited in number and extent of development than in C_4 grasses. C_4 grasses are distinguished by their gas-impermeable and photosynthetic BS cells, which are specialized for the decarboxylation of C_4 acids, received from mesophyll (M) cells, and refixation by Rubisco of released CO_2. A gas-impermeable barrier (suberin in grasses) surrounding the BS limits or prevents the entry of oxygen and the leakage of CO_2. Our limited understanding of the developmental regulation of the features of the BS in C_4 grasses, including the development of abundant specialized BS chloroplasts, presents an obstacle to the engineering of C_4 rice. It is currently difficult to suggest specific strategies without first increasing our knowledge in this area. How can the massive BS chloroplast development required for C_4 physiology be stimulated? The literature on efforts to introduce maize or other C_4 BS-specific and M-specific genes into rice suggests that there are more significant "*trans*" regulatory differences between the BS cells of C_3 and C_4 grasses than between their M cells, at least as regards the soluble C_4 and C_4-like enzymes. Possibly, the stimulation of BS chloroplast development, together with the immediate proximity to vein-associated developmental and physiological signals, would in itself be sufficient to produce BS-specific activities. Another BS feature in need of much fundamental knowledge is the regulation of suberin synthesis and deposition. At what level is the deposition of an impermeable suberin layer in the wall regulated? Are developing BS cells exposed to particular developmental or environmental signals in C_4 species? How much does gas impermeability rely on additional specializations of the secondary wall? The "-omic" profiling and systems biology strategy described later in this paper might provide a means of identifying the regulators of the pathways that produce these BS characteristics. It should at very least be capable of providing us with rice BS-specific promoters and transcription factors, as well as a complete list of genes that are coexpressed during the formation of these features.

Grasses with files of distinctive cells (DC) might be a resource for information on BS development. The development and physiology of C_4 grasses with DC files instead of BS cells exhibit unconventional but potentially useful features. Certain species within the genus *Arundinella* develop DC files in place of the BS-surrounded minor veins that are found in most C_4 grasses (Dengler et al 1990, 1996, Dengler and Dengler 1990). Their major and intermediate veins are surrounded with typical C_4 BS cells, however. In most ways characterized to date, the DC files appear to be functionally equivalent to BS cells, but without an adjacent vein. Photosynthate formed in DC files moves laterally to veins instead of being loaded directly. It has been postulated that the DC files form at sites patterned as minor veins, but at which the vascular development that would normally occur has degenerated in evolution (Dengler et al 1997). The signals (vein-associated?) at these leaf sites for making BS or DC are still at the sites but are interpreted for DC development. This system bears further study for the identification of the developmental signals that stimulate the formation of C_4-type BS cells, without the complication of associated vascular development. The lack of genetic and molecular tools for analysis of these species is an obstacle. However, if *Arundinella* is similar enough at a molecular level to rice, maize, or another grass with genomic resources, it may be possible to profile *Arundinella* DC, BS, and M cell transcripts and proteins on platforms from those systems, or even to test candidate regulators from *Arundinella* in transgenic rice.

Plasmodesmatal density to support increased BS–M flux

Whether for movement of C_4 intermediates or for enhanced flux of photosynthate to veins, an improvement in plasmodesmatal density is likely to be needed for either a two-cell or single-cell C_4 pathway in rice. Rice lacks the high density of plasmodesmata that facilitates the fluxes of metabolites between BS and M cells in C_4 grasses such as sugar cane and maize. In addition, a highly active C_4 rice is likely to require the high plasmodesmatal density between BS cells and adjacent vascular tissue that is generally observed in C_4 grasses (Laetsch 1974, Evert et al 1977, 1978, Robinson-Beers and Evert 1991, Botha 1992, Dengler and Nelson 1999), even if it is engineered with a single-cell C_4 scheme. Plasmodesmatal position and density are regulated throughout plant development, and may guide the spatial patterning of many features. Developmental changes in the positioning and frequency of plasmodesmata have been described in several systems, particularly as regards the apparent role of plasmodesmata in the channeling of communication between regions and cell layers during embryogenesis and organogenesis. It is possible that the regulation of plasmodesmatal placement has a direct or indirect effect on the full photosynthetic development of the BS, since the two features are generally associated (Dengler and Nelson 1999). The discovery that macromolecules such as transcription factors and RNAs can pass through the symplastic route regulated by plasmodesmata has intensified their study. Our current knowledge of the structure and function of plasmodesmata has grown as a result of mechanistic studies of viral movement, although this process generally includes significant modifications to their normal properties. However, much more needs to be learned, particularly about their developmental regulation in the grasses. This is likely to come by testing of paradigms from *Arabidopsis* developmental studies.

Regulation of compartmentalization of enzyme activities in BS and M

Why is the development of chloroplasts and photosynthetic metabolism far more extensive in the BS of C_4 grasses than in C_3 grasses? How are the complementary patterns of accumulation of C_4 photosynthetic enzymes achieved in BS and M cells of C_4 leaves? A growing body of evidence suggests that maize M-specific genes but not BS-specific genes find the appropriate endogenous signals in transgenic rice to retain their cell specificity of transcription in transgenic rice plants. The compartmentalization of C_4 activities in BS and M cells, as well as the cell-specific transcription, *cis*-factors and *trans*-factors, mRNA stability, activation, and other properties that contribute to this have been much reviewed elsewhere (Edwards et al 2001, Leegood 2002). One body of work of particular relevance to an evaluation of the re-regulation of rice is the analysis of maize mutants with defects in BS development. In *golden2* (*g2)* mutants, the development of BS chloroplasts throughout the leaf and of M chloroplasts in "C_3" (vein-distal) regions of the leaf sheath is disrupted (Cribb et al 2001, Rossini et al 2001). Vein-proximal "C_4" mesophyll cells are normal. The *G2* gene encodes a transcriptional regulator expressed in photosynthetic tissue in a light-enhanced manner. A second gene, *ZmGlk1*, is expressed in C_4 M cells in light, and is proposed to have a role in producing specialized photosynthetic development in those cells, whereas *G2* is responsible for specialized PS development in BS cells. In the absence of *G2*, C_4 enzymes accumulated in both BS and M cells nonspecifically. Rice orthologs for both genes have been identified and are further described in an accompanying article (Langdale et al, this volume). It is currently unknown whether regulators of photosynthetic development, such as *G2*, are permissive or instructive—necessary or sufficient—to the localization of C_4 activities. The localization of activities is dependent on illumination, on spatial regulation at multiple levels (transcription, translation, activity, etc.), and certainly on the underlying anatomical context and associated signals.

The C_4 syndrome and systems biology

From this brief consideration of four complex features of development and anatomy that distinguish C_4 grasses, it should be clear that the engineering of C_4 rice with these features on a gene-product-by-gene-product basis would be very challenging, since many interrelated multigene features are involved. Might it be possible instead to cause rice to reshape itself into a C_4 or C_4-like plant, using its endogenous genetic resources, by altering the developmental and physiological regulators of entire traits? The C_4 pathway and other plant metabolic pathways have already been profitably investigated with "systems" approaches that evaluate the effects of experimental perturbation on the fluxes and concentrations of all related metabolites, to identify control points in pathways (Hammer et al 2004). We should be able to extend this approach to the developmental and environment-response pathways that produce the C_4 syndrome, and to identify and modify the key regulators. At present, we can only infer that such high-order regulators exist, but the means are also emerging rapidly to identify them and the networks they regulate. Emerging tools for altering the expression of multiple target genes *in trans*, such as synthetic microRNAs, place this strategy within the realm of discussion, if enough fundamental information about the target networks can be gained. This approach would integrate and augment the detailed knowledge we now have of the C_4-related genes, enzymes, and metabolites with the comprehensive data sets we could gather by profiling all transcripts, proteins, and metabolites in the leaf cell types of rice and C_4 grasses under a variety of developmental, environmental, and physiological conditions. To what extent do the same networks of interactions exist in rice as in maize, despite the obvious differences in output?

Profiles, perturbations, networks, and models

Since the C_4 syndrome appears to rely on a regulative network of interdependent gene products and metabolites, it is a promising subject for a systems biology "discovery" approach. A systems approach is designed to be broad and unbiased, to permit the discovery of "emergent" properties and relationships within the entire system that might not be revealed in hypothesis-driven experimentation that is targeted at specific genes, proteins, activities, or metabolites (Ideker et al 2001a, Raikhel and Coruzzi 2003). By evaluating all components of the system when it is perturbed by genetic mutations or environmental treatments, computational approaches are able to find and group coordinately induced changes and to infer networks of relationships that can then be tested. Models of the networks and their regulation can be refined iteratively based on the results of the experiments they suggest. With the rice (and soon maize) genome completely sequenced and with constantly improving annotation, it now makes sense to build the "-omics" data sets from BS and M cells and their precursors in rice and maize that will permit this computational discovery approach. The comprehensive and multilevel nature of a systems approach necessitates a multilab collaboration to collect genome-wide data on the transcriptomes, proteomes, protein-protein interactions, enzymatic activities, and metabolites for BS and M cells. The initial efforts for data-gathering would be significant and, in the case of proteomics, protein-protein interactions, and metabolomics, would rely on methodologies that are emerging or under development. Equally important in this effort will be the informatics and computational expertise that links information at all levels, derives the networks and other emergent properties of the system, and proposes testable models that can incorporate and predict the effects of specific modifications on the entire system.

Several such systems approaches have begun to examine complex metabolic and developmental systems in plants. To date, most of the effort has focused on *Arabidopsis* because of the concentration of genomic, proteomic, and other tools in that system, and the mass of available public data. However, analogous resources are rapidly becoming available in rice and maize. Examples of systems seeking to integrate transcriptional, proteomic, and metabolomic data and tools for pathways can be viewed at VirtualPlant (www.virtualplant.org), MapMan (http://gabi.rz-berlin.mpg.de/projects/MapMan/), PlaNet (http://mips.gsf.de/projects/plants/PlaNetPortal/index_html), and VPIN (http://vpin.ncgr.org/). All of these projects seek to make it possible to detect and model networks of interactions. To date, the most successful implementations of this comprehensive systems biology approach have been in yeast, in which data at all levels, including protein-protein interactions, have been gathered (e.g., Ideker et al 2001b). However, the means are increasingly available to apply this to model plant systems.

The production of some of the public data sets that could serve a systems approach to the re-regulation of rice into a C_4 scheme is under way. The next section describes transcript profiling for the BS and M cells of rice seedling leaves.

Transcriptional profiling of rice BS, mesophyll, and other leaf cell types

As a part of a project to produce a comprehensive transcriptional atlas of rice cell types (http://plantgenomics.biology.yale.edu/riceatlas), we have generated transcriptional profiles of a variety of cell types from developing rice leaves (second leaf of 5–7-d seedling), and we are currently profiling cell types from more mature leaves. To produce profiles of specific cell types, tissue at the required stage of growth is harvested from plants grown under controlled conditions of illumination, temperature, and humidity. Tissue is immediately treated with a precipitative fixative (cold 100% acetone), processed under vacuum through histological preparative steps, and embedded in paraffin blocks. RNA is well preserved through all histological steps, as judged by size measurements (Agilent Bioanalyzer). RNA probes from fixed, embedded, and sectioned tissue exhibit a correlation coefficient of >0.9 compared to RNA from the same fresh tissue, when hybridized to whole-genome microarrays, as described below, suggesting that histological steps do not significantly alter the RNA population. The paraffin blocks can serve as a tissue archive that can be sampled over a period of months without significant degradation of RNA. To obtain individual cell-type samples, microtome sections of 8 μm are prepared from the blocks, mounted on slides, and deparaffinized under RNAase-free conditions. Using a laser microdissection microscopy system (Arcturus Veritas or Pixcell IIe, Molecular Devices, Mountain View, California), target cells were visually identified, harvested onto thermoplastic film using the near-infrared laser of the instrument, and visually inspected for purity.

In some cases, the cell harvest was edited on the film by ablation of contaminants with a UV laser. When 500–1,000 cells of the desired type were harvested onto the capture film, RNA was extracted from the film and evaluated for quantity (RiboGreen fluorescence) and quality (Agilent Bioanalyzer). For most leaf cell types, approximately 20 ng of total RNA could be recovered from 1,000 cells. A 15-ng aliquot of total RNA was subjected to two rounds of linear amplification of polyA-containing mRNAs (Arcturus Ribo-Amp-OA system) to produce 10–20 μg of aRNA, which was amino-allyl labeled for use in microarray hybridizations.

The transcript population was profiled by hybridization to a spotted 70-mer-long oligo array with 58k features, representing approximately 80% of all rice genes and gene models, as currently annotated (TIGR version 4 rice annotation). This microarray platform and the oligo set from which it was constructed are described in detail elsewhere (Ma et al 2005). The oligo set, which was derived from a combination of japonica and indica sequence information, is mapped on the TIGR Rice Genome Browser. To assure that expression intensities for individual genes can be compared across all cell-type data sets in the rice atlas, each of the four biological replicates for a particular cell type was hybridized in parallel with an aliquot of a Common Reference RNA sample (rice tissue culture RNA) that produced a significant signal at >70% of the array sites. Along with dye-swaps, this permits the progressive normalization of the entire database, producing normalized intensity values for each gene that can be directly compared among all cell types. Details of the statistical quality control and data normalization for the cell-type atlas will be published elsewhere.

Relevant to the evaluation of C_4 resources in rice, the transcriptional atlas has robust data sets for BS, mesophyll, "vein" (combination of phloem, developing xylem, and associated parenchymal cells), epidermal pavement, bulliform, and guard cells in the developing rice leaf. The profiling of BS and M cells from the C_4 grass maize has been independently performed on maize EST arrays (Thomas Brutnell, unpublished data), thus permitting at least preliminary comparisons between maize and rice BS, M, and other profiles. Within the rice cell-type profiles for BS and M cells, where the correct gene can be identified, the intensities of specific C_4-related gene transcripts are consistent with published literature. For example, the four rice NADP-malic enzyme genes were characterized by Chi et al (2004) to have distinct expression patterns in leaf, panicle, and root. Two of the genes were expressed in all three locations to different extents (*OschlMe1* = *Os01g9320* and *OscytMe1* = *Os01g52500*), one in all locations to a small extent (*OscytMe3* = *Os01g54030*), and one highly expressed but only in roots (*OscytMe2* = *Os05g09440*). The first of these (*Os01g9320*) is likely to be the C_4-like gene. Our rice cell profiles are in general agreement with this organ distribution, but with the added detail that the C_4-like gene is much more highly expressed in leaf cell types than in root and that it is roughly equally expressed in rice BS, M, and vascular tissue (unpublished results).

Distribution of C_4-related transcripts in rice BS, M, and other leaf cells

In addition to C_4 pathway genes, the rice transcriptional atlas (http://plantgenomics.biology.yale.edu/riceatlas) contains data on the BS-M-vascular distribution and intensity of expression for many other genes relevant to the C_4 syndrome, including auxin-related (Aux/IAA family proteins, ARF family proteins, etc.), suberin synthesis, photosystem proteins, and others. At this point, the interpretation of the data is limited by the current state of gene annotation in rice and maize (see below), and results should be considered as preliminary. For some pathways, data are not available for genes encoding every step, either because they have not yet been identified and annotated for rice or because they are not optimally represented in the oligo set on the current microarray platform. One notable observation to date is that the putative orthologs of many or most of the genes associated with the patterning of veins in *Arabidopsis* are expressed in rice below the threshold of detection. Whether this represents a fundamental biological difference or simply a limitation of the technology or the developmental stage of sampling is currently unknown.

What next for a systems approach?

To realize the potential of a systems biology approach to the development and regulation of the C_4 pathway, the following steps would be needed.

More robust transcript profiling of rice BS, M, and other leaf cells types

This should include sampling at various stages in leaf development, different illumination, temperature, and other conditions relevant to C_4, and should include selected rice mutants, varieties, and relatives that exhibit variation in photosynthetic or developmental properties. All profiling should be done on a second-generation (or later) microarray platform that optimizes the detection of every gene and that benefits from improvements in rice gene models and annotation. It should be noted that the annotation of the rice genome is largely the result of automatic analysis thus far. Ongoing manual curation to incorporate data from the international research community is needed to correct errors and to add biological detail, as is continuing in the *Arabidopsis* and maize communities. This is of particular significance to the evaluation of genes for C_4-related enzymes, which are nearly all members of multigene families in rice and maize. A significant number of biochemical pathways, including photosynthetic enzymes, are being curated by Gramene (www.gramene.org/pathway/, Pankaj Jaiswal, personal communication); the first version of their Rice Cyc pathway browser has just been released. As this curation is ongoing and large amounts of data are produced, it is essential that the rice research community embrace the controlled vocabularies regulated by the Plant Ontology Consortium (www.plantontology.org) and Gene Ontology Consortium (www.geneontology.org), to assure that all data related to particular genes, plants, developmental stages, environmental conditions, etc., can be found in computer queries.

Transcript profiling of BS and M cells of maize and other C_4 grasses

This should include relevant maize mutants such as *g2*. It may be particularly revealing to profile *Arundinella* distinctive cells, along with their BS and M cells, if a rice, maize, or other profiling platform can be found to be appropriate for the species. If not, it may be possible to evaluate particular transcripts by means of RT-PCR (reverse transcription polymerase chain reaction) or a related method.

Extend profiling efforts to proteomics (abundance and activities), protein-protein interactions, and metabolites

A significant proteomic effort has been initiated for maize BS and M cells (Majeran et al 2005), and this should be extended to rice and to the developmental and environmental conditions described above. Laser microdissection or other technologies should make it possible to perform similar analyses in rice and other species for which it is not possible to perform cell separations. The cell-specific analysis of metabolites would be challenging or impossible with current methods for isolating BS and M cells, although some progress is being made for metabolite profiling of cells laser-microdissected from fresh frozen plant sections (Schad et al 2005). Ideally, the protein and metabolite profiling should be performed on the same materials as for transcript profiling, or at least on materials grown and sampled under identical controlled conditions, to permit subsequent linking of information at all levels.

Computational integration and modeling of data at the RNA, protein, and metabolite levels

The C_4 system should provide a creative challenge to experts in modeling and informatics, since we already know that important controls exist at levels of transcription, posttranscription, translation, posttranslation, enzymatic activity, and more, all with distinct cell specificity for these controls. Are the relationships and networks revealed in C_4 grasses (and *Arundinella*?) effectively the same as those revealed in rice, but with different regulation? How are the regulatory networks different between BS and M cells, in both C_4 and C_3 backgrounds? Are the pathways involved in vein initiation and patterning, BS development, and plasmodesmatal development different between *Arabidopsis* and monocots? Between maize and rice? Proposed models of relationships and control points would provide the targets for experimentation, such as knockouts or other alterations to the candidate regulators, followed by selective profiling to evaluate the effects of that perturbation on all levels. Success with the analysis in this manner may provide us with the key targets for using rice's own resources in a C_4 or C_4-like manner. Along the way, this approach should provide a tremendous resource for molecular tools (cell-specific promoters, protein tags, and markers) and for fundamental biology (monocot venation patterning, regulation of BS cell differentiation, etc.) for the C_4 syndrome.

References

Benckova E, Michniewicz M, Sauer M, Teichmann T, Seifertova D, Jurgens G, Friml J. 2003. Local, efflux-dependent auxin gradients as a common module for plant organ formation. Cell 115:591-602.

Berleth T, Sachs T. 2001. Plant morphogenesis: long-distance coordination and local patterning. Curr. Opin. Plant Biol. 4:57-62.

Botha CEJ. 1992. Plasmodesmatal distribution, structure and frequency in relation to assimilation in C_3 and C_4 grasses in southern Africa. Planta 187:348-358.

Chi W, Yang J, Wu N, Zhang F. 2004. Four rice genes encoding NADP malic enzyme exhibit distinct expression profiles. Biosci. Biotechnol. Biochem. 68:1865-1874.

Cribb L, Hall LN, Langdale JA. 2001. Four mutant alleles elucidate the role of the G2 protein in the development of C_4 and C_3 photosynthesizing maize tissues. Genetics 159:787-797.

Dengler NG, Dengler RE, Grenville DJ. 1990. Comparison of photosynthetic carbon reduction (Kranz) cells having different ontogenetic origins in the C_4 NADP-malic enzyme grass *Arundinella hirta*. Can. J. Bot. 68:1222-1232.

Dengler NG, Dengler RE, Hattersley PW. 1985. Differing ontogenetic origins of PCR (Kranz) sheaths in leaf blades of C_4 grasses (Poaceae). Am. J. Bot. 72:284-302.

Dengler NG, Donnelly PM, Dengler RE. 1996. Differentiation of bundle sheath, mesophyll and distinctive cells in the C_4 grass, *Arundinella hirta* (Poaceae). Am. J. Bot. 83:1391-1405.

Dengler NG, Nelson T. 1999. Leaf structure and development in C_4 plants. In: Sage RF, Monson RK, editors. C_4 plant biology. San Diego, Calif. (USA): Academic Press. p 133-172.

Dengler NG, Woodvine MA, Donnelley PM, Dengler RE. 1997. Formation of the vascular pattern in developing leaves of the C_4 grass *Arundinella hirta*. Int. J. Plant Sci. 158:1-12.

Dengler RE, Dengler NG. 1990. Leaf vascular architecture in the atypical NADP-malic enzyme grass *Arundinella hirta*. Can. J. Bot. 68:1208-1221.

Deyholos MK, Cordner G, Beebe D, Sieburth LE. 2000. The *SCARFACE* gene is required for cotyledon and leaf vein patterning. Development 127:3205-3213.

Dimitrov P, Zucker SW. 2006. A constant production hypothesis guides leaf venation patterning. Proc. Natl. Acad. Sci. USA 103:9363-9368.

Edwards GE, Franceschi VR, Voznesenskaya EV. 2004. Single-cell C_4 photosynthesis versus the dual-cell (Kranz) paradigm. Annu. Rev. Plant Biol. 55:173-196.

Edwards GE, Furbank RT, Hatch MD, Osmond CB. 2001. What does it take to be C_4? Lessons from the evolution of C_4 photosynthesis. Plant Physiol. 125:46-49.

Evert RF, Eschrich W, Heyser W. 1977. Distribution and structure of the plasmodesmata in mesophyll and bundle-sheath cells of *Zea mays* L. Planta 136:77-89.

Evert RF, Eschrich W, Heyser W. 1978. Leaf structure in relation to solute transport in phloem loading in *Zea mays* L. Planta 138:279-294.

Hammer GL, Sinclair TR, Chapman SC, van Oosterom E. 2004. On systems thinking, systems biology, and the *in silico* plant. Plant Physiol. 134: 909-911.

Häusler RE, Hirsch HJ, Kreuzaler F, Peterhänsel C. 2002. Overexpression of C_4-cycle enzymes in transgenic C_3 plants: a biotechnological approach to improve C_3-photosynthesis. J. Exp. Bot. 53:591-607.

Ideker T, Galitski T, Hood L. 2001a. A new approach to decoding life: systems biology. Annu. Rev. Genomics Hum. Genet. 2:343-372.

Ideker T, Thorsson V, Ranish JA, Christmas R, Buhler J, Eng JK, Bumgarner R, Goodlett DR, Aebersold R, Hood L. 2001b. Integrated genomic and proteomic analyses of a systematically perturbed metabolic network. Science 292:929-934.

Kellogg EA. 1999. Phylogenetic aspects of the evolution of C_4 photosynthesis. In: Sage RF, Monson RK, editors. C_4 plant biology. San Diego, Calif. (USA): Academic Press. p 411-444.

Kramer EM. 2004. PIN and AUX/LAX proteins: their role in auxin accumulation. Trends Plant Sci. 9:578-582.

Laetsch WM. 1974. The C_4 syndrome: a structural analysis. Annu. Rev. Plant Physiol. 25:27-52.

Leegood RC. 2002. C_4 photosynthesis: principles of CO_2 concentration and prospects for its introduction into C_3 plants. J. Exp. Bot. 53:581-590.

Ma L, Chen C, Liu X, Jiao Y, Su N, Li L, Wang X, Cao M, Sun N, Zhang X, Bao J, Li J, Pedersen S, Bolund L, Zhao H, Yuan L, Wong GK, Wang J, Deng XW, Wang J. 2005. A microarray analysis of the rice transcriptome and its comparison to *Arabidopsis*. Genome Res. 15:1274-1283.

Majeran W, Cai Y, Sun Q, van Wijk KJ. 2005. Functional differentiation of bundle sheath and mesophyll maize chloroplasts determined by comparative proteomics. Plant Cell 17:3111-3140.

Matsuoka M, Furbank RT, Fukayama H, Miyao M 2001. Molecular engineering of C_4 photosynthesis. Annu. Rev. Plant Physiol. Plant Mol. Biol. 52:297-314.

Miyao M. 2003. Molecular evolution and genetic engineering of C_4 photosynthetic enzymes. J. Exp. Bot. 54:179-189.

Nelson T, Dengler NG. 1997. Leaf vascular pattern formation. Plant Cell 9:1121-1135.

Raikhel NV, Coruzzi GM. 2003. Achieving the *in silico* plant: systems biology and the future of plant biological research. Plant Physiol. 132:404-409.

Robinson-Beers K, Evert RF. 1991. Ultrastructure of and plasmodesmatal frequency in mature leaves of sugarcane. Planta 184:291-306.

Rolland-Lagan A-G, Prusinkiewicz P. 2005. Reviewing models of auxin canalization in the context of leaf vein pattern formation in *Arabidopsis*. Plant J. 44:854-865.

Rossini L, Cribb L, Martin DJ, Langdale JA. 2001. The maize *golden2* gene defines a novel class of transcriptional regulators in plants. Plant Cell 13:1231-1244.

Runions A, Fuhrer M, Lane B, Federl P, Rolland-Lagan A-G, Prusinkiewicz P. 2005. Modeling and visualization of leaf venation patterns. ACM Trans. Graph. 24:702-711.

Sachs T. 1981. The control of patterned differentiation of vascular tissues. Adv. Bot. Res. 9:152-255.

Sachs T. 1991. Cell polarity and tissue patterning in plants. Development 91 Suppl.:83-93.

Sage R. 2004. The evolution of C_4 photosynthesis. New Phytol. 161:341-370.

Scarpella E, Marcos D, Friml J, Berleth T. 2006. Control of leaf vascular patterning by polar auxin transport. Genes Dev. 20:1015-1027.

Schad M, Mungur R, Fiehn O, Kehr J. 2005. Metabolic profiling of laser microdissected vascular bundles of *Arabidopsis thaliana*. Plant Methods 1:2. (Online journal: volume 1, article 2.)

Smith LG, Hake S, Sylvester AW. 1996. The tangled-1 mutation alters cell division orientations throughout maize leaf development without altering leaf shape. Development 122:481-489.

Taylor WC. 2000. C_4 rice: What are the lessons from developmental and molecular studies? In: Sheehy JE, Mitchell PL, Hardy B, editors. Redesigning rice photosynthesis to increase yield. Amsterdam (Netherlands): Elsevier. p 87-96.

Tsiantis M, Brown MIN, Skibinski G, Langdale JA. 1999. Disruption of auxin transport is associated with aberrant leaf development in maize. Plant Physiol. 121:1163-1168.

Turner S, Sieburth L. 2002. Vascular patterning. In: Somerville CR, Meyerowitz EM, editors. The *Arabidopsis* book. Rockville, Md. (USA): American Society of Plant Biologists. www.aspb.org/publications/arabidopsis.

Ueno O, Kawano Y, Wakayama M, Takeda T. 2006. Leaf vascular systems in C_3 and C_4 grasses: a two-dimensional analysis. Ann. Bot. (Lond.) 97:611-621.

Notes

Authors' addresses: T. Nelson, S.L. Tausta, N. Gandotra, T. Liu, T. Ceserani, M. Chen, Y. Jiao, L. Ma, and X.-W. Deng, Department of Molecular, Cellular, and Developmental Biology; N. Sun, M. Holfold, N. Li, and H. Zhao, Center for Statistics and Bioinformatics, Yale University, P.O. Box 208104, New Haven CT 06520-8104, email: timothy.nelson@yale.edu.

Toward C_4 rice: learning from the acclimation of photosynthesis in the C_3 leaf

E.H. Murchie and P. Horton

Previous studies of photosynthesis made in the field under irrigated conditions have shown that there are several potential sites of limitation at the leaf level, including light saturation of photosynthesis in upper leaves, mid-morning depression of photosynthetic capacity, and photoinhibition of photosynthesis (reduction in quantum yield). The response of rice photosynthesis in the field to a high CO_2 concentration (measured in a leaf chamber) indicates the central importance of Rubisco chemistry and photorespiration in limiting the assimilatory capacity and potential biomass production of rice under tropical conditions in the field. In this paper, we discuss the relationship between C_3 photosynthesis and rice leaf morphology and how this research may be incorporated into a program to produce a rice plant with C_4 features.

Rice crops typically demand high inputs of N fertilizer to achieve high grain yield and this is reflected in the Rubisco concentrations observed in field-grown rice leaves. We have described an inconsistent relationship in the field between Rubisco content and *in situ* rates of rice leaf photosynthesis in some genotypes and postulated the role of Rubisco in forming part of an N store for later remobilization to the grain. Rice leaf morphology (in this case thickness and area) is a feature of rice crops important for canopy efficiency and integrity, photosynthetic rate, and N content. However, the relationship among leaf thickness, N content, and photosynthesis is not clear. We have adopted a number of lines of research that explore the factors responsible for leaf thickness determination in rice. First, using differences in morphology induced by acclimation to irradiance, we suggest that this results from a signal provided by mature leaves. We postulate that these changes are a "fine-tuning" of cellular morphology and that the establishment of Kranz anatomy in rice may not require such signals. Second, we are exploiting genotype differences and rice mutant collections. The exploitation of new mutant resources for rice will be essential if the goal of C_4 rice is to be achieved. Although high-throughput screening of rice mutant populations is still largely impracticable, this may not apply to IRRI's IR64 deletion mutant collection.

Keywords: acclimation, leaf morphology, leaf photosynthesis, leaf thickness, P_{max}

An increase in the rate of biomass production of rice crop canopies in tropical regions would form the basis for new and unprecedented improvements in rice grain yield potential (Mann 1999, Sheehy 2000). With an ever-increasing global population, this would be a critical step toward the future prevention of suffering and environmental degradation of resource-poor communities in rice-dependent developing nations. Since the Green Revolution, improvements in rice yield potential have come from traits such as harvest index (HI) and total biomass production (Peng et al 2000). However, HI and leaf area index (LAI) are now believed to be optimized and any further increases in yield potential will have to involve an increase in total crop biomass only. To achieve this, the rate of canopy photosynthesis per unit radiation intercepted will need to increase (Sheehy 2000). This characteristic greatly influences the radiation conversion factor (RCF) or radiation-use efficiency described by Mitchell et al (1998). Assuming that canopy architecture is optimal for light penetration and that LAI is sufficiently high, the improvement in carbon assimilation must arise from the leaf level, that is, photosynthetic rate per unit area of leaf. The rate of leaf photosynthesis has not been used successfully in plant breeding in the past, partly because of the poor correlation between leaf photosynthesis and yield and partly because of the focus on the improvement of other important agronomic characteristics.

It is in fact highly unlikely that the photosynthetic rate of a small area of a single leaf at a given point in time will enable the prediction of grain yield. Too many biological processes are occurring at different spatial and temporal scales to make this possible. A suggested analogy would be trying to predict leaf photosynthesis from the activity of a single mesophyll chloroplast or granal stack: leaf ultrastructure is simply too heterogeneous. We must understand how leaf photosynthetic measurements are integrated into canopy photosynthetic rates. This will arise from knowledge of how the photosynthetic operation of all leaves at each life cycle phase is integrated into the formation of a canopy that supports high grain yield. Recent literature shows that interest is growing in the role of photosynthesis in crop yield improvement (Long et al 2006, Horton 2000, Mann 1999, Edwards 1999). Such interest has a firm basis in experimentation. For example, it is clear that increases in yield associated with growth in elevated CO_2 (e.g., Ziska et al 1997) result from an increased rate of photosynthesis, providing direct evidence that improvement of photosynthetic efficiency can give rise to improved yield potential.

A detailed survey of available literature suggested that the RCF of rice is one of the lowest for C_3 crops (Mitchell et al 1998). The reasons for this may include the high LAI of rice and the high rates of photorespiration caused by oxygenation reaction of Rubisco under warm tropical conditions. Suppression of photorespiration in rice crops could theoretically occur by engineering of Rubisco protein or by introducing the C_4 pathway into rice leaves and there has perhaps been a greater focus placed on the latter. As we shall discuss in this paper, these are not the only options available: there are other possibilities for improving the RCF of rice crops, which require further research. However, the C_4 option remains the most immediately attractive because the substantially higher photosynthetic capacity, RCF, and yield potential can already be seen in existing C_4 crop species such as maize.

The characteristic of interest to the improvement of photosynthesis in the context of yield potential is the light-saturated rate of photosynthesis under ambient CO_2 conditions (P_{max}). In fact, surprisingly few studies of photosynthesis attempt to quantify the expression of P_{max} in tropical field conditions. One of the most important features surrounding P_{max} is its relationship with amounts of light experienced in field conditions. Irradiance (photosynthetically active radiation—PAR) can reach over 2,000 $\mu mol\ m^{-2}\ s^{-1}$ in the field (measured on a horizontal surface) and this is sufficient to completely saturate photosynthesis in rice leaves. In fact, it can be estimated that around 40% of the photons absorbed by the leaf in full tropical sunlight are in excess of carbon assimilatory requirements (Murchie et al 1999, 2002a). This was amply demonstrated by diurnal measurements of chlorophyll fluorescence parameters (ϕPSII, the operational quantum efficiency of photosystem II) and the amount of photochemical quenching.

The "expression" of P_{max} in the field is dependent on a plethora of regulatory factors. One way of altering P_{max} is to grow plants at different irradiances (acclimation). We showed in a field experiment that the ability to alter P_{max} with increasing growth irradiance saturates at relatively low irradiance (Murchie et al 2002a). Above this point, acclimation occurs to enhance photoprotection, not photosynthesis. This demonstrates clearly that acclimation to full tropical conditions has not been attained. If the maximum Rubisco capacity and activity of the leaf has been reached, this saturation point may be unavoidable. If not, there may be unknown developmental processes in place that are curtailing the full expression of P_{max} in rice (Horton and Murchie 2000).

With this in mind, we have conducted several such studies that suggested that a number of limitations are imposed on the rate of C_3 photosynthesis in the field in tropical conditions, some unexpected, some not. These limitations have been discussed previously (Horton 2000, Horton and Murchie 2000) but here we will summarize and update this work and discuss its relevance for the proposal to improve rice photosynthesis by introducing a C_4 mechanism. We will describe work undertaken to analyze the role of leaf morphology in acclimation of C_3 rice photosynthesis and how this may complement the C_4 goal.

Losses in photosynthesis in the tropical environment

Carbon assimilation rate in a C_3 leaf is determined by a combination of extrinsic and intrinsic factors: temperature, irradiance, supply of water and CO_2, the capacity of the leaf (amounts of photosynthetic components per unit leaf area), and the efficient export of assimilate. In fact, high light-saturated rates of photosynthesis will ultimately be determined genetically, and be dependent upon sufficient content of photosynthetic proteins per unit leaf area. Under otherwise optimal conditions, photosynthetic rate is frequently observed to be below the potential maximum, and this is caused by events such as stomatal closure, inhibition of Rubisco activity, and lowered quantum yield. These processes may be prevalent under field conditions where fluctuations in environmental variables such as irradiance can be large. We have previously referred to this as a "loss" of photosynthesis since it does not necessarily represent "damage" to photosynthetic components but rather a set of highly regulated responses to unfavorable abiotic conditions whose benefits remain somewhat speculative.

We have carried out several field experiments to measure photosynthetic rates in tropical rice and identified areas in which photosynthetic losses may occur (described below). If measurements are made in such a way as to accommodate factors such as leaf age and canopy position, it is clear that the relationship between irradiance and CO_2 assimilation rate shows a high frequency of "underperformance" of leaf photosynthesis in the field. Relationships of this type are more often described for natural populations of plants under severely limiting environmental conditions rather than a crop under optimal conditions for growth and high yield (Cheesman et al 1991).

1. Leaf temperature and photorespiration

Experiments were carried out at the IRRI farm during the dry seasons of 1997-2001. Daytime air temperature typically varied between 28 and 35 °C and maximum irradiance reached a maximum of 2,000 μmol m^{-2} s^{-1}. At these temperatures, significant photosynthetic losses arise from increased photorespiratory flux, and CO_2 response measured in the field predicts a 40% loss due to photorespiration (Murchie et al 1999, Leegood and Edwards 1996). It is clear that Rubisco oxygenation is a significant hindrance to attaining high CO_2 assimilation rates under these conditions, affirming the value of seeking a C_4 option in rice crops.

2. Diurnal depressions in photosynthesis

Diurnal variation in the capacity of photosynthesis is relatively common in plants. Specifically, the decline in photosynthesis, usually during the mid-part of the day, has been recorded, with individual stimuli being suggested as the cause, such as high light, vapor pressure deficit, and carbohydrate accumulation. However, in many cases, mechanisms may be dependent on multiple factors. Significantly, there is evidence for genetic variation in the extent of midday depression in rice, and it was even associated with changes in grain yield (Black et al 1995).

In rice grown at IRRI, depression was observed not at midday but mid-morning (Murchie et al 1999). We suggested that this was related to the upright posture of rice leaves: as a result, solar movements cause exposure to high irradiance of one side of the leaf in the morning and the other in the afternoon, meaning that the highest irradiance received on the surface of the leaf in fact occurs mid-morning. It is currently unclear what the mechanism of photosynthetic depression is in rice. Assimilation versus C_i (internal leaf CO_2 concentration) curves made in the morning and compared with those made in the afternoon indicate a slightly reduced carboxylation capacity (initial slope of the curve) and reduced capacity for regeneration of ribulose bisphosphate (CO_2-saturated rate) (Fig. 1). A lowered stomatal conductance was observed and it is possible that the high leaf temperatures and irradiance resulted in stomatal closure, although there was no measured high vapor pressure deficit. Feedback inhibition via the accumulation of photosynthate is possible though Murchie et al (2002a) showed that carbohydrates had not accumulated to high concentrations at this time of morn-

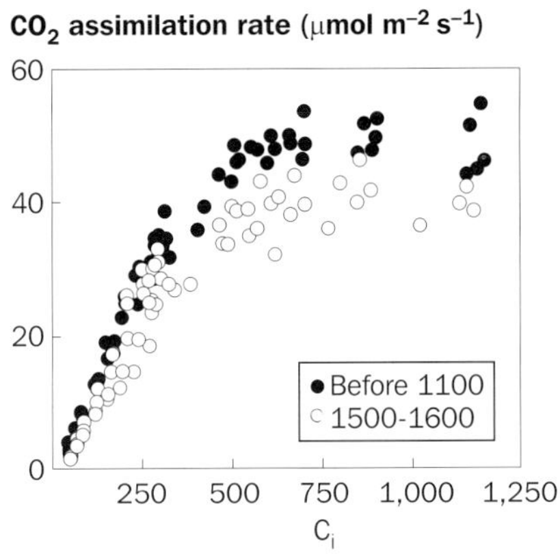

Fig. 1. Characterizing midday depression in rice leaves in the field. CO_2 assimilation versus internal CO_2 concentration (C_i, μL L^{-1}) of IR72 rice leaves. Measurements were made in the field using a Licor 6400 IRGA (Nebraska) at an air temperature of 31 °C and PAR of 1,700 μmol m^{-2} s^{-1}. Two times of day were used: prior to midday depression (up to 1100) and after midday depression (1500-1600).

ing. It is possible that feedback inhibition occurred as a direct result of sustained high irradiance. The irradiances reached in the field are substantially higher than that needed by the rice leaf to saturate photosynthesis (Murchie et al 1999, 2002a). There are a number of ways in which an imbalance may be attained between the amount of light energy absorbed and the capacity for assimilation and transduction (Horton and Murchie 2000). For example, the rates of electron transport caused by assimilatory and photorespiratory flux are likely to be extremely high. Over-reduction of the electron transport system can result in the induction of a number of stress-related signaling pathways (e.g., Pfannschmidt 2003) and this may lead to the inhibition of photosynthetic activity and/or the closure of stomata. It may be of great importance to establish the mechanism of midday depression and whether it is widespread in rice, since it can result in an estimated loss of 30% of daily leaf photosynthesis (Murchie et al 1999).

3. Photoinhibition

The thylakoid membrane possesses a number of processes that serve to balance the excitation level of the chlorophyll pigments with capacity for assimilation. One of these is nonphotochemical quenching (NPQ). At sufficiently high irradiances, this regulatory mechanism operates to quench excited chlorophyll, dissipating excess excitation energy and preventing an over-reduction of the electron transport chain (Horton et al 1994). Most of this consists of high-energy state quenching that is capable of relaxing to a nonquenched form within a time period of minutes. A slowly-relaxing form can sometimes be observed that can take hours or days to relax and is often termed "photoinhibition." When a leaf is in a "quenched" or dissipative state, the result is a lowered quantum yield of photosynthetic CO_2 assimilation and a lowered value of F_v/F_m (Murchie et al 1999). Photoinhibition may also result from damage to photosynthetic reaction centers. By definition, therefore, a reduced F_v/F_m will reduce leaf CO_2 assimilation only in low light: it will not cause a reduction in the light-saturated rate of photosynthesis. Under the high irradiance in conditions that test yield potential, it is therefore considered that F_v/F_m does not limit canopy photosynthesis (Horton 2000). However, there has been recent interest in situations where light is heterogeneous over space and time. Zhu et al (2004) modeled leaf photosynthesis within a crop canopy. Models incorporated the established principle that a leaf with a lowered F_v/F_m will, on transfer from high to low irradiance, possess a lowered assimilation rate. This study predicted a 17% reduction in canopy photosynthesis (C_3) at 30 °C. It may be important, therefore, that significant photoinhibition was observed in "optimal" conditions for rice growth in the field at IRRI (Murchie et al 1999). This particular type of photoinhibition was acute during the middle hours of the day but was totally removed by the following morning. If the total daily radiation load was increased by forcing rice leaves into a horizontal position, then the amount of photoinhibition increased. It is predicted that under extremely unfavorable conditions such as drought, photoinhibition may present a greater limitation to photosynthesis.

It is possible to move now to an empirical testing of these ideas, using rice plants that possess altered dynamics of NPQ. It is of great interest that genotype variation in photoinhibition was also observed (Murchie et al 1999). However, rice plants that have been transformed to overexpress or underexpress genes such as *PsbS* that are involved in the regulation of NPQ will be of great interest.

Photosynthesis, Rubisco, and leaf N

Can leaf Rubisco content predict photosynthetic rate in the field?

The literature shows that the predominant factor that limits photosynthetic rate in rice leaves under controlled growth conditions in which irradiance is close to saturation, and in which CO_2 concentrations are at ambient levels, is the content (and activity) of Rubisco (Makino et al 1985, von Caemmerer and Farquhar 1981). In the field, this will apply to localities and seasons where light may be consistently high. However, in many localities, temporary cloud cover will cause photosynthesis in the upper canopy to fluctuate between light saturation and light limitation. Additionally, a large proportion of canopy photosynthesis arises from light-limited leaves in the lower canopy. So, although Rubisco content is imperative when considering leaf photosynthetic capacity, one would also predict from models of C_3 photosynthesis that Rubisco amount may not limit photosynthesis in the field to the same extent as seen in controlled-environment conditions. Such ideas are important in the context of reducing N inputs into crop systems. We present two lines of evidence from the field suggesting that, in the field, P_{max} may become "uncoupled" from Rubisco amount.

1. Acclimation of photosynthesis
   The relationship between leaf Rubisco content and incident irradiance was tested in the field by imposing partial shading over rice plants (Murchie et al 2002a). Interestingly, Rubisco contents were reduced and P_{max} did not decline by the same extent, indicating an excess capacity of Rubisco within rice leaves in the field. There was also a decline in the capacity for photoprotective processes, providing further support for the existence of excess excitation energy under full sunlight.
2. Rubisco loss during leaf aging
   The photosynthetic function of Rubisco could be compromised by the onset of leaf senescence during the reproductive phase of the life cycle. The decline in P_{max} and Rubisco content of flag leaves during the grain-filling phase was measured in IR72 and several NPT lines that differed in their rates and patterns of grain filling (Murchie et al 2002b). IR72 showed a good correlation between P_{max} and Rubisco content. In the NPT, Rubisco content declined in a manner dependent on leaf age with no alteration in P_{max}, indicating an excessive capacity for Rubisco in these lines. This may be a significant observation: the NPT was "designed" to possess a number of morphological features that support high grain yield, that is, fewer, thicker leaves and tillers with a higher leaf N content and often a higher Rubisco content per unit leaf area. The reduction in Rubisco in all lines generally coincided with a period just after the rapid phase of grain filling and represents a remobilization of N to the developing grain.

In a second, related experiment, the relationship among photosynthetic capacity and leaf N and leaf protein content was tested in IR72 and an NPT line from the point of full leaf expansion through to senescence (Fig. 2). Maximum Rubisco and leaf N contents were higher in the NPT than in IR72. A simultaneous decline in Rubisco and P_{max} was observed for the two genotypes, consistent with a breakdown of leaf protein and remobilization. However, when Rubisco is plotted against P_{max}, it is clear that the two genotypes differed in their efficiency of use of Rubisco. For a given Rubisco content, the P_{max} of NPT was lower (Fig. 2). The reasons for this remain unclear; suggested limitations to photosynthesis could include increased stomatal or mesophyll resistance or feedback limitation by photosynthate. Perhaps most importantly, the extent of this variation in "Rubisco-use efficiency" among rice genotypes is actually unknown. The data from Murchie et al (1999, 2002b) suggest that this may be a common feature of NPT lines from this period. Zhang et al (2003) in similar work suggest that during rice leaf senescence Rubisco and chlorophyll amounts decline before total leaf N declines.

From this, we suggest that the relationship between Rubisco amounts and *in situ* photosynthesis can be uncoupled in field conditions, and that the relationship between the two is dependent on plant type. This may not be particularly startling, but it becomes significant when we consider that Rubisco is intricately involved in the N-use efficiency of the plant: it is a significant N sink within the rice plant, composing up to

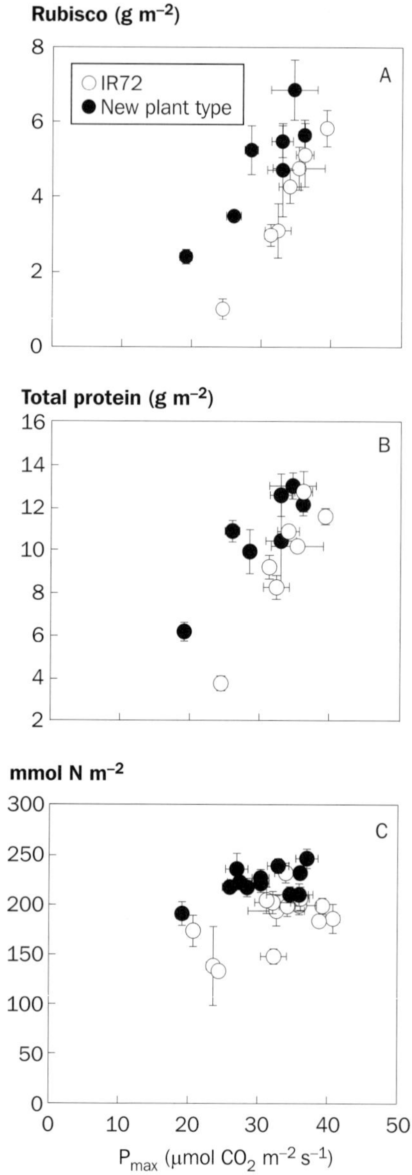

Fig. 2. The relationships of Rubisco content per unit leaf area (A), total protein per unit leaf area (B), and total N per unit leaf area (C) with light-saturated photosynthetic rate (P_{max}) measured on flag leaves of plants grown in the field at IRRI at 350 $\mu L\ L^{-1}$ CO_2. The variation in all characteristics was made by sampling leaves during the process of leaf aging: this was from the point of full leaf expansion to 40 days old. Means and standard error are shown. Results of correlation: A, IR72, r = 0.99, n = 7, $P<0.001$; A, NPT, r = 0.89, n = 7, $P<0.01$; B, IR72, r = 0.93, n = 7, $P<0.01$; B, NPT, r = 0.89, n = 7, $P<0.01$; C, IR72, r = 0.57, n = 14, $P<0.05$; C, NPT, r = 0.55, n = 12, $P>0.05$.

35% of total leaf protein. It is important to consider the role of the leaf photosynthetic apparatus in the N economy of the rice plant.

High yields of rice since the Green Revolution have been dependent upon high rates of application of N fertilizer. It has been argued that a reduction in the amount of Rubisco (while maintaining P_{max}) would improve N-use efficiency. However, leaf N is often correlated with photosynthesis (Peng et al 1995), and it has been suggested that an increase in photosynthesis can be achieved only by applying more fertilizer. Both of these views ignore the key role of Rubisco as a reservoir of remobilizable N that is needed to sustain grain yield. Rice canopies need to accumulate substantial amounts of N in order to support high rates of canopy photosynthesis, but also grain protein synthesis following senescence and protein remobilization. It has been argued that the provision of N for grain production is a principal factor in determining canopy architecture (Sinclair and Sheehy 1999). Approximately half of grain N must be derived from leaf N; it can be calculated that an LAI of 7 is needed for the purpose of storing N before transport to the grain. A significant amount of N in the lower leaves of rice canopies will exist in the form of photosynthetic proteins. To prevent senescence and to reduce the costs of respiratory maintenance, good light distribution is essential (Sinclair and Sheehy 1999). Canopies of many modern varieties of rice (such as NPT) possess upright leaves that permit good penetration through to lower layers. However, high N contents can be correlated with higher maintenance respiration and the cost of this to the plant during periods of low irradiance is unknown.

One of the outcomes of introducing C_4 features into rice is to reduce the amounts of Rubisco. C_4 plants tend to have a lower N per unit leaf area, so the effects on plant N economy need to be carefully evaluated in order to maintain grain protein content and quality. This may be compensated by a number of strategies such as altering leaf area, improving efficiency of remobilization, and considering alternative N storage locations such as the stem tissue.

Toward C_4 rice: the relationship between photosynthesis and leaf morphology

The previous section has discussed the relationships between leaf photosynthesis and N content and how grain yield is dependent upon leaves as a source of both carbon and N. The development of recent plant types seems to have reflected this dual role. For example, the NPT mentioned in this study in comparison with IR72 have fewer, larger, and thicker upright leaves that possess a higher Rubisco and N content and permit a greater amount of light through to lower layers of the canopy. The unusual feature is that these plants had a lower P_{max} (Fig. 2). This leads to interesting questions about the nature of the relationship among leaf thickness, N, and photosynthetic capacity. Clearly, it is in the interests of optimizing biomass production rates for these features to act in a mutually beneficial manner. Leaf thickness does not always correlate with photosynthetic capacity: it is a relatively crude measurement that does not take into account subtle features of leaf ultrastructure, such as mesophyll cell area, cell size, and stomatal density. However, these features are often associated because a high leaf thickness is needed to supply the high N and protein content per unit leaf area

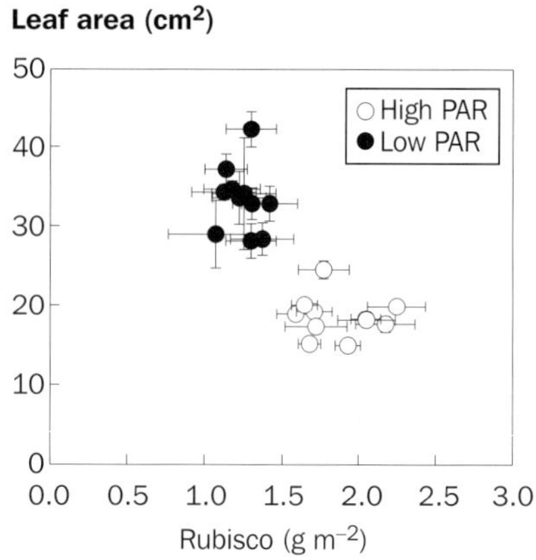

Fig. 3. The relationship between leaf area and Rubisco content per unit leaf area for ten historical rice varieties developed at IRRI. Plants were grown at Sheffield University at a high PAR (48.8 mol m^{-2} day^{-1} and a maximum of 1,500 μmol m^{-2} s^{-1}) and low PAR (10.8 mol m^{-2} day^{-1} and a maximum of 300 μmol m^{-2} s^{-1}). Leaf used was leaf nine. Means and standard error of means are shown.

needed for high photosynthetic rates. We suggest that one of the determining factors for thickness in rice leaves may be the capacity for N storage.

Another factor to consider is leaf area: a survey of historical IRRI cultivars (Peng et al 2000) that have been released since 1960 demonstrated a large variation in area per leaf, but this was not correlated with P_{max} or Rubisco per unit leaf area (Fig. 3). We would assume from this that photosynthesis per leaf is not constant but dependent on features of whole-plant development that regulate the formation of a particular canopy morphology. An inverse correlation between leaf thickness and leaf area can sometimes be observed.

It is therefore important to determine the factors, both genetic and otherwise, that determine leaf thickness and area in rice plants. Although progress in our understanding of the molecular processes of leaf development is rapidly improving, the consideration of leaf thickness as an important agronomic trait seems under-represented. As a result, we have started work that explores three aspects of leaf morphology and photosynthesis in rice. This has parallels that may be of use to the proposal to introduce C_4 characteristics into rice leaves.

Determining leaf thickness in rice leaves: the role of mature leaves

We have used acclimation of rice leaf morphology to irradiance as a model for leaf thickness development studies: when rice plants are grown under low irradiance (LL), leaves are thinner and larger in area than those grown under high irradiance (HL) (Murchie et al 2005). This is a typical example of sun/shade leaf acclimation.

The thicker leaves also had higher Rubisco, protein, and P_{max}. Examination of leaf sections revealed that this was not due to an alteration in cell number or differentiation of a palisade and spongy mesophyll as in dicotyledons but rather a change in cell size only.

Transfer of plants between HL and LL conditions also resulted in an alteration in the morphology of leaves developed subsequently. However, it was necessary to transfer plants prior to any significant division or expansion within the leaf primordium (Fig. 4). The developing leaf does not respond to changed conditions unless the change occurs before significant cell division and expansion of the leaf in question. Therefore, leaf thickness is established very early in leaf development, consistent with the fact that all rice leaf cell division and cellular elongation/expansion takes place in a zone just a few cm long above the meristem tissue (Hoshikawa 1989). This zone of growth is positioned typically at the base of the rice plant and relatively isolated from external stimuli. We questioned whether the zone of growth was able to receive irradiance signals from the external environment in a quantitative manner: to test this, we covered the stems and roots of rice with a light-proof black plastic material during growth. The seedlings were covered at a point in development that preceded the division and expansion of the leaf-five primordium. The amount of light available to the leaf sheath and base of the plant decreased by more than 95%; however, leaves that developed in this region were still able to acclimate to the light environment. This suggests that the zone of division and elongation does not receive and respond to light signals in a quantitative manner (Fig. 4).

The role of mature leaves in the determination of acclimation of leaf morphology is well established. Lake et al (2002) showed that stomatal index and stomatal densities were affected by the light and CO_2 concentration supplied to mature leaves of *Arabidopsis thaliana*. They postulated that the developing leaf used cues received from mature leaves to provide information about the environment in which it was growing. These long-distance signals are still unknown, but the possibility of hormonal signals and photosynthate has been discussed (Lake et al 2002). There is also important evidence that the formation of sun/shade leaf morphology is dependent on similar cues: Yano and Terashima (2001, 2004) used *Chenopodium album* to show that cellular division within the mesophyll to form a sun leaf-type ultrastructure was related to light supplied to mature leaves only.

These features of acclimation were tested in rice plants within the laboratory (Fig. 4). It was considered likely that photosynthate supply would be necessary for the formation of a thick mesophyll layer within rice leaves. Table 1 shows data from an experiment where plants were grown under low light (which would normally induce formation of shade-type thin leaf) with ambient and elevated levels of CO_2. Elevated CO_2 stimulated higher rates of gas exchange and higher growth rate, indicating higher availability of photosynthate for leaf growth. However, leaf thickness was unaltered and we conclude that there is either a signal in addition to carbohydrate supply required or a much higher level of photosynthesis in mature leaves is necessary.

Another feature of rice leaf morphology examined was stomatal density. Counts were made of rice stomata from plants growing in conditions of high and low CO_2

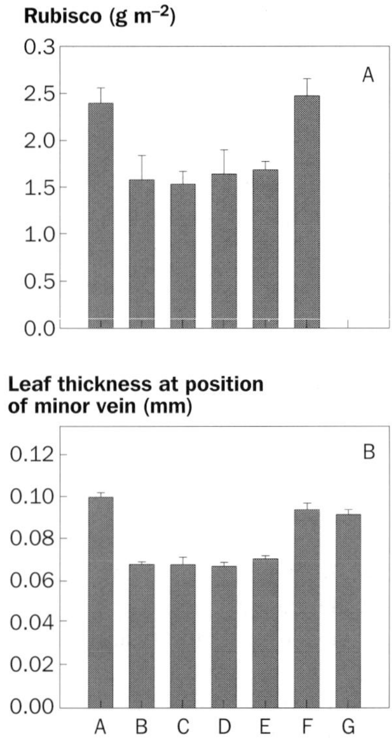

Fig. 4. Rubisco content (A) and leaf thickness (B) in leaf number five of plants transferred from LL (low light) to HL (high light) at points in time that incrementally precede the point of full leaf extension. Thickness was measured approximately halfway between the mid-rib and leaf margin. Distance was measured across a minor vein (not adjacent to a major vein). Plants were maintained at HL (A) and LL (B) throughout or grown under LL and then transferred to HL at the point of full leaf extension (C), 4 days before the point of full leaf extension (D), 8 days before the point of full leaf extension (just before emergence from the leaf sheath (E), and 14 days before full leaf extension (at this point, leaf three is approaching full leaf extension) (F). The final point (F) occurred before the formation of the zone of rapid division and elongation of leaf five. (G) Leaf thickness in plants grown with the stem/sheath enclosed in black plastic light-proof material. Means and standard error of means are shown, n = 8.

Table 1. The influence of elevated CO_2 on photosynthate production in leaf three and morphological features of leaf five. IR72 plants were grown in a controlled-environment chamber at low light (8.6 mol m^{-2} day^{-1}, 200 μmol m^{-2} s^{-1}), 28/20 °C. CO_2 assimilation rate was measured by infrared gas analysis at the same irradiance (200 μmol m^{-2} s^{-1}). Thickness of leaf five was not affected by growth in elevated CO_2. Stomata per unit leaf area (abaxial surface) of leaf five were measured after growth in high light (34.6 mol m^{-2} day^{-1}, 800 μmol m^{-2} s^{-1}). Means and standard error of means are shown, n = 8.

Item	Growth [CO_2] 400 μL L^{-1}	Growth [CO_2] 1,000 μL L^{-1}
Assimilation rate of leaf three at 200 μmol m^{-2} s^{-1}	11.88 ± 0.54	17.47 ± 0.85
Thickness of leaf five (μm)	0.70 ± 0.01	0.73 ± 0.01
Stomata mm^{-2} of leaf five	203 ± 12	163 ± 6

and irradiance. Under high irradiance, growth of rice plants in high CO_2 resulted in an increase in leaf area and also a reduction in the number of stomata per unit leaf area (Table 1). It would be critical to know whether this results from a long-distance signal.

Clues from acclimation studies

The last section described evidence of the operation of long-distance signals within the rice plant. Characteristics affected include mesophyll cell size and alterations in the frequency of certain cell types such as stomatal guard cells and pavement cells. To move from C_3 morphology to C_4 morphology (Kranz anatomy) would require a radical alteration in the arrangement of mesophyll cells within the rice leaf, that is, mesophyll cells would be reduced to a single layer surrounding the bundle sheath. It would seem unlikely at first that such a radical alteration to leaf cellular arrangement requires these (relatively) subtle long-distance signaling effects.

It is worth noting, however, that the work by Yano and Terashima (2001, 2004) indicated clearly that periclinal cell division within the mesophyll cell layer in *C. album* is indeed influenced by long-distance signals. This would imply that the signaling processes involved in the acclimation to irradiance can potentially interact with the cell types of interest in the creation of a potential C_4 plant. However, in initial studies, we concluded that such cell divisions in rice during acclimation were minimal (Murchie et al 2005), which agrees with data by Makino et al (1997). A survey of rice genotypes that provides an analysis of mesophyll cell patterning in terms of number of cell layers and cell sizes would be of great use. There is currently great interest in establishing the genetic regulation of leaf development, and it is hoped that an understanding of the factors affecting mesophyll cell layer development within rice will soon be achieved.

Another feature of the rice leaf is stomatal development. The processes that give rise to stomatal patterning in rice differ in some ways from those of *Arabidopsis*: stomata in rice are formed in rows or files (Hoshikawa 1989) in a longitudinal direction along the rice leaf. An important point to note is that files normally consist of single or double rows of stomatal cells, and they always form alongside veins (Hoshikawa 1989). We have observed extremely high frequencies of stomata covering the mid-rib of rice leaves (Murchie 2005, unpublished results). There has been great interest and progress in elucidating the mechanism for stomatal and epidermal cell development processes and the genes involved (e.g., Lake et al 2002, Bird and Gray 2003). It may therefore be important to consider the mechanisms regulating the formation and frequency of stomatal files alongside those concerned with veins and vein spacing. C_4 plants typically have a much smaller distance between adjacent veins, with a single mesophyll cell layer surrounding the bundle sheath. C_4 rice would probably need to show a similar arrangement to reduce path length and allow the efficient concentration of Rubisco into adjacent bundle sheath cells. Mesophyll cell mutants are known in other plant species. Several mutants of stomatal cell patterning are available for *Arabidopsis* but not for rice. It is possible that the altered vein spacing of Kranz anatomy also results in an altered stomatal cell patterning: the stomatal density of rice is in fact much higher than that of wheat or maize. It has been suggested that this high density compensates for the smaller size and potential aperture of rice stomata (Peng 2000) and allows a higher stomatal conductance in flooded, irrigated conditions. It is difficult to predict the impact of this type of arrangement for C_4 rice, but, assuming that the regulation of stomatal aperture is appropriate, it is unlikely to be deleterious.

The search for novel rice leaf morphology mutants

We have described how an understanding is needed of the relationships among protein and N content, photosynthetic capacity, and leaf morphology in rice leaves. One way of achieving this is to study the physiology and biochemistry of rice genotypes. However, a fundamental understanding will come only from analysis at the genetic level and it is here that we are in a position to take advantage of recent developments. In particular, the use of rice mutant resources should prove invaluable. A comprehensive survey of the phenotypic rice mutants available to date (almost 2,000 genes) is provided by Kurata et al (2005).

The sequencing of the rice genome by both private and public means has established rice as a model monocotyledonous organism, and the genomic resources that have been available for *Arabidopsis* research are now becoming available to the rice research community. The "knocking out" of genes by insertional mutagenesis is the best way of elucidating function. A growing number of such rice mutant collections use different insertion elements (Hirochika et al 2004, Wu et al 2005). The advantage of this strategy is in reverse genetic analysis, in which the gene that has been knocked out is easily identifiable. One way in which rice mutant collections differ greatly from their *Arabidopsis* counterparts is in forward genetic screens, in which mutant plants are grown in large numbers and screened for a phenotype, sometimes after imposing a change in their environmental conditions. So far, it has been largely impractical to screen rice insertional mutant collections in this way, mainly because of the resources required to produce and ship large quantities of rice seed. It has been much easier to provide such services for *Arabidopsis*, which requires less growth space and produces a greater quantity of smaller seed in a fraction of the time.

The International Rice Research Institute has focused on the large-scale production of deletion mutant lines by chemical and irradiation mutagenesis (www.iris.irri.org). Currently, 50,000 independent lines are at the M_4 stage, meaning that this is one of the most advanced collections available. The IR64 line was chosen because it is the most widely grown variety in the tropics and carries many valuable agronomic traits related to yield, architecture, and tolerance of stress. Tagged insert mutants have been the tool of choice due to the ease with which the gene of interest can be identified. On the other hand, more effort has normally been required to identify deletion mutant genes. However, this view is changing: analysis of deletion mutations is becoming more advanced, allowing both forward and reverse genetic analysis of these collections (Gong et al 2004, Hirochika et al 2004, Wu et al 2005). Detection of deletions can be carried out using oligomer chips, and it is suggested that this may become a general technique to identify deletions when genome-wide oligomer chips become available (Wu et al 2005).

The IR64 deletion mutant collection at IRRI currently offers some advantages. First, the deletions are likely to knock out more than one gene. The rice genome sequence indicates that 22% of the genes in rice are tandemly duplicated, suggesting that some functional redundancy may be present (Hirochika et al 2004). Two or more genes may have to be removed in order to discover their function. Second, the collection is advanced and resources are sufficient to allow the analysis of relatively large numbers of seeds. Third, the collection has been characterized for a variety of visible traits, making it an ideal resource to use for a forward genetic screen of morphological traits. Fourth, IR64 is an indica, whereas the genotypes used for insertional mutagenesis are normally japonica, which is more amenable to these transformation methods.

Around 1,000 lines from the IR64 deletion mutant collection that already show altered leaf shape, size, and chlorophyll content have been identified (www.iris.irri.org). These will be further screened for alterations in leaf thickness and leaf ultrastructure. Those that show differences in leaf thickness and/or cell size and number will be analyzed for their photosynthetic responses and N content under a variety of conditions. In this way, we hope to identify some of the genetic components involved in determining the relationships between leaf thickness and photosynthesis in rice, described above.

This gives us a great opportunity to search for traits useful in the quest for a rice leaf with C_4 traits. We will extend our screen to look for ultrastructural features that would prove useful in identifying which genes determine the C_3 or C_4 leaf anatomy, notably vein spacing and frequency, bundle sheath cell size, mesophyll cell number and arrangement, epidermal cell structure and arrangement, and sensitivity of photosynthesis to low and high CO_2 and/or low O_2.

Rice leaves are in fact well suited to this type of rapid throughput analysis, being rigid and silicized so that very thin sections can be cut transversely by hand and examined fresh under a microscope with no need for fixing and staining (Murchie et al 2005). This alone will provide most of the information for a screen of leaf ultrastructure and vein spacing.

Conclusions

1. Data from tropical field experiments support the need to suppress the oxygenation reaction of Rubisco in photosynthesis to increase biomass accumulation rates.
2. Losses to photosynthesis in the field such as photoinhibition and midday depression may still apply to C_4 rice, offering further scope for improvement.
3. Since one of the outcomes of C_4 rice is likely to be a reduced Rubisco content, the function of Rubisco as a significant store of N should be taken into account. Although C_4 rice should have a lower N requirement, there may need to be compensation to maintain grain protein content.
4. Using rice as a model, it is suggested that features of morphological acclimation of C_3 leaves in cell types such as mesophyll and stomata may help to point the way toward the requirements for C_4 leaf formation.
5. Rice mutant collections will be used to search for novel rice mutants with leaf ultrastructural features useful in the development of C_4 rice.

References

Bird S, Gray J. 2003. Signals from the cuticle affect epidermal cell differentiation. New Phytol. 157:9-23.

Black CC, Tu Z-P, Counce PA, Yao P-F, Angelov MN. 1995. An integration of photosynthetic traits and mechanisms that can increase crop photosynthesis and grain production. Photosynth. Res. 46:169-175.

Cheesman JM, Clough BF, Carter DR, Lovelock CE, Eong OJ, Sim RG. 1991. The analysis of photosynthetic performance in leaves under field conditions: a case study using *Bruguiera* mangroves. Photosynth. Res. 29:11-22.

Edwards G. 1999. Tuning up crop photosynthesis. Nature Biotechnol. 17:22-23.

Gong J-M, Waner DA, Horie T, Li SL, Horie R, Abid KB, Schroeder JI. 2004. Microarray-based rapid cloning of an ion accumulation deletion mutant in *Arabidopsis thaliana*. Proc. Natl. Acad. Sci. USA 101:15404-15409.

Hirochika H, Guiderdoni E, An G, Hsing Y, Eun M-Y, Han C-D, Upadhyaya N, Ramachandran S, Zhang Q, Pereira A, Sundaresan V, Leung H. 2004. Rice mutant resources for gene discovery. Plant Mol. Biol. 54:325-334.

Horton P, Ruban AV, Walters RG. 1994. Regulation of light harvesting in green plants. Plant Physiol. 106:415-420.

Horton P, Murchie EH. 2000. C_4 photosynthesis in rice: some lessons from studies of C_3 photosynthesis in field-grown rice. In: Sheehy JE, Mitchell PL, Hardy B, editors. Redesigning rice photosynthesis to increase yield. Proceedings of the Workshop on the Quest to Reduce Hunger: Redesigning Rice Photosynthesis, 30 Nov-3 Dec 1999, Los Baños, Philippines. Makati City (Philippines): International Rice Research Institute, and Amsterdam (The Netherlands): Elsevier. p 127-144.

Horton P. 2000. Prospects for crop improvement through the genetic manipulation of photosynthesis: morphological and biochemical aspects of light capture. J. Exp. Bot. 51:475-485.

Hoshikawa K. 1989. The growing rice plant: an anatomical monograph. Tokyo (Japan): Nobunkyo.

Kurata N, Miyoshi K, Nonomura K-I, Yamazaki Y, Ito Y. 2005. Rice mutants and genes related to organ development, morphogenesis and physiological traits. Plant Cell Physiol. 46(1):48-62.

Lake JA, Woodward FI, Quick P. 2002. Long-distance CO_2 signalling in plants. J. Exp. Bot. 367:183-193.

Leegood RC, Edwards G. 1996. Carbon metabolism and photorespiration: temperature dependence in relation to other environmental factors. In: Baker NR, editor. Photosynthesis and the environment. Dordrecht (Netherlands): Kluwer Academic Publishers. p 101-121.

Long S, Zhu X-G, Naidu SL, Ort DR. 2006. Can improvement in photosynthesis increase crop yields? Plant Cell Environ. 29:315-330.

Makino A, Mae T, Ohira K. 1985. Photosynthesis and ribulose-1,5-bisphosphate carboxylase/oxygenase in rice leaves from emergence through senescence: quantitative analysis by carboxylation/oxygenation and regeneration of ribulose 1,5-bisphosphate. Planta 166:414-420.

Makino A, Sato T, Nakano H, Mae T. 1997. Leaf photosynthesis, plant growth and nitrogen allocation in rice under different irradiances. Planta 203:390-398.

Mann CC. 1999. Genetic engineers aim to soup up crop photosynthesis. Science 283(5400):314-316.

Mitchell PL, Sheehy JE, Woodward FI. 1998. Potential yields and the efficiency of radiation use in rice. IRRI Discussion Paper Series No. 32. Manila (Philippines): International Rice Research Institute. 62 p.

Murchie EH, Chen Y-Z, Hubbart S, Peng S, Horton P. 1999. Interactions between senescence and leaf orientation determine *in situ* patterns of photosynthesis and photoinhibition in field-grown rice. Plant Physiol. 119:553-563.

Murchie EH, Hubbart S, Chen Y-Z, Peng S, Horton P. 2002a. Acclimation of rice photosynthesis to irradiance under field conditions. Plant Physiol. 130:1999-2010.

Murchie EH, Yang J, Hubbart S, Horton P, Peng S. 2002b. Are there associations between grain-filling rate and photosynthesis in the flag leaves of field-grown rice? J. Exp. Bot. 53:2217-2224.

Murchie EH, Hubbart S, Peng S, Horton P. 2005. Acclimation of photosynthesis to high irradiance in rice: gene expression and interactions with leaf development. J. Exp, Bot. 56:449-460.

Peng S, Cassman KG, Kropff MJ. 1995. Relationship between leaf photosynthesis and nitrogen content of field-grown rice in the tropics. Crop Sci. 35:1627-1630.

Peng S. 2000. Single leaf and canopy photosynthesis of rice. In: Sheehy JE, Mitchell PL, Hardy B, editors. Redesigning rice photosynthesis to increase yield. Proceedings of the Workshop on the Quest to Reduce Hunger: Redesigning Rice Photosynthesis, 30 Nov-3 Dec 1999, Los Baños, Philippines. Makati City (Philippines): International Rice Research Institute, and Amsterdam (The Netherlands): Elsevier. p 213-228.

Peng S, Laza RC, Visperas RM, Sanico AL, Cassman KG, Khush GS. 2000. Grain yield of rice cultivars and lines developed in the Philippines since 1966. Crop Sci. 40:307-314.

Pfannschmidt T. 2003. Chloroplast redox signals: how photosynthesis controls its own genes. Trends Plant Sci. 8:33-37.

Sheehy JE. 2000. Limits to yield for C_3 and C_4 rice: an agronomist's view. In: Sheehy JE, Mitchell PL, Hardy B, editors. Redesigning rice photosynthesis to increase yield. Proceedings of the Workshop on the Quest to Reduce Hunger: Redesigning Rice Photosynthesis, 30 Nov-3 Dec 1999, Los Baños, Philippines. Makati City (Philippines): International Rice Research Institute, and Amsterdam (The Netherlands): Elsevier. p 39-52.

Sinclair TR, Sheehy JE. 1999. Erect leaves and photosynthesis in rice. Science 283:1455-1456.

von Caemmerer S, Farquhar GD. 1981. Some relationships between the biochemistry of photosynthesis and the gas-exchange of leaves. Planta 153:376-387.

Wu J-L, Wu C, Lei C, Baraoidan M, Bordeos A, Madamba MRS, Ramos-Pamplona M, Mauleon R, Portugal A, Ulat VJ, Bruskiewich R, Wang G, Leach J, Khush G, Leung H. 2005. Chemical and irradiation induced mutants of indica rice IR64 for forward and reverse genetics. Plant Mol. Biol. 59:85-87.

Yano S, Terashima I. 2001. Separate localisation of light signal perception for sun or shade type chloroplast and palisade tissue differentiation in *Chenopodium album*. Plant Cell Physiol. 42:1303-1310.

Yano S, Terashima I. 2004. Developmental process of sun and shade leaves in *Chenopodium album* L. Plant Cell Environ. 27:781-793.

Zhang C, Peng S, Laza RC. 2003. Senescence of top three leaves in field-grown rice plants. J. Plant Nutr. 26:2453-2468.

Zhu X-G, Ort DR, Whitmarsh J, Long SP. 2004. The slow reversibility of photosystem II thermal energy dissipation on transfer from high to low light may cause large losses in carbon gain by crop canopies: a theoretical analysis. J. Exp. Bot. 55:1167-1175.

Ziska LH, Numuco O, Moya T, Quilang J. 1997. Growth and yield responses of field-grown tropical rice to increasing carbon dioxide and air temperature. Agron. J. 89:45-53.

Notes

Authors' addresses: E.H. Murchie, Agricultural and Environmental Sciences Division, Sutton Bonington Campus, University of Nottingham, Leicestershire, LE12 5RD, UK; P. Horton, Department of Molecular Biology and Biotechnology, Firth Court, Western Bank, University of Sheffield, S10 2TN, UK.

Wild species of *Oryza*: a rich reservoir of genetic variability for rice improvement

D.S. Brar and J.M. Ramos

The genus *Oryza*, to which cultivated rice (*O. sativa*) belongs, has 22 wild species (2n=24, 48) representing 10 genomes. These wild species show tremendous diversity in plant morphology, life cycle (perennial, annual), growth habit, light requirement (full sun, full shade, partial shade, etc.), including agronomic traits (resistance to biotic and abiotic stresses). Wild species are an important reservoir of useful genetic variability to broaden the gene pool of rice for tolerance of major biotic and abiotic stresses, for cytoplasmic diversification, and to introgress yield-enhancing loci/QTLs. Low crossability, increased sterility, reduced recombination, and linkage drag limit the introgression of genes from wild species into rice. Recent advances in tissue culture and molecular markers have facilitated alien introgression in rice. At IRRI, using direct crosses, embryo rescue, anther culture, molecular markers, and flourescence *in situ* hybridization (FISH), a series of interspecific hybrids, alien introgression lines, monosomic alien addition lines (MAALs), and chromosome segmental substitution lines (CSSLs) have been produced. Genes for resistance to brown planthopper (BPH), bacterial blight (BB), blast, tungro, acid sulfate soils, and iron toxicity have been introgressed from AA, BBCC, CC, CCDD, EE, and FF genomes into rice. Some of the introgressed genes (*Bph10, Bph18, Xa21, Pi-9*) introgressed from wild species have been mapped and used in marker-assisted selection (*Xa21, Bph18*). Some of the breeding lines with genes introgressed from wild species have been released as varieties. Opportunities exist in wild species germplasm for novel genetic variability for traits such as C_4 or C_3-C_4 intermediates. Once variability for C_4 traits is identified in *Oryza*, it should be possible to exploit the transfer of such traits following wide hybridization techniques already used successfully for the transfer of resistance to biotic and abiotic stresses and thus enhance the photosynthetic efficiency and yield of rice by modifying it from C_3 to C_4.

Keywords: C_3-C_4 intermediates, wild species, genetic variability, alien introgression, molecular markers

Rice is an important cereal and a source of calories for more than one-third of the world population. More than 90% of rice is produced and consumed in Asia. During the last few decades, world rice production has more than doubled from 257 million tons to 600 million tons. This has mainly been achieved through the application of the principles of classical Mendelian genetics and conventional plant breeding methods. To meet the growing need of the human population, 40% more rice is needed by 2020, with less land, water, labor, and chemicals. Further, several biotic stresses (bacterial blight, blast, tungro, stem borer, planthopper) and abiotic stresses (drought, submergence, salinity, iron toxicity, P deficiency) continue to reduce rice productivity. To overcome these constraints, there is thus an urgent need to increase rice productivity. Besides many approaches for the genetic enhancement of rice, one new option is to explore developing C_4 rice.

Wild species of *Oryza* (2n=24, 48) with 10 genomes, although of inferior, grassy appearance, are an important reservoir of useful genetic variability to broaden the gene pool of rice for tolerance of major biotic and abiotic stresses, for cytoplasmic diversification, and to introgress yield-enhancing loci/QTLs (Table 1). IRRI maintains more than 100,000 accessions of cultivated rice and a large collection of different wild species in the gene bank (Table 1). Wild species of *Oryza* are widely distributed across Asia, Africa, Latin America, and Australia. These species show tremendous diversity in life cycle (perennial, annual), growth habit, light requirement (full sun, full shade, partial shade, etc.), including agronomic traits (resistance to biotic and abiotic stresses) and in growth in terms of height, leaf, stem, panicle, and seed characteristics (Vaughan 1994). Opportunities exist in the wild species germplasm for novel genetic variability for traits such as C_4 or C_3-C_4 intermediates. Naturally-occurring species intermediate between C_3 and C_4 plants have been reported in the genera *Panicum*, *Moricandia*, and others (Brown and Hattersley 1989). Four species of the genus *Flaveria* have been identified as intermediate C_3-C_4 plants based on leaf anatomy, photosynthetic carbon dioxide compensation point, inhibition of photosynthesis by oxygen, and activities of C_4 enzymes (Ku et al 1983). Imaizumi et al (1997) examined photosynthetic carbon metabolism of spikelets in rice. In a ^{14}C pulse-^{12}C study of photosynthetic carbon dioxide fixation, about 35% and 25% of ^{14}C fixed in rice lemmas was incorporated initially into 3-phosphoglycerate (3-PGA) and C_4 acids, respectively. This suggests that lemmas participate mainly in C_3-type photosynthetic metabolism, but that lemmas may also participate in the metabolism of C_4 acids to some extent. Preliminary results indicate that some of the wild species may have some of the anatomical features of C_4 plants (J. Sheehy, personal communication). Once variability for C_4 traits is identified in *Oryza*, it should be possible to exploit this and transfer it into commercial cultivars and thus enhance the photosynthetic efficiency and yield of rice by modifying it from C_3 to C_4.

Table 1. Chromosome number, genomic composition, and potential useful traits of *Oryza* species. Modified from Brar and Khush (2006).

Species	2n	Genome	Number of accessions[a]	Distribution	Useful or potentially useful traits[b]
***O. sativa* complex**					
O. sativa L.	24	AA	102,785	Worldwide	Cultigen
O. glaberrima Steud.	24	A^gA^g	1,656	West Africa	Cultigen; tolerance of drought, acidity, and iron toxicity; resistance to RYMV, African gall midge, nematodes, and weed competitiveness
O. nivara Sharma et Shastry	24	AA	1,130	Tropical and subtropical Asia	Resistance to grassy stunt virus
O. rufipogon Griff.	24	AA	858	Tropical and subtropical Asia, tropical Australia	Resistance to BB and tungro virus, tolerance of aluminum and soil acidity, source of CMS
O. breviligulata A. Chev. et Roehr. (*O. barthii*)	24	A^gA^g	214	Africa	Resistance to GLH, and BB, drought avoidance
O. longistaminata A. Chev. et Roehr.	24	A^lA^l	203	Africa	Resistance to BB, nematodes, drought avoidance
O. meridionalis Ng	24	A^mA^m	46	Tropical Australia	Elongation ability, drought avoidance
O. glumaepatula Steud.	24	$A^{gp}A^{gp}$	54	South and Central America	Elongation ability, source of CMS
***O. officinalis* complex**					
O. punctata Kotschy ex Steud.	24, 48	BB, BBCC	59	Africa	Resistance to BPH and zigzag leafhopper
O. minuta J.S. Pesl. ex C.B. Presl.	48	BBCC	63	Philippines and Papua New Guinea	Resistance to BB, blast, BPH, GLH, tolerant of Shb
O. officinalis Wall ex Watt	24	CC	265	Tropical and subtropical Asia, tropical Australia	Resistance to thrips, BPH, GLH, WBPH, BB, and stem rot
O. rhizomatis Vaughan	24	CC	19	Sri Lanka	Drought avoidance, rhizomatous
O. eichingeri A. Peter	24	CC	29	South Asia and East Africa	Resistance to BPH, WBPH, and GLH
O. latifolia Desv.	48	CCDD	40	South and Central America	Resistance to BPH, high biomass production
O. alta Swallen	48	CCDD	6	South and Central America	Resistance to striped stem borer, high biomass production

Continued on next page

Table 1 continued.

Species	2n	Genome	Number of accessions[a]	Distribution	Useful or potentially useful traits[b]
O. grandiglumis (Doell) Prod.	48	CCDD	10	South and Central America	High biomass production
O. australiensis Domin.	24	EE	36	Tropical Australia	Resistance to BPH and BB, drought avoidance
***O. meyeriana* complex**					
O. granulata Nees et Arn. ex Watt	24	GG	24	South and Southeast Asia	Shade tolerance, adaptation to aerobic soil
O. meyeriana (Zoll. et Mor. ex Steud.) Baill.	24	GG	11	Southeast Asia	Shade tolerance, adaptation to aerobic soil
***O. ridleyi* complex**					
O. longiglumis Jansen	48	HHJJ	6	Irian Jaya, Indonesia, and Papua New Guinea	Resistance to blast and BB
O. ridleyi Hook. F.	48	HHJJ	15	South Asia	Resistance to BB, stem borer, and whorl maggot
Unclassified					
O. brachyantha A. Chev. et Roehr.	24	FF	19	Africa	Resistance to BB, yellow stem borer, leaf-folder, and whorl maggot, tolerance of laterite soil
O. schlechteri Pilger	48	HHKK	1	Papua New Guinea	Stoloniferous
Related genera			15	–	–

[a]Accessions maintained at T.T. Chang Genetic Resources Center, International Rice Research Institute, Philippines. [b]BPH = brown planthopper, GLH = green leafhopper, WBPH = white-backed planthopper, BB = bacterial blight, Shb = sheath blight, CMS = cytoplasmic male sterility, RYMV = rice yellow mottle virus.

Strategies for gene transfer from wild species into rice

Some of the steps followed at IRRI for alien gene transfer are described below.

1. Search for useful genetic variability for target traits in wild species germplasm.
2. Produce interspecific hybrids through direct crosses and embryo rescue.
3. Develop introgression lines through backcrossing with the recurrent rice parent.
4. Evaluate alien introgression lines for target traits under natural hot spots and artificial conditions in the screenhouse, field, and laboratory.
5. Develop species-specific libraries through representational difference analysis (RDA) and characterize alien introgression using molecular markers and tagging of introgressed alien genes for use in marker-assisted selection (MAS) and the location of introgressed alien segments using genomic *in situ* hybridization.
6. Construct alien chromosome segmental substitution lines (CSSL) for mapping genes/QTLs and use in functional genomics.

Alien gene transfer and varietal development using wild species of *Oryza*

We have established strong collaboration with national agricultural research and extension systems (NARES) and advanced research institutes (ARI). Hybrids have been produced between rice and wild species representing each of the 10 genomes of *Oryza* across crossability barriers using embryo rescue techniques, including the production of a large number of advanced alien introgression lines. Many novel genes have been transferred with a wide spectrum of resistance to brown planthopper (BPH), bacterial blight (BB), blast, and tungro virus and tolerance of acidity and iron toxicity, including new cytoplasmic male sterile (CMS) sources from many wild species representing AA, BBCC, CC, CCDD, EE, and FF genomes into elite breeding lines of rice (Table 2, see Brar and Khush 1997, 2002, 2006, Multani et al 2003). These elite alien introgression lines have been used in rice breeding programs in India, China, Indonesia, the Philippines, Bangladesh, and Vietnam. New sources of resistance to yellow stem borer, sheath blight, and black streak dwarf virus (BSDV) have been identified from wild species. Resistance to BSDV, BPH, and BB is being introduced to broaden the gene pool of japonica cultivars. A large number of introgression lines derived from *O. sativa* × *O. glaberrima* are under evaluation for transfer of weed competitive ability and tolerance of iron toxicity, P deficiency, and drought. Of the several elite alien introgression lines developed, some have been released as varieties by NARES. Recently, three varieties were released in Vietnam and the Philippines from crosses of rice with AA genome wild species (Table 3).

Table 2. Genes for tolerance of biotic and abiotic stresses transferred at IRRI from wild species into rice.

Donor wild species	Genome	Traits transferred
O. nivara	AA	Grassy stunt virus resistance
Different species	AA, BBCC, CC, CCDD, EE	Brown planthopper resistance
Different species	AA, CC, BBCC, CCDD, EE, FF	Bacterial blight resistance
O. minuta	BBCC	Blast resistance
O. australiensis	EE	Blast resistance
O. rufipogon	AA	Tungro virus resistance
O. rufipogon	AA	Acid sulfate soil tolerance
O. glaberrima[a]	AA	Iron toxicity tolerance
Alien introgression lines under evaluation		
O. longistaminata	AA	Yellow stem borer resistance
O. glaberrima[a]	AA	Nematode resistance, iron and aluminum toxicity and drought tolerance, brown planthopper resistance
O. rufipogon	AA	Elongation ability (in deep water); rice black streak dwarf virus, bacterial blight, and tungro resistance

[a]Cultivated African rice.

Table 3. IRRI breeding lines with genes introgressed from wild species released as varieties.

Trait	Wild species	Variety released	Country
Grassy stunt virus resistance	*O. nivara*	Many	Rice-growing countries in Asia
Brown planthopper resistance	*O. officinalis*	MTL 98, MTL 103, MTL 105, MTL 114	Vietnam
Tungro virus resistance	*O. rufipogon*	Matatag 9	Philippines
Acid sulfate soil tolerance	*O. rufipogon*	AS996	Vietnam
High yield, brown planthopper resistance	*O. longistaminata*	NSICRc112	Philippines

Molecular mapping of introgression alien genes/QTLs

Some of the alien genes for BB, blast, and BPH resistance introgressed from wild species have been tagged with molecular markers. Two genes (*Bph10* and *Bph 18(t)*) introgressed from *O. australiensis* have been tagged (Ishii et al 1994, Jena et al 2006). QTLs for BPH resistance introgressed from *O. minuta* and for aluminum toxicity tolerance introgressed from *O. rufipogon* have been identified. One QTL located on chromosome 3 is conserved across cereal genomes (Nguyen et al 2003). One of the genes, *Xa21* from *O. longistaminata*, has been used by NARES in India, China, and the Philippines in gene pyramiding and MAS for BB resistance (Sanchez et al 2000). The *Xa21* gene has been incorporated through MAS in commercial hybrids released in China.

Molecular characterization of alien introgression and homoeologous pairing using GISH

We have carried out extensive molecular analysis of alien introgression lines derived from crosses of *O. sativa* × *O. glaberrima*, *O. sativa* × *O. rufipogon*, and *O. sativa* × distantly related genomic types (CC, BBCC, EE, FF, and GG). The results indicate frequent introgression from the AA genome wild species, including from *O. rufipogon*. However, limited introgression occurs from distant genomes of *Oryza*. It is interesting to note that small segments are introgressed from wild species through crossing-over: a possible cause of clean gene transfer and rapid recovery of recurrent phenotype in wide crosses of rice.

GISH protocols have been successfully used to characterize parental genomes and extra chromosomes in monosomic alien addition lines (MAALs). MAALs have been established from seven wild species (CC, BBCC, CCDD, EE, FF, GG, and HHJJ genomes) (Brar and Khush 2002, 2006). GISH was used successfully to characterize homoeologous pairing among several genomes: A × C, A × E, A × BC, A × F, A × HJ, and A × G. Autosyndetic pairing (chromosome pairing within the genome of rice) is suggestive of duplications in the *sativa* genome. Genome-specific clones from *O. minuta* have been developed for large-scale analysis of alien introgression lines using microarrays.

Development of new genetic resources for functional genomics

CSSLs and near-isogenic alien introgression lines are being developed from *O. rufipogon*, *O. longistaminata*, and *O. glaberrima* in the genetic background of *O. sativa* and are being characterized using simple sequence repeat (SSR) markers. These are valuable genetic resources to map genes/QTLs governing tolerance of biotic and abiotic stresses and for functional genomics.

Future outlook

Wild species are an important reservoir of useful genes and offer opportunities to broaden the gene pool of both indica and japonica rice through the introgression of genes for resistance to major diseases and insects and tolerance of various abiotic stresses, including the transfer of yield-enhancing loci/QTLs and weed competitive ability into elite breeding lines. The introgressed alien genes with a broad spectrum of resistance have become a valuable addition for marker-assisted selection. The production of BAC libraries from 10 different genomic types of wild species (Ammiraju et al 2006) have become a novel genetic resource for the fine-mapping of genes/QTLs and in understanding evolutionary relationships in *Oryza*. The search for C_4 and C_3-C_4 intermediates in wild species of *Oryza* using different anatomical, biochemical, physiological, and molecular approaches is important. Once such sources for C_4 activity become available, it should be possible to introgress these novel traits from wild species into rice using strategies followed for the introgression of other traits as listed in the previous sections. MAS could further facilitate the transfer of different components of C_4-ness into elite breeding lines.

References

Ammiraju JSS, Lou M, Goicoechea JL, Wang W, Kudrna D, Mueller C, Talag J, Kim H, Sisneros NB, Blackmon B, Fang E, Tomkins J, Brar D, Mackill D, McCouch S, Kurata N., Lambert G, Galbrait DW, Arumuganathan K, Rao K, Walling JG, Gill N, Yu Y, SanMiguel P, Soderlund C, Jackson S, Wing RA. 2006. The *Oryza* bacterial artificial chromosome library resource: construction and analysis of 12 deep-coverage large-insert BAC libraries that represent the 10 genome types of the genus *Oryza*. Genome Res. 16:140-147.

Brar DS, Khush GS. 2002. Transferring genes from wild species into rice. In: Kang MS, editor. Quantitative genetics, genomics and plant breeding. Wallingford (UK): CAB International. p 197-217.

Brar DS, Khush GS. 2006. Cytogenetic manipulation and germplasm enhancement of rice (*Oryza sativa* L.). In: Singh RJ, Jauhar PP, editors. Genetic resources, chromosome engineering and crop improvement. II. Boca Raton, Fla. (USA): CRC Press. p 115-158.

Brar DS, Khush GS. 1997. Alien introgression in rice. Plant Mol. Biol. 35:35-47.

Brown RH, Hattersley PW. 1989. Leaf anatomy of C_3-C_4 species as related to the evolution of C_4 photosynthesis. Plant Physiol. 91:1543-1550.

Ishii T, Brar DS, Multani DS, Khush GS. 1994. Molecular tagging of genes for brown planthopper resistance and earliness introgressed from *Oryza australiensis* into cultivated rice, *O. sativa*. Genome 37:217-221.

Imaizumi N, Samejima M, Ishihara K. 1997. Characteristics of photosynthetic carbon metabolism of spikelets in rice. Photosynth. Res. 53:75-82.

Jena KK, Jeung JU, Lee JH, Choi HC, Brar DS. 2006. High-resolution mapping of a new brown planthopper (BPH) resistance gene, *Bph18*(t), and marker-assisted selection for BPH resistance in rice (*Oryza sativa* L.). Theor. Appl. Genet. 112:288-297.

Ku MSB, Monson RK, Littlejohn RB, Nakamoto H, Fisher DB, Edwards GE. 1983. Photosynthetic characteristics of C_3-C_4 intermediate *Flaveria* species. Plant Physiol. 71:944-948.

Multani DS, Khush GS, delos Reyes BG, Brar DS. 2003. Alien genes introgression and development of monosomic alien addition lines from *Oryza latifolia* Desv. to rice, *O. sativa*. Theor. Appl. Genet. 107:395-405.
Nguyen BD, Brar DS, Buu BC, Nguyen TV, Pham LN, Nguyen HT. 2003. Identification and mapping of the QTL for aluminum tolerance introgressed from a new source, *Oryza rufipogon* Griff., into indica rice, *Oryza sativa* L. Theor. Appl. Genet. 106:583-593.
Sanchez AC, Brar DS, Huang N, Li Z, Khush GS. 2000. Sequence tagged site marker-assisted selection for three bacterial blight resistance genes in rice. Crop Sci. 40:792-797.
Vaughan DA. 1994. The wild relatives of rice: a genetic resources handbook. Los Baños (Philippines): International Rice Research Institute. 137 p.

Notes

Authors' address: International Rice Research Institute, Los Baños, Philippines.

C_4 rice: a plant breeder's perspective

P.S. Virk and S. Peng

Rice is one of the most important cereal crops in the world. Rice production has been increasing at a steady pace since the adoption of modern high-yielding rice cultivars. However, demand for rice continues to increase and so does the number of rice consumers. It is estimated that we need to produce an extra 200 million tons of rice by 2025 to meet the demand. To achieve this objective, we have to breed rice varieties with higher yield potential. Several breeding strategies are being employed at IRRI for developing rice varieties with higher yield potential, such as empirical breeding, heterosis breeding, ideotype breeding, and wide hybridization. Modern approaches such as the development of C_4 rice for increasing the yield potential of rice are being investigated. To maximize the benefit of C_4-ness in rice, breeders should target associated yield component traits. Furthermore, rice breeders can play a major role in transferring C_4-ness (once available) from donor or engineered genotypes into commercially acceptable backgrounds.

Keywords: empirical breeding, ideotype breeding, heterosis breeding, wide hybridization

Rice is the staple food for the largest number of people on Earth and it has a special significance in Asia, where about 90% of the rice is produced and consumed. According to one estimate, global rice production must reach 800 million tons from the 608 million tons in 2004 to meet demand in 2025. Hence, there is an urgent need to investigate ways to enhance the yield potential of rice to meet the global rice requirement for the additional 192 million tons of rice. Therefore, the average yield of irrigated rice varieties must increase in tropical rice lands from 5 to 8.5 t ha^{-1} (Peng et al 1999). To achieve the enhanced yield, rice varieties with yield advantages must be developed to meet the goal of increased rice production. Several breeding strategies for increasing the yield potential of rice are currently being used at IRRI, such as empirical selection, ideotype breeding, heterosis breeding, and wide hybridization.

Empirical breeding

This is a time-tested approach and it has two phases. Variability is created through hybridization, followed by empirical phenotypic selection for target traits such as growth duration, lodging resistance, disease and insect resistance, and physicochemical grain characteristics. Elite lines developed through empirical breeding are evaluated for yield and other agronomic traits and superior lines are released as varieties. It has been estimated that, on average, around a 1% annual increase in yield has occurred since the development of the first improved semidwarf variety, IR8, when all historical and newly developed varieties were grown under current conditions (Peng et al 1999). Small incremental improvements in yield are expected to continue for the foreseeable future following this strategy.

Ideotype breeding

Scientists at IRRI in the late 1980s postulated that the plant type of semidwarf varieties might limit further improvement in their yield potential. These genotypes produce a large number of unproductive tillers and have excessive leaf area that may cause mutual shading and thus a reduction in canopy photosynthesis and sink size (Dingkuhn et al 1991). Most of these varieties have high tillering capacity and small panicles. A large number of unproductive tillers, which limit sink size and increase lodging susceptibility, were identified as the major constraint to yield improvement in these varieties. Furthermore, simulation models predicted that a 25% increase in yield potential was possible by modifying certain traits of the semidwarf plant type (Dingkuhn et al 1991).

These factors prompted IRRI scientists to propose modifications to the high-yielding indica plant type in the late 1980s in order to increase yield potential. The proposed new plant type (NPT) or ideotype (The ideotype was defined by Donald [1968] as an idealized plant type with a specific combination of characteristics favorable for photosynthesis, growth, and grain production based on knowledge of plant and crop physiology and morphology. He anticipated that it would be more efficient to define a plant type that was theoretically efficient and then breed for this ideotype [Hamblin 1993]) has low tillering capacity (8 to 10 tillers); few unproductive tillers; 200 to 250 grains per panicle; a plant height of 90 to 100 cm; thick and sturdy stems; leaves that are thick, dark green, and erect; a vigorous root system; 100 to 130 days' growth duration; and increased harvest index (Peng et al 1994, Khush 1995). This ideotype became the "new plant type (NPT)" highlighted in IRRI's strategic plan (IRRI 1989). The goal was to develop a new plant type with yield potential 20–25% higher than that of existing semidwarf varieties of rice in the tropical environment during the dry season.

Hybridization at IRRI was undertaken in 1990. Since most bulu varieties are tall in stature, they were crossed with semidwarf breeding line Sheng Nung 89-366 obtained from Shenyang Agricultural University, China. The selected donors were crossed and breeding lines with the proposed ideotype were selected. Since then, more than 2,500 crosses have been made, 150,000 pedigree lines were produced, breeding lines with the desired morphological ideotype traits were selected, and more than 500 NPT-tropical japonica (TJ) lines have been evaluated in observational yield trials. The NPT-TJ lines based on tropical japonica germplasm were developed in less than 5 years. The first batch of NPT-TJ lines had large panicles, few unproductive tillers, thick stems, and large and dark green flag leaves, but grain yield of these lines was not encouraging. This was attributed to low biomass production and poor grain filling. Reduced tillering capacity contributed to low biomass production (Khush and Peng 1996). Less biomass production was also associated with poor grain filling. Fortunately, we traced the problem of unfilled grains to certain donor parents and were able to overcome this problem by including only parents with a high percentage of filled grains in the subsequent hybridization programs (Virk and Khush 2003). As a result, the yield of later NPT-TJ lines was higher than that of the previous NPT-TJs and the indica check variety. For example, the best NPT-TJ line outyielded the check variety (IR72) by 9.5% in an observational field trial conducted at IRRI during the 1998 dry season.

Several NPT-TJ lines were shared with Yunnan Academy of Agricultural Sciences, China. After evaluation in their local conditions, the Chinese rice breeders, between 2000 and 2003, released three NPT-TJ varieties, Dianchao 1, Dianchao 2, and Dianchao 3, developed from two of IRRI's NPT-TJ lines, namely, IR64446-7-10-5 and IR69097-AC2-1. The NPT-TJ lines perform very well, primarily in temperate areas where disease pressure is low, and the preference is for sticky-cooking and bold-grain types.

Further improvement of the new plant type

An increase in the tillering capacity of the NPT-TJ lines was envisaged to increase biomass production. Second, most of the NPT-TJ lines lacked resistance to tropical diseases and insects as the parents used for developing these lines were susceptible. For example, there were no donors for resistance to brown planthopper and tungro virus in the tropical japonica germplasm. Third, farmers and consumers in the tropical rice-growing countries prefer varieties with long and slender grains and with intermediate amylose content.

In 1995, modern high-yielding indica varieties or elite lines were included in the hybridization program. With this, the development of improved NPT lines began. Since these are derivatives from crosses between indica and japonica germplasm, we shall call them NPT-IJs. This was necessary to increase the biomass, incorporate genes for resistance to tropical diseases and insects, and change grain appearance and quality. More than 500 NPT-IJ lines have been evaluated in observational yield trials. During 2001, several NPT-IJ lines outyielded the best indica check variety by up to 30% in breeders' replicated yield trials (Khush et al 2001). Encouraged by the superior performance of the NPT-IJs, we included them in more rigorous yield testing during 2002 and 2003. In the 2002 dry and wet seasons, several NPT-IJ lines significantly outyielded check variety IR72 (Table 1). The NPT-IJ lines approached the yield bar-

Table 1. Grain yield and some yield components of new plant type lines grown at IRRI during the 2002 dry and wet seasons.

Genotype	Grain yield ($t\ ha^{-1}$)	Panicles per m^2	Spikelets per panicle
2002 dry season			
IR72 (check)	7.80	450	83.4
IR71700-247-1-1-2	9.75	470	120.7
IR72158-16-3-3-1	9.69	328	145.7
IR72967-12-2-3	9.59	343	133.5
LSD[a] (0.05)	0.85	31	8.8
2002 wet season			
IR72 (check)	5.35	404	81.5
IR71700-247-1-1-2	6.51	407	106.2
IR72164-348-6-2-2-2	6.41	303	122.4
IR73711-130-1-3-1	5.73	282	115.3
LSD (0.05)	0.61	25	9.2

[a]LSD (0.05) is the least significant difference for comparing the trait mean between breeding lines at $P = 0.05$.

rier of 10 t ha^{-1} (Peng et al 2004). In the 2003 dry season, IR72967-12-2-3 was the top-yielding NPT-IJ line. It yielded 10.16 t ha^{-1}, which was significantly higher than the yield of the indica check variety.

The yield increase over check variety IR72 was attributed to larger panicles with a large number of spikelets, more biomass, and a higher harvest index, through the introduction of genes from elite indica parents into the NPT-TJ lines. Recently, in the Philippines, an IRRI-bred NPT-IJ line has been recommended as a commercial variety.

Heterosis breeding

Hybrid rice research in China started in 1964 and the first commercial hybrid rice was released in 1976. Hybrid rice showed a yield advantage of about 15–20% over the best inbred varieties. In China, close to 50% of the rice area is under hybrid rice cultivation. Five South and Southeast Asian countries, India, the Philippines, Vietnam, Indonesia, and Bangladesh, have also released rice hybrids for commercial cultivation. These hybrids show 15–20% standard heterosis, and are developed primarily using indica germplasm.

However, Chinese hybrids perform poorly when grown in tropical Asian countries. Hence, IRRI initiated hybrid rice research to develop hybrids for tropical countries. Breeding for rice hybrids began in 1980 at IRRI by developing suitable parental lines through the use of wild abortive (WA) cytoplasm from China. Through continuous breeding efforts, various CMS lines were developed at IRRI. One of the male sterile lines, IR58025A, has been used widely in developing commercial hybrids in several tropical rice-growing countries.

The increased yields of hybrids are due to their increased biomass, higher grain number per panicle, and slightly higher grain weight. All the hybrids developed at IRRI have been between indica varieties. However, genetic diversity between parents is generally related to the magnitude of heterosis. Our results showed that various groups of hybrids yielded in the order tropical japonica (TJ)/indica (I) > indica/indica > tropical japonica/tropical japonica (Khush et al 1998, Virk et al 2003). However, many of the TJ/I F_1s showed varying degrees of spikelet sterility because of interaction at the S-5 locus, which causes female gametes carrying the japonica allele to be eliminated (Ikehashi and Araki 1984). However, varieties with wide compatibility (WC) genes carry a neutral allele at the S-5 locus (Ikehashi and Araki 1984, 1986). Our experience showed that the WC gene is widely distributed in TJ germplasm and offers an opportunity for use in the development of parental lines for hybrid breeding (Virk et al 2003).

Developing NPT parental lines for hybrid rice breeding

A breeding program to identify maintainer and restorer lines in the TJ germplasm started at IRRI in 1995. However, fertility restorers in the TJ germplasm could not be found. On the other hand, maintainer frequency of the TJ lines was 70% as against 5% in indicas. The apparent lack of restoration ability in the TJ germplasm indicated that these lines couldn't be used as pollen parents of an I/TJ hybrid. Hence, it was sensible to convert TJ lines into CMS lines. In fact, male sterile lines in the TJ background have been produced; however, the phenotypic acceptability of such CMS lines is rather low and, more importantly, their outcrossing rate is also very low. A continued degeneration of the panicles in certain TJ genotypes even in the advanced backcross generations was observed. Besides, TJ lines do not possess the desired level of disease or insect resistance and they possess short and bold grains. It appears that the original TJ CMS lines will also be of limited use for developing commercial TJ/I hybrids. However, the original NPT lines have been improved for these limitations.

Prospects of using improved NPT lines for hybrid rice breeding

Improved NPT lines (I-TJ derivatives) are good restorers. Also, their maintainer frequency is almost 3 times higher than that of indicas. Thus, they have great potential to be used as pollen parents in hybrid rice breeding. Unlike original TJ lines, they possess long and slender grains with intermediate amylose content. Also, their level of resistance to various diseases and insects is more like that of indicas. All these features make them excellent candidates for developing commercial rice hybrids. Efforts are under way to determine the level of heterosis of hybrids between I-TJ NPT and indica lines.

Two-line hybrids using NPT-TJ lines

Usage of TJ lines in the development of commercial I/TJ rice hybrids using the three-line system is limited. Therefore, a breeding program to incorporate the thermosensitive genetic male sterility (TGMS) system in TJ lines began. Crosses were made between indica TGMS lines with NPT-TJ lines possessing the WC gene to incorporate a TGMS gene into them. Advanced male sterile lines with *tms2* and *tms3* genes are now available. We are evaluating two-line hybrids in replicated yield trial nurseries. Incorporation of TGMS genes in I-TJ NPT lines, which would be useful in developing future two-line hybrids, is under way.

Constraints in using hybrid rice

There are several major constraints to the large-scale adoption of hybrid rice technology:

- The need to buy fresh hybrid seed, from a credible source, for every planting season.
- The high cost of hybrid seed.
- The need to establish a seed production infrastructure.
- Other factors such as poor grain quality of some hybrids, inconsistent performance of the first set of released hybrids, and inconsistent seed yields have also been identified (Virmani and Kumar 2004).

Wide hybridization

Wild species of rice are phenotypically inferior to cultivated ones. However, evidence is accumulating suggesting that wild species have hidden genetic variability that can be exploited to improve plant productivity traits, including yield. With the help of molecular markers, it is possible to identify and transfer desirable QTLs from wild species into elite breeding lines (Tanksley and Nelson 1996). In particular, *Oryza rufipogon* alleles at two marker loci, RM5 (*yld1-1*) on chromosome 1 and RG256 (*yld2-1*) on chromosome 2, were found to be associated with enhanced yield (Xiao et al 1996). Xiao et al (1998) and Moncada et al (2001) have also reported QTLs with trait-improving alleles originating from phenotypically inferior wild species. Preliminary observations at IRRI on progenies derived from crosses between a tropical japonica NPT line with *O. longistaminata* and IR64 × *O. rufipogon* support these observations.

C_4 rice

Various breeding strategies are available for yield enhancement. However, none of the methods described above can meet global rice demand by 2050. Improving photosynthesis for enhancing yield potential offers exciting possibilities for rice scientists.

Scenario 1: Use of C_4 or C_3-C_4 intermediates from wild rice types

This would require screening of wild species germplasm for traits potentially indicative of C_4-ness or targeting C_3-C_4 intermediates. Accessions possessing desirable traits

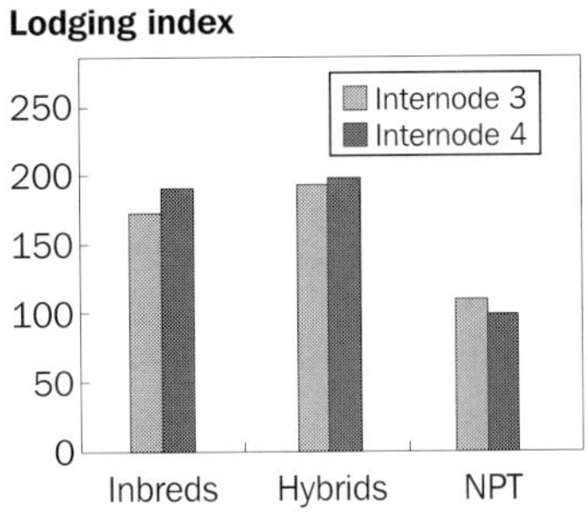

Fig. 1. Lodging index (lower lodging index is desirable) for inbreds, hybrids, and NPT lines at 3rd and 4th internodes.

could then be used as donor parents in various cross combinations with cultivated rice (for details, see chapter by D.S. Brar and J.M. Ramos).

Scenario 2: Genetically engineered C_4 rice

Protocols to transfer a transgenic allele from a donor line to various popular varieties are available, if C_4 rice were to be genetically engineered. These could range from conventional backcrossing and selection to molecular marker-aided backcrossing. From the breeders' point of view, the first step would be to assess the effect of the C_4 transgene on various agronomic traits such as resistance to pests, emergence of new rice pests, physical properties of rice grains, and cooking and eating quality. Second, it would be important to evaluate the amount of expression of the transgene in different genetic backgrounds and identify the most promising transgenic event.

Associated traits

Elite lines in the breeding stream have some traits thought to be important in harnessing the full potential of C_4 rice once developed. For example, we have been able to breed lines in the indica background using tropical japonica germplasm, which have a lower lodging index score (or higher degree of tolerance of lodging) than indica inbreds or hybrids (Fig. 1). Similarly, we have bred lines with bigger sink size. For example, the panicle size of newly bred lines is larger and the number of spikelets per panicle is higher (Fig. 2). Finally, we should remember that rice is grown in diverse ecosystems, namely, irrigated, rainfed, upland, and flood-prone, and, to maximize the potential of C_4 rice, one should prioritize a particular ecosystem, at least to start with.

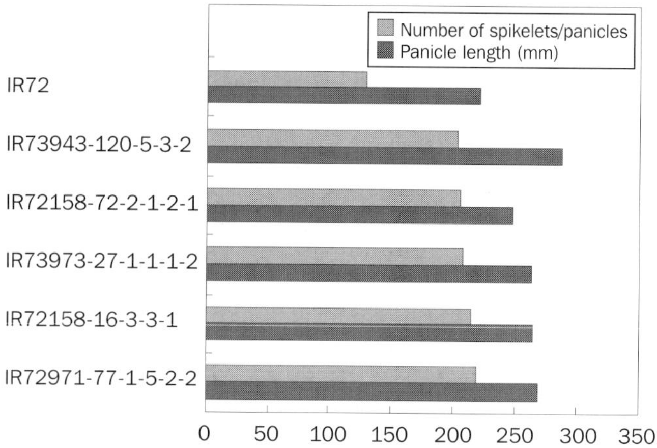

Fig. 2. Panicle length and number of spikelets/panicles for some selected lines.

References

Dingkuhn M, Penning de Vries FWT, De Datta SK, van Laar HH. 1991. Concepts for a new plant type for direct seeded flooded tropical rice. In: Direct seeded flooded rice in the tropics. Manila (Philippines): International Rice Research Institute. p 17-38.

Donald CM. 1968. The breeding of crop ideotypes. Euphytica 17:385-403.

Hamblin J. 1993. The ideotype concept: useful or outdated? In: International Crop Science I, Crop Science Society of America, Madison, Wis., USA. p 589-597.

Ikehashi H, Araki H. 1984. Varietal screening of compatibility types revealed in F_1 fertility of distant crosses in rice. Jpn. J. Breed. 34:304-313.

Ikehashi H, Araki H. 1986. Genetics of F_1 sterility in remote crosses of rice. In: Rice genetics. Manila (Philippines): International Rice Research Institute. p 119-130.

IRRI. 1989. IRRI towards 2000 and beyond. Los Baños (Philippines): International Rice Research Institute. p 36-37.

Khush GS. 1995. Breaking the yield frontier of rice. GeoJournal 35:329-332.

Khush GS, Aquino RC, Virmani SS, Bharaj TS. 1998. Using tropical japonica germplasm to enhance heterosis in rice. In: Virmani SS, Siddiq EA, Muralidharan K, editors. Advances in hybrid rice technology. Proceedings of the 3rd International Symposium on Hybrid Rice, 14-16 November 1996, Hyderabad, India. Manila (Philippines): International Rice Research Institute. p 59-66.

Khush GS, Peng S. 1996. Breaking the yield frontier of rice. In: Reynolds MP, Rajaram S, McNab A, editors. Increasing yield potential in wheat: breaking the barriers. El Batán (Mexico): International Maize and Wheat Improvement Center. p 36-51.

Khush GS, Virk PS, Evangelista A, Romena B, Pamplona A, Lopena V, Dela Cruz N, Peng S, Cruz CV, Cohen M. 2001. Germplasm with high yield potential. In: 2001 annual report. Plant Breeding, Genetics, and Biochemistry Division. International Rice Research Institute, Los Baños, Philippines. p 4-5.

Moncada P, Martinez CP, Borrero J, Chatel M, Gauch H Jr, Guimaraes E, Tohme J, McCouch SR. 2001. Quantitative trait loci for yield and yield components in an *Oryza sativa* × *Oryza rufipogon* BC_2F_2 population evaluated in an upland environment. Theor. Appl. Genet. 102:41-52.

Peng S, Cassman KG, Virmani SS, Sheehy J, Khush GS. 1999. Yield potential trends of tropical rice since the release of IR8 and the challenge of increasing rice yield potential. Crop Sci. 39:1552-1559.

Peng S, Khush GS, Cassman KG. 1994. Evaluation of a new plant ideotype for increased yield potential. In: Cassman KG, editor. Breaking the yield barrier. Proceedings of a Workshop on Rice Yield Potential in Favorable Environments. Manila (Philippines): International Rice Research Institute. p 5-20.

Peng S, Laza R, Visperas RM, Khush GS, Virk PS. 2004. Rice: progress in breaking the yield ceiling. New directions for a diverse planet: Proceedings of the 4th International Crop Science Congress, Brisbane, Australia, 26 Sep.-1 Oct. 2004. Published on CD. Online at www.cropscience.org.au.

Tanksley SD, Nelson JC. 1996. Advanced backcross QTL analysis: a method for the simultaneous discovery and transfer of valuable QTLs from unadapted germplasm into elite breeding lines. Theor. Appl. Genet. 92:191-203.

Virk PS, Khush GS. 2003. New plant type in rice. In: Proceeding of the National Symposium on Priorities and Strategies for Rice Research in High Rainfall Tropics. 10-11 October 2002, Pattambi, India. p 7-16.

Virk PS, Khush GS, Virmani SS. 2003. Breeding strategies for enhancing heterosis in rice. In: Virmani SS, Mao CX, Hardy B, editors. Hybrid rice for food security, poverty alleviation, and environmental protection. Proceedings of the 4th International Symposium on Hybrid Rice, 14-17 May 2002, Hanoi, Vietnam. Los Baños (Philippines): International Rice Research Institute. p 21-28.

Virmani SS, Kumar I. 2004. Development and use of hybrid rice technology to increase rice productivity in the tropics. Int. Rice Res. Notes 29(1):10-19.

Xiao J, Grandillo S, Ahn SN, McCouch SR, Tanksley SD, Li J, Yuan L. 1996. Genes from wild rice improve yield. Nature 384:223-224.

Xiao J, Li J, Grandillo S, Ahn SN, Yuan L, Tanksley SD, McCouch SR. 1998. Identification of trait-improving quantitative trait loci alleles from a wild rice relative, *Oryza rufipogon*. Genetics 150:899-909.

Notes

Authors' address: International Rice Research Institute, DAPO Box 7777, Metro Manila, Philippines.

From allele engineering to phenotype

P. Hervé

The delivery of improved rice varieties with C_4-like yield advantages will require a precise modeling of the genome. Interestingly, the novelty of C_4 photosynthesis appears to have evolved from a key reservoir of genes present also in C_3 plants through gene duplication and neofunctionalization. Both specific patterns of gene expression and kinetic properties of the enzymes together with anatomical changes are observed in C_4 plants. Improved photosynthesis in rice can probably be achieved by engineering alleles involved in biochemical pathways and plant development. Key limiting steps in C_3 photosynthesis have also been envisaged as a possible strategy to lead to yield increases. From these results, I believe that a dual approach based on directed evolution and allele engineering could boost the yield potential of rice. Here, we discuss briefly some possible options from allele engineering to phenotype.

Keywords: photosynthesis, rice, genome, transgenic

From the agricultural perspective, C_4 photosynthesis is contemplated as a powerful engine leading to high yield performance, in particular in the context of a changing environment and suboptimal field conditions such as drought and high air temperature at the canopy level. An improvement in photosynthesis to increase yield is particularly relevant in the context of an elevated CO_2 environment (Long et al 2006). Equipped with all C_3 photosynthesis mechanisms, C_4 plants exhibit an optimized and efficient carbon metabolism based on C_4 dicarboxylic acids (aspartate, oxaloacetate, and malate) that leads to high photosynthetic performance, and highly efficient use of nitrogen and water. Importantly, one major feature of C_4 photosynthesis is the lack of oxygenase activity catalyzed by Rubisco (photorespiration). But only a limited number of crops such as maize, sugar cane, and sorghum possess C_4 advantages. Other major crops such as soybean and rice are classified as C_3 crops and the introduction of C_4-like mechanisms may drastically increase photosynthesis efficiency of these species, which could lead to higher yield performance under different environments. Key genes involved in the C_4 pathways probably evolved from a set of existing alleles in ancestral C_3 plants that did not function in photosynthesis (Monson 2003). Higher promoter activity and cell-specific or organ-specific expression are a clear difference between C_3 and C_4 genes. The enzymes in C_3 and C_4 plants differ distinctly in kinetics, subcellular localization, and regulatory properties. An approach that would switch rice photosynthesis to a C_4 format thus appears challenging. Alternatively, an increase in the amount of enzymes in the regenerative phase of the C_3 cycle has been envisaged as a possible target for improving carbon fixation in C_3 plants (Raines 2003).

As recently reviewed by Raines (2006), three major transgenic approaches have been assessed to reduce photorespiration: Rubisco protein engineering, genetic manipulation of the C_2 photorespiratory cycle, and the introduction of CO_2 concentrating mechanisms into C_3 plants. So far, direct protein engineering of Rubisco has not been sufficiently explored to be a successful strategy. Despite a loss of 25% carbon, the C_2 photorespiratory cycle is probably required to sustain photosynthesis in C_3 plants by regenerating ADP and NADP under normal conditions. As a consequence, direct manipulation of this pathway does not seem straightforward. The introduction of an alternative and complete glycolate catabolic cycle into chloroplasts in order to release CO_2 directly within the chloroplast is under investigation. In past years, it has been demonstrated that transgenic rice possessing a single C_4 enzyme did not increase photosynthesis efficiency despite a modified carbon metabolism (for review, Matsuoka et al 2001, Häusler et al 2002, Miyao 2003). The yield advantage of the single C_4 gene approach for rice has never been shown. It is possible, however, that a beneficial physiological advantage can be detectable only under certain conditions and phenotypic evaluation of transgenic material often lacks robust and reliable methods. The insertion of multiple genes with different promoter-gene combinations has also been envisaged, but so far the yield advantage of these transgenic plants has not been demonstrated. Complete leaf anatomy changes that are required for true C_4 mechanisms still need to be addressed even if such a morphological adaptation is perhaps not a prerequisite for improving photosynthesis in C_3 plants. From these previous attempts, it is clear that spatial localization of the proteins is crucial, and targeted expression of the genes in the appropriate cell or tissue compartment is required.

Recent improvements in genome engineering, sequencing of rice genomes, functional genomics, proteomics, and phenotyping tools now offer exciting possibilities to boost the photosynthetic efficiency of rice. In addition, the available genetic resources for rice appear to be an important and valuable reservoir of key genes. In this chapter, we discuss the different components of possible scientific strategies. Although biosafety and regulatory issues are important, they are not discussed here.

Candidate genomic regions

Extensive functional genomics studies in rice (including the available mutant collections) have greatly enhanced our capacity to identify candidate genome regions for particular traits and have accelerated the discovery of promoters specific to cells or tissues and other *cis*-regulatory sequences. The availability of several plant genome sequences including rice and the use of bioinformatics tools also trigger a better reading of the genome structure. The introduction of C_4-like mechanisms in rice has been attempted by targeting limiting steps or introducing alternative pathways thanks to the overexpression of key genes. Although it could be a valuable approach, there are clear limitations and, so far, it has not been successful in enhancing yield. Interestingly, the *cis*-regulatory sequences in the C_4 genes confer both cell-specific and high expression. This is well illustrated by the significant increase in PEPC and PPDK activity that was achieved by introducing in rice the whole maize genomic region, including promoter region, exons, introns, and the terminator (Ku et al 1999, Fukayama et al 2001). In other words, a strategy based on the whole genomic region (functional gene in its chromosomal context) rather than cDNA is probably preferable even though the size of the DNA sequence may be limiting. For the genes already present in rice, the lack of some regulatory elements rather than negative regulatory mechanisms is probably a limiting factor toward C_4 photosynthesis. Engineering the expression of the native rice genes, rather than introducing large genomic fragments, could be an attractive strategy. The possible options and current technical limitations are discussed below. Transcription factor genes or other noncoding regulatory sequences that are involved in the photosynthetic mechanisms or regulatory network could be other interesting targets.

The size and the number of genes to be engineered is a key element in defining technical strategies and road maps toward a biotech product, as described below.

Genome engineering and directed evolution strategies

Genome engineering of cereal crops has been significantly improved during the last decade, in particular with the development of robust and reliable *Agrobacterium*-mediated protocols. A range of indica and japonica rice can now be genetically engineered using these available protocols (Hervé and Kayano 2006, Hiei and Komari 2006). The insertion of large fragments as mentioned above, however, could be technically difficult with some genotypes because of low transformation efficiency. Interestingly, other methods using new gene delivery vehicles are under development and may be available on a medium term. A limited number of genes contained in a single medium-sized cassette (20–30 kb) could easily be transferred in rice by available methods. If large genomic regions (more than 30 kb) or multiple genes are to be engineered, three major strategies can be envisaged. Large fragments can be inserted in modified vectors and introduced in one locus of the rice genome. Multiple genes could also be stacked in one single recipient or locus in a step-by-step manner via site-specific recombination technology (Ow 2005). Such an approach offers the advantage of selecting the best transgenic events after each round of gene transfer. Several other technologies have been recently developed in the private sector. For example, the proprietary chromosome-based engineering system, called ACE platform, designed by Agrisoma Inc., could be useful for multiple gene delivery and expression using standard technologies. For rice, this system could be evaluated by creating an ACE platform on rice chromosome 9 or 10 (Agrisoma Inc., personal communication). A different system based on an artificial chromosome has been developed by the company Chromatin Inc. Alternatively, several independent events each containing one or two genes can be generated in parallel and combined by crossing. The advantage of this approach is the ease of transformation, the ability to screen the best events independently prior to crossing, and the possibility of multiple independent crossing combinations. Marker-assisted backcrossing can now be used to facilitate and accelerate the necessary breeding steps.

A very interesting and powerful approach could be based on the modulation of the expression of native C_3 genes that may lead to C_4 features in rice. Although site-specific recombination or targeted insertion has been demonstrated in rice (Terada et al 2002), there are still technical limitations to using such strategies on a short term. Nevertheless, the use of the zinc finger nuclease (ZFN) approach (Wright et al 2005, D. Voytas, personal communication) or modified transactivation system (O. Coltsaftis, personal communication) could be valuable for a limited number of genes. For example, the ZFN-based technology offers a tool for targeted genomic manipulation in humans, plants, and insects. Zinc finger nucleases are synthetic proteins consisting of an engineered zinc finger DNA-binding domain covalently linked to an endonuclease domain. They can induce double-stranded breaks in targeted DNA sequences that promote both homologous and nonhomologous recombination and thereby promote targeted genomic manipulation of genomic loci. The main technical challenge of ZFN technology is the design of specific zinc finger domains targeted to any genomic locus of interest. The other example, based on the available GAL4 transactivation system, may increase the expression of a native gene while preserving spatial and temporal specificity of the native promoter (O. Coltsaftis, personal communication). In short, the concept is based on the targeted insertion of an activator cassette (using the GAL4: UAS system) between the promoter and the coding region of the gene of interest, which would allow the expression of the native gene in larger amounts by a positive feedback loop triggered by the activator. These technologies could be used to induce targeted modification in both *cis*-regulatory regions of C_4 genes and the coding region to trigger a new gene expression pattern or protein features or both.

Finally, all published results reported the overexpression of different genes to enhance photosynthesis. It may be particularly useful and powerful to envisage strategies based on down-regulation or a combination of down-regulation and overexpression to drive decisive physiological changes. In other words, one should investigate several and new mechanisms for gene switching to trigger changes.

Directed evolution of proteins using random mutagenesis or gene shuffling and high-throughput screening is now a reality to improve protein catalytic properties for therapeutic and industrial enzymes. It is obvious that Rubisco enzyme with an enhanced CO_2/O_2 specificity would improve photosynthesis. Both existing natural variation of the enzyme and directed evolution may be beneficially used to boost photosynthesis in rice (Galmes et al 2005, Yu et al 2005). It is interesting to note that species growing in a dry and hot environment may exhibit different specificity of Rubisco. It would be valuable to analyze the natural variation of wild species, which could lead to the identification of better natural Rubisco.

The hypothesis that C_4 genes and mechanisms may have evolved from an existing reservoir of genes in ancestral C_3 plants offers the exciting challenge of accelerating the evolution of C_3 crops toward C_4. The major factors that may enhance such a process remain to be elucidated. It is unlikely that growing rice in extreme conditions would rapidly induce C_4 features but improved photosynthesis could be acquired. It is questionable, however, whether the presence of one or a limited number of C_4 mechanisms combined with specific growing conditions would accelerate such an evolution. One strategy may thus consist of growing transgenic rice plants with C_4 features in different environments and screening for newly acquired C_4 features. To our knowledge, such an approach has not been attempted so far but a single gene may affect plant metabolism so significantly that it leads to adjustments of interconnected pathways and to a new equilibrium. It remains to be seen whether a single gene that confers a new metabolic property may also affect the mutation rate of genes that encode enzymes involved in the same pathways. A very exciting challenge would be to identify a gene or a few genes that may trigger major changes at both the biochemical and developmental level under specific environmental conditions.

Whatever the general approach is, the choice of germplasm will be an important decision. Although a major constraint is the time that is required to develop an improved variety, the yield gain to be expected may equip this cultivar with major advantages over new released cultivars in the next 10–20 years. The material should be suitable for breeding and genome engineering. An elite germplasm is not necessarily the most preferred recipient since new improved cultivars will be made available in the meantime. More importantly, the material should be well known by breeders in order to facilitate its use. As indicated above, marker-assisted backcrossing greatly facilitates the transfer of useful genomic regions from one donor cultivar to a recipient. With the existing technology, IR64 could be the preferred indica recipient because of a recent breakthrough transformation method for that cultivar (Hervé, unpublished data) and its wide use by breeders and farmers since its release.

Phenotyping

The final product of any breeding strategy is a phenotype with a desired feature. The road map of a crop biotech product includes an essential breeding step and robust phenotypic screening. Furthermore, the directed evolution strategy and screening of wild species would also require very efficient phenotyping tools. So far, the limited success rate of genome engineering strategies is partly due to a lack of reliable phenotyping together with a limited scale of experiments. With the genome engineering methods currently used, the first step requires the capability of generating a large number of independent events. The second step needs a large population of a limited number of events. Although the only valid result is crop productivity under field conditions, large-scale field trials with transgenic plants will be achieved for only a limited number of events. Robust early screening tools are thus absolutely required to both reduce the population size of transgenic events to be analyzed and to select the most likely successful events for field trials. This is particularly relevant if multiple independent strategies are tackled in parallel. There is thus a need to define the criteria and standard methods to be used for C_4-ness.

It is important that early screenings be technically applicable to a large number of independent events (with a limited number of plants per event), which is a prerequisite for unraveling a gene effect versus phenotypic changes due to somaclonal or position effects. Another type of screening must be applied to a larger number of plants of only a set of selected events. As mentioned above, early screening of transgenic material cannot be performed in the field but is restricted to confined environments such as greenhouses or contained screenhouses. Since the environmental conditions are very different from those in the field, the early screening methodology must be adapted. In the case of the introduction of C_4 genes, quantitative expression profiling of the transgenic events for a set of genes could be used since the higher expression of these genes is probably a prerequisite for success. Metabolic profiling tools could also be applied in order to identify the events showing significant changes in carbon metabolic pathways. Such profiles could also detect any unexpected changes in secondary metabolism. A single assay to measure Rubisco specificity would also be very useful as a systematic screening tool. A second step of the screening could be based on preliminary agronomic performance, photosynthesis, and leaf nitrogen content. This could probably be performed in a containment facility. Yield performance of a very limited number of selected events will be assessed ultimately in field trials.

Conclusions

The challenge of and opportunities for C_4 rice are obvious. As discussed briefly in this chapter, there are several possible approaches toward an improved photosynthesis for rice. The scientific strategies will depend on both the number of genomic regions to be engineered and the size of these regions. The most currently feasible road map could be based on single or double transformation using exogenous genes with combinations by crossing. The transgenic plants could be grown under particular environmental conditions to trigger additional changes. There are promising and exciting strategies for targeted locus engineering that may be used on a medium term if they are shown to be successful in rice. Since the genes that are required for C_4 exist in rice, it is theoretically possible to switch the expression of these genes to a C_4 mode. Whatever the strategy is, the different phenotypic procedures should be standardized in the short term in order to define the scale and feasibility of the whole road map toward a product available to farmers.

References

Fukayama H, Tsuchida H, Agarie S, Nomura M, Onodera H, Ono K, Lee BH, Hirose S, Toki S, Ku MSB, Makino A, Matsuoka M, Miyao M. 2001. Significant accumulation of C_4-specific pyruvate, orthophosphate dikinase in a C_3 plant, rice. Plant Physiol. 127:1136-1146.

Galmés J, Flexas J, Keys AJ, Cifre J, Mitchell RAC, Madgwick PJ, Haslam RP, Medrano H, Parry MAJ. 2005. Rubisco specificity factor tends to be larger in plant species from drier habitats and in species with persistent leaves. Plant Cell Environ. 28:571-579.

Häusler RE, Hirsch HJ, Kreuzaler F, Peterhansel C. 2002. Overexpression of C_4-cycle enzymes in transgenic C_3 plants: a biotechnological approach to improve C_3-photosynthesis. J. Exp. Bot. 53(369):591-607.

Hiei Y, Komari T. 2006. Improved protocols for transformation of indica rice mediated by *Agrobacterium tumefaciens*. Plant Cell Tissue Org. Cult. 85(3):271-283.

Hervé P, Kayano T. 2006. Japonica rice varieties (*Oryza sativa* Nipponbare and others). In: Wang K, editor. *Agrobacterium* protocols. Second edition. Volume I. Totowa, N.J. (USA): Humana Press. p 213-222.

Ku MSB, Agarie S, Nomura M, Fukayama H, Tsuchida H, Ono K, Hirose S, Toki S, Miyao M, Matsuoka M. 1999. High-level expression of maize phospho*enol*pyruvate carboxylase in transgenic rice plants. Nat. Biotechnol. 17:76-80.

Li L, Liu YG, Xu XP, Li B. 2003. Efficient linking and transfer of multiple genes by a multigene assembly and transformation vector system. Proc. Natl. Acad. Sci. USA 100:5962-5967.

Long SP, Zhu XG, Naidu SL, Ort DR. 2006. Can improvement in photosynthesis increase crop yields? Plant Cell Environ. 29:315-330.

Matsuoka M, Furbank RT, Fukuyama H, Miyao M. 2001. Molecular engineering of C_4 photosynthesis. Annu. Rev. Plant Physiol. Plant Mol. Biol. 52:297-314.

Miyao M. 2003. Molecular evolution and genetic engineering of C_4 photosynthetic enzymes. J. Exp. Bot. 54(381):179-189.

Monson RK. 2003. Gene duplication, neofunctionalization, and the evolution of C_4 phtosynthesis. Int. J. Plant Sci. 164(3 Suppl.):543-554.

Ow DW. 2005. Transgene management via multiple site-specific recombination systems. In Vitro Cell. Dev. Biol. Plants 41(3):213-219.

Raines CA. 2003. The Calvin code revisited. Photosynth. Res. 75:1-10.

Raines CA. 2006. Transgenic approaches to manipulate the environmental responses of the C_3 carbon fixation cycle. Plant Cell Environ. 29:331-339.

Terada R, Urawa H, Inagaki Y, Tsugane K, Iida S. 2002. Efficient gene targeting by homologous recombination in rice. Nat. Biotechnol. 20:1030-1034.

Yu GX, Park BH, Chandramohan P, Geist A, Samatova NF. 2005. An evolution-based analysis scheme to identify CO_2/O_2 specificity-determining factors for ribulose 1,5-bisphosphate carboxylase/oxygenase. Protein Engin. Design Select. 18(12):589-596.

Notes

Author's address: International Rice Research Institute, Los Baños, Philippines.

Section 5:

Setting up the consortium

C_4 rice: brainstorming from bioinformaticians

R. Bruskiewich and S. Wanchana

With the sequencing of the rice genome and pending sequences of several C_4 cereals (i.e., maize and sorghum), functional genomics analysis may elucidate the critical genetic differences contrasting C_4 and C_3 photosynthetic biochemical and anatomical systems. Recent genomics findings may also deepen our understanding of the evolution of the C_4 syndrome, pointing at the evolutionary processes involved and possibly the plasticity of the rice genome for conversion to C_4 photosynthesis. Bioinformatics systems and methodology serve as critical glue for cross-linking such genomic information into an interpretable framework.

Keywords: crop structural and functional genomics, bioinformatics, evolution

The genome (of an organism) is the entire DNA content of a cell, including all of the genes and all of the intergenic regions.[1] The field of genomics strives to characterize genome structure (by sequencing) and function (by experimental and bioinformatics analysis).

A general paradigm of modern germplasm-based crop research is that high-throughput characterization of the structure and function of genomes of diverse germplasm (including specialized collections such as mapping populations and mutants) can reveal by functional annotation (obtained from forward and reverse genetics, gene expression experiments, and similar techniques) many specific targets ("genes") important in many plant traits. Such genes become the targets for high-throughput molecular characterization of germplasm to identify genetic polymorphism ("alleles") causally associated with a desirable phenotype. Genes and alleles with positive trait values can be subsequently applied to crop improvement using gene transfer technologies, such as marker-assisted selection (MAS) and transgenic transformation.

[1] A definition borrowed, with slight modification, from Brown (1999).

What is the role of bioinformatics?

Bioinformatics is the discipline integrating the tools and techniques of mathematics and statistics, computing science, information technology, and the natural sciences to capture, analyze, store, represent, and disseminate biological information from a variety of sources to generate results that guide further experimentation or provide valuable insights about biology. As such, bioinformatics is a critical tool for genomics data management and analysis, given the explosive growth in the volume and complexity of genomic data.[2]

The application of informatics to crop research is not a new activity. The International Rice Research Institute (IRRI) already had an IBM computer storing and analyzing crop data back in the early days of the Institute. However, the onset in the 1990s of high-throughput genome sequencing of plant genomes, the application of new laboratory technologies for probing gene function, and the rapid expansion of the Internet and online databases established bioinformatics as an essential tool for modern crop research (Bruskiewich et al 2006b).

Bioinformatics research and development now offer a rich combination of protocols, tools, databases, and computing infrastructure that can be applied to help answer biological research questions, often at a significant savings of time and laboratory resources. Bioinformatics can integrate information across a diverse collection of crop data about germplasm, genotype, phenotype, cellular expression (of transcripts, proteins, and metabolites), growth characteristics, applied treatments, and environmental conditions. These data are cross-linked and interpreted to develop a complete picture of genome function from DNA sequence, through RNA and protein structures, into biochemical and cell structural characteristics interacting with the plant's environment, giving rise to the observable phenotype of the germplasm.

Of special interest here is the task of linking the impact of specific variations in germplasm DNA sequence (genotype) on all these components, resulting in the variation of morphology or behavior (phenotype) that one observes among distinct germplasm varieties in the field. Significant progress is anticipated in the near future to characterize such molecular variation in a wide range of representative rice germplasm (McNally et al 2006).

Crop (plant) genome sequencing

The first genome sequence of a model plant, *Arabidopsis thaliana* (Meinke et al 1998), was completed in December of 2000 (Pennisi 2000).

Draft sequences of rice were completed in 2002 (Yu et al 2002, Goff et al 2002), with completed chromosome 1 (Sasaki et al 2002) and chromosome 4 (Feng et al 2002),

[2] Database repositories of one common biological data type, sequence data, alone have increased at least 10-fold over the past five years, to more than 100 billion base pairs. See www.ncbi.nlm.nih.gov/Genbank/index.html.

and a finished genome published in 2005 (IRGSP 2005), with the genome annotation available online (Ohyanagi et al 2006, Yuan et al 2003).

Genome sequencing of gene-rich regions of larger crop genomes such as maize and sorghum is advancing rapidly and becoming publicly available (Chan et al 2006, Bedell et al 2005). Although the large size of some crop genomes may discourage complete sequence characterization, the available crop genome sequence resources and associated functional genomics data already provide a rich foundation for comparative studies (Paterson et al 2005). For comparative genomics among species, for example, see the comparison between wild species of rice (OMAP, www.omap.org; Wing et al 2005).

General observations about sequenced plant genomes

Gene identification is proving to be an especially challenging task in crop genomes. Early estimates of gene count suggested a larger complement of rice (monocot)-specific genes; however, this has subsequently been subject to considerable debate. In addition, in recent years, increased evidence for noncoding microRNA fragments has complicated the picture. The nature of whole-genome expression is a story remaining to be fully told.

Some things are clear. There is significant evidence for ancient whole and segmental genomic duplications, followed by gene loss or divergence (Yu et al 2005). Specific gene families often exhibit differential expansion in different clades (Rabinowicz et al 1999). Most eukaryote (especially plant) genomes are found to be replete with repetitive transposable elements. These latter sequence elements stir up the pot quite a bit (Morgante 2006).

- They serve as hot spots for homologous recombination, creating unequal crossing over, to create tandem gene duplications. Homologous recombination also generates significant structural rearrangements. These may be the norm rather than the exception, even within a species (at the micro-synteny level) and most certainly between species.
- They knock out genes: transposons can insert to create new gene structures and may even introduce novel genome regulatory signals such as *cis*-element binding sites and chromatin matrix attachment sites.

Novel genes can arise from transposable element insertion in animals and sometimes such novel genes can have DNA binding character with pleiotropic regulatory impact within the genome (Cordaux et al 2006, Jordan 2006). Although equivalent examples in plant genomes remain to be discovered, some observations suggest the possibility that this process also occurs in plants (Elrouby and Bureau 2000, 2001).

Framework for gene discovery

Accurate and complete functional annotation of the genetic function of genomes remains a daunting task. Efforts to characterize gene function are largely driven by a paradigm of "intersecting evidence" using experimental results about genes from the following features.

- *Position.* Quantitative trait locus results link segregating chromosome markers with plant traits.
- *Function.* Experimental or *in silico* analysis of biological sequences, combined with genetic dissection of rice mutants, can reveal the specific role of candidate genes in biological processes.
- *Expression.* Gene expression experiments at the transcript, protein, and metabolic level can associate the expression of specific genes with specific processes, both constitutively and under conditional treatments.
- *Selection.* Analysis of genetic resource collections (through association genetics) and bulk population selection experiments can identify genomic regions in linkage disequilibrium with traits or alleles conserved under selection pressure in short-term breeding screens. This again provides additional support for the role of specified loci and alleles in pertinent biological processes and phenotypes.
- *Crop models.* This a relatively new source of evidence, but it is expected that linking gene systems information with whole-crop models will help support or refute gene candidacy in specific field-level processes, such as photosynthetic yield.

A key objective of contemporary crop bioinformatics research and development is to provide effective information systems for capturing and integrating such intersecting experimental evidence, allowing researchers to use intersection sets of gene candidates to narrow their focus down to a few key candidate gene loci or alleles for further cost-effective laboratory or field validation.

Perspectives in modern biology

In contemplating the task of engineering C_4 into rice using genomics, one can view the challenge from various perspectives: evolutionary, molecular biology, and systems biology.

Evolutionary perspective

> "*... I submit that all these remarkable findings make sense in the light of evolution: they are nonsense otherwise. ... Seen in the light of evolution, biology is, perhaps, intellectually the most satisfying and inspiring science. Without that light it becomes a pile of sundry facts, some of them interesting or curious but making no meaningful picture as a whole....*"[3]

[3] Quote attributed to the American evolutionary biologist Theodosius Dobzhansky (1900-75) in his article for the *American Biology Teacher*, March 1973.

Genomes can use only what DNA is in front of them at any one time, or have DNA transferred to them from outside by reproductive or infection processes. Genome changes are generally incremental (a single or a small number of mutations or rearrangements) except under rare circumstances (e.g., viable whole-genome duplications or hybridizations, hyperactive infection of transposons). Some genome mutations can be irreversible (e.g., chromosomal rearrangements, insertions and deletions of sequence), driving a unidirectional genetic drift process.

Incremental steps (by genome mutation) that increase reproductive fitness may or may not get fixed in a species (population). Intermediates need to be fit enough only to survive (reproductively), until they make the transition to a state of significant competitive advantage for a given environment. Nature is patient: it has millions of years to experiment. Evolutionary selection acts on large, genetically diverse populations over many generations.

For example, for the predecessor of a C_4 species, one can make a crude estimate of the opportunity: perhaps 10^6 hectares of land (a small fraction of the world's land area) can host up to maybe 10^4 plants per hectare. This gives approximately 10^{10} meioses per year over 10^6 years, equaling 10^{16} meioses. This is a big number. Even with the suggested observed plant mutation rate of perhaps only about 10^{-6} mutations per gene locus per generation, one might still expect bountiful opportunities for mutation and selection in the system. C_4 appears to be "exapted"[4] from C_3 species that evolved to "local maxima" of fitness in hot, dry, high-light, low-CO_2 environments. The C_3 to C_4 evolution has clearly involved many changes. All of these changes would need to have been applied to the successful lineage in each C_4 species. This seems like a big task, but C_4 has evolved independently several times, so the process cannot be too special. One also suspects that these changes would mainly have been modest DNA changes to existing genes (with perhaps a few gene duplications).

Comparative genomics of biochemistry, development, and regulation across C_3/C_4

In the era of whole-genome sequencing, we can now conceivably guess what those critical changes were, through the application of comparative genomics analysis of a C_3 genome and a C_4 genome. Of course, the target species of our C_4 rice project, *Oryza sativa,* is a sequenced monocot C_3 genome. Fortunately for us, the gene-rich regions of two suitable C_4 monocot genomes, maize and sorghum, are swiftly being sequenced.

Already, comparative analysis of the chloroplast extra-nuclear genomes of these plants has been undertaken (Wakasugi et al 2001). A more comprehensive comparative analysis of the transcriptional and posttranslational regulation and metabolic activity of gene orthologs of C_4 enzymes across nuclear and chloroplast genomes of these species is certainly feasible.

---

[4] An "exaptation" is a biological adaptation where the biological functon currently performed by the adaptation was not the function performed while the adaptation evolved under earlier pressures of natural selection. See http://en.wikipedia.org/wiki/Exaptation.

These comparisons can focus on several salient features of the C_4 system:

- The biochemical enzymes in the C_4 syndrome[5] in comparison to their C_3 orthologs.
- The developmental trajectory of leaf tissues and cells, in particular
  - — genes controlling leaf cell differentiation and proliferation: the mesophyll cell (MC)–bundle sheath cell (BSC) dichotomy; MC proliferation relating to vein spacing;
  - — genes controlling the presence or absence and internal localization of chloroplasts in cells: character of chloroplasts within MC versus BSC; centripetal versus centrifugal positioning of chloroplasts.
- Genes controlling cell-cell channel proteins, plasmodesmata numbers, and cell-wall barriers that modulate metabolite or CO_2 penetration across MC-BSC boundaries.
- Regulatory systems of all of the above.

The potential of *Arabidopsis thaliana* is not to be ignored in the light of the above comparisons, since, even though it is a dicotyledon, research on this model plant is swiftly probing all the boundaries of fundamental plant biology. Although monocots have clear differences from dicots, a common Magnoliophyta (Angiospermae or flowering plant) ancestry bespeaks of significant common biology that can point the way to the genes involved in fundamental processes of interest in the C_4 rice project.

Preliminary molecular phylogenetic trees of the peptide sequences of various biochemical enzymes involved in C_4 processes—for example, phospho*enol*pyruvate carboxylase (PEPC), pyruvate orthophosphodikinase (PPDK), NADP-malic enzyme (ME), NADP-malate dehydrogenase (MDH), phospho*enol*pyruvate carboxykinase (PEPCK), and the small subunit (RBCS) of ribulose-1,5-bisphosphate carboxylase/oxygenase (Rubisco)—are informative (Wanchana, personal communications, data not shown). There are some gene duplications (i.e., a C_4-specific PEPC in a phylogenetic tree outlier branch) and some apparent sequence-level differences introducing phylogenetic distance between C_3 and C_4 versions of the same gene (e.g., in RBCS and possibly PPDK and MDH).

Molecular biology perspective

Figure 1 is a basic road map for the molecular scope of biology. It is beyond the scope of this paper to dwell upon it—it is self-explanatory. The take-home message is that a full understanding of any biological system requires a complete cataloguing of the relationship between sequence variation at the DNA level and the observed functional variation at the molecular level: in terms of the stability and function of molecular structure, biochemical activity, and gene (product) regulation.

[5] Too numerous to list here, but practitioners in the field largely know what they are.

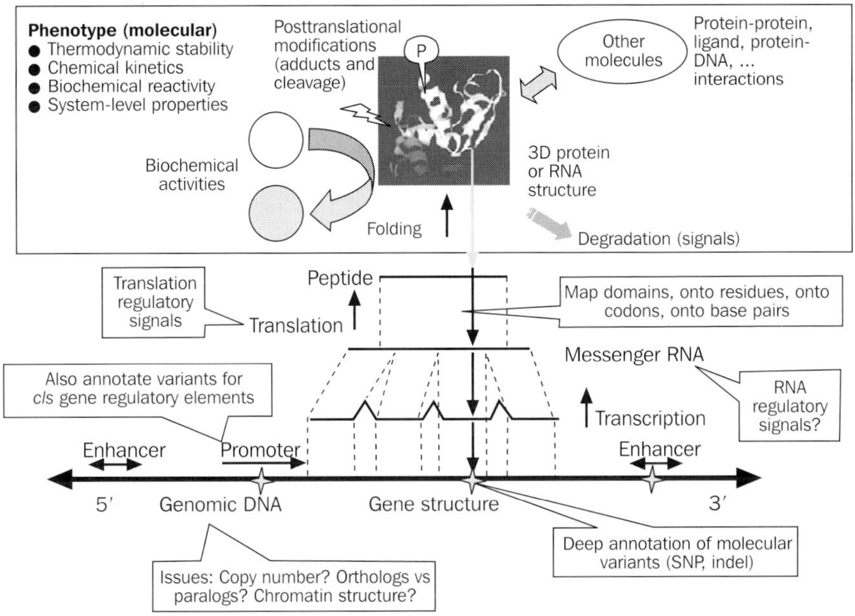

Fig. 1. From genome to molecular phenotype through the central dogma of biology (DNA to RNA to protein).

Systems biology perspective

Plants, like all organisms, which are often quite aptly called "living systems," cannot simply be reduced to their component parts (genes, molecules). Rather, all such parts synergistically combine together into complex systems.

It is not surprising, therefore, that after taking organisms apart by genome sequencing and biochemical analysis, scientists are driven to try to put those systems back together, using bioinformatics, in order to gain a more realistic perspective on the interactions and emergent properties inherent in living organisms. This "systems biology" perspective may embrace a range of systems:

- *Regulatory networks:* signal transduction, genetic regulation
- *Metabolic networks:* biochemical fluxes
- *Structural networks:* morphology (e.g., chloroplast positioning)
- *Physiological systems:* whole-plant development and function

Understanding the C_4 photosynthetic syndrome will require special attention to this perspective. It involves a complex biochemical and anatomical structure, including both nuclear and organellar genomes.[6]

[6] See the laboratory site http://chloroplast.net of the Nagoya Plant Genome group led by Masahiro Sugiura for the chloroplast view of this complexity, in particular the biochemistry of C_4 carbon fixation illustrated at http://chloroplast.net/nge/map/m01.html.

It is pertinent to note that evolution acts on whole living systems, not really just their component parts. For systems too, optimization is a global task, not local. This is already apparent in C_4 in the apparent trade-off between increased catalytic turnover of Rubisco and reduced affinity for CO_2 (Leegood 2002). Although reduced affinity for CO_2 may seem, at first consideration, to run contrary to the goal of increasing CO_2 fixation rate, such decreased affinity is not a problem since the C_4 syndrome is an optimization at the whole-system level, providing for a local compartmental increase in the level of CO_2 that saturates the enzyme, allowing the increased turnover rate of CO_2 to be of greater benefit.

Some brainstorming

In this section, we will brainstorm about how we might further apply genomics, sequencing information, and bioinformatics to the problem of introducing C_4 photosynthesis into rice.

Naïve idea 1—accelerated evolution: screen rice (and maize?) mutant pools for C_4 traits

Significant mutant genomic stocks are now available for rice (Hirochika et al 2004, Wu et al 2005). These stocks are generated by chemical or radiation mutagenesis from interesting backgrounds, usually elite rice genotypes of relatively high agronomic (IR64, Tainung 67) or scientific (Nipponbare, the IRGSP sequenced genome) importance. Some of these stocks exhibit startling morphological diversity: there is a field of Tainung 67 mutants in Taiwan that seem like an apparent model of rapid reverse evolution, with enormous visible morphological diversity from one source background derived from sodium azide (NaN_3) mutagenesis (Jeng et al 2003).

Perhaps, too, some mutation breeding and random crossing of pools of mutant germplasm will generate novel allelic combinations for high-throughput phenotypic characterization for C_4-like features.

Exploration of the pool of available mutants in *Arabidopsis thaliana* and *Zea mays*, and perhaps other relevant species, to identify mutants with disruption of key biochemical and anatomical components likely to be involved in C_4, is also likely to be profitable. The former species is of interest given the wide extent of fundamental developmental biology already known. The latter is, of course, a relevant cereal monocot relative of rice hosting the C_4 syndrome itself.

A short bioinformatics exercise illustrates the potential value of these mutant stocks. One interesting trait to examine in C_4 is the positioning of chloroplasts in leaf cells. As outlined earlier in this conference, there is an interesting *A. thaliana* mutant, the *Chloroplast Unusual Positioning 1* (CHUP1) mutant. Using The Arabidopsis Resource (TAIR; www.arabidopsis.org) online database to identify the Genbank sequence (NP_189197), then searching the IRGSP Rice Annotation Project genome database (Ohyanagi et al 2006), identifies a homologous rice locus (Os12g0105300) in the IRGSP assembly. Searching one of the available rice mutant databases (TOS 17 at NIAS; Hirochika et al 2004) for mutants hitting this gene locus reveals a rich allelic

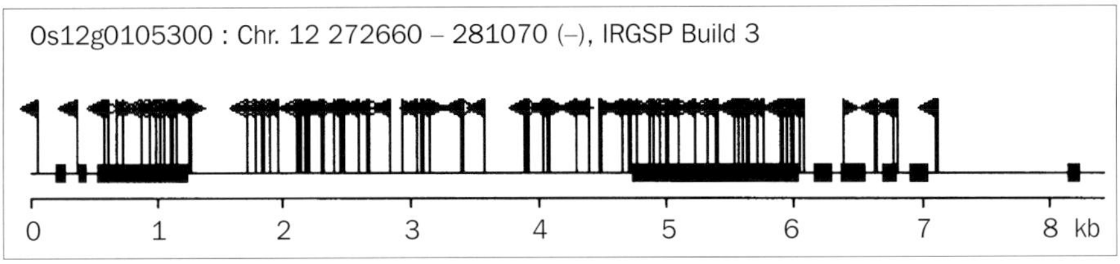

Fig. 2. TOS 17 rice mutants for *Arabidopsis thaliana* CHUP1 mutant. This diagram represents the sequence rice gene ortholog (*Os12g0105300*), along a scale in kilobase pairs (kb), extracted from the IRGSP database. The squares represent the exons of the gene. The triangles represent TOS 17 insertions along the gene where a mutation has been identified in a mutant recorded in the TOS 17 database (http://tos.nias.affrc.go.jp/).

series of available mutant stocks with inserts into this gene (Fig. 2). The phenotype of these mutants is telling: generally dwarf and of low fertility. It would be natural now, of course, to examine the leaf chloroplast phenotype of one of the mutants under a microscope.

Naïve idea 2—study gene and regulatory changes due to transposable element insertions?

Sequenced plant genomes, including rice, exhibit a considerable number and diversity of repetitive sequences due to the insertion of T-DNA and retro-transposable elements from diverse families (IRGSP 2005). Such a large background of transposon activity may have played a significant role in the evolution of complex systems such as the C_4 syndrome.

One can illustrate the possibilities with a small exploratory analysis. This analysis is not scientifically rigorous, but it does illustrate the possibilities.

The *cis*-element GATAAG, also known as an "I-box," is found in several promoters of genes involved in C_4 (see Table 2 from Sheen 1999) and is generally suggested to be involved in light-regulated and/or leaf-specific gene expression of photosynthetic genes. Interestingly, the tomato I-box binding factor, LeMYBI, encodes a member of a class of Myb-like proteins that binds specifically to the I-box sequence of the RBCS1, RBCS2, and RBCS3A promoters, that is, to the GATAAG in the promoter of the RBCS genes that code for the small subunit of Rubisco (Rose et al 1999). Myb genes are a family of transcription factors widely distributed in the higher plants, making up one of the largest known families of regulatory proteins, and are thought to have considerable latitude for evolutionary modulation of binding and activation specificity for diverse biochemical pathways (Dias et al 2003). Although somewhat tangential to our small bioinformatics analysis here, it is interesting to mention that there is some evidence that the insertion of transposable elements can affect the expression of Myb genes themselves (Pooma et al 2002).

A search for the GATAAG motif in the 623 "characterized" repeats of the TIGR Maize Repeat Database (http://maize.tigr.org/repeat_db.shtml, accessed 15 July 2006) reveals that 226 (or about 36%) of the repeat sequences contain the GATAAG motif. Incidentally, other motifs of interest are also observed, such as the CCAAT *cis*-binding motif in 367 (59%) of the sequences.

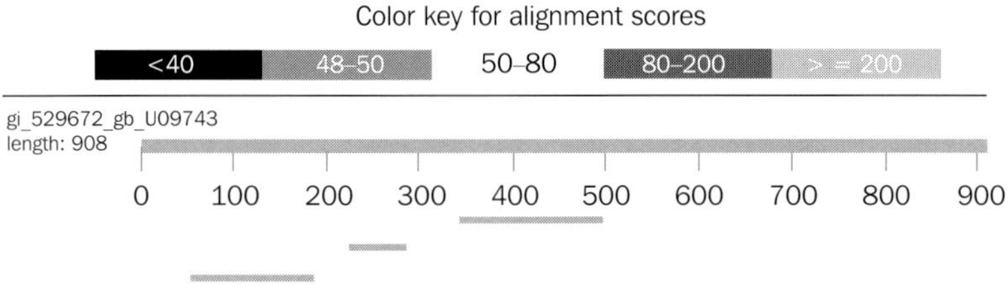


```
                                                                          Smallest
                                                                            sum
                                                                    High  probability
Sequences producing high-scoring segment pairs:                    score  P(N)      N


ZRSiTETN00T0008 gi|1840117|gb|U85626.1|ZMU85626 Zea mays ...       191  0.75       1
ZRSm0T0T00201446 family_583_C8                                     138  0.96       1
ZRSgTERT00T13633 0GJAJ83TV||nt17-836|ZRSiTERT00T0212 puta...       174  0.998      1

>ZRSiTETN00T0008 gi|1840117|gb|U85626.1|ZMU85626 Zea mays PIF-12 transposable
 Score = 191 (34.7 bits), Expect = 1.4, P = 0.75
 Identities = 103/166 (62%), Positives = 103/166 (62%), Strand = Plus / Plus
Query:   342 GAGGCGGAAT-TGC--CATGAATTCGAGGTAGCTAAGTAAG-GCGCATGCATATATATGC 397
             || |||||||| ||   | |||  || | | ||   ||||    | |  | |  ||  | ||
Sbjct:  2170 GAAGCGGAATCTGTTTCCTGATTTAG-GTTAAGAAAGTTTCAGAGACTTCTAATTCAGGC 2228
Query:   398 CAAA-AATTC-T--ACTGTCACTTTCCAATTTCAATGCGCTGCCAAACAAGCCATCCTGG 453
             | || |||   |  |  || | ||||||||| | ||    ||||||||||  || ||
Sbjct:  2229 CTAAGAATGGATGGAAAGTTAGTTTCCAATT-CTATAGTTTGCCAAACAAATCAGCCCAA 2287
Query:   454 AAACTGACTTG-AATTCAGCCCAATTCTGTAGATCCAAACAGGGCC 498
              || | | ||  |||||||||||||||      |  ||||| |||||
Sbjct:  2288 GAATTAAATTCCAATTCAGCCCAATTCCAATCAAACAAAC-GGGCC 2332

>ZRSm0T0T00201446 family_583_C8
 Score = 138 (26.8 bits), Expect = 3.3, P = 0.96
 Identities = 46/63 (73%), Positives = 46/63 (73%), Strand = Minus / Plus
Query:   285 CTATTCTATTTTTCTACAAAAAATAATTACCCCAGCCTCAGCTATTTATGAA-AAATATA 227
             |||||| ||||||  || |||||||||||     ||||  | ||   || | || ||||||
Sbjct:    61 CTATTC-ATTTTTGAACTAAAAATAATTAAG--AGCCCTAACTTATTGTAAAGAAATATT 117
Query:   226 TGA 224
             |||
Sbjct:   118 TGA 120

>ZRSgTERT00T13633 0GJAJ83TV||nt17-836|ZRSiTERT00T0212 putative retrotransposon;
 Score = 174 (32.2 bits), Expect = 6.3, P = 0.998
 Identities = 90/141 (63%), Positives = 90/141 (63%), Strand = Minus / Plus
Query:   186 TATCAGTCATCACGAAAGAAAGGT-AAATCTA-AAACTCCTCCTAAAAATGTTATTTAAT 129
             | |||  ||     || | |||  | ||||||  ||| | || || ||||||||||| | |
Sbjct:   404 TTTCAAACAGGGGGAGA-AAATTTCAAATCTTGAAAATGCTTCTCAAAATGTTATTCATT 462
Query:   128 TACCCA--AC-ATGCCTAAAACTGCATTTAAG-GATATCTACTATGAAAGGTATTGCAAA 73
             ||||     || ||     ||||   | || ||  || || ||| |  ||||    |||||||
Sbjct:   463 TACCTTTGACTATTTGCAAAAAGACTTTGAAAAGA-ATTTAC-A--AAAGAATTTGCAAA 518
Query:    72 ATAACACACTAAATG-TGCCA 53
             | |||| |  || || ||| |
Sbjct:   519 AAAACAAAACAAGTGGTGCAA 539
```

Fig. 3. Basic Local Alignment Search Tool (BLAST; Altschul et al 1990) nucleotide alignment to the TIGR maize repeat database (http://maize.tigr.org/repeat_db.shtml), of the *Zea mays* Rubisco small subunit promoter sequence, retrieved as Genbank entry U09743.1 from the National Center for Biotechnology Information (NCBI; www.ncbi.nlm.nih.gov). The pairwise sequence alignment suggests the presence of retrotransposable element inserts that could potentially influence promoter activity.

A representative *Zea mays* Rubisco small subunit promoter sequence retrieved from NCBI (ZMRBCSM3S1, Genbank accession number U09743.1) was used as a query sequence to search the TIGR BLAST sequence alignment server (http://tigrblast.tigr.org/tgi_maize/index.cgi, accessed 15 July 2006). Three transposable element entries in the maize repeat database were hit (ZRSiTETNOOT0008, ZRSmOTOT00201446, and ZRSgTERTOOT13633; Fig. 3). The alignments are not particularly highly significant but only suggestive, perhaps ancient.

Table 1. PLACE database search of putative transposable element subsequences of the *Zea mays* Rubisco small subunit promoter sequence retrieved from NCBI (Genbank entry # U09743.1).

Factor or site name	Signal sequence	Place entry	No. of hits
CAATBOX1	CAAT	S000028	5
CACTFTPPCA1	YACT	S000449	5
GT1CONSENSUS	GRWAAW	S000198	5
DOFCOREZM	AAAG	S000265	4
GATABOX	GATA	S000039	4
TATABOX5	TTATTT	S000203	3
WRKY71OS	TGAC	S000447	3
-300ELEMENT	TGHAAARK	S000122	2
CARGCW8GAT	CWWWWWWWWG	S000431	2
CCAATBOX1	CCAAT	S000030	2
EECCRCAH1	GANTTNC	S000494	2
GT1GMSCAM4	GAAAAA	S000453	2
GTGANTG10	GTGA	S000378	2
POLLEN1LELAT52	AGAAA	S000245	2
SEF4MOTIFGM7S	RTTTTTR	S000103	2
WBOXHVISO1	TGACT	S000442	2
WBOXNTERF3	TGACY	S000457	2
-10PEHVPSBD	TATTCT	S000392	1
AGMOTIFNTMYB2	AGATCCAA	S000444	1
ANAERO3CONSENSUS	TCATCAC	S000479	1
ARR1AT	NGATT	S000454	1
BOXIINTPATPB	ATAGAA	S000296	1
L1BOXATPDF1	TAAATGYA	S000386	1
MYBST1	GGATA	S000180	1
POLASIG2	AATTAAA	S000081	1
POLASIG3	AATAAT	S000088	1
RAV1AAT	CAACA	S000314	1
ROOTMOTIFTAPOX1	ATATT	S000098	1
RYREPEATBNNAPA	CATGCA	S000264	1
RYREPEATGMGY2	CATGCAT	S000105	1
RYREPEATLEGUMINBOX	CATGCAY	S000100	1
SEBFCONSSTPR10A	YTGTCWC	S000391	1
SITEIIATCYTC	TGGGCY	S000474	1
SORLIP2AT	GGGCC	S000483	1
TATABOXOSPAL	TATTTAA	S000400	1

Extracting the subsequences of the promoter aligning with those repeat elements and searching the PLACE *cis*-element database (Higo et al 1999, Prestridge 1991) reveal a number of interesting putative *cis*-regulatory signals (Table 1). Unfortunately, GATAAG is not obviously contained in this list (that motif appears to be farther downstream in the promoter, circa nucleotides 692 to 697), although several closely related motifs are listed (e.g., GATABOX and MYBST1).

Perhaps of slightly greater interest are *cis*-elements noted with the highest number of hits, listed here with the definitions and representative citations as recorded within the PLACE database (Higo et al 1999).

- **CAATBOX1:** CAAT promoter consensus sequence thought to be involved in tissue-specific regulation of genes (Shirsat et al 1989).
- **CACTFTPPCA1:** tetranucleotide (CACT) is a key component of Mem1 (mesophyll expression module 1) found in the *cis*-regulatory element in the distal region of the phospho*enol*pyruvate carboxylase (ppcA1) of the C_4 dicot *Flaveria trinervia* (Gowik et al 2004).
- **GT1CONSENSUS:** which is the GT-1 binding site in many light-regulated genes (Zhou 1999).
- **DOFCOREZM:** DE core site required for binding of Dof proteins in maize; Dof proteins are DNA binding proteins, with presumably only one zinc finger, and are unique to plants; four cDNAs encoding Dof proteins, Dof1, Dof2, Dof3, and PBF, have been isolated from maize; PBF is an endosperm-specific Dof protein that binds to a prolamin box; maize Dof1 enhances transcription from the promoters of both cytosolic orthophosphate kinase (CyPPDK) and a nonphotosynthetic PEPC gene; maize Dof2 supressed the C_4 PEPC promoter (Yanagisawa 2000).
- **GATABOX:** GATA box; GATA motif in CaMV 35S promoter; binding with ASF-2; three GATA box repeats were found in the promoter of petunia chlorophyll a/b binding protein, *Cab22* gene; required for high-level, light-regulated, and tissue-specific expression; conserved in the promoter of all LHCII type I Cab genes (Reyes et al 2004).

Thus, in the regions of the promoter spanning sequences exhibiting apparent (though weak) homology with transposable element sequences, there appear to be a significant number of interesting *cis*-element motifs pertinent to the regulation of the gene in question. So, it is conceivable that transposable element insertion played a role in the formation of this and other C_4 promoters. However, the evidence is not conclusive at this point.

Transcriptional activation or suppression is conjectured to involve regulatory modules of a combinatorial set of cooperatively binding transcription factors expressed in a tissue-specific manner and binding multiple *cis*-element sites, in particular orientations, and separated by specific distances, with defined protein-protein interactions—another complex (local) system.

How such complex regulatory modules are created over the time scale of plant evolution is difficult to ascertain. Genome sequences are full of putative *cis*-elements, most of which are unlikely to be of biological significance because of the lack of such an important context. A deeper understanding of the evolution of (plant) genome regulation awaits considerable additional research.

A (really) naïve idea 3—get a system of homologous recombination to work in rice

We conclude here with one final, very naïve idea.

Since we know the rice genome sequence, and can perhaps identify the kinds of *cis*-regulatory and functional sequence changes needed for C_4 (say, from a study of maize), could we eventually attempt site-specific homologous recombination of specific target (regulatory, coding) sequences into the target genome, in this case rice?

This has not, of course, yet been achieved in rice (it is hard enough to get this working in simpler eukaryotes such as yeast) but perhaps it is an idea to consider as biotechnology progresses—see Kumar et al (2006) for inspiration.

Other pertinent tools and initiatives

Researchers wishing to apply genomics and bioinformatics toward the problem of creating C_4 rice have various additional public resources to draw upon.

International Crop Information System (ICIS)

One of the key tasks of crop bioinformatics is the systematic documentation of germplasm and field data for germplasm. This is not a new challenge; therefore, not too surprisingly, suitable databases and software already exist to meet this need.

The International Crop Information System (ICIS; www.icis.cgiar.org; Fox and Skovmand 1996) is one example of such a computerized database system and suite of tools for general, integrated management and use of genealogy, nomenclature, evaluation, and characterization data for a wide range of crops. One useful feature of ICIS is that it is a public open-source software collaboration involving several Consultative Group on International Agricultural Research (CGIAR) centers and non-CG partners in Australia (University of Queensland, University of Western Australia), Canada (Agriculture and Agri-foods Canada), the Netherlands (a private seed company, Nunhems), and Singapore (Bayer hybrid rice seed division). Thus, the underlying computer code is free for the asking, hosted on a software project development site for agricultural projects called "CropForge" (http://cropforge.org), and it is being customized and enhanced by a wide public community of developers and end users.

ICIS is successfully deployed for many nonrice crops: wheat, barley, maize, common bean, chickpea, cowpea, sugar cane, potato, sweet potato, and several vegetable crops. From the perspective of this meeting on C_4 in rice, the International Rice Information System (IRIS; www.iris.irri.org; McLaren et al 2005, Bruskiewich et al 2003) is of specific interest as the largest curated public rice installation of ICIS.

International Rice Functional Genomics Consortium

In the area of rice functional genomics, several resources are available. In January 2003, representatives of 18 institutions from 10 countries and two international agricultural research centers established an International Rice Functional Genomics Consortium (IRFGC; www.iris.irri.org/IRFGC). The proposed mandate of the IRFGC is to coordinate research in the postsequencing "functional genomics" era by exploring ways to consolidate international rice functional genomics resources and to build common strategies.

The consortium strives to encourage sharing and consolidation of several useful resources for gene characterization of rice in the areas of genomic stocks, expression arrays, proteomics, and bioinformatics. For example, one specific bioinformatics resource under development is the establishment of a "Rice MOBY Network" that will provide easier integrated access to globally distributed information of the consortium.

Generation Challenge Program

A second pertinent international initiative is the Generation Challenge Program (GCP; www.generationcp.org), an international research consortium established in 2003, with a projected lifetime of 10 years, hosted by the CGIAR and involving global partners from international agricultural research centers, advanced research institutes, and national agricultural research and extension systems. The research themes of the GCP are directed to crop improvement through genomics and comparative biology across species, as well as molecular and phenotypic characterization of genetic resources to discover valuable alleles for crop improvement.

The GCP's primary trait focus is drought tolerance across a range of crops. Since the C_4 syndrome confers several features enhancing the performance of plants under hot, dry conditions, the study of C_4 is perhaps of some relevance to this research consortium.

One major research and development subprogram of the GCP focuses on crop informatics. This bioinformatics subprogram is striving to develop global public standards for crop information management, including a comprehensive crop scientific domain model (see http://pantheon.generationcp.org; Bruskiewich et al 2006a) and a platform of tools for accessing and analyzing information available from the globally networked databases of GCP partners and other pertinent external data sources. All of the software technology of the GCP, such as ICIS, is open-source and will be available on CropForge. This platform and network are envisioned to specifically empower crop researchers to perform the kinds of integrated genomics and related data management and analyses discussed in this paper.

IRRI-CIMMYT Alliance in research informatics

Perhaps pertinent to the task of engineering C_4 into rice is an alliance between IRRI and CIMMYT in the area of research informatics. The resulting Crop Research Informatics Laboratory (CRIL) has as one of its goals the development of a comparative crop resource for rice, maize, and wheat. CRIL efforts to align maize, a C_4 system, against rice will be of special interest to the C_4 rice project.

Summary

Introducing C_4 photosynthesis into rice is a significant challenge. Comparative genomics and bioinformatics can help identify molecular targets for modification or transfer into pertinent genetic backgrounds to achieve the biochemical and morphological changes required for the goal. Bioinformaticians could perhaps also help in a practical way by constructing a consolidated online repository of biological data and information about C_4 photosynthesis to facilitate collaboration and sharing of knowledge.

References

Altschul SF, Gish W, Miller W, Myers EW, Lipman DJ. 1990. Basic local alignment search tool. J. Mol. Biol. 215:403-410.

Bedell JA, Budiman MA, Nunberg A, Citek RW, Robbins D, Jones J, Flick E, Rohlfing T, Fries J, Bradford K, McMenamy J, Smith M, Holeman H, Roe BA, Wiley G, Korf IF, Rabinowicz PD, Lakey N, McCombie WR, Jeddeloh JA, Martienssen RA. 2005. Sorghum genome sequencing by methylation filtration. PLoS Biol. 3(1):e13.

Brown TA. 1999. Genomes. Oxford (UK): Bios Scientific Publishers. p 2.

Bruskiewich R, Davenport G, Hazekamp T, Metz T, Ruiz M, Simon R, Takeya M, Lee J, Senger M, McLaren CG, Van Hintum T. 2006a. The Generation Challenge Programme (GCP): standards for crop data. OMICS 10(2):215-219.

Bruskiewich R, Metz T, McLaren CG. 2006b. Bioinformatics and crop information systems in rice research. Int. Rice Res. Notes 31(1):5-12.

Bruskiewich R, Cosico A, Eusebio W, Portugal A, Ramos LR, Reyes T, Sallan MAB, Ulat VJM, Wang X, McNally KL, Sackville Hamilton R, McLaren CG . 2003. Linking genotype to phenotype: The International Rice Information System (IRIS). Bioinformatics 19(Suppl.1):i63-i65.

Chan AP, Pertea G, Cheung F, Lee D, Zheng L, Whitelaw C, Pontaroli AC, SanMiguel P, Yuan Y, Bennetzen J, Barbazuk WB, Quackenbush J, Rabinowicz PD. 2006. The TIGR maize database. Nucl. Acids Res. 34(Database issue):D771-776.

Cordaux R, Udit S, Batzer MA, Feschotte C. 2006. Birth of a chimeric primate gene by capture of the transposase gene from a mobile element. Proc. Natl. Acad. Sci. USA 103(21):8101-8106.

Dias AP, Braun EL, McMullen MD, Grotewold E. 2003. Recently duplicated maize R2R3 Myb genes provide evidence for distinct mechanisms of evolutionary divergence after duplication. Plant Physiol. 131(2):610-620.

Elrouby N, Bureau TE. 2001. A novel hybrid open reading frame formed by multiple cellular gene transductions by a plant long terminal repeat retroelement. J. Biol. Chem. 276(45):41963-41968.

Elrouby N, Bureau TE. 2000. Molecular characterization of the *Abp1* 5′-flanking region in maize and the Teosintes. Plant Physiol. 124:369-377.

Feng Q, Zhang Y, Hao P, Wang S, Fu G, Huang Y, Li Y, Zhu J, Liu Y, Hu X, Jia P, Zhang Y, Zhao Q, Ying K, Yu S, Tang Y, Weng Q, Zhang L, Lu Y, Mu J, Lu Y, Zhang LS, Yu Z, Fan D, Liu X, Lu T, Li C, Wu Y, Sun T, Lei H, Li T, Hu H, Guan J, Wu M, Zhang R, Zhou B, Chen Z, Chen L, Jin Z, Wang R, Yin H, Cai Z, Ren S, Lv G, Gu W, Zhu G, Tu Y, Jia J, Zhang Y, Chen J, Kang H, Chen X, Shao C, Sun Y, Hu Q, Zhang X, Zhang W, Wang L, Ding C, Sheng H, Gu J, Chen S, Ni L, Zhu F, Chen W, Lan L, Lai Y, Cheng Z, Gu M, Jiang J, Li J, Hong G, Xue Y, Han B. 2002. Sequence and analysis of rice chromosome 4. Nature 420(6913):316-320.

Fox PN, Skovmand B. 1996. The International Crop Information System (ICIS): connects genebank to breeder to farmer's field. In: Cooper M, Hammer GL, editors. Plant adaptation and crop improvement. Wallingford (UK): CAB International. p 317-326.

Goff SA, Ricke D, Lan TH, Presting G, Wang R, Dunn M, Glazebrook J, Sessions A, Oeller P, Varma H, Hadley D, Hutchison D, Martin C, Katagiri F, Lange BM,Moughamer T, Xia Y, Budworth P, Zhong J, Miguel T, Paszkowski U, Zhang S,Colbert M, Sun WL, Chen L, Cooper B, Park S, Wood TC, Mao L, Quail P, Wing R,Dean R, Yu Y, Zharkikh A, Shen R, Sahasrabudhe S, Thomas A, Cannings R, Gutin A,Pruss D, Reid J, Tavtigian S, Mitchell J, Eldredge G, Scholl T, Miller RM,Bhatnagar S, Adey N, Rubano T, Tusneem N, Robinson R, Feldhaus J, Macalma T, Oliphant A, Briggs S. 2002. A draft sequence of the rice genome (*Oryza sativa* L. ssp. japonica). Science 296(5565):92-100.

Gowik U, Burscheidt J, Akyildiz M, Schlue U, Koczor M, Streubel M, Westhoff P. 2004. *cis*-regulatory elements for mesophyll-specific gene expression in the C_4 plant *Flaveria trinervia*, the promoter of the C_4 phospho*enol*pyruvate carboxylase gene. Plant Cell 16:1077-1090.

Higo K, Ugawa Y, Iwamoto M, Korenaga T. 1999. Plant *cis*-acting regulatory DNA elements (PLACE) database. Nucl. Acids Res. 27(1):297-300.

Hirochika H, Guiderdoni E, An G, Hsing YI, Eun MY, Han CD, Upadhyaya N, Ramachandran S, Zhang Q, Pereira A, Sundaresan V, Leung H. 2004. Rice mutant resources for gene discovery. Plant Mol. Biol. 54(3):325-334.

IRGSP (International Rice Genome Sequencing Project). 2005. The map-based sequence of the rice genome. Nature 436:793-800.

Jeng TL, Tseng TH, Wang CS, Chen CL, Sung JM. 2003. Starch biosynthesis enzymes in developing grains of rice cultivar Tainung 67 and its sodium azide-induced rice mutant. Field Crops Res. 84:261-269.

Jordan IK. 2006. Evolutionary tinkering with transposable elements. Proc. Natl. Acad. Sci. USA 103(21):7941-7942.

Kumar S, Allen GC, Thompson WF. 2006. Gene targeting in plants: fingers on the move. Trends Plant Sci. 11(4):159-161.

Leegood RC. 2002. C(4) photosynthesis: principles of CO(2) concentration and prospects for its introduction into C(3) plants. J. Exp. Bot. 53(369):581-590.

McNally KL, Bruskiewich R, Mackill D, Buell CR, Leach JE, Leung H. 2006. Sequencing multiple and diverse rice varieties: connecting whole-genome variation with phenotypes. Plant Physiol. 141:1-6.

McLaren CG, Bruskiewich RM, Portugal AM, Cosico AB. 2005. The International Rice Information System: a platform for meta-analysis of rice crop data. Plant Physiol. 139(2):637-642.

Meinke DW, Cherry MJ. Dean C. Rounsley SD, Koornneef M. 1998. *Arabidopsis thaliana*: a model plant for genome analysis. Science 282:662-682.

Morgante M. 2006. Plant genome organisation and diversity: the year of the junk! Curr. Opin. Biotechnol. 17(2):168-173.

Ohyanagi H, Tanaka T, Sakai H, Shigemoto Y, Yamaguchi K, Habara T, Fujii Y, Antonio BA, Nagamura Y, Imanishi T, Ikeo K, Itoh T, Gojobori T, Sasaki T. 2006. The Rice Annotation Project Database (RAP-DB): hub for *Oryza sativa* ssp. japonica genome information. Nucl. Acids Res. 34(Database issue):D741-744.

Paterson AH, Freeling M, Sasaki T. 2005. Grains of knowledge: genomics of model cereals. Genome Res. 15(12):1643-1650.

Pennisi E. 2000. Plants join the genome sequencing bandwagon. Science 290:2054-2055.

Pooma W, Gersos C, Grotewold E. 2002. Transposon insertions in the promoter of the *Zea mays a1* gene differentially affect transcription by the Myb factors P and C1. Genetics 161(2):793-801.

Prestridge DS. 1991. SIGNAL SCAN: a computer program that scans DNA sequences for eukaryotic transcriptional elements. CABIOS 7:203-206.

Rabinowicz PD, Braun EL, Wolfe AD, Bowen B, Grotewold E. 1999. Maize R2R3 Myb genes: sequence analysis reveals amplification in the higher plants. Genetics 153(1):427-444.

Reyes JC, Muro-Pastor MI, Florencio FJ. 2004. The GATA family of transcription factors in *Arabidopsis* and rice. Plant Physiol. 134:1718-1732.

Rose A, Meier I, Wienand U. 1999. The tomato I-box binding factor LeMYBI is a member of a novel class of Myb-like proteins. Plant J. 20(6):641-652.

Sasaki T, Matsumoto T, Yamamoto K, Sakata K, Baba T, Katayose Y, Wu J, Niimura Y, Cheng Z, Nagamura Y, Antonio BA, Kanamori H, Hosokawa S, Masukawa M, Arikawa K, Chiden Y, Hayashi M, Okamoto M, Ando T, Aoki H, Arita K, Hamada M, Harada C, Hijishita S, Honda M, Ichikawa Y, Idonuma A, Iijima M, Ikeda M, Ikeno M, Ito S, Ito T, Ito Y, Ito Y, Iwabuchi A, Kamiya K, Karasawa W, Katagiri S, Kikuta A, Kobayashi N, Kono I, Machita K, Maehara T, Mizuno H, Mizubayashi T, Mukai Y, Nagasaki H, Nakashima M, Nakama Y, Nakamichi Y, Nakamura M, Namiki N, Negishi M, Ohta I, Ono N, Saji S, Sakai K, Shibata M, Shimokawa T, Shomura A, Song J, Takazaki Y, Terasawa K, Tsuji K, Waki K, Yamagata H, Yamane H, Yoshiki S, Yoshihara R, Yukawa K, Zhong H, Iwama H, Endo T, Ito H, Hahn JH, Kim HI, Eun MY, Yano M, Jiang J, Gojobori T. 2002. The genome sequence and structure of rice chromosome 1. Nature 420(6913):312-316.

Sheen J. 1999. C_4 gene expression. Annu. Rev. Plant Physiol. Plant Mol. Biol. 50:187-217.

Shirsat A, Wilford N, Croy R, Boulter D. 1989. Sequences responsible for the tissue specific promoter activity of a pea legumin gene in tobacco. Mol. Gen. Genet. 215:326-331.

Tsudzuki J, Tsudzuki T, Wakasugi T, Kinoshita K, Kondo T, Itho Y, Sugiura M. n.d. Comparative analysis of the whole chloroplast genomes from rice, maize and wheat. In: Sugiura M, Olemueller R, Obokata J, editors. Endocytobiology and eukaryotic organelles. Berlin (Germany): Logos Verlag. (In press.)

Wakasugi T, Tsudzuki T, Sugiura M. 2001. The genomics of land plant chloroplasts: gene content and alteration of genomic information by RNA editing. Photosynth. Res. 70(1):107-118.

Wing RA, Ammiraju JS, Luo M, Kim H, Yu Y, Kudrna D, Goicoechea JL, Wang W, Nelson W, Rao K, Brar D, Mackill DJ, Han B, Soderlund C, Stein L, SanMiguel P, Jackson S. 2005. The *Oryza* map alignment project: the golden path to unlocking the genetic potential of wild rice species. Plant Mol. Biol. 59(1):53-62.

Wu JL, Wu C, Lei C, Baraoidan M, Bordeos A, Madamba MRS, Ramos-Pamplona M, Ramil Mauleon, Portugal A, Ulat VJ, Bruskiewich R, Wang G, Leach J, Khush G, Leung H. 2005. Chemical- and irradiation-induced mutants of indica rice IR64 for forward and reverse genetics. Plant Mol. Biol. 59(1):85-97.

Yanagisawa S. 2000. Dof1 and Dof2 transcription factors are associated with expression of multiple genes involved in carbon metabolism in maize. Plant J. 21:281-288.

Yu J, Wang J, Lin W, Li S, Li H, Zhou J, Ni P, Dong W, Hu S, Zeng C, Zhang J, Zhang Y, Li R, Xu Z, Li S, Li X, Zheng H, Cong L, Lin L, Yin J, Geng J, Li G, Shi J, Liu J, Lv H, Li J, Wang J, Deng Y, Ran L, Shi X, Wang X, Wu Q, Li C, Ren X, Wang J, Wang X, Li D, Liu D, Zhang X, Ji Z, Zhao W, Sun Y, Zhang Z, Bao J, Han Y, Dong L, Ji J, Chen P, Wu S, Liu J, Xiao Y, Bu D, Tan J, Yang L, Ye C, Zhang J, Xu J, Zhou Y, Yu Y, Zhang B, Zhuang S, Wei H, Liu B, Lei M, Yu H, Li Y, Xu H, Wei S, He X, Fang L, Zhang Z, Zhang Y, Huang X, Su Z, Tong W, Li J, Tong Z, Li S, Ye J, Wang L, Fang L, Lei T, Chen C, Chen H, Xu Z, Li H, Huang H, Zhang F, Xu H, Li N, Zhao C, Li S, Dong L, Huang Y, Li L, Xi Y, Qi Q, Li W, Zhang B, Hu W, Zhang Y, Tian X, Jiao Y, Liang X, Jin J, Gao L, Zheng W, Hao B, Liu S, Wang W, Yuan L, Cao M, McDermott J, Samudrala R, Wang J, Wong GK, Yang H. 2005. The genomes of *Oryza sativa*: a history of duplications. PLoS Biol. 3(2):e38.

Yu J, Hu S, Wang J, Wong GK, Li S, Liu B, Deng Y, Dai L, Zhou Y, Zhang X, Cao M, Liu J, Sun J, Tang J, Chen Y, Huang X, Lin W, Ye C, Tong W, Cong L, Geng J, Han Y, Li L, Li W, Hu G, Huang X, Li W, Li J, Liu Z, Li L, Liu J, Qi Q, Liu J, Li L, Li T, Wang X, Lu H, Wu T, Zhu M, Ni P, Han H, Dong W, Ren X, Feng X, Cui P, Li X, Wang H, Xu X, Zhai W, Xu Z, Zhang J, He S, Zhang J, Xu J, Zhang K, Zheng X, Dong J, Zeng W, Tao L, Ye J, Tan J, Ren X, Chen X, He J, Liu D, Tian W, Tian C, Xia H, Bao Q, Li G, Gao H, Cao T, Wang J, Zhao W, Li P, Chen W, Wang X, Zhang Y, Hu J, Wang J, Liu S, Yang J, Zhang G, Xiong Y, Li Z, Mao L, Zhou C, Zhu Z, Chen R, Hao B, Zheng W, Chen S, Guo W, Li G, Liu S, Tao M, Wang J, Zhu L, Yuan L, Yang H. 2002. A draft sequence of the rice genome (*Oryza sativa* L. ssp. indica). Science 296(5565):79-92.

Yuan Q, Ouyang S, Liu J, Suh B, Cheung F, Sultana R, Lee D, Quackenbush J, Buell CR. 2003. The TIGR rice genome annotation resource: annotating the rice genome and creating resources for plant biologists. Nucl. Acids Res. 31(1):229-233.

Zhou DX. 1999. Regulatory mechanism of plant gene transcription by GT-elements and GT-factors. Trends Plant Sci. 4:210-214.

Notes

Authors' address: Crop Research Informatics Laboratory, International Rice Research Institute, DAPO Box 7777 Metro Manila, Philippines.

Acknowledgments: Phylogenomic analysis was done by Samart Wanchana, a GCP-funded postdoc, using analysis tools from K. Sjölander (University of California, Berkeley, USA; http://phylogenomics.berkeley.edu). We discussed ideas relating to transposable element genome evolution and regulatory impact with T. Bureau (McGill University, Canada) and D. Hoen, his student. S. Wanchana is supported by a postdoctoral fellowship from the Generation Challenge Program.

Surveying the possible pathways to C_4 rice

P.L. Mitchell and J.E. Sheehy

The conference on "Supercharging the Rice Engine," held at the International Rice Research Institute in July 2006, discussed various ways of constructing C_4 rice, given its desirability for substantially increasing rice yields along with better use of water and nitrogen. It was agreed to set up a consortium to stimulate and coordinate research to produce C_4 rice. The organization of the consortium, stages of work, proof of concept, and the need for sustained funding are outlined. The two main pathways to C_4 rice are with Kranz anatomy, as in maize, or as a single-cell system. The features of each system are summarized, as currently understood, and questions for research are identified. Both pathways deserve attention and research on each will benefit the other; there is no clear single route to success at present. Other possibilities for improving photosynthesis are mentioned but only C_4 rice brings the whole package of high productivity plus better use of resources that is necessary to help alleviate hunger and poverty in countries dependent on rice.

Our short-term goal is to demonstrate that the problems faced in constructing C_4 rice are solvable, and our long-term goal is to produce C_4 rice. In this chapter, we outline the coordination of research for C_4 rice, review briefly some of the areas of ignorance from the science of C_4 photosynthesis, and try to map the pathways to success. We say "pathways" because it is too soon to identify a single route to success and parallel approaches are appropriate, at least in the early stages. We attempt to draw conclusions from the work presented by the various authors. All of them are interested in the science of photosynthesis and how it might be applied to make improvements to crop plants and in particular to rice. The Kranz C_4 system might be the ultimate goal because it appears to be the most efficient, but it has been argued that single-cell systems could be simpler to install, offering the possibility of more rapid progress. Research on one system will likely inform research on the other. There are still many discoveries to be made, especially in the area of the genetics of the control of leaf anatomy. Nonetheless, it cannot be too strongly stated that we are not trying to invent a new photosynthetic mechanism; rather, we wish to imitate in rice Nature's achievements in C_4 plants.

The background to C_4 rice

In 1999, a conference was held at IRRI called "The Quest to Reduce Hunger: Redesigning Rice Photosynthesis," and the proceedings were issued as the book *Redesigning Rice Photosynthesis to Increase Yield* (Sheehy et al 2000). Several methods for improving rice photosynthesis were discussed and it was clear that only a rice with C_4 photosynthesis offered the prospect of a substantial gain in yield potential, together with better use of water and nitrogen. The possibility of producing a C_4 rice was discussed at length. Responses ranged from skeptical to enthusiastic, but the enthusiasts were daunted by the size and complexity of the task, largely because of a lack of knowledge of the genes controlling leaf anatomy or ways to accelerate their discovery. A coordinated program of research did not emerge from the conference, mostly because funding for such a long-term and risky venture was impossible. However, the potential benefits of C_4 rice ensured that the idea resurfaced at intervals (e.g., Leegood 2002, Surridge 2002) and was often referred to in passing (e.g., Edwards et al 2001, Sage 2004).

Continuing progress, with some surprises, in the understanding of C_4 photosynthesis (Mitchell, this volume), developments in molecular biology and genetic engineering, and increasingly clear recognition of the need for a substantial increase in rice productivity (IRRI 2006) contributed to the decision to hold a conference at IRRI in July 2006, called "Supercharging the Rice Engine." Here, the atmosphere of the meeting was more positive. The case for a C_4 rice (Mitchell and Sheehy, this volume) was largely accepted, participants arrived with new ideas for how the task might be started (e.g., Sage, this volume; Hibberd, this volume), generously shared their relevant knowledge and experience (e.g., Burnell, this volume), and contributed to lively discussions. There was enthusiasm and optimism for the task, along with recognition of the length and mountainous terrain of the pathway to C_4 rice, that is, the difficult methods and long time scale of reaching that ultimate objective. A clear sense of purpose emerged and it was resolved to set up the Consortium for C_4 Rice with a determination to overcome the difficulty of obtaining sustained funding (Table 1).

One disputable point to settle initially is what is meant by C_4 photosynthesis and, in particular, by C_4 rice. Often, the inclusion of a gene related to C_4 photosynthesis in rice is sufficient for authors reporting their work to call the modified plants "C_4 rice" when in fact there is no identifiable mechanism concentrating carbon around Rubisco.

Photosynthesis by the C_4 pathway can be defined by the following combination of characteristics: high activity of phospho*enol*pyruvate carboxylase (PEPcase); initial fixation of carbon dioxide in four-carbon acids followed by refixation by Rubisco; an increased concentration of carbon dioxide around Rubisco, which leads to much reduced photorespiration and a low carbon dioxide compensation point; and usually increased photosynthesis and growth in warm environments (above 20 °C) compared with C_3 plants. Plants with these characteristics are undoubtedly C_4 plants. It is not sufficient to find high activity of PEPcase, or a large flux of carbon dioxide fixed by

Table 1. Synthesis of responses of participants to some pertinent questions asked at the conference "Supercharging the Rice Engine" in July 2006.

1. **Is now the time to get serious about C_4 rice?**
   Universal YES, because
   (a) the case for increasing productivity has been made,
   (b) the knowledge is available or is becoming available, and
   (c) techniques are advancing.

2. **So why now?**
   It will be a long time (15 years) before reaching the ultimate goal; thus, the sooner we start, the better.

3. **Given the results with C_3 plants transgenic with C_4 enzymes, does the evidence support the view that Kranz anatomy is also required?**
   The majority say yes, that Kranz anatomy is required in addition to C_4 enzymes added to a C_3 plant. An authoritative minority point out that it may be possible to have effective single-cell C_4 in exceptional cases.

4. **Will further study of single-cell C_4 plants be relevant to constructing C_4 rice with highly productive photosynthesis?**
   Almost all say yes, that further study of single-cell C_4 will be useful, for example, on enzyme function and regulation, or on whether and exactly how carbon dioxide is concentrated in a single-cell system.

5. **Given the ultimate aim of 50% greater yield potential from substantially improved photosynthesis, are there ways of achieving this apart from the C_4 route?**
   No consensus. Various alternatives are mentioned: modifying Rubisco, recapturing photorespired carbon dioxide, finding a different carbon-concentrating mechanism, or making C_3 photosynthesis less sensitive to feedback by carbohydrate accumulation.

6. **So, what are the steps? Sequential? Parallel?**
   There are two contrasting strategies: (a) Kranz C_4 and (b) single-cell C_4. It was recommended that both strategies be pursued in parallel, at least initially. Steps in each strategy will include screening for variation (natural or mutant-induced) and transformation, for both structural and functional changes.

7. **What degree of coordination is required? Who will do this?**
   Coordination by a steering committee based at IRRI to avoid duplication or omission of tasks is preferred.

8. **What will this cost? Who will support it?**
   Start-up support can be provided by IRRI for activities here, but determined and even managed by participants. Examples are $1–2 million a year, or at least $10 million; but few ventured a cost. Sources of funding: IRRI for seed money and later directing funds from large donors; national funding agencies, especially for nonrice work or basic science work.

Continued on next page

Table 1 continued.

9.	**What will success (acceptable progress) look like in three years, or in five years?** Notably varied responses that are hard to summarize; no consensus. For five years, they range from (least optimistic) introduction of a simple single-cell C_4 cycle, knowing whether anatomy can be altered, to (very optimistic) first series of transgenic rice available for testing the principle, solid evidence of C_4-like function in rice. A number of criteria are proposed for three and five years, and for single-cell and Kranz systems. Clearly, definitions of progress and proof of concept will require detailed discussion by the steering committee.

PEPcase, because this is a routine occurrence in C_3 plants in particular tissues or circumstances, for example, in seed pods of rape (canola), where the high flux of respired carbon dioxide can be captured in an organ that does not have much gas exchange with the atmosphere (King et al 1998). All C_3 plants possess PEPcase for topping up the Krebs cycle as intermediates are drawn off, and for other purposes in regulating the biochemistry in the cell, and these roles are sometimes referred to as "housekeeping." Good measures of C_4-ness are the carbon dioxide compensation point (C_4 typically 0–5 ppm; C_3 40–60 ppm), sensitivity of photosynthesis to low oxygen concentration (C_4 insensitive, C_3 increased photosynthesis as the oxygenase reaction of Rubisco is prevented), $\delta^{13}C$ because the biochemical pathways discriminate between the isotopes differently (C_4 typically –9 to –18‰; C_3 –20 to –34‰), and labeling patterns of metabolites with ^{14}C (in C_4, first four-carbon acids, then metabolites of the Calvin cycle, then triose phosphate, glucose, fructose, and sucrose).

What is wanted in C_4 rice is substantially increased photosynthesis, growth, biomass, and grain yield, along with the benefits of reduced use of water and nitrogen (Mitchell and Sheehy, this volume). As a rule of thumb, photosynthesis, growth, and yield as good as those of maize are required, and an experiment has shown that, grown side by side, maize is about 50% better than rice in these properties (Sheehy et al, this volume). This will be achieved primarily through the increased concentration of carbon dioxide around Rubisco, along with minimal leakage of carbon dioxide out of that compartment so that quantum yield is larger than the C_3 value, which includes the effects of photorespiration. Rice with increased activity of C_4 enzymes, readily produced nowadays (Ku et al 1999, Wang et al 2006, Jiao, this volume), cannot meet these criteria and cannot claim to be "C_4 rice." Certainly, some experimental rice has some C_4 features but it is not yet the C_4 rice that is the ultimate goal, with greatly enhanced productivity, able to contribute to the alleviation of hunger and poverty in rice-dependent regions of the world.

Coordinating research for C_4 rice

Success in this venture will require sustained and coordinated effort that harnesses the ingenuity and expertise of many scientists based in institutions around the world. The organization that emerged from the conference in 2006 is a Consortium for C_4

Rice. This is not a consortium with its own funds to disburse, unlike other consortia associated with IRRI such as the Rice-Wheat Consortium or the Consortium for Unfavorable Rice Environments. In general, the role for the consortium is to stimulate cross-fertilization of ideas among members, to promote novel collaborative research, and to set out a plan for research and coordinate it so that there are minimal gaps or overlaps. What follows in this section is based on the assumption that the consortium will be successful in proving that a sustained C_4 rice project is a risk worth taking.

The structure of the consortium was envisaged as a steering committee, scientific advisers, and ad hoc working groups for particular tasks. Communication would be by email for the most part, with opportunities for meetings at large scientific assemblies that members were attending. The steering committee should consist of around six members, representing a range of scientific disciplines, and wide geographical coverage as far as possible. Membership of the steering committee will need to be reviewed periodically as the project develops. In the future, external reviewers will be needed so it may be helpful to have in mind several prominent researchers with knowledge of C_4 photosynthesis who are not involved elsewhere in the consortium. The members of the consortium will be associate members of IRRI.

A small secretariat is required at IRRI to serve the steering committee, keep a database of members, and set up and maintain a Web site. If necessary, the Web site can have a private section accessible only to members of the consortium. The Web site can be freestanding, for example, www.c4rice.org, with links from IRRI (www.irri.org) and from the rice portal that is being developed. The Web site is the ideal place for a bibliography, technical discussions and exchanges of experience, and frequently asked questions (especially for the public). A newsletter, perhaps quarterly, sent by email could report progress and help members keep in touch, highlight new resources on the Web site, and so on. There is still value in a paper document for publicity, and a brochure or leaflet could be produced with expertise from IRRI.

The consortium is open to new members who share the values of the consortium, and who wish to contribute to the C_4 rice project. Some thought will need to be given to the scope of research. Does the consortium have a place for other ways of improving photosynthesis, such as redirecting the photorespiratory pathways (Kebeish et al 2007, Peterhänsel, this volume) or improving Rubisco?

An early task for the steering committee will be strategic planning, setting out the vision, values, mission, and so on, aligning the consortium with the IRRI Strategic Plan (2006). The values could include statements about collaboration and how intellectual property (IP) rights will be set up, if required. The tradition at IRRI, and in the Consultative Group on International Agricultural Research (CGIAR) generally, is to produce and make available breeding lines and finished cultivars as an international public good. There is a tendency now in the CGIAR to establish protection of IP, in the interests of making it freely available to resource-poor farmers. The steering committee will need to consider this carefully, especially as the consortium is a loose collaboration of members, without a formal structure for funding and control. Again, we proceed in the belief that the humanitarian benefits flowing from the success of the C_4 project would be so large that goodwill would ensure that IP problems would not be insurmountable obstacles.

A certain amount of seed money will be available from IRRI; apart from this, the project relies at present on national agencies funding the work of members of the consortium. In the future, the consortium, through IRRI, can apply for funds from international donors. The seed money from IRRI is part of the strategy to fund a few long-term, risky ventures, and shows donors that IRRI is committed to the project. If thought helpful, IRRI can write letters of support for applications to national agencies. Collaboration is available in terms of access to materials in the genebank, expertise in growing rice experimentally, and perhaps field experiments at IRRI. Visits to IRRI are also encouraged. Workers from less developed countries (Ph.D. students or post-doctoral fellows) could be partly funded by IRRI because this support contributes to the building of research capacity on return to their home countries.

Experience with previous difficult projects at IRRI has shown the value of having more than one donor, and of being aware that at later stages there must be a clear focus, not multiple strands of research. Another foreseeable problem is breaks in funding. If work started for three years, in the expectation of substantial further funding to follow, say for 12–15 years for full development, there could be a gap of months or a year between the end of one grant and the start of another. Although some equipment can be put aside or stored, staff and their expertise cannot, nor can growing space be left unused or reserved indefinitely, so there will be a need for continuity funding, to bridge the gap.

A possible scheme of four stages for the C_4 rice project is given in Table 2. Each stage could have proof of concept as a substage within it. Some stages could be skipped, for example, if the knowledge is already there, or if it is convenient to do the work in rice straight away (skip Stage 2).

Proof of concept is seen as an essential early phase of any long-term and risky project such as this one. Donors will need assurance that each step along the pathway to C_4 rice can be shown to work before further funds are applied to improving that step or starting the next one. There is a trade-off between the duration of a proof of concept phase and the end-point, that is, the harder the task or the more tightly defined the end-point, the longer the time required for proof of concept. Researchers vary in their starting points. Some have projects running that are relevant to C_4 rice or could be directed that way; others have ideas to try but no funding, staff, or facilities in place. Obtaining funding from national agencies and then setting up the research takes 12–18 months.

Three to five years appeared to be a widely accepted duration for proof of concept; IRRI has provided three years of seed money. Proof of concept should flow from investigations of the degree of C_4-ness in the wild relatives of rice, yield comparisons of C_3 and C_4 crops grown alongside each other, the maximum crop yields that can be achieved when there are no carbon dioxide limitations to growth, the identification of mutants of vein spacing in maize or sorghum, and comparisons of numbers of plasmodesmata between mesophyll and bundle sheath cells in rice and *Echinochloa*. The steering committee will need to consider carefully the end-points of the proof

Table 2. A suggested scheme for the C_4 rice project in four stages.

Stage 1: Foundation
Acquire the basic scientific information and understanding.
Examples Looking for variation in C_4 characters in rice and near relatives (other species of *Oryza*, other genera closely related).
Control of development of leaf anatomy.
Regulation and control of properties and activities of genes for the C_4 enzymes.

Stage 2: Demonstration of principle
Show that the scientific information and understanding can be made to work, in any convenient species.
Examples Single-cell system.
Changed leaf anatomy toward Kranz anatomy.
Refinement of the above to produce marked (i.e., nontrivial, greater than 10%) increase in photosynthesis, plant growth, and yield.

Stage 3: Application to rice
Show that the scientific information and understanding can be made to work in rice.
Changed metabolism.
Changed anatomy (if required).
Refinement of the above to produce marked (i.e., nontrivial, greater than 10%) increase in photosynthesis, plant growth, and yield.

Stage 4: Production of usable cultivars
Show that C_4 rice can be used in breeding to produce locally adapted cultivars with all the usual advanced traits: resistance to pests and diseases, good grain quality, responsive to fertilizer, and resistance to lodging.

of concept phase to ensure that they are realistic in the agreed time period and also effective in establishing the concept so that further funding follows.

Kranz anatomy and C_4 photosynthesis

Kranz anatomy—enlarged bundle sheath cells filled with dark green chloroplasts appearing as a wreath around each vascular bundle—was discovered long before C_4 photosynthesis. Once the mechanism of C_4 photosynthesis had been elucidated, Kranz anatomy and those metabolic pathways were seen as essential to each other, neither anatomy nor biochemistry making sense alone. Consequently, when considering making C_4 rice, C_4 photosynthesis with Kranz anatomy is the obvious system, all the more so since it is particularly widespread in the grass family with 11 independent origins (Sage 2004).

It is clear in principle how the concentration of carbon dioxide can be raised around Rubisco at the site of decarboxylation of four-carbon acids in the chloroplast of the bundle sheath cell. As far as we know, the conductance to carbon dioxide of the chloroplast envelope is high, as for chloroplasts generally, but the chloroplasts are clumped together in one part of the cell, which probably minimizes the net leakage from any one chloroplast. Suberin occurs in the cell wall of the bundle sheath cells (or the mestome sheath cells) of all grasses and may be a significant barrier to carbon dioxide diffusion out of the cell, but C_4 dicots (e.g., *Flaveria*) lack this feature (Dengler and Nelson 1999). The importance of the close coupling of mesophyll cell adjacent to bundle sheath cell is evident from the greatly increased frequency of plasmodesmata between these cells in C_4 plants (Leegood, this volume) and from the observation that, in maize, mesophyll cells not adjacent to a bundle sheath cell remain in the default, C_3, condition (Langdale and Nelson 1991). The absence of a more distributed C_4 system, that is, three or four mesophyll cells between bundle sheath cells, suggests that exchanges between adjacent mesophyll and bundle sheath cells are crucial to the success of highly productive C_4 plants.

Thus, C_4 photosynthesis with Kranz anatomy, as in maize, has increased quantum yield (because of minimal leakage of carbon dioxide from the bundle sheath cell) and increased the maximum rate of leaf photosynthesis (because the concentration of carbon dioxide is raised around Rubisco) compared with C_3 plants. It seems the best-known system to copy for C_4 rice. Unfortunately, although the features of C_4 photosynthesis that are desired for rice can be identified, we have as yet insufficient knowledge about their development, and the set of genes that are involved and how they are regulated. Much is known about the enzymes and their genes (e.g., Gowick, this volume), a little about genes controlling anatomy (Nelson, this volume), but not enough to control expression of the genes precisely. Given the apparent ease of evolution of C_4 photosynthesis (Sage, this volume), it is possible that there are only a few key regulatory changes from which all the other developmental sequences follow. It is clear that C_4 biochemistry and anatomy have been added on to the basic, C_3, system. Where there are closely related C_3 and C_4 species, it will be possible to search for the crucial differences between them in their genomes, using modern techniques of molecular biology as well as looking for mutants with the methods of classical genetics.

Numerous questions arise. Most are familiar indicators of long-standing ignorance, for example, vein spacing. Answers to these questions will contribute to making C_4 rice with Kranz anatomy.

1. *Comparative genomics of C_3 and C_4 plants.* Clearly, the differences between C_3 and C_4 plants must derive from differences in their genomes. Are pairs of closely related C_3 and C_4 species to be found, in which the genomes can be examined for these key differences? The C_3 and C_4 subspecies of *Alloteropsis semialata* (Ripley et al 2007) appear to be suitable at first sight, but there are difficulties. The two taxa are certainly closely related, but the C_3 is probably a revertant from a C_4 ancestor, containing C_4 genes that somehow are not functioning properly; and the current C_4 plants form a polyploid series that would complicate genetic analysis (unpublished data of D.G. Ibrahim, C.P. Osborne, and T.A. Burke).
2. *Vein spacing.* The importance of closely spaced minor veins, so that each mesophyll cell is adjacent to a bundle sheath cell, is clearly understood. The close spacing arises from prolongation of the phase of vein initiation in C_4

grasses (Nelson, this volume). What controls the duration of vein initiation? Can it be manipulated? There used to be a mutant of *Panicum maximum* with veins spaced more widely (Fladung 1994) that did not persist in cultivation; will it be worth trying to produce it again?

3. *Bundle sheath cells.* The bundle sheath occurs in C_3 plants, sometimes conspicuously, as in rice, where it appears in transverse sections of a leaf as a necklace of large, rounded cells around the vascular bundle. The cells contain a few chloroplasts but otherwise appear empty. What exactly is the function of the bundle sheath in C_3 plants? The natural speculation is that it facilitates or controls the movement of substances between the vascular tissue and the mesophyll. If so, is this role maintained in C_4 plants where the bundle sheath has been commandeered for another purpose?
4. *Distinctive cells.* In *Arundinella*, the files of distinctive cells that function as bundle sheath cells develop from ground tissue, unlike bundle sheath cells proper (Wakayama et al 2003). This indicates that neither closeness to vascular tissue nor origination from procambial tissue is essential for bundle sheath functions. How do certain cells in the leaf ground tissue, in longitudinal files, switch to become distinctive cells? How does it come about that files are neatly inserted between the minor veins to ensure that each mesophyll cell is next to either a distinctive cell or a bundle sheath cell?
5. *Plasmodesmata.* Little seems to be known in detail about plasmodesmata in rice leaves; it is assumed that they are typical of a C_3 plant. The much more abundant (three to five times) plasmodesmata connecting mesophyll and bundle sheath cells in C_4 plants are thought to be essential for the high fluxes of metabolites between the two cells. How do plasmodesmata develop? Are they solely formed soon after cell division, as the primary cell wall is deposited, or can new plasmodesmata appear between mature cells? What controls the number of plasmodesmata? How do C_4 plants achieve the high number between mesophyll and bundle sheath cells?
6. *Translocators.* The different or modified translocators in the chloroplast envelope of C_4 plants have been reviewed by Leegood (this volume). Clearly, much more needs to be discovered about these translocators and how they can be manipulated. Have they all been identified? How are the properties of the same type of translocator altered in a C_4 plant compared with related C_3 plants? Can the genes be identified and manipulated?
7. *Conductance of bundle sheath cells to carbon dioxide.* In C_4 plants, decarboxylation of four-carbon acids occurs in the bundle sheath chloroplasts, but, since the chloroplast envelope is assumed to have its usual high conductance to carbon dioxide, it is the bundle sheath cell wall that is thought to be the barrier to carbon dioxide diffusion, thus minimizing leakage. What exactly is the role of suberin in the cell walls? Where suberin does not occur, how is a low conductance to carbon dioxide achieved?

8. *Location of chloroplasts.* The centripetal or centrifugal distribution of chloroplasts in bundle sheath cells, according to C_4 subtype, is a striking feature of leaf sections under the light microscope. What exactly is the significance of this location in the cell? How would it be accomplished in C_4 rice?

Single-cell C_4 photosynthesis

There is no longer any doubt that C_4 photosynthesis can be carried out in a single cell, so *a priori* a single-cell system for C_4 rice is worth considering. However, single-cell C_4 photosynthesis is not a uniform category but includes diverse mechanisms. Type 1 is the system found in *Hydrilla verticillata* (Bowes, this volume) and *Egeria densa*, where the C_4 pathways are induced by conditions that produce high photorespiration, and carbon dioxide is concentrated, if at all, within the chloroplasts that previously conducted C_3 photosynthesis. Type 2 are the systems where there is specialization of structure of the photosynthetic cells in order to separate the locations of initial and final fixation of carbon dioxide (Edwards, this volume). In *Borszczowia aralocaspica* (Type 2a), the cells are elongated radially around the vascular bundles with initial fixation at the distal end and final fixation in the chloroplasts clustered at the proximal end. Superficially, the appearance is like Kranz anatomy but without a cell wall separating the two locations. (If the reclassification of *Borszczowia aralocaspica* Bunge as *Suaeda aralocaspica* (Bunge) Freitag & Schütze is accepted [Schütze et al 2003], then it will be convenient to use *Borszczowia* as a common name, much as *Arabidopsis* is becoming used for *Arabidopsis thaliana*.) Type 2b is found in *Bienertia cycloptera*, where initial fixation occurs in the peripheral cytosol and final fixation in chloroplasts in a central portion of cytoplasm suspended in the vacuole.

From the point of view of productivity, there are two questions to ask of the *Hydrilla* system. Is there a higher concentration of carbon dioxide inside the chloroplast when the C_4 pathway is operating? Is the quantum yield for C_4 photosynthesis higher or lower than for C_3? The chloroplast envelope is expected to have a high conductance for carbon dioxide so that it diffuses inward under C_3 photosynthesis as rapidly as possible. Therefore, its diffusion outward, if the concentration inside is raised by decarboxylation in the C_4 mechanism, would be equally facilitated. The high leakage has an energy cost since two ATP are required to regenerate phospho*enol*pyruvate from pyruvate, and this would be evident in the quantum yield. The only values known are from Spencer et al (1994), where the C_4 value was half the C_3 one, but these were measurements apparently made only in passing. A more definitive answer is required.

A leaky system with low quantum yield may well be good enough for *Hydrilla* in its extreme conditions (very low concentrations of carbon dioxide and bicarbonate in the water, high concentration of oxygen, high temperature, and high light), ensuring limited, rather costly, carbon gain and giving it an advantage over C_3 competitors. It is probably also significant that *Hydrilla* is a submerged plant so that loss of carbon dioxide by diffusion from the plant is slower than it would be in air. These points make it unlikely that the *Hydrilla* system is worth imitating in rice, where high productivity is the main requirement, unless there are unexpected answers to the questions posed above. Otherwise, the main interest is in the occurrence of C_4 isoforms of the enzymes and their regulation as the C_4 mechanism is induced. Insights here may be useful for making C_4 rice.

It has been speculated that the conductance to carbon dioxide of the chloroplast envelope could be changed, by altering the abundance of aquaporins (Bowes et al, this volume; von Caemmerer et al, this volume). If a decrease in carbon dioxide conductance is part of the inducible C_4 mechanism in *Hydrilla*, this would make the system look more attractive for rice, if the chloroplast conductance can be controlled. Several questions come to mind. Is the conductance of the chloroplast envelope altered during induction of C_4 photosynthesis? Presumably, if there is subsequent reversion to C_3 photosynthesis, then the reverse change in conductance must occur. Are aquaporins the only molecules of interest? What genes are involved and could they be manipulated to produce chloroplasts with altered conductance to carbon dioxide?

Type 2 single-cell C_4 photosynthesis occurs in plants of dry, saline habitats, apparently of low productivity and with modest rates of photosynthesis (von Caemmerer 2003). But there is some evidence (Edwards, this volume) that *Borszczowia* may have higher rates of photosynthesis (30 μmol CO_2 m^{-2} s^{-1}) than previously published. There are no single-cell C_3-C_4 intermediates, not surprisingly since single-cell C_4 is a recent discovery and probably not common even when its full extent is known. Consequently, there is no guidance from intermediates as to how the single-cell system might have evolved. Similarly, it is less obvious how rice could be screened for any tendency toward single-cell C_4 features; presumably, there would be reduced photorespiration and a lowered carbon dioxide compensation point but anatomical changes would be subtle. We do not know what intermediate stages to look for, in contrast to C_4 photosynthesis with Kranz anatomy. Overall, Type 2 single-cell C_4 offers little inspiration for making C_4 rice, except perhaps through mechanisms that isolate chloroplasts to one end of the cell. It appears to be the 40 μm of vacuole between the locations of initial and final fixation that ensures a raised concentration of carbon dioxide around Rubisco in *Borszczowia* (von Caemmerer 2003).

Artifical single-cell C_4 photosynthesis constitutes a third type; one case is known (Burnell, this volume) but others may be designed in the future. The Burnell system uses phospho*enol*pyruvate carboxykinase (PEPCK) inside the chloroplast to decarboxylate oxaloacetate (OAA) and produce carbon dioxide for Rubisco and phospho*enol*pyruvate (PEP) that diffuses back to the cytosol. The first transgenic rice with PEPCK had limited operation of a C_4 cycle but photosynthesis not significantly different statistically from untransformed rice (Suzuki et al 2000). Further work (Suzuki et al 2006) included adding the maize gene for phospho*enol*pyruvate carboxylase (PEPcase), to be expressed in the cytosol and stimulate that part of the cycle. Although there was a flux of carbon from carbon dioxide to four-carbon acids, in these plants photosynthesis was about 10% lower than in control plants, and there were abnormalities in the chloroplasts. The lack of appropriate transporters for OAA and PEP across the chloroplast envelope, or lack of sufficient capacity, was suggested as one reason for no improvement in photosynthesis. The Burnell system can fairly be called C_4, rather than C_4-like, because when fully developed it will have typical C_4 values for carbon dioxide compensation point and $\delta^{13}C$.

Given its novelty and ambition, progress so far with the Burnell system can be seen as a success. Rice transgenic with PEPCK and PEPcase, separately and together, has been constructed and grown with photosynthetic rate the same as, or slightly reduced from, that of the untransformed control. This stage is equivalent to an intermediate C_3-C_4 in the evolution of C_4 photosynthesis. The plant can survive and is only marginally worse than the wild type; any tendency to improved photosynthesis will be selected for strongly, leading to further development and optimization of the C_4 mechanism. The same question arises for the Burnell system as for *Hydrilla*: How can a high concentration of carbon dioxide around Rubisco be maintained in a chloroplast where the envelope, adapted for C_3 photosynthesis, must have a high conductance for carbon dioxide? If this problem can be solved, and also the translocators across the chloroplast envelope, perhaps here is a pathway to C_4 rice. The original work used an experimentally convenient japonica cultivar, Nipponbare. As a first step, would it be worth introducing the same genes to an indica rice, for example, IR64, to see if better photosynthesis resulted, and as a base for further improvement?

Outlook

It is not clear which pathway to C_4 rice will be successful, or whether an alternative, non-C_4, method of substantially improving photosynthesis will turn out to be worthwhile, at least as an interim solution. The several non-C_4 possibilities include introduction of Rubisco with improved specificity for carbon dioxide (Zhu et al 2004), introducing empirically a cyanobacterial enzyme for a carbon-concentrating mechanism (Raines 2006), recapturing carbon dioxde from photorespiration (Kebeish et al 2007), or simply enhancing the regeneration of ribulose 1,5-bisphosphate (Raines 2006). Some of these procedures improve photosynthesis only when light-saturated, or only when light-limited. It is worth reiterating that the attraction of the full C_4 system is not only the high productivity and yield but also the better use made of water and nitrogen; no non-C_4 solution offers this complete package of benefits.

One dichotomy that emerges from recent work is between imitating a natural system or introducing a novel one. The idea of C_4 rice was prompted by the comparison with maize, so copying the NADP-ME subtype of C_4 photosynthesis appeared attractive. It is evident now that only a natural system carefully copied to provide the precise coordination of leaf anatomy (or cellular compartmentation) and biochemistry is likely to be successful. On the other hand, a novel C_4 system (Suzuki et al 2000, 2006; Burnell, this volume) or a novel method of recapturing photorespired carbon dioxide (Kebeish et al 2007) is simpler to introduce and has given early success. Thus, it appears that producing novel biochemistry may be better than imitating badly a finely tuned natural system, at least in the short term (say, 5–10 years). Ultimately, the full C_4 system, its high productivity and economy as amply demonstrated by C_4 crops and weeds, remains desirable, and should be the objective of the Consortium for C_4

Rice. Which pathways lead there, and how long the journey will take, will become apparent in the coming decades.

References

Dengler NG, Nelson T. 1999. Leaf structure and development in C4 plants. In: Sage RF, Monson RK, editors. C_4 plant biology. London (UK): Academic Press. p 133-172.

Edwards GE, Furbank RT, Hatch MD, Osmond CB. 2001. What does it take to be C_4? Lessons from the evolution of C_4 photosynthesis. Plant Physiol. 125:46-49.

Fladung M. 1994. Genetic variants of *Panicum maximum* (Jacq.) in C_4 photosynthetic traits. J. Plant Physiol. 143:165-172.

IRRI. 2006. Bringing hope, improving lives: strategic plan 2007-2015. Los Baños (Philippines): International Rice Research Institute. 61 p.

Kebeish R, Niessen M, Thiruveedhi K, Bari R, Hirsch H-J, Rosenkranz R, Stäbler N, Schönfeld B, Kreuzaler F, Peterhänsel C. 2007. Bypassing photorespiration in the chloroplast improves photosynthesis, reduces photorespiration, and increases biomass production in transgenic *Arabidopsis* plants. Nature Biotechnol. (In press.)

King SP, Badger MR, Furbank RT. 1998. CO_2 refixation characteristics of developing canola seeds and silique wall. Aust. J. Plant Physiol. 25:377-386.

Ku MSB, Agarie S, Nomura M, Fukayama H, Tsuchida H, Ono K, Hirose S, Toki S, Miyao M, Matsuoka M. 1999. High-level expression of maize phospho*enol*pyruvate carboxylase in transgenic rice plants. Nature Biotechnol. 17:76-80.

Langdale JA, Nelson T. 1991. Spatial regulation of photosynthetic development in C_4 plants. Trends Genet. 7:191-196.

Leegood RC. 2002. C_4 photosynthesis: principles of CO_2 concentration and prospects for its introduction into C_3 plants. J. Exp. Bot. 53:581-590.

Raines C.A. 2006. Transgenic approaches to manipulate the environmental responses of the C_3 carbon fixation cycle. Plant Cell Environ. 29:331-339.

Ripley BS, Gilbert ME, Ibrahim DG, Osborne CP. 2007. Drought constraints on C_4 photosynthesis: stomatal and metabolic limitations in C_3 and C_4 subspecies of *Alloteropsis semialata*. J. Exp. Bot. (In press.)

Sage RF. 2004. The evolution of C_4 photosynthesis. New Phytol. 161:341-370.

Schütze P, Freitag H, Weising K. 2003. An integrated molecular and morphological study of the subfamily Suaedoideae Ulbr. (Chenopodiaceae). Plant Syst. Evol. 239:257-286.

Sheehy JE, Mitchell PL, Hardy B, editors. 2000. Redesigning rice photosynthesis to improve yield. Amsterdam (The Netherlands): Elsevier and Makati City (Philippines): International Rice Research Institute. Proceedings of a workshop The Quest to Reduce Hunger: Redesigning Rice Photosynthesis, 30 November-3 December 1999, at IRRI. 293 p.

Spencer WE, Teeri J, Wetzel RG. 1994. Acclimation of photosynthetic phenotype to environmental heterogeneity. Ecology 75:301-314.

Surridge C. 2002. The rice squad. Nature 416:576-578.

Suzuki S, Murai N, Burnell JN, Arai M. 2000. Changes in photosynthetic carbon flow in transgenic rice plants that express C_4-type phospho*enol*pyruvate carboxykinase from *Urochloa panicoides*. Plant Physiol. 124:163-172.

Suzuki S, Murai N, Kasaoka K, Hiyoshi T, Imaseki H, Burnell JN, Arai M. 2006. Carbon metabolism in transgenic rice plants that express phospho*enol*pyruvate carboxylase and/or phospho*enol*pyruvate carboxykinase. Plant Sci. 170:1010-1019.

von Caemmerer S. 2003. C_4 photosynthesis in a single C_3 cell is theoretically inefficient but may ameliorate internal CO_2 diffusion limitations of C_3 leaves. Plant Cell Environ. 26:1191-1197.

Wakayama M, Ueno O, Ohnishi J. 2003. Photosynthetic enzyme accumulation during leaf development of *Arundinella hirta*, a C_4 grass having Kranz cells not associated with veins. Plant Cell Physiol. 44:1330-1340.

Wang Q, Zhang Q, Fan D, Lu C. 2006. Photosynthetic light and CO_2 utilization and C_4 traits of two novel super-rice hybrids. J. Plant Physiol. 163:529-537.

Zhu X-G, Portis AR, Long SP. 2004. Would transformation of C_3 crop plants with foreign Rubisco increase productivity? A computational analysis extrapolating from kinetic properties to canopy photosynthesis. Plant Cell Environ. 27:155-165.

Notes

Authors' addresses: P.L. Mitchell, Department of Animal and Plant Sciences, University of Sheffield, Sheffield S10 2TN, U.K.; J.E. Sheehy, Crop and Environmental Sciences Division, IRRI, Philippines.

Acknowledgments: Use of facilities at the Department of Animal and Plant Sciences and support from Professor F.I. Woodward are gratefully acknowledged by P.L.M.

Index

Page numbers followed by the letter *t* refer to tables; numbers followed by the letter *f* refer to figures.

A

Alga(e), red, 69, 70, 276, 299, 300
Alloteropsis semialata, 406
Alternanthera, 179, 180, 181*f*, 182, 199
Alternanthera pungens, 180, 181*f*
Alternanthera sessilis, 180, 181*f*
Amaranthus, 60, 69, 260
Amaranthus edulis, 86, 88, 90, 99*t*, 107, 131*f*, 265
Amaranthus hypochondriacus, 220, 221
Amaranthus retroflexus, 102
anaplerosis, 57, 58, 165*f*, 279, 284
aquaporin, 111, 197*t*, 205, 275, 289
Arabidopsis, 32, 66, 67, 71, 83, 84, 87, 117, 119, 120, 123, 163, 166*t*-168, 170, 179, 186-189, 212, 217-228, 240, 254, 267*f*, 268, 319, 322, 324, 326, 328, 329, 346, 348, 382, 386, 388, 389*f*, 408
Arabidopsis thaliana, 67, 178, 186, 188, 217, 219, 221, 343, 382, 386, 388, 389*f*, 408
Arundinella, 60, 322, 323, 328, 329, 407
Arundinella hirta, 318
aspartate aminotransferase, 64, 184, 197*t*
auxin, 319, 320-322, 328

B

barrier to leakage, 62, 104, 112, 205, 304, 305, 322, 406, 407 (*see* leakage of carbon dioxide, leakiness)
Benson–Calvin cycle, 86-89, 249 (*see* Calvin cycle)
bicarbonate, 62, 65, 97, 101, 111, 180, 182, 203, 239, 240, 241, 408
Bienertia cycloptera, 61, 62*t*, 253, 265, 408
Bienertia sinuspersici, 253, 258, 259, 262, 263, 266
Borszczowia, 62-64, 408, 409
Borszczowia aralocaspica, 61, 62*t*, 253, 408 (*see Suaeda aralocaspica*)
BSC, 21, 22*t*, 277, 280, 282, 288, 289, 290, 386 (*see* bundle sheath cell)
bundle sheath cell(s), 21, 22*t*, 30*t*, 31, 55, 58-61*t*, 64, 66, 70, 72, 81-84, 87, 96, 99*t*, 104, 111, 112, 127-131, 133, 134, 136, 138, 140, 141, 145, 161, 175, 176, 183-188, 195, 201, 204, 205, 210, 212, 217, 220, 221, 223-227, 235-237, 239-242, 245, 250, 251, 253-255, 260-262, 277, 298, 346, 347, 386, 404-408 (*see* BSC

and bundle sheath tissue)
bundle sheath tissue, 58, 195, 201-203, 205 (*see* bundle sheath cell)

C

^{14}C labeling experiments, 302
^{14}C pulse, 56, 352
^{14}C pulse-chase, 155
C_3-C_4 intermediate(s), 3, 21, 55, 60, 61*t*, 72, 86, 177, 180, 181, 184, 195, 198, 199, 200*f*, 202, 203, 206, 237, 243, 266, 267*f*, 318, 351, 358, 366, 409
C_3-like, 58, 60, 122, 200, 260, 300
C_4-like, 57, 58, 60, 68, 122, 164, 165, 167-170, 177, 180, 181, 183, 185, 186, 195, 200, 203, 204, 210, 212*f*, 220, 242, 297-301*t*, 303, 322, 325, 327, 329, 371, 372, 388, 402*t*, 410
C_4-ness, 18, 20, 21, 358, 361, 366, 375, 402, 404
C_4 acid decarboxylase, 235, 250
C_4 lineages, 60, 65, 66, 71, 118, 195, 198-200, 205, 206*f*, 219, 220, 277, 278, 287, 385
C_4 photosynthesis, defined, 400
C_4 rice, defined, 402
C_4 sub-type(s), 86, 222, 250, 408, 410 (*see* NAD-ME subtype, NADP-ME subtype, PEPCK subtype)
  C_4 decarboxylation type(s), 82, 132*f*
  groups of C_4 plants, 128
CA, 31, 57, 64, 65, 82, 146-148, 151, 153*f*, 154, 164, 169, 170, 236, 239-241, 275, 281*t*, 282*f*, 288, 290, 297, 301, 303-306 (*see* carbonic anhydrase)
Calvin cycle, 57, 58, 86, 89, 90, 118, 139, 197*t*, 203, 224, 402 (*see* Benson–Calvin cycle)
canopy architecture, 10, 14, 15, 32, 334, 341
canopy morphology, 342
canopy structure, 70
carbon-concentrating mechanism(s), 64, 297, 401*t*, 410 (*see* CCM; CO_2-concentrating mechanism; Rubisco, concentrate CO_2 around)
carbon dioxide compensation point(s), 20, 57, 68, 72, 83, 102, 109, 133, 147, 154, 164, 165, 166*t*, 203, 263, 264*t*, 266, 280, 281*t*, 288, 352, 400, 402, 409, 410
carbonic anhydrase(s), 31*t*, 32, 57, 62*t*, 68, 82*f*, 97, 145, 146, 148*t*, 153*f*, 163, 164, 169, 197*t*, 203, 217, 235-239, 244, 275, 282, 288, 297, 301, 303, 304, 306, 307 (*see* CA)
carbon isotope discrimination, 95-101, 102*f*, 104, 106, 107*f*, 108
carrier(s), 84, 85 (*see* translocator, transporter)
CCM(s), 275-282, 289, 290, 297, 299, 300-304, 306, 307 (*see* carbon-concentrating mechanism; CO_2-concentrating mechanism; Rubisco, concentrate CO_2 around)
cell-specific expression, 184, 259, 261 (*see* cell-specific transcription)
cell-specific transcription, 140, 184, 186, 188, 220, 250, 259, 261, 324, 329, 371, 373
Chenopodiaceae, 61, 199, 249-251, 252*f*, 255, 256*f*, 258, 262, 264*t*, 265, 298
chenopods, 61, 63, 199, 260, 265, 268, 277
Chlamydomonas reinhardtii, 278, 304
chloroplast (*see* plastid, proplastid)
  agranal chloroplast(s), 224, 279, 289
  agranal morphology, 134
  biogenesis, 117, 120, 121
  centrifugal, 254, 255, 256*f*, 266, 268, 298, 386, 408
  centripetal, 104, 254-257, 266, 268, 298, 386, 408
  development, 66, 117-121, 123, 186, 187, 322

dimorphic, 249, 253, 259-261, 267*f*, 268, 277
endoplasmic reticulum, 261, 304
envelope(s), 31*t*, 63, 68, 82*f*, 83-85, 102, 104, 109, 110*f*, 111, 112, 170, 261, 275, 289, 290, 405, 407-410
granal, 120, 224, 275, 290
location, 251 257, 408
monomorphic, 258, 268
peripheral reticulum, 85
stroma, 98, 100, 169, 238, 240, 242, 243, 288, 304, 306
chloroplast-targeted protein, 117 (*see* nuclear-localized protein, transit peptide)
cis-factor, 324
cis-acting, 242
cis-element(s), 184, 221, 383, 389, 391, 392
cis-regulatory, 183-186, 372-374, 391-393
Cleome, 23, 67, 73, 112, 189, 199, 217, 218, 221, 226, 228
Cleome gynandra, 67, 201, 217, 224
Cleome sparsifolia, 201, 207*f*
climate change, 4, 22, 27, 34, 37, 50 (*see* global warming)
CO_2-concentrating mechanism, 95, 96, 103, 109, 111, 128, 153, 178, 203, 249, 250, 262, 275, 276 (*see* carbon-concentrating mechanism; CCM; Rubisco, concentrate CO_2 around)
CO_2 conductance, 30*t*, 95-97, 100, 103, 109, 111, 275, 277, 288, 289, 405, 407-410 (*see* CO_2 permeability, leakage of carbon dioxide, resistance to CO_2)
CO_2 permeability, 100, 102, 106, 112, 277, 289, 290, 386 (*see* barrier to leakage, leakage of carbon dioxide, leakiness)
compartmentalization, 119, 204, 258, 283, 317, 318, 324 (*see* compartmentation, localization)
compartmentation, 90, 176, 182, 223, 225, 235, 238, 239, 262, 264, 266, 268, 277, 301, 410 (*see* compartmentalization)
Consortium for C_4 Rice, 23, 34, 400, 402, 410
crop duration, 4, 19, 29*t*, 34 (*see* growth duration)
cyanobacteria, 32, 64, 81, 117, 167, 179, 277, 303, 304, 306, 307, 410
$\delta^{13}C$, 21, 301*t*

D

DC, 59, 60, 318, 322, 323 (see distinctive cell)
distinctive cell(s), 59, 60, 317, 318, 323, 328, 407 (*see* DC)
down-regulation, 240, 241, 275, 279, 280, 288, 289, 291, 374 (*see* transgene)

E

Echinochloa, 16, 18, 20, 21, 28, 58, 72, 404
Echinochloa colona, 21
Echinochloa glabrescens, 16, 18, 20, 28
ectopic expression, 168
efficiency
carboxylation, 145, 147, 154
energetic, 277
light use, 146 (*see* efficiency, radiation-use)
nitrogen-use, 50, 95, 96, 108, 176, 196, 204, 208, 250, 279, 291, 339, 341, 371
photochemical, 137, 138, 145, 150, 156
radiation-use, 3, 7, 18, 27, 31, 128, 334 (*see* efficiency, light use; radiation conversion factor)
Rubisco-use, 339
water-use, 5, 95, 96, 108, 196, 263,

264t, 266, 278 (*see* WUE)
Egeria, 62, 278, 284
Egeria densa, 61, 62*t*, 408
Eleocharis, 58, 60
Eleocharis vivipara, 58
evolution
accelerated, 388
C_4, 56, 60, 61, 64, 65, 71, 72, 86, 87, 121, 175-180, 182, 183, 186, 195, 197-201, 203-208, 210-212, 218, 220, 233, 251, 318, 381, 385, 406, 410
directed 195, 205, 210, 371, 373-375
rapid, 72
stepwise, 177*f*, 318

F

F_1 hybrid, 57 (*see* hybrid rice, super-hybrid rice)
Flaveria, 23, 65, 73, 99*t*, 107*f*, 112, 127, 129, 131, 136*f*, 137, 138*t*, 139*f*, 140*f*, 141, 167, 169, 175, 177-185, 187-189, 199-202, 204, 206, 219-221, 228, 238-240, 243, 245, 260, 267, 268, 284, 288, 291, 352, 358, 392, 406
Flaveria bidentis, 65, 99*t*, 107, 108*f*, 127, 129-131, 135-141, 169, 182, 184-186, 189, 200*f*, 221, 240, 243, 288
Flaveria brownii, 177*f*, 180, 181, 200, 204, 238, 243
Flaveria linearis, 177*f*
Flaveria pringlei, 177*f*, 180, 181*f*, 184, 185, 267*f*
Flaveria pubescens, 177*f*, 180, 181*f*
Flaveria robusta, 200*t*, 201
Flaveria trinervia, 167, 169, 177*f*, 180, 181*f*, 183-187, 220, 392
flood-prone rice, 367
futile cycling, 30*t*, 31*t*, 32, 68, 84, 197, 258, 277, 305, 322 (*see* overcycling)

G

GDC, 60, 61*t*, 72, 87, 90, 171*f*, 182, 186, 195, 197*t*, 200*f*, 202, 203, 211, 212, 262, 265 (*see* glycine decarboxylase)
gene duplication(s), 61*t*, 65, 72, 119, 200, 281, 347, 371, 383, 385, 386
genetic engineering, 27, 28, 33, 34, 56, 64, 145, 160, 163, 167, 222, 279, 367, 371, 373, 400 (*see* transformation, transgene, transgenic plant)
genome annotation, 5, 33, 123, 212, 222, 328, 346, 347, 371, 373, 381, 383, 393
genomic(s), 222, 226, 227, 243, 323, 326, 346, 353, 355, 357, 358, 372-376, 381-384, 388, 393-395, 406 (*see* rice genome)
global warming, 48, 249 (*see* climate change)
glycerate-3-P, 82*f*, 83, 85, 86, 88-90 (*see* PGA, phosphoglyceric acid)
glycine, 61*t*, 88-90, 265, 267*f*
glycine decarboxylase, 60, 61*t*, 72, 86*f*, 90, 171*f*, 183, 186, 195, 197*t*, 202, 211, 262 (*see* GDC)
Glycine max, 278 (*see* soybean)
glycolate, 87, 170, 171, 250, 262, 266, 267, 276, 300, 372 (*see* P-glycolate)
Green Revolution, 3, 4, 43, 45, 334, 341
Griffithsia, 69
Griffithsia monilis, 276
growth duration, 28, 29*t*, 32, 362 (*see* crop duration)

H

harvest index, 7, 29*t*, 334, 362, 364
HCO_3-, 167-169, 181*f*, 218, 220, 276, 281, 288, 297, 300*t*-307 (*see* bicarbonate)
heterosis, 146, 361, 364, 365 (*see* hybrid

rice)
HI (*see* harvest index)
hybrid rice, 45, 46*f*, 146, 156, 160, 364-366, 393 (*see* F_1 hybrid, super-hybrid rice)
Hydrilla, 62-64, 70, 258, 275, 277, 278, 280-291, 408, 410
Hydrilla verticillata, 61, 62*t*, 258, 275, 276, 298, 408

I
indica, 43, 123, 279, 327, 347, 358, 362, 363-367, 373, 375, 410
interveinal distance, 58, 82, 176, 317, 320 (*see* vein density, vein spacing)
IR64, 22*t*, 333, 347, 366, 375, 388, 410
IR72, 8, 9*f*, 12, 13*f*, 16, 18-21, 279, 337*f*, 339-341, 345*t*, 363, 364, 368*f*
irrigated rice, 6, 9*f*, 32, 361
isoform, 176, 178, 182, 281-284, 286, 287

J
japonica, 123, 156, 279, 327, 347, 355, 358, 363, 365-367, 373, 410
javanica, 123 (*see* tropical japonica)

K
Kitaake, 147*f*, 152-157, 160*f*
Kranz anatomy, 8, 21, 30-32, 55, 57, 58, 62-64, 69, 128, 135, 160, 176-178, 203, 204, 210-212, 238, 239, 249, 250-254, 256*f*, 257, 265, 266, 298, 317, 333, 345, 346, 399, 401*t*, 405, 406, 408, 409
Kranz-like anatomy, 195, 203
Kranz-type C_4, 30*t*, 109, 112, 128, 140, 141, 253, 256*f*, 259-262, 265, 266, 399, 405, 406, 409 (*see* two-cell C_4)
Krebs cycle, 57, 402

L
LAI, 8, 12*f*, 15, 334, 341 (*see* leaf area index)
laser microdissection, 145, 161, 317, 326, 329
leaf area, 8, 10, 11, 14, 15, 85, 95, 98, 100*f*, 102*f*, 103, 106, 111, 224, 262, 264, 265*t*, 334, 335, 339-342, 345, 362
leaf area index, *8, 11, 33* (*see* LAI)
leaf development, 66, 73, 84, 108, 204, 211, 217, 218, 223-227, 317, 319, 321, 328, 342, 343, 345
leaf morphology, 333, 335, 341-343, 346
leaf sheath(s), 57, 59, 70, 118, 324, 343, 344*f*
leaf thickness, 82, 176, 197t, 224, 333, 341-343, 344*f*, 347
leakage of carbon dioxide, 30*t*, 31, 32, 61*t*, 62, 68, 104, 107, 111, 112, 129, 130*f*, 141, 197, 210, 258, 264, 275, 277, 288, 290, 298, 304-306, 322, 402, 405-408 (*see* barrier to leakage, leakiness, CO_2 conductance)
leakiness, 63, 96, 97, 104-108, 111, 127, 128, 130-135, 138
localization, 57, 86, 195, 197, 202-204, 211, 260, 261, 283, 319-322, 324, 372, 386 (*see* compartmentalization)
lodging, 362, 367, 405*t*

M
malate dehydrogenase, 146, 147*f*, 164*f*, 197*t*, 217, 282*f*, 386 (*see* NADP-dependent malate dehydrogenase)
malic enzyme, 164-166*t*, 169 (*see* ME, NAD-dependent malic enzyme, NADP-dependent malic enzyme)
marker-assisted selection, 351, 355, 357, 358, 367, 374, 375, 381 (*see* MAS)
MAS, 31 (*see* marker-assisted selection)
MDH (*see* malate dehydrogenase)
mesophyll/bundle sheath interface, 103,

111, 112
mestome sheath, 59, 60, 406
microarray(s), 226, 317, 326-328, 358
millet(s), 196, 236
(*see Panicum*)
Miscanthus, 30*t*
mitochondria, 61*t*, 87, 90, 104, 134, 170, 184, 222, 225, 241, 244, 250, 251, 254, 262, 263*f*, 265-268, 277
mutagenesis, 40, 122, 180, 284, 346, 347, 374, 388
mutagenized populations, 122, 195
mutant collection(s), 222, 333, 346-348, 372
mutation, 120, 122, 202, 254, 265, 265, 375, 385, 388, 389*f*
gain of function, 122, 188
loss of function, 119, 122, 188

N

NAD-dependent malic enzyme, 59, 84, 88, 218, 250, 277, 298 (*see* malic enzyme, NAD-ME)
NAD-ME, 59, 86, 87, 90, 104, 127, 129-133, 135, 140, 141, 218, 250, 251, 254, 255f, 262, 263f, 265, 267*f*, 268, 277, 298, 306 (*see* NAD-dependent malic enzyme)
NADP-dependent malate dehydrogenase, 57, 82, 88, 89, 278, 282*f*, 386 (*see* NADP-MDH, malate dehydrogenase)
NADP-dependent malic enzyme, 31*t*, 57, 82*f*, 88, 89, 127, 145, 146, 147*f*, 163, 167, 168, 178, 218, 238, 240, 250, 275, 282*f*, 285, 298, 327, 386 (*see* malic enzyme, NADP-ME)
NADP-MDH, 57, 64, 82*f*, 278, 283, 289, 290, 386 (*see* NADP-dependent malate dehydrogenase)
NADP-ME, 31, 57, 59, 64-66, 68, 82*f*, 84, 89, 103, 127, 129-135, 140, 145, 146, 155, 168, 178, 179, 203, 218, 220, 222, 224-226, 240-242, 250-252*f*, 254, 255*f*, 257, 258, 266, 275, 278-281, 283, 285, 286, 289, 290, 298, 306, 410 (*see* NADP-dependent malic enzyme)
Neurachne, 199
new plant type, 8, 362, 363, 364*t* (*see* NPT)
Nicotiana tabaccum, 280 (*see* tobacco)
Nipponbare, 57, 388, 410
NPT, 8, 9*f*, 16, 339-341, 362, 367*f* (*see* new plant type)
nuclear-localized protein, 117 (*see* chloroplast-targeted protein, transit peptide)

O

Orcuttia, 257
Orcuttia viscida, 257, 267*f*, 268
organ-specific expression, 371
orthologous gene, 66, 87, 177, 178, 188, 221
Oryza alta, 22*t*, 353*t*
Oryza australiensis, 22*t*, 354*t*, 356*t*, 357
Oryza barthii, 22*t*, 353*t*
Oryza longistaminata, 21, 22*t*, 353*t*, 356*t*
Oryza rufipogon, 20, 353*t*, 356*t*, 357, 358, 366
overcycling, 85, 290 (*see* futile cycling)
overexpression, 156, 158, 165-170, 178, 179, 186, 189, 279, 288, 290, 338, 373, 374

P

P-glycolate, 163, 276 (*see* glycolate)
panicle, 7, 8, 9*t*, 19, 29*t*, 31, 56, 149, 150*t*, 327, 352, 362, 364, 365, 367
Panicum, 352
Panicum coloratum, 135
Panicum maximum, 21, 82, 83, 89, 242
Panicum miliaceum, 87, 127, 135, 184, 225

Panicum milioides, 22
PCK, 99*t*, 164, 168, 169 (*see* PEP carboxykinase, PEPCK, phospho*enol*pyruvate carboxykinase)
PCK type(s), 104, 183 (*see* PEPCK subtype)
PEP carboxykinase, 84, 88, 164*f*, 168, 237-239, 241 (*see* PCK, PEPCK, phospho*enol*pyruvate carboxykinase)
PEP carboxylase, 20, 32, 82*f*, 86, 88, 103, 109, 164*f*, 179, 197*t*, 204, 238, 241, 244, 250 (*see* PEPC, PEPcase, phospho*enol*pyruvate carboxylase)
PEPC, 16, 73, 82*f*, 88-90, 96, 97, 101, 106, 107, 109, 145-151, 153-158, 160, 161, 164*f*-170, 175, 177*f*-182, 220, 224, 241-244, 250, 257-259, 262-267, 275-279, 281-284, 287, 288, 290, 298, 299, 303, 305, 373, 386, 392 (*see* PEP carboxylase, PEPcase, phospho*enol*pyruvate carboxylase)
PEPcase/PEPCase, 21, 31*t*, 56-58, 60, 61*t*, 62-68, 200*f*, 203, 204, 400, 402, 409 (*see* PEP carboxylase, PEPC, phospho*enol*pyruvate carboxylase)
PEPC inhibitor, 58, 106, 154, 155, 264, 265, 282
PEPCK, 31*t*, 59, 64, 65, 68, 84, 87-89, 218, 235, 236, 238, 241-244, 250, 251, 255*f*, 266, 278, 290, 298, 299, 303, 305, 306, 386, 409 (*see* PCK, PEP carboxykinase, phospho*enol*pyruvate carboxykinase)
peroxisomes, 250, 254, 262
PGA, 127, 130, 131, 132*f*, 133, 138, 140, 155, 163, 202, 352 (*see* glycerate-3-P, phosphoglyceric acid)
phospho*enol*pyruvate carboxykinase (*see* PCK, PEP carboxykinase, PEPCK)
phospho*enol*pyruvate carboxylase, 31*t*, 56, 62*t*, 96, 145-150, 153*f*, 158*f*, 163, 164, 175, 176, 178, 184, 217, 235, 275, 276, 282*f*, 283, 297, 298, 386, 392, 400, 409 (*see* PEP carboxylase, PEPC, PEPcase)
phosphoglyceric acid, 127, 163, 202, 250, 302, 304, 352 (*see* glycerate-3-P, PGA)
photon yield, 281, 304 (*see* quantum yield)
photorespiration, 5, 20, 27, 29*t*, 30*t*, 32, 55, 57, 59, 61*t*-63, 68, 70-72, 87, 90, 129, 130*f*, 133, 163, 164, 168, 170, 171*f*, 175, 176, 183, 186, 195, 196, 202*f*, 204-208, 210-212, 249, 250, 262, 266, 275, 276, 280, 288, 333, 334, 336, 337, 371, 372, 400, 402, 408-410
photosynthesis
  acclimation, 90, 278, 333, 335, 342, 343, 345, 348
  light-limited, 70, 338, 410
  light-saturated photosynthesis, 27, 32, 70, 147, 158, 149*t*, 262, 333, 335, 338, 340f, 410
photosystem I, 120, 129, 131, 133, 197*t* (*see* PSI)
photosystem II, 31*t*, 66, 120, 129-131, 133, 134, 136, 150, 179, 226, 304, 335 (*see* PSII)
phylogenetic analysis 177, 221, 286, 287, 386
plasmodesmata, 30*t*, 81-83, 111, 176, 197*t*, 203, 211, 217, 221, 226, 227, 237, 245, 317, 318, 323, 324, 329, 386, 404, 406, 407 (*see* symplastic connection)
plastid(s), 33, 83, 84, 117, 123, 257, 298, 299, 303-305 (*see* chloroplast, proplastid)
  envelope, 84, 85, 290, 303, 305
  stroma, 85, 104, 240, 243, 303, 305
pleiotropy, 66, 73, 90, 168-170, 321, 383
PPDK, 31*t*, 56, 57, 61*t*, 64, 65, 68, 82*f*,

145-147*f*, 150*t*, 164*f*, 169, 179, 204, 218-220, 224, 235, 236-239, 241-245, 250, 258, 260-263*f*, 268, 278, 279, 281*t*-283, 287, 289, 290, 298, 303, 373, 386 (*see* pyruvate, orthophosphate dikinase; Pi dikinase; PYR/Pi dikinase)
productivity, 5, 19, 28-30*t*, 32, 37, 38, 40-42, 48, 50, 95, 108, 150, 160, 161, 163, 170, 196, 209, 236, 249, 275, 291, 297, 352, 366, 375, 399-402, 408-410
promoter(s), 67, 68, 165, 171*t*, 179, 183-188, 202, 219-221, 224, 242, 243, 280, 371-374, 389-392
  bundle sheath-specific, 183, 188
  cell-specific, 329
  mesophyll-specific, 183, 202
promoter-gene combinations, 372
proplastid, 121, 123 (*see* chloroplast, plastid)
proteomic(s), 188, 228, 227, 325, 326, 329, 372, 394
PSI, 120, 150 (*see* photosystem I)
PSII, 30*t*, 31*t*, 120, 145, 150, 156, 279 (*see* photosystem II)
pump
  bicarbonate, 111
  C_4, 95, 133, 264, 275, 290
  CO_2, 95, 104, 111, 141, 168, 175, 176, 202*f*, 203, 205, 212, 260
pyruvate kinase, 235, 303
pyruvate, orthophosphate dikinase, 31*t*, 56, 145, 146, 147*f*, 179, 183, 217, 218, 235, 236, 278, 282*f*, 287*f*, 386 (*see* PPDK; pyruvate, Pi dikinase; PYR/Pi dikinase)
pyruvate, Pi dikinase, 88, 89, 238, 245, 250 (*see* PPDK; pyruvate, orthophosphate dikinase; PYR/Pi dikinase)
PYR/Pi dikinase, 164*t* (*see* PPDK; pyruvate, orthophosphate dikinase; pyruvate, Pi dikinase)

Q

quantum requirement, 131, 132*f*, 133, 134 (*see* quantum yield)
quantum yield(s), 14, 15, 29-32, 58, 62-64, 127, 128, 134, 138, 158, 264, 275, 282, 333, 335, 338, 402, 406, 408 (*see* photon yield, quantum requirement)

R

radiation conversion factor, 18, 27, 334 (*see* efficiency, radiation-use; RCF)
rainfed rice, 32, 367
RCF, 334 (*see* radiation conversion factor)
regulatory genes, 5
regulatory mechanism, 161, 241, 337
regulatory network, 183, 317, 373
resistance to CO_2, 63, 64, 68, 70, 95, 96, 103, 104, 250, 255, 257, 264, 265 (*see* CO_2 conductance, leakage of carbon dioxide)
reverse transcription polymerase chain reaction, 329 (*see* RT-PCR)
ribulose 1,5-bisphosphate, 31*t*, 32, 55, 56, 163, 171f, 175, 249, 386, 410 (*see* ribulose bisphosphate, RuBP)
ribulose bisphosphate, 123, 139, 250, 276, 282f, 297, 298, 304, 336 (*see* ribulose 1,5-bisphosphate, RuBP)
rice genome, 5, 33, 123, 212, 328, 346, 347, 371, 373, 381, 393 (*see* genome annotation)
RT-PCR, 58-60, 127, 156, 157, 203, 285, 329 (*see* reverse transcription polymerase chain reaction)
Rubisco (*see* C_4 photosynthesis, defined; carbon-concentrating mechanism; CO_2-concentrating mechanism)
  concentrate CO_2 around, 27, 30*f*, 31, 63, 68, 70, 170, 197, 249, 276, 301, 304, 400, 402, 409, 410
  inhibition by oxygen, 166*t*, 178,

275-277, 280-282, 288, 352
oxygenase reaction, 29, 32, 61, 87, 96, 163, 164, 170, 202, 275, 276, 334, 336, 348, 371, 402
RuBP, 32, 88, 110*f*, 130, 250, 262, 276 (*see* ribulose 1,5-bisphosphate, ribulose bisphosphate)
RUE, 27, 28, 29*t*, 30*t*, 31*t* (*see* efficiency, radiation-use)

S

screening phenotypes, 21, 27, 28, 33, 90, 188, 195, 210, 211, 223, 227, 333, 366, 374-376, 401*t*
semidwarf, 4, 146, 362
serine, 61*t*, 87, 88, 167, 171, 180-182, 202, 204, 241, 283, 284
single-cell C_4, 30*t*, 31, 55, 56, 61-63, 64, 68, 70, 95, 96, 109, 111, 112, 128, 141, 222, 223, 235, 249, 253, 257, 258, 260-268, 275, 277, 278, 280, 288, 297-299, 306, 323, 399, 401, 401*t*, 402*t*, 405, 408, 409
sink, 3, 7, 19, 31, 140, 146, 319, 339, 362, 367
sorghum, 71*t*, 84, 89, 123, 166*t*, 196, 236, 240, 260, 371, 381, 383, 385, 404
Sorghum bicolor, 86, 99*t*, 107, 131*f*, 135, 183, 284
source, 3, 31, 146, 160, 319
soybean, 33, 278, 291, 371 (*see Glycine max*)
specificity for carbon dioxide, 30*t*, 61*t*, 68-70, 276, 374
spikelet, 6, 19, 57, 67, 365
spinach, 86, 208*f*, 225, 236
Spinacia, 138, 139*f*
Spinacia oleracea, 137, 138*t*, 139*f*, 141
stomatal conductance, 17, 18*t*, 32, 98, 147, 153, 154, 249, 264, 336, 346
Suaeda, 262, 263
Suaeda aralocaspica, 253, 258, 260, 264t, 265, 267*f*, 408 (*see Borszczowia aralocaspica*)
suberin, 57, 99*t*, 104, 111, 164, 227, 277, 322, 323, 328, 406, 407 (*see* suberization)
suberization, 222, 226, 227 (*see* suberin)
sugar cane, 30*t*, 47, 71*t*, 82*f*, 196, 210, 236, 323, 371, 393
super-hybrid rice, 57, 146, 160 (*see* F_1 hybrid, hybrid rice)
symplastic connection (*see* plasmodesmata)

T

targeted expression, 372
Thalassiosira, 297, 301*t*, 303
Thalassiosira pseudonana, 301*t*-303
Thalassiosira weisflogii, 278, 299, 301*t*, 302
thylakoid membrane, 120, 140, 337
thylakoid stacking, 118, 179, 222, 224, 226
tiller(s), 4, 5, 7-10, 339, 362, 363
tissue-specific expression, 134, 183, 392
tobacco, 32, 56, 69, 85, 87, 110, 160, 163, 166*t,* 167-170, 186, 187, 208, 209, 218-221, 224, 227, 280, 289 (*see Nicotiana tabaccum*)
TP (*see* transit peptide)
trans-factor, 183, 184, 217, 220, 242, 324
transformation, 40, 41, 170, 182, 209, 210, 212, 224, 238, 239, 241, 347, 373-376, 381, 401*t* (*see* genetic engineering, transgene, transgenic plant)
transgene, 33, 67, 121, 122, 367 (*see* down-regulation, expression, genetic engineering, transformation, transgenic plant, up-regulation)
transgenic plant(s), 64, 65, 67, 68, 73, 107, 119, 145-161, 178, 179, 239, 240, 242, 243, 288, 290, 323, 324, 372, 375, 376, 402*t*, 409 (*see* ge-

netic engineering, transformation, transgene)
transit peptide, 65, 68, 220, 235, 240, 242, 261, 285 (*see* chloroplast-targeted protein)
translocator, 83-85, 163, 164*t*, 169, 183, 187, 225, 250, 289, 407, 410 (*see* carrier, transporter)
transpiration, 16-18*t*, 32, 176, 201, 218
transporter, 83, 84, 87, 219*f*, 225, 243, 244, 275, 289-291 (*see* carrier, translocator)
tropical japonica, 363, 365-367 (*see* javanica)
two-cell C_4, 177, 178, 183, 184 (*see* Kranz-type C_4)

U
Udotea, 278
Udotea flabellum, 278, 298
upland rice, 32, 64, 123, 367
up-regulation, 87, 241, 275, 279, 280, 283, 285 (*see* transgene)
Urochloa panicoides, 89, 99*t*, 168, 235, 237, 238, 241

V
vein density, 8, 109, 201, 211, 320, 321 (*see* interveinal distance, vein spacing)
vein spacing, 30*t*, 58, 72, 73, 195, 200, 201, 223, 346-348, 386, 404, 406 (*see* interveinal distance, vein density)

W
wheat, 3, 37, 43, 46, 48, 71*t*, 196, 346, 393, 394
wild relative (*see* wild rice, wild species)
wild rice, 3, 20-22, 123, 366 (*see* wild species)
wild species, 122, 351, 352, 355-358, 366, 374, 375, 383 (*see* wild relative, wild rice)
WUE, 196, 264, 278 (*see* efficiency, water-use)

Y
yield(s) 8, 146, 196, 361-364
yield potential, 27-29*t*, 34, 334, 335, 338, 361, 362, 366, 371, 400, 401*t*

Z
Zea mays, 18*t*, 66, 99*t*, 107, 127, 129, 131*f*, 135*f*, 181*f*, 255*f*, 277, 388, 390, 391*t*